Soil Degradation and Restoration in Africa

Advances in Soil Science

Series Editors
Rattan Lal
B.A. Stewart

For more information about this series, please visit:
https://www.crcpress.com/Advances-in-Soil-Science/book-series/CRCADVSOILSCI

Soil Degradation and Restoration in Africa

Edited by
Rattan Lal and B. A. Stewart

CRC Press
Taylor & Francis Group
Boca Raton London New York

CRC Press is an imprint of the
Taylor & Francis Group, an **informa** business

CRC Press
Taylor & Francis Group
6000 Broken Sound Parkway NW, Suite 300
Boca Raton, FL 33487-2742

First issued in paperback 2021

ISBN 13: 978-1-03-209135-8 (pbk)
ISBN 13: 978-1-138-10331-3 (hbk)

Library of Congress Cataloging-in-Publication Data

Names: Lal, R., editor. | Stewart, B. A. (Bobby Alton), 1932- editor.
Title: Soil degradation and restoration in Africa / editor(s): Rattan Lal,
B.A. Stewart.
Other titles: Advances in soil science (Boca Raton, Fla.)
Description: Boca Raton, FL : CRC Press, Taylor & Francis Group, 2019. |
Series: Advances in soil science
Identifiers: LCCN 2019009205 | ISBN 9781138103313 (hardback : alk. paper)
Subjects: LCSH: Soil degradation--Africa. | Soil restoration--Africa. |
Soils--Africa.
Classification: LCC S625.A4 S65 2019 | DDC 631.4/76--dc23
LC record available at https://lccn.loc.gov/2019009205

Visit the Taylor & Francis Web site at
http://www.taylorandfrancis.com

and the CRC Press Web site at
http://www.crcpress.com

Publisher's Note
The publisher has gone to great lengths to ensure the quality of this reprint but points out that some imperfections in the original copies may be apparent.

Contents

Preface

In 2017, the Food and Agriculture Organization of the United Nations (FAO) estimated that the percentage of undernourished people in sub-Saharan Africa (SSA) increased from 20.8% (200 million) in 2015 to 22.7% (224 million or about one in four persons in SSA) in 2016, which is also 25% of the 815 million food-insecure people in the world. Over and above the adverse effects of a harsh and uncertain climate, characterized by pedological and agronomic droughts and prevalence of heatwaves during the crop-growing seasons, low and stagnating agronomic yields are also attributed to severe problems of soil degradation and land desertification throughout the SSA. The issue of soil degradation and undernourishment is intricately interlinked with that of growing food demand because of the increase in population. The present population of 856 million is projected to increase to 2 billion by 2050 and 4 billion by 2100. It is also widely feared that 40% of Africa's growing population may be prone to hunger and malnutrition during the 21st century. Furthermore, the absolute number of food-insecure population is likely to increase with the increase in population. Among the vulnerable regions of food and nutritional insecurity in Africa are the West African Sahel and the Horn of Africa. Most farmers in SSA are small landholders (<2 hectare), and farm sizes will continue to decrease with the projected increase in rural population. In addition to degraded and depleted soils, low agronomic yields are also due to the prevalence of extractive farming practices characterized by the negative nutrient (N, P, K, Ca, Mg and micronutrients) budget at the continental scale, which has been occurring for decades. The average annual rate of fertilizer use in Africa of 13–20 kg/ha is about 10% of the global average rate. Most resource-poor farmers can neither afford the expensive farm input nor are sure about their effectiveness because of the uncertain and changing climate and degraded and depleted soils. The problem is confounded by the severe problem of soil degradation and land desertification. Among major processes of soil degradation are accelerated soil erosion by water and wind, soil compaction and densification, depletion of soil organic matter content and of essential plant nutrients, salinization, elemental imbalance, soil pollution and contamination.

The Montpellier Panel (2014) reported that in Africa soil degradation affects 65% of cropland, 30% of grazing land, and 20% of forest land. Soil degradation (physical, chemical, biological, and ecological) is caused by neglecting soils and taking these finite but fragile and crucial resources for granted. The Montpellier Panel reported that neglecting the health of Africa's soil will lock the continent into food and nutritional insecurity for generations to come. Indeed, even the yield potential of elite varieties and species (of crops, animals, and trees) cannot be realized until grown on healthy soils. Despite its severe impact on agronomic yield, food/nutritional security, and human wellbeing, there are no credible estimates of the extent and severity of the problem, temporal and spatial variations, cause–effect relationship, and strategies to reverse the degradation trends for restoring the health and ecosystem services provisioned by the soil resources. The adverse effects of drought, especially agronomic and pedologic ones, are exacerbated by degraded and depleted soils, particularly where the surface layer has been truncated by accelerated erosion caused by water and wind. Erosion-prone soils, depleted of their organic carbon and clay contents because colloidal materials constitute light fractions of low density and are preferentially removed by water and wind, have low water- and nutrient-retention capacity and thus low use efficiency of these essential constituents.

The Green Revolution, which brought about a quantum leap in food production in Asia and Central America during the 1960s and 1970s, bypassed SSA because of soil degradation and the fact that essential inputs (irrigation, fertilizers) were not available to resource-poor farmers. Consequently, the agronomic yield of cereals (corn, sorghum, millet) has stagnated between 1 and 1.5 Mg/ha since the 1960s. Indeed, food and nutritional security in SSA can only be met, and the Sustainable Development Goals realized, if degraded soils and desertified ecosystems are restored through a soil-centric approach to modernize agriculture. Soils of managed ecosystems, especially those of agroecosystems, need to be restored and managed sustainably and prudently.

This 15-chapter book discusses the issue of soil degradation in SSA and the causes, consequences, and strategies of soil restoration and rehabilitation for advancing food and nutritional security and strengthening other ecosystem services. The book also provides case studies on soil degradation for different regions, as well as a range of processes and land use. Additionally, it contains chapters on recommended practices and policies for minimizing the risks of soil degradation and promoting the restoration policies.

The editors thank all the authors for their outstanding contributions and for sharing their knowledge and experience with the global soil science community. Preparation of the manuscripts, involving collation and synthesis of the literature and interpretation of the data from context-specific situations, is a time-consuming process that requires dedication and commitment. Thanks are also due to the editorial staff of Taylor & Francis for their timely help and prompt response to numerous questions and queries from the editors and authors. Special thanks are due to the office staff of the Carbon Management and Sequestration Center of the Ohio State University for providing support for the flow of manuscripts between authors and editors and for making valuable contributions. In this context, special thanks and appreciation are due to Ms. Laura Conover, Ms. Janelle Watts, and Ms. Terese Phinney, who formatted the text and prepared the final submission. While it is a major challenge to list all those who made direct and indirect contributions toward the completion of this book, thanks are due to everyone who supported the completion of this volume. It is important to build upon the contributions of all those who study the process, factors, and causes of soil degradation in SSA and share the knowledge contained in this volume with others from around the world.

Rattan Lal
B. A. Stewart

Editors

Rattan Lal, Ph.D., is Distinguished University Professor of Soil Science and Director of the Carbon Management and Sequestration Center, Ohio State University, USA, and Adjunct Professor at the University of Iceland, Reykjavik. His current research focus is on climate-resilient agriculture, soil carbon sequestration, sustainable intensification, enhancing use efficiency of agroecosystems, and sustainable management of soil resources of the tropics. He received an honorary degree of Doctor of Science from Punjab Agricultural University (2001), India; the Norwegian University of Life Sciences, Ås, (2005); Alecu Russo Balti State University, Moldova (2010); Technical University of Dresden, Germany (2015); University of Lleida, Spain (2017); and Gustavus Adolphus College, St. Peter, Minnesota. He was President of the World Association of Soil and Water Conservation (1987–1990), the International Soil Tillage Research Organization (1988–1991), the Soil Science Society of America (2005–2007), and is President of International Union of Soil Science (2017–2018). He was a member of the US National Climate Assessment and Development Advisory Committee (NCADAC) (2010–2013), a member of the SERDP Scientific Advisory Board of the US Department of Energy (US-DOE) (2011–), Senior Science Advisor to the Global Soil Forum of Institute for Advanced Sustainability Studies, Potsdam, Germany (2010–), a member of the Advisory Board of Joint Program Initiative of Agriculture, Food Security and Climate Change (FACCE-JPI) of the European Union (2013–2016), and Chair of the Advisory Board of the Institute for Integrated Management of Material Fluxes and Resources of the United Nations University (UNU-FLORES), Dresden, Germany (2014–2017). Professor Lal was a lead author of IPCC (1998–2000). He has mentored 110 graduate students and 54 postdoctoral researchers, and has hosted 174 visiting scholars. He has authored/co-authored 907 refereed journal articles, has written 20 books and edited/co-edited 71 books. For three years (2014, 2015, 2016), Thomson Reuters listed him among the world's most influential scientific minds and as having citations of publications among the top 1% of scientists in agricultural sciences. He is the recipient of the 2018 GCHERA World Agriculture Prize, 2018 Glinka World Soil Prize, and 2019 Japan Prize.

B. A. Stewart is Director of the Dryland Agriculture Institute and Distinguished Professor of Agriculture at West Texas A&M University, USA. He is a former Director of the USDA Conservation and Production Laboratory at Bushland, Texas; past President of the Soil Science Society of America; and a member of the 1990–1993 Committee on Long-Range Soil and Water Policy, National Research Council, National Academy of Sciences. He is a fellow of the Soil Science Society of America, American Society of Agronomy, and Soil and Water Conservation Society. Dr. Stewart is a recipient of the USDA Superior Service Award and of the Hugh Hammond Bennett Award of the Soil and Water Conservation Society. He is also an honorary member of the International Union of Soil Sciences since 2008 and was inducted into the USDA Agriculture Research Service Science Hall of Fame in 2009. Dr. Stewart is very supportive of education and research on dryland agriculture. The B.A. and Jane Ann Stewart Dryland Agriculture Scholarship Fund was established in West Texas A&M University in 1994 to provide scholarships for undergraduate and graduate students with a demonstrated interest in dryland agriculture.

Contributors

Wulf Amelung
Institute of Crop Science and Resource
 Conservation
Soil Science and Soil Ecology
University of Bonn
Bonn, Germany

George Ayaga
Kenya Agricultural and Livestock Research
 Organization
Food Crop Research Institute - Alupe Centre
Nairobi, Kenya

Mohamed Badraoui
National Institute of Agronomic Research
Rabat, Morocco

Laurent Barbiero
IRD-CNRS-UPS-OMP
Geoscience Environment Toulouse
Berlin, Toulouse

Twaha Ali Ateenyi Basamba
School of Agricultural Science
Makerere University
Kampala, Uganda

Mateete A. Bekunda
International Institute of Tropical
 Agriculture
The World Vegetable Center
Arusha, Tanzania

Rachid Bouabid
National School of Agriculture
Meknes, Morocco

Marcia J. Bunge
Gustavus Adolphus College
St. Peter, Minnesota
USA

Chris C. Du Preez
Department of Soil, Crop and Climate
 Sciences
University of the Free State
Bloemfontein, South Africa

Peter Ebanyat
School of Agricultural Science
Makerere University
Kampala, Uganda

Claude Hammecker
IRD, UMR,
Eco&Sols
Montpellier, France

Cornie W. Van Huyssteen
Department of Soil, Crop and Climate Sciences
University of the Free State
Bloemfontein, South Africa

Alice A. Katusabe
School of Agricultural Science
Makerere University
Kampala, Uganda

E. Kotzé
Department of Soil, Crop and Climate Sciences
University of the Free State
Bloemfontein, South Africa

Rattan Lal
Carbon Management and Sequestration Center
The Ohio State University
Columbus, Ohio
USA

Alpha P. Mtakwa
Department of Soil and Geological Sciences
College of Agriculture
Sokoine University of Agriculture
Morogoro, Tanzania

Peter W. Mtakwa
Department of Soil and Geological Sciences
College of Agriculture
Sokoine University of Agriculture
Morogoro, Tanzania

Sybrand Jacobus Muller
Department of Geography and
 Environmental Studies
Stellenbosch University
Stellenbosch, South Africa

Faith Milkah Wakonyo Muniale
Masinde Muliro University of Science and
Technology
Kakamega, Kenya

Patrick Musinguzi
School of Agricultural Science
Makerere University
Kampala, Uganda

Giregon Olupot
School of Agricultural Science
Makerere University
Kampala, Uganda

Emmanuel Opolot
School of Agricultural Science
Makerere University
Kampala, Uganda

Amit Roy
International Fertilizer Development Center
Muscle Shoals, Alabama
USA

Ernest Semu
Department of Soil and Geological Sciences
Sokoine University of Agriculture
Morogoro, Tanzania

Hussein B. Shelukindo
PO-RALG
Dodoma, Tanzania

Darryl D. Siemer
ISU Nuclear Engineering
Idaho Falls, Idaho
USA

Bal Ram Singh
Environmental Sciences and Natural Resource
Management
Norwegian University of Life Sciences
Ås, Norway

Hennie A. Snyman
Department of Animal, Wildlife and Grassland
Sciences
University of the Free State
Bloemfontein, South Africa

Brahim Soudi
Hassan II Agronomic and Veterinary Institute
Rabat, Morocco

Hamisi J. Tindwa
Department of Soil and Geological Sciences
Sokoine University of Agriculture
Morogoro, Tanzania

Ndelilio N. Urio
Department of Animal Science and Aquaculture
College of Agriculture
Sokoine University of Agriculture
Morogoro, Tanzania

David A.N. Ussiri
Carbon Management and Soil Sequestration
Center
The Ohio State University
Columus, Ohio
USA

Adriaan Van Niekerk
Department of Geography and Environmental
Studies
Stellenbosch University
Stellenbosch, South Africa

Kennedy Were
Kenya Agricultural and Livestock Research
Organization
Food Crops Research Institute -
Kabete Centre
Nairobi, Kenya

1 Soil Degradation in Sub-Saharan Africa

Challenges and Opportunities for Restoration

David A.N. Ussiri and Rattan Lal

CONTENTS

1.1 INTRODUCTION

Soil is one of world's most important non-renewable resources essential to all life forms on Earth. It provides the physical medium, chemical environment, and biological setting for water, nutrients, air, and heat exchange for organisms. Soils also provide ecosystem services such as climate regulation and food production by supporting production of food, feed, fiber, wood, clean water, and clean air, and they provide the basis of livelihood for millions of people across the world (MEA 2005). Soil is a natural habitat that regulates the environment and also responds to pressures imposed upon it. Other soil functions include the recycling and purification of water, and the provision of mechanical support for structures such as buildings and other installations. Soils influence the hydrological processes such as infiltration, deep percolation, drainage, streamflow, and water storage. They regulate the exchange of energy, water, and gas within the lithosphere–hydrosphere–biosphere–atmosphere system. Despite being a basic resource on which life depends, soil is degrading in many parts of the world. Land and soil degradation occur when the potential productivity associated with land becomes unsustainable, or when land is no longer able to perform its environmental moderation function within an ecosystem, commonly due to human activities. Degradation is generally preventable through the understanding and remediation of underlying root causes (Vlek et al. 2008). It is commonly caused by mismanagement or overexploitation of land resources, including vegetation clearance; nutrient depletion; overgrazing; inappropriate irrigation; excessive use of agrochemicals; urban sprawl; pollution; or other direct impacts, such as mining, quarrying, and compaction by heavy machinery. Soil can be degraded over time both qualitatively (e.g., nutrient depletion, acidification, salinization) and quantitatively (e.g., soil erosion). Major types of soil degradation processes fall under: (i) physical degradation; (ii) chemical degradation; and (iii) biological degradation (Table 1.1). Among physical degradation processes, soil erosion is the most visible and widespread form of degradation since loss of topsoil by erosion has deleterious effects on the productive potential as well as ecological and ecosystem well-being (Lal 2003). Global estimates suggest that 1,965 Mha (mega hectare) of land have been degraded, of which 1,642 Mha have been eroded by water and wind (Oldeman 1991; Bridges and Oldeman 1999). In Africa, 494 Mha out of the total land area of 2,966 Mha have been degraded (Figure 1.1).

Sub-Saharan Africa (SSA) is a large region of 24.6 million km^2, with a wide range of soil types and land uses (Dewitte et al. 2013). The SSA landscape is a mosaic of land resources including forests and woodland, grassland, arable land, mountains, and dryland. Forests and woodland occupy 6.7 million km^2, and arable land is estimated at 8.1 million km^2, of which about 2 million km^2 is under cultivation (UNEP 2013, 2016). SSA has a wide range of soils and climate. The soils range from stony and shallow to deeply weathered soils that are capable of recycling and supporting large amounts of biomass. The climate of SSA ranges from arid (rainfall < 500 mm yr^{-1}) to humid (rainfall $> 1,300$ mm yr^{-1}). The soils of the region fall under Alfisols, Andisols, Aridisols, Entisols, Inceptisols, Oxisols, Spodosols, Ultisols, and Vertisols. Land is central to development in Africa since the livelihoods of nearly 50% of the population (almost 70% in Eastern Africa) depend on agriculture (NEPAD 2013). Except for the Rift Valley area that benefited from volcanic rejuvenation, most soils of SSA have inherently poor fertility because they largely originate from Precambrian rocks that are very old (hundreds of millions to billions of years) and lack volcanic rejuvenation of mineral nutrients. The soils of the continent have also undergone various cycles of weathering, erosion, and leaching, leaving them poor in nutrients (Bationo et al. 1998; 2006). About 8% of land is of high-quality soil that is relatively free of natural constraints for agriculture and can be effectively managed (Jones et al. 2013), 34% is of medium or low potential with at least one major constraint for agriculture, and 55% is unsuitable for any kind of agriculture except nomadic grazing (Eswaran et al. 1997; Bationo et al. 2006; Vlek et al. 2008).

Land degradation, low input use, population pressure, and climate change have locked the majority of the farming communities of SSA in poverty and food insecurity, with their earnings below US \$1.50 per day (Vlek et al. 2010). With limited resources to invest in land management, the continued pressure on resources exacerbates soil degradation by processes such as erosion by tillage, erosion by water and/or wind, soil compaction, soil nutrient depletion, acidification, and loss of organic matter (OM) (Table 1.1). Soils of croplands in SSA are under threat from a wide range of human activities, and soil degradation in its diverse forms is a fundamental and persistent problem throughout the SSA region. Since the observed impacts are gradual, soil degradation is often ignored. However, it is a major development issue that causes pressure on land, extreme poverty, and migration in many parts of SSA (Dewitte et al. 2013). Consistent studies covering wide geographical areas of SSA are limited, but some studies suggest that about 500 Mha of Africa (about 16% of the continental land area) are affected by different types of degradation (FAO 1995; Jones et al. 2013). If non-productive land (deserts, salt pans, mountains, and lakes) are discounted, some form of soil degradation affects over 22% of the African continent (Jones et al. 2013). About 10% of SSA is estimated to be very severely degraded with an additional 15% to be severely degraded (Table 1.2); further, about 5 Mha are considered to be irreclaimable and beyond restoration (Jones et al. 2013). It is estimated that about 65% of agricultural land area in SSA is degraded (Vlek et al. 2008; Scherr 1999). Information derived

TABLE 1.1

Major Types of Soil Degradation and the Conditions under Which They Occur

Category	Processes	Parent material/ topography	Socioeconomic drivers and predominant climate
Physical	Soil erosion by water	Slope	Deforestation, overgrazing and improper grazing, and tillage practices generally in humid to semi-arid regions.
	Soil erosion by wind	Vegetation cover loss	Soil and vegetation disturbance or bio-crust by tillage, poor grazing management, and trafficking commonly in semi-arid to arid regions.
	Soil compaction	Clayey soils	Heavy machinery traffic, grazing. Common in clayey soils in humid regions.
	Soil erosion by tillage	Hilly landscapes	Continuous cultivation especially with tillage upslope and downslope.
	Surface sealing	Low organic matter, sandy and silty soils	Compaction, excessive tillage in cropland, and urbanization.
	Reduced water storage capacity	Low organic matter content	Common in areas where soil depth is reduced by erosion, pore space reduced by compaction, and water-holding capacity reduced by loss of OM.
Chemical	Acidification	Highly weathered soils	Excessive N fertilization, leaching of cations, sulfur, and N oxidation. Common in humid regions.
	Salinization	Shallow water table	Shallow water table, excessive irrigation, and removal of native vegetation can cause salinization in arid to semi-arid regions.
	Alkalization/ dispersion	Excessive monovalent cations, mixing calcareous subsoil and topsoil	Poor quality irrigation water, tillage, and loss of perennial vegetation.
	Nutrient depletion	Low inherent fertility, nutrient mining	Low input agriculture, overgrazing, and excessive forest harvesting.
	Toxic contamination	—	Urbanization, mining, industrial waste disposal, or spillage.
Biological	Soil organic matter depletion	Sandy texture, steep slopes, deep water table	Loss/reduced vegetation, excessive tillage, lack of sufficient organic amendments and plant residues, excessive biomass removal (by harvest, fire or grazing), and erosion of sloping surface (by tillage, wind or water). Occurs in regions with high temperature and limited rainfall.
	Loss of biological diversity		Monocropping, deforestation, and poorly managed grazing are generally associated with low soil biodiversity.
	Loss of plants, soil animals, and microbial biomass	Shallow bedrock, root limiting subsoil layers (fragipan, cemented layer, calcic horizons, Al toxicity)	Reduced plant growth and subsequent low litter and root exudates addition limits C for food web, exposure to extremes of temperature and dryness by plant litter removal, and destruction of macropores, aggregates and other soil habitat by tillage, compaction and/or soil erosion.

Source: Modified from Weil and Brady (2016).

from remote sensing observation over a 20-year time slice (1982–2003) indicated that the areas that show a consistent and significant decline in Normalized Difference Vegetation Index (NDVI) (or land degradation) over time amount to around 2.13 million km^2 or 10% of the SSA land area (Vlek et al. 2008; Vlek et al. 2010). This corresponds with the fraction of SSA with very severe land degradation symptoms in the Global Assessment of Human-induced Soil Degradation (GLASOD) assessment

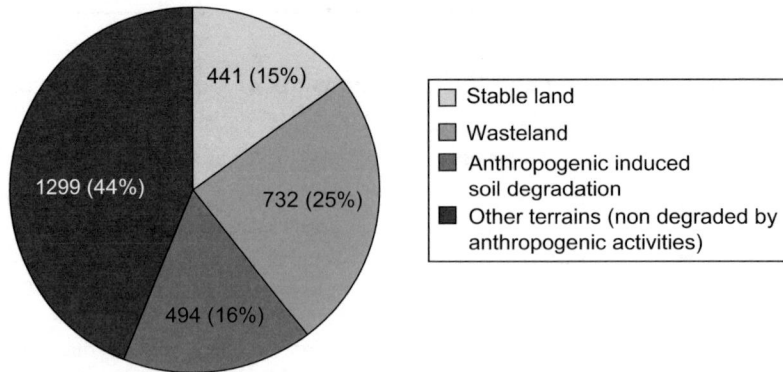

FIGURE 1.1 Proportion of Africa's land area (Mha) that is degraded. Data summarized from Oldeman (1991) and Bridges and Oldeman (1999).

TABLE 1.2
Degradation Severity in SSA Compared to Global Extent

Land degradation severity	Sub-Saharan Africa (%)	World (%)
None	33	35
Light	24	18
Moderate	18	21
Severe	15	20
Very severe	10	6
Total (light—very severe)	65	65

Source: Modified from Oldeman (1991)

of the late 1980s, and it is more likely that these areas are additive, suggesting a dwindling of land resources in SSA. Since the 1950s, Africa has lost about 20% of its soil productivity irreversibly due to degradation (Dregne 1990). Soil degradation in Africa is generally anthropogenic—and mostly agriculture-related. Human population growth, inappropriate land use, poor soil management, intrinsic characteristics of fragile soils in diverse agroecological zones, deforestation, and overgrazing are some of human-induced factors responsible for soil degradation in SSA. About 6 Mha of productive land are lost every year due to land degradation (Bationo et al. 2006).

Demand for land has been increasing over the years due to a rising population and a high population density. For example, the population density increased from 31 persons per km^2 in 2005 to 40 persons per km^2 in 2015 (UNDESA 2015). Agriculture plays a major role in the SSA economy. It employs more than 50% of the total labor force and generates an average of 15% of the gross domestic product (GDP) for SSA economies (OECD/FAO 2016). Despite the region's high dependence on land and its resources by a majority of people, land productivity is low due to inherently low soil fertility, low fertilizer input, and also declining fertility resulting from a steady negative nutrient imbalance (Bationo et al. 2006). Agricultural productivity in SSA is low compared to that for Asia and Latin America (Benin et al. 2011), and many countries in the region produce only 25% of their crop yield potential (Byerlee and Deininger 2013). The 20th-century Green Revolution that revolutionized agriculture in Latin America and Asia bypassed SSA because of the ecological challenges and political climate particular to Africa. The type and degree of soil constraints in the region vary widely (Table 1.3). About 40% of soils in SSA are inherently low in nutrient reserves with 10% weatherable minerals, while 25%

TABLE 1.3

Prevalence of Soil Constraints in Sub-Saharan Africa Based on the Fertility Capability Classification System

Soil constraint	Area in SSA (Mha)	Percentage of land in SSA (%)*
Low nutrient capital reserves	942.1	39.94
Steep slope (>30%)	55.6	2.36
Al toxicity	588.3	24.94
High P fixation	200.4	8.49
Poor drainage	160	6.78
High leaching potential	425.1	18.02
Calcareous soil (high pH)	158.1	6.70
Salinity	19.1	0.81
Alkalinity	52.1	2.21
Shrink-swell	132.7	5.62
Allophone	2.8	0.12
Total	2358.8	

Source: Modified from Sanchez et al. 2003; HarvestChoice 2010).

Notes: *Soil may be affected by multiple constraints; therefore, percentage areas may not add up to 100%.

are prone to aluminum (Al) toxicity, and 18% have low buffering capacity (Sanchez et al. 2003). The prevalence of hunger in SSA is pervasive and rising. In many countries, this is attributed to adverse climatic conditions resulting in poor crop harvests and loss of livestock (FAO 2017). The SSA region is experiencing a decline in overall per capita food production. The objective of this review is to highlight the causes, extent, and impact of soil degradation on food security in SSA.

1.2 DRIVERS OF SOIL DEGRADATION IN SUB-SAHARAN AFRICA

The drivers and underlying causes of soil degradation in SSA can be grouped into two categories (Figure 1.2): (i) those due to natural causes and biophysical processes—such as aridity or dry-land distribution, intrinsic soil quality, climatic variables, terrain and landscape position, vegetation cover, and soil biodiversity; and (ii) human-induced causes—such as land use and soil management, including deforestation, overgrazing, tillage practices, socioeconomic factors (e.g., land tenure, markets and infrastructure, population pressure, institutional support, income and human health, poverty, demographic change), and political factors (e.g., incentives and political stability) (Eswaran et al. 2001). The SSA region is one of the hotspots of soil erosion-related soil degradation (Lal 2001) and is vulnerable to desertification (Lal 1995).

1.2.1 BIOPHYSICAL PROCESSES

Some of the natural causes of soil degradation are factors related to climate, topography, soil, and vegetation. These factors include rainfall distribution, quantity, and intensity; frequent floods and tornadoes; storms and high-velocity winds; leaching in humid regions; drought; steep slopes; land terrain and relief; soil erodibility characteristics; species; patchiness; and density of vegetation (Figure 1.2). The majority of the agricultural land in SSA falls in semi-arid regions. About 66% of Africa is classified as desert or dryland—this category includes arid, semi-arid, and sub-humid areas (UNEP 2015). Due to the fragility of dryland soils, these areas are more susceptible to land degradation and desertification. Inherent soil quality is the ability of soil to perform its functions of agricultural production and enable it to respond to sustainable management. Soil resilience is the

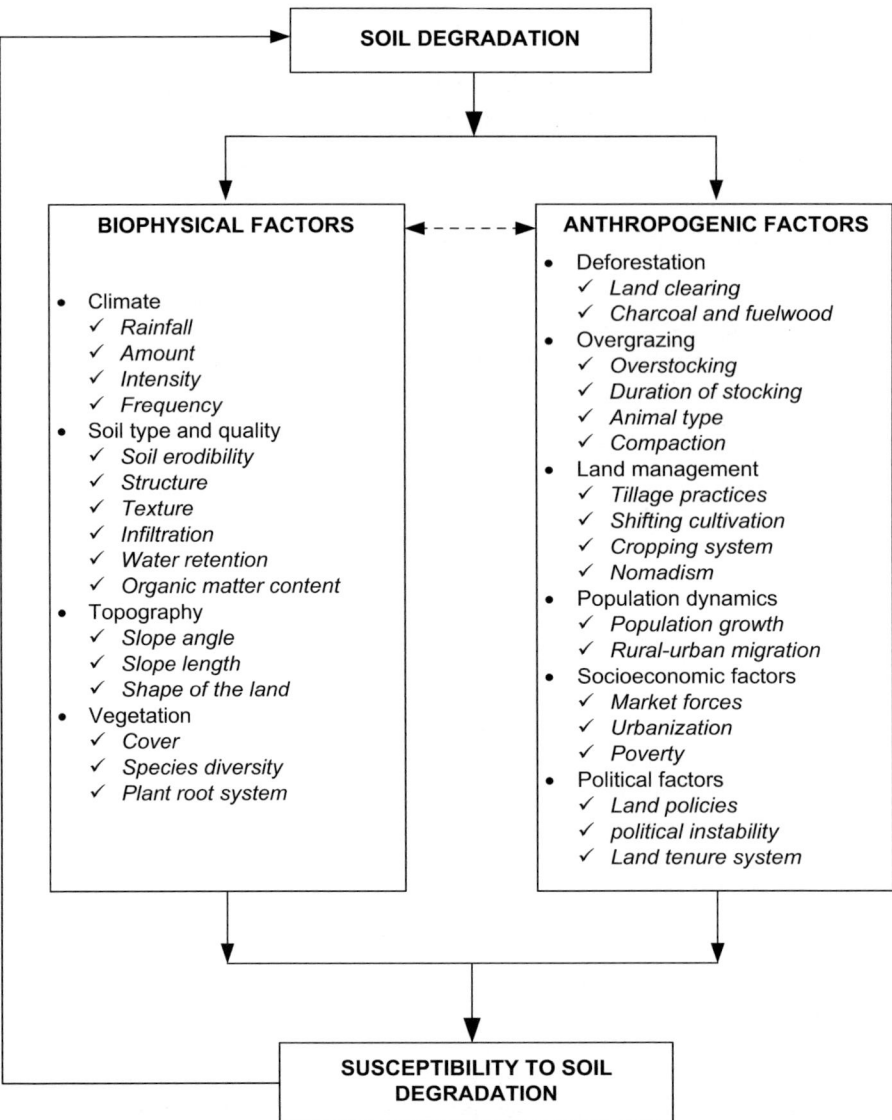

FIGURE 1.2   Interaction of biophysical and anthropogenic factors during the soil degradation process.

ability of soil to revert to the earlier state after it is degraded by mismanagement or change in land use. Land with low soil resilience remains irreversibly degraded. A large proportion of soils of SSA are of poor quality. Agricultural production in SSA is generally hampered by the predominance of inherently low soil fertility and fragile ecosystems that cannot support intensive agriculture (Jones et al. 2013). Soil resilience is an important underlying condition contributing to degradation in SSA (UNEP 2015). In addition to soil resilience, climate variability and climate change are other underlying drivers of soil degradation that figure prominently among underlying causes for the land and soil degradation in SSA. Depending on their inherent characteristics and the climate, soils vary from highly resistant to highly vulnerable. Most parts of SSA experience large variations in rainfall within and between the years, and they are subject to frequent storms that bring extreme flooding as well as extended drought periods, which contribute to soil erosion and degradation (UNEP 2013). Global warming and the associated changes in soil temperature and water regime affect all soil processes. Changes in rainfall and wind regimes associated with climate change will also lead to changes in the rates of soil erosion by water and wind. Similarly, the changing hydrological regimes and soil temperatures affect soil nutrient dynamics and properties, soil organic matter (SOM) decomposition rate, nutrient release, and plant metabolic processes.

1.2.2 HUMAN-INDUCED FACTORS

Although both biophysical and human-induced processes contribute to soil erosion and soil degradation, the occurrence of widespread soil degradation in SSA is attributed to anthropogenic factors, the root cause being the dominant low-input agricultural systems on land with moderate to poor potential. Other factors include: land misuse; cultivation of marginal lands such as semi-arid land, land with steep slopes, or shallow soils in harsh and highly variable climates; and subsistence farming by resource-poor farmers (Lal 2000; Kiage 2013). Marginal land cultivation is the outcome of rapid population growth that has led to the shortage of prime agricultural land in many parts of SSA. Anthropogenic drivers of soil degradation in SSA generally accelerate the natural degradative processes, and they are the outcome of depletive anthropogenic activities and their interaction with natural environments. Some anthropogenic pressures and interventions that have caused soils in SSA to degrade include deforestation, overgrazing, agriculture land mismanagement, and overexploitation of land and its vegetative cover. It is estimated that 121 Mha (65%) of cropland and 243 Mha (31%) of pasture and rangeland in Africa have been degraded as a result of cropland mismanagement and overgrazing, respectively (Table 1.4). Desertification is land degradation that occurs in arid, semi-arid, and sub-humid areas (drylands). It involves loss of biological and economic productivity, and complexity in woodland, pasture, and cropland. It is caused by various factors including climatic variations (droughts, aridity, and irregular and intense precipitation regimes) and unsustainable human activities (deforestation, overgrazing, overcultivation, soil structure deterioration, and poor irrigation practices). Degradation in drylands creates desert-like conditions, an irreversible change of land to a state where it can no longer be restored to its original use (UNCCD 2011). The affected land can no longer support vegetation. Climatic conditions and intensive agriculture make many parts of SSA particularly vulnerable to desertification. Desertification has extremely serious socioeconomic consequences and can ultimately cause the destabilization of societies and the migration of human populations. Approximately 26% of Africa, mostly in SSA, is vulnerable to the desertification processes, and nearly 47% of Africa is characterized as desert (Jones et al. 2013). With desertification, soils lose their structure and fertility, affecting crop yields and vegetation for livestock browsing, with serious impacts to the local and national economies and further impoverishment to communities already suffering from poverty and food insecurity (Jones et al. 2013; UNEP 2015).

1.2.2.1 Deforestation

Deforestation is one of the leading causes of soil erosion and a significant cause of degradation and desertification in SSA. Forest lands are generally cleared to create land for agricultural purposes

TABLE 1.4

Soil Degradation by Land Use in Africa

Land use	Total area (Mha)	Degraded Area (Mha)	Degraded (%)
Agricultural land	187	121	65
Permanent pasture	793	243	31
Forest and woodland	683	130	19
All land uses	1663	494	30

Source: Modified from Oldeman (1991).

(crops and livestock production), timber harvesting for large-scale commercial use, or fuelwood for small-scale subsistence reasons. Other reasons for deforestation include road construction, urban development, etc. (Oldeman 1991). Biomass represents about 80% of the energy source in many countries in SSA and a significant direct cause of soil degradation. Removal of vast amounts of trees and bushes, mainly for fuel, has left large areas bare and susceptible to erosion and degradation (ECA 2007; UNEP 2015). According to the 2015 Forest Resource Assessment by the Food and Agriculture Organization of the United Nations' (FAO), Africa is losing 0.5% of forest area annually, and, from 1990 to 2015, about 82 Mha (11.6%) of forest land were converted to other land uses (FAO 2015a; Ussiri and Lal 2017). Major drivers of deforestation in SSA are population growth and the associated demand for cropland expansion, as well as the demand for biomass fuel. Cases of logging activities for making charcoal and firewood collection, both of which contribute to the decimation of tree cover, are widespread in many parts of SSA (Braimoh 2006). Similarly, tree clearing for firewood is a crucial factor in soil degradation of the Sahel (Gorse and Steeds 1985). It is estimated that, over the last several decades, deforestation or domestic exploitation of wood for fuel or other domestic uses has accounted for more than 17% of soil degradation in SSA (Cherlet et al. 2018).

1.2.2.2 Population Growth

The increasing population pressure of people and livestock is one of the human-related causes of soil degradation in SSA. Increasing population and population density are potential contributors to land degradation and can be intensified with increased demand for crop and livestock production. Pressures on land contribute to soil degradation when intensified crop and livestock production is not accompanied by increased soil conservation measures to prevent exceeding the land's carrying capacity. For example, the SSA population grew from 100 million people in 1900 to 970 million in 2015, at annual growth rates of 2.2% in 1900–2005 and 2.56% in 2010–2015 and 2.75% in 2015 (UNDESA 2017). Many countries in SSA are experiencing fast population growth with high fertility rates. Africa has experienced rapid population growth of about 300% since 1962. At the same time, the area under agriculture increased by 59%. Each cultivated hectare supported 1.91 people in 1962, but by 2009 1 hectare supported 4.55 people (UNEP 2015; Jones et al. 2013). Africa's current population of 1.2 billion people is projected to reach 2.6 billion people by 2050, of which 2.12 billion people will be residing in the SSA region (UNDESA 2017). Similarly, SSA accounts for approximately 14% of the world's livestock resources (Otte and Chilonda 2002), and more than half of all ruminant livestock are kept in arid and semi-arid zones. Such rapid population growth leads to intensified stresses on natural resources—water, land, forest, and pasture in rangeland ecosystems (Balasubramanian et al. 2007). The demand for cropland expansion leads to expansion into more marginal conditions unsuitable for agriculture. The stress on natural resources and ecosystems often translates into overexploitation through overcultivation, deforestation, and overgrazing, and also poor land management practices. The consequences of stresses on ecosystems are soil erosion and degradation that perpetuate food insecurity and poverty in SSA. There is a general consensus that population pressure has disrupted the traditional farming systems of shifting cultivation and temporary resource use, particularly in the semi-arid environments (Kiage 2013). Shifting cultivation involves clearing a bush or forested area, farming the land for a period of several years, and thereafter abandoning it, permitting natural regeneration. Typically, abandoned land would be left fallow for up to a decade. Population increases have disrupted the fallow arrangement for natural recovery that existed under shifting cultivation (de Rouw and Rajot 2004) with the inevitable outcome of soil erosion and degradation.

1.2.2.3 Socioeconomic and Political Factors

Socioeconomic and political factors such as land tenure, marketing institutional support, income, human health, poverty, incentives, population density, land tenure, and political stability are also important causes of soil degradation in SSA. Interactions between natural and human society activities generally determine the success or failure of resource management. Poverty is a driver of soil degradation when farmers and herders resort to inappropriate land management practices such as farming in marginal lands, overgrazing, overstocking in marginal grazing lands, and the elimination of fallow periods. The SSA region has a disproportionately large share of low-income countries compared to other regions, and Africa on the whole has a very low level of economic development. Other socioeconomic factors include trade patterns that encourage short-term exploitation of land for export crops, nutrient mining, shortage of labor, the presence of conflict that prevents soil conservation, and land tenure arrangements. The land tenure system often determines how land is managed and used, and is generally considered as one of the primary drivers of soil degradation (Rohde et al. 2006; Thomas et al. 2000). A land governance system that provides secure rights to land resources offers an incentive for investments in land management that favors sustainable land-use and soil conservation practices, resulting in increased land productivity. Similarly, secure land rights and the presence of land titles are often associated with greater long-term land investment as well as allowing market transactions. For example, in smallholder systems in Eastern Africa, investments in soil fertility management are more likely when there is security of tenure or land ownership (Mafongoya et al. 2006). Generally, policies that raise the farm-gate crop prices play a critical role in encouraging good land management strategies, especially for those with tenure, since these provide farmers with both resources and incentives.

1.2.2.4 Overgrazing

Overgrazing decreases protective cover and leaves a soil vulnerable to water and wind erosion. Overgrazing is the major cause of soil erosion in the rangelands of different climates in SSA environments (Podwojewski et al. 2011). Trampling also causes soil compaction and reduced infiltration or loosening of the soil surface, depending on type of livestock involved. Livestock grazing and pastoralism in SSA have often been viewed as a critical factor in the interaction between agriculture and the natural resource base, and overstocking has long been blamed for the cause of extensive soil degradation in rangeland areas. Overgrazing is especially damaging to soils in marginal areas, on sandy soils, when the livestock is of only one species, and when there are especially high stock densities (UNEP 2013). It is estimated that overgrazing is responsible for nearly 50% of soil degradation in dryland SSA, followed by that with poor agricultural management practices (24%), deforestation and vegetation removal (14%), and overexploitation of soils (13%) (UNEP 2013, 2016). Successful implementation of destocking is difficult in many SSA rangelands since the ownership of a large herd is considered a sign of wealth and status within the community. For example, in Tanzania, officials viewed large herd size and overgrazing as major causes of land degradation and so attempted to enforce mandatory destocking and introduced zero-grazing of improved dairy cows for milk. Yet, livestock were simply moved to other areas, rather than having their numbers reduced, thereby transferring the problem to different locations and also leading to increased malnutrition (Dejene et al. 1997). A lack of understanding of the social, cultural, and economic roles of livestock most likely led to misguided solutions that did not have the intended effect but had overall negative consequences. There is increasing evidence that climate, rather than overgrazing, is the key cause of land degradation in the rangelands of SSA. Climate change is likely therefore to exacerbate the problem of soil degradation.

1.2.2.5 Mismanagement of Agriculture

Cropland mismanagement refers to the improper management of agricultural land through variety of practices that fail to conserve and improve soil quality, maintain vegetative land cover, protect soils from water and wind erosion, and degrade soils through overexploitation or pollution. These include insufficient or excessive use of fertilizers, shortening the fallow periods in shifting cultivation, poor irrigation practices, lack of soil conservation practices and erosion control measures, and the use of heavy farm machinery when the soil is fragile and susceptible to compaction (Oldeman 1991; Jones et al. 2013). Overall, human activities and how land is managed have a major impact on soil characteristics and the provision of soil functions in large parts of SSA, and under increased population density, many of these impacts exacerbate soil degradation.

1.3   TYPES OF DEGRADATION IN SUB-SAHARAN AFRICA

According to GLASOD, two categories of soil degradation processes include (i) soil erosion by water and/or wind forces, and (ii) *in situ* soil deterioration by physical, chemical, and biological processes (Oldeman 1991). These can be grouped into degradation by the displacement and transport of soil material by water and/or wind, physical, chemical, and biological soil degradation (Table 1.1). The underlying geographical conditions, natural factors, and anthropogenic activities such as deforestation, overgrazing, soil management practices, and socioeconomic conditions that are conducive to overexploitation of land resources make much of Africa highly vulnerable to soil degradation. Soils can be altered physically, chemically, or biologically as the result of natural processes. However, anthropogenic activities play a significant role in accelerating many of the natural processes, causing human-induced degradation of soils. Human-induced soil degradation processes vary according to land use. Deforestation and overgrazing are two types of land mismanagement that deprive soils of their vegetative cover and have negative impacts on soil resources (Vlek et al. 2010). In forested areas of the humid tropics, degradation occurs due to deforestation. Overgrazing is the prime form of soil degradation on grasslands, and the greater impact occurs in arid–semi-arid regions of SSA, commonly on soils with better soil and terrain conditions.

1.3.1   PHYSICAL DEGRADATION OF SOILS

Important physical properties linked to soil quality, fertility and productivity include soil texture, structure, porosity, bulk density, soil water, and air and soil temperature. Except soil texture, all other properties are readily altered by soil management practices. Physical degradation of soil includes compaction, sealing and crusting, and waterlogging. About 4% of soil degradation in Africa is caused by physical degradation (Table 1.5). Soil compaction is generally caused by the

TABLE 1.5
Severity of Anthropogenic-Induced Soil Degradation in Africa
(million hectares)

Degradation type	Area (Mha)				
	Light	**Moderate**	**Strong**	**Extreme**	**Total**
Water erosion	57.5	67.4	98.3	4.2	227.4 (46%)
• Loss of topsoil	53.9	60.5	86.6	3.8	204.9
• Terrain deformation	3.6	6.9	11.7	0.4	22.5
Wind erosion	88.3	89.3	7.4	—	186.5 (38%)
• Loss of topsoil	79.1	84.2	7.4	—	170.7
• Terrain deformation	9.2	5.1	—	—	14.3
• Overblowing	—	—	0.5	1.0	1.5
Chemical degradation	26.0	27.0	8.6	—	61.5 (12%)
• Loss of nutrients	20.4	18.8	6.2	—	45.1
• Salinization	4.2	7.7	2.4	—	14.8
• Acidification	1.1	0.3	<1	—	1.5
• Pollution	—	0.2	—	—	0.2
Physical degradation	1.8	8.1	8.8	—	18.7 (4%)
• Compaction	1.4	8.0	8.8	—	18.2
• Waterlogging	0.4	0.1	—	—	0.5
Total degraded area	173.5 (35%)	191.8 (38%)	123 (25%)	5.2 (1%)	494.2

Source: Data compiled from Oldeman 1991; Bridges and Oldeman 1999.

use of heavy machinery on soils with low structural stability. In agriculture, soil compaction due to heavy machinery can alter infiltration, drainage, bulk density, and other physical properties of the soil as a result of reduced porosity. This can restrict the movement of water and gases through the soil and cause high surface runoff, which may lead to significant water erosion. It is estimated that about 18 Mha are compacted in Africa (Table 1.5). Compaction is particularly evident across Sahel, South Africa, and Zambia (Jones et al. 2013). The physical degradation of soil is mainly due to improper soil management. For example, excessive soil tillage breaks down soil aggregates, thus rapidly decomposing organic matter, loosening the soil, and making it vulnerable to wind and water erosion. The poor management of both grazing and tillage can lead to compaction of surface or subsurface soil layers (Collins et al. 2001), resulting in reduced infiltration.

1.3.2 SOIL EROSION

Erosion is a process of detachment of soil particles and their transport and deposition at distant places by natural agents such as water, wind, tillage, glaciers, and gravity. It is the loss of topsoil through the destructive action of water, wind, gravity, and human actions of tillage, especially when the vegetation cover has been removed. Soil erosion is the most serious form of soil degradation and fertility decline (Tamene and Vlek 2008) because it involves the removal of the most fertile topsoil where SOM and plant nutrients are concentrated. It is one of the major forms of land degradation in SSA, with a severe threat to the environment and serious negative impacts on agricultural productivity. The loss of topsoil by soil erosion can cause the deterioration of the physical, chemical, and biological properties of the soil, loss of nutrients and overall reduction of the fertility of soils, loss of soil productivity, and cropland loss (Obalum et al. 2012). In SSA, the reduction in crop yields on eroded soil occurs through the removal of plant nutrients with eroded sediments, exposure of root-toxic and poorly aerated subsoil, soil structure deformation, surface sealing and crusting, reduced seed emergence, and reduced infiltration (Lal 1995). Erosion damage can cause significant economic loss. It is estimated that, in the past, erosion has caused a yield reduction ranging from 2% to 40% (Lal 1995). Cultivation on steep slopes, clearing of vegetation (especially leaving land bare between cultivation cycles), and poorly managed grazing are the primary factors accelerating soil erosion in SSA (Tamene and Vlek 2008). On steep slopes, hand or animal traction tillage moves the soil preferentially in the easier downslope direction (Kimaro et al. 2005). Poorly managed grazing in pastureland can also contribute significant amounts of sediment downstream. For example, 17% of suspended sediment loss from the Kaleya catchment in southern Zambia that reached downstream originated from grazing (Collins et al. 2001).

Studies on the effect of soil erosion on crop yield on field and simulated erosion in SSA are summarized in Table 1.6. The adverse impacts on agronomic productivity are due to the decline in land and soil quality, generally considered as on-site effects. Soil erosion also causes damage off-site such as fluvial sediment deposition, reservoir sedimentation, river channel silting, and enhanced flooding (Mullan 2013). High rates of topsoil loss contribute to downstream sedimentation and the degradation of local and regional water bodies. Reservoirs act as a large sediment trap, and siltation causes rapid decline in the lifespan of the reservoir, threatening the sustainability of inland water storage. For example, in Tigray, Ethiopia, reservoirs designed to improve water access with a 20-year lifespan lost half of their storage capacity in only 5 years due to sedimentation (Tamene et al. 2006). Similarly, the Angereb domestic water supply dam in northern Ethiopia, with a lifespan of 25 years, was losing its water storage capacity by 3.26% annually between 1997 and 2007 due to sediment deposition from erosion within the catchment area (Haregeweyn et al. 2012). Similarly, dam siltation rates ranging between 1% and 4% annually have been reported in Zimbabwe and Tanzania (DFID 2004). Soil erosion also has negative impacts on the infrastructure and water quality (Vrieling 2006; Obalum et al. 2012). Heavy metal contaminants, fertilizers, herbicides, and pesticides contained in sediments can pollute water bodies, cause eutrophication, and also affect carbon (C), nitrogen (N), and phosphorus (P) cycling (Li and Fang 2016; Issaka and Ashraf 2017).

TABLE 1.6

Soil Erosion–Productivity Relationship in Sub-Saharan Africa

Soil loss (cm)	Yield reduction (%)	Country and climate	Soil order	References
Field erosion studies				
Maize (*Zea mays* L.) as a crop				
0.0024	26.9	Zimbabwe, semi-arid	Alfisol	Lal 1995
0.0080	0.1513	Nigeria, sub-humid	Alfisol	Lal 1981
0.0080	0.1720	Nigeria, sub-humid	Alfisol	Lal 1981
Pearl millet (*Pennisetum americanum* L.) as a test crop				
0.0928	51.6	Burkina Faso, semi-arid	Aridisol	Lal 1995
Desurfacing experiments				
Maize as a test crop				
2.5–12.5	23–56	Nigeria, sub-humid	Alfisol	Lal 1976
5–20	30.5–100	Nigeria, humid	Ultisol	Mbagwu et al. 1984
5	15–69.7	Nigeria, sub-humid	Alfisol, Inceptisol, Ultisol	Obalum et al. 2012
10–20	39–81.2	Nigeria, sub-humid	Ultisol	Lal 1995
2.5–7.5	50–100	Cameroon, humid	Alfisol	
5–20	47–63	Burkina Faso, semi-arid	Ultisol	
3.6	23–95	Nigeria, sub-humid	Ultisol	Obi et al. 2005
5–20	56–95.5	Nigeria, sub-humid	Oxisol	Oyedele and Aina 2006
5–20	17–76	Nigeria, sub-humid	Alfisol	Salako et al. 2007
Cowpeas (*Vigna anguiculata* L.) as a test crop				
5–20	1.5–80.5	Nigeria, sub-humid	Alfisol	Mbagwu et al. 1984
5–20	62–70.6	Nigeria, humid	Ultisol	Mbagwu et al. 1984
Cassava (*Manihot esculentus* C.) as a test crop				
10–20	35.7–53.7	Nigeria, sub-humid	Alfisol	Mbagwu et al. 1984

In areas where soil is shallow and land is sloping, such as the Ethiopian highlands (Tamene and Vlek 2007), erosion can cause the irreversible loss of soil and land degradation, leading to land abandonment. Increases in water and wind erosion result from land-use changes that alter the vegetative cover of the land (e.g., when forest or grassland is converted to cropland). Because of ongoing climate change, frequent storms and extended droughts are likely to become common in some parts of SSA with a significant impact on soil erosion. In the Horn of Africa, agriculture production is largely rainfed. Droughts, erratic rainfall distribution, and increased temperatures affect crop production significantly. For example, the drought of 2011 in the region caused the migration of people to other, less-affected areas in order to access food, water, and feed. In the settlement locations, large-scale deforestation occurred, driven by the demand for charcoal, which left land exposed and prone to erosion (Terefe 2012).

Erosion causes soil loss, particularly from the surface, but sometimes large masses of soil may be lost—as in landslides or riverbank erosion. The erosion process brings changes both on-site and at the deposition site. In addition to on-site soil loss, erosion moves sediment and nutrients off the land, causing significant yield loss and loss of soil productivity while also creating widespread water pollution problems in rivers and lakes. The nutrients impact water quality largely through eutrophication. Geological or natural erosion that takes place in undisturbed landscapes by natural forces, without the influence of human activities, is of little concern from the point of view of soil quality because of its low rate, and soil loss can generally be offset by soil formation. Human actions during the exploitation of land, water, vegetation, and soil resources such as deforestation, overgrazing, soil tilling, and shifting cultivation accelerate soil erosion beyond the tolerance limit. They reduce soil quality, damage the land, and reduce crop yield on-site.

Changes in land use are widely recognized as capable of greatly accelerating soil erosion, and it is recognized that erosion in excess of soil production would eventually result in decreased agricultural potential (Montgomery 2007). In the United States, a soil loss range of 2–12 Mg ha^{-1} yr^{-1} is considered tolerable, depending on soil type (Schertz 1983). Accelerated erosion occurs at an alarming rate that reduces soil quality and crop yield. Globally, soil erosion remains the most extensive problem, and is estimated to be responsible for 80% of the degraded areas (Oldeman 1991; den Biggelaar et al. 2004). Estimates of the global annual soil loss by water erosion range between 20 and 37 Pg (Pg = petagram = 10^{15}g = billion metric ton = Gt) (FAO 2015b; Walling 2008, 2009; Borrelli et al. 2017), and wind mobilizes an additional 2 Pg yr^{-1} as dust (FAO 2015b). Borrelli et al. (2017) estimated a decadal increase of 2.5% in global soil loss from 2001 to 2012, driven by spatial changes of land use. Erosion is prevalent in the eastern highlands, humid, sub-humid, and semi-arid regions of SSA. The severity of erosion in these regions is attributed to high population density, overgrazing and uncontrolled grazing, excessive stocking rate, soils that are highly prone to erosion due to the harsh climate and intensive farming, and a combination of soil profile characteristics and climatic factors that renders them highly susceptible to severe erosion. Assessments of agronomical impact of soil erosion indicate that Africa, especially SSA, is more vulnerable to economic losses due to soil erosion (den Biggelaar et al. 2004).

1.3.2.1   Water Erosion

Soil erosion by water involves the detachment of soil particles from soil mass, primarily by raindrops and transport by runoff and flowing water. Water erosion is the most common type among soil degradation processes, accounting for about 56% of degraded soils globally (Oldeman 1991; Bridges and Oldeman 1999). In Africa, about 227 Mha (46% of degraded soils) are affected by water erosion (Table 1.5). In addition to soil loss, surface wash and sheet erosion remove considerable amount of nutrients from the topsoil, leading to loss of soil quality. In some cases, rills and gullies are formed as a result of large amounts of water movement on susceptible terrain. Water erosion is particularly destructive in humid tropical regions of Africa where the convergence of deforestation, population pressure and torrential rain episodes can lead to annual soil losses >50 Mg ha^{-1} yr^{-1} (FAO 1995; UNEP 2015). Northern Africa, Madagascar, and South Africa experience the most severe water erosion. Erosion selectively detaches the colloidal fractions of soil (clay and OM, which is more easily eroded) and transports them off-site by runoff. These fractions are important to soil fertility, aggregation, structural stability, and favorable pore size distribution. The process of water erosion is influenced by soil properties, slope of the land, vegetation cover, and rainfall amount and intensity. A meta-analysis of soil erosion rates across the globe (García-Ruiz et al. 2015) identified general trends that are also applicable to SSA: (i) erosion rates tend to increase with increasing slope below 0.2 mm^{-2}; (ii) erosion rates increase with the increase in mean annual rainfall and are greater for mean annual rainfall between 1,000 and 1,400 mm yr^{-1}; (iii) land use/land cover have a determining effect in erosion rates and agriculture activities are generally associated with the highest erosion rates compared to other land uses (García-Ruiz et al. 2015). Water erosion is prominent in the sloping landscapes and is the most widespread process leading to topsoil loss and land degradation. It occurs all over the world, varying in intensity and scope according to climatic and physical conditions as well as human activities (Oldeman 1991). Assessment of soil erosion in the Volta and Nile basins, regions covering areas of 105,000 and 2.9 million km^2, respectively indicated an average soil loss of 35 Mg ha^{-1} yr^{-1} for the Volta basin in West Africa and 75 Mg ha^{-1} yr^{-1} for the Nile basin (Tamene and Le 2015). As evidenced in many regions of Africa, water erosion is the greatest factor in limiting soil productivity and impeding agriculture enterprise in humid and sub-humid tropical regions of SSA (Dregne 1990).

1.3.2.2   Wind Erosion

Wind erosion, which occurs when strong winds blow across soils with relatively dry surface layers, is a serious problem in arid and semi-arid regions where vegetation is sparse, rainfall is low, and temperature is high. It is more evident in areas where the annual rainfall is below 600 mm and the dry season lasts for longer than six months. Generally, coarse-textured soils are more susceptible to

wind erosion than are fine-textured soils. Anthropogenic activities that remove the protective vegetation such as tree cutting, overgrazing, and plowing cause or exacerbate wind erosion (Oldeman
1991). About 38% of soil degradation in Africa is caused by wind erosion (Table 1.5), which significantly reduces soil productivity and crop yield due to the loss of the most fertile part of the soil.
Desertification is an ultimate result of wind erosion. The Sahel region and parts of Southern Africa
are among the most severely affected by erosion in the SSA region (Jones et al. 2013). Wind erosion
causes a considerable loss of soil and associated nutrients in the Sahel.

1.3.3 CHEMICAL DEGRADATION OF SOILS

Chemical degradation of soil refers to the undesirable changes in soil chemical properties that
lead to a decline in soil quality as a result of human intervention. It includes the loss of nutrients,
salinization, acidification, aluminum toxicity, and pollution (Table 1.1). The loss of nutrients is the
most important form of chemical degradation in Africa. Nutrient outputs have exceeded inputs
for decades across SSA, exhausting soil nutrient pools. Approximately 62 Mha (12%) of degraded
soils in Africa result from chemical degradation, of which 45 Mha are affected by the loss of nutrients (Table 1.5; Osman 2014). Unlike physical degradation, chemical soil degradation is not easily observed by the naked eye. Increasingly, global analyses are recognizing that such less visible
changes in soil properties are critical in affecting crop yields (Mueller et al. 2012).

Nutrient depletion is a major setback for low-input agriculture and is recognized as the primary
cause of loss of soil fertility in SSA cropland soils. It is recognized as a fundamental biophysical
cause of food insecurity among small-scale farmers in the region (Sanchez et al. 1997) since food
production in the tropics and subtropics of Africa is mostly rainfed and generally relies on available
soil nutrient pools. The general cause of depletion is the negative balance between output (harvesting, burning, and leaching) and input (through fertilizers, manure, and crop residues) of nutrients
and OM. A study of the nutrient balance of the key agroecological regions of Africa estimated that,
during 2002–2004 cropping seasons, about 85% of farmlands in Africa had nutrient mining rates
exceeding 30 kg NPK (nitrogen, phosphorus, potassium) ha^{-1} yr^{-1}, and 40% of these had rates >60
kg NPK ha^{-1} yr^{-1} (Henao and Baanante 2006). The highest rates of nutrient depletion (>100 kg
NPK ha^{-1} yr^{-1}) occur in Eastern Africa—Rwanda, Burundi, and Malawi—where fertilizer use is
rather low and the loss of nutrients through soil erosion is high (Henao and Baanante 2006). Most
countries in Eastern and Southern Africa experience high nutrient depletion rates resulting from:
(i) high population density and continuous cultivation with low or no nutrients addition; (ii) hilly
and mountainous terrain rendering land susceptible to erosion; and (iii) soils that are still fertile
and have a lot to lose in terms of plant nutrients supply (Stoorvogel and Smaling 1990; Stoorvogel
et al. 1993). In Sudano-Sahelian zones, continuous cultivation, high OM decomposition rates, and
wind erosion are the major paths for nutrient loss. In the equatorial forest zone, with high rainfall
and a favorable soil moisture regime, water erosion, leaching, reduced fallow, and residue burning
are the major nutrient loss paths (Buerkert and Hiernaux 1998). Effective soil fertility management remains a major challenge in SSA (Onduru et al. 2007). In addition to nutrient depletion, low
nutrient holding capacity, high acidity, aluminum toxicity, and low organic matter are some of the
chemical-related constraints to soil productivity in SSA. These constraints are exacerbated by overexploitation through continuous cropping and low rates of nutrient application.

For decades, nutrient removal across SSA has exceeded input, exhausting soil nutrient pools
(Table 1.7). The lack of application of required nutrients is causing nutrient depletion and the reduction of agricultural productivity in most agricultural areas in SSA. Other factors contributing to
nutrient depletion are soil erosion for P and leaching of P and N. Partial nutrient budgets are often
used in Africa to evaluate management practices that promote nutrient surpluses or deficits (Henao
and Baanante 1999; Cobo et al. 2010). Results have indicated that most systems in Africa
have negative N and K balances, while P was less noteworthy (Table 1.7; Henao and Baanante 1999;
Cobo et al. 2010), and broadly supports the claim of nutrient mining in SSA and across the continent

TABLE 1.7

Annual Nutrient Depletion and Average Fertilizer Use (2011–2015) in Agricultural Soils for Selected Countries of Southern Africa

Country	Nutrient depletion (kg ha⁻¹ year⁻¹)			Fertilizer use (kg ha⁻¹ year⁻¹)		
	N	P	K	N	P	K
Benin	−16	−2	−11	3.1	1.0	2.0
Ethiopia	−47	−7	−32	11.9	4.1	1.0
Cameroon	−21	−2	−13	4.4	0.6	2.4
Ghana	−35	−4	−20	5.1	2.1	3.6
Kenya	−46	−1	−36	21.9	4.2	2.4
Malawi	−67	−10	−48	20.2	2.2	3.0
Tanzania	−32	−5	−21	4.3	0.8	0.5
Mozambique	−23	−4	−19	4.2	0.3	0.5
Zambia	−13	−1	−12	34.9	3.7	3.2
Zimbabwe	−27	2	−26	11.1	3.0	3.2
Botswana	−2	0	−2	52.6	0.9	2.0
Rwanda	−60	−11	−61	3.9	1.2	1.2

Source: Nutrient depletion data modified from Henao and Baanante (1999) fertilizer use data from FAOSTAT (2018).

(Sanchez et al. 1997; Hartemink 2006). In many SSA farming systems, certain soils suffer from nutrient depletion even if the whole farm or farming community does not. This pattern of nutrient depletion has been documented in many studies that show how nutrients are transported from out of fields and transported to fields near the homestead in the form of crops harvested and animal manure deposited (Amede et al. 2001).

Soils in SSA are also experiencing declining cation exchange capacity (CEC), cation imbalances, and declining soil pH (which can lead to Al toxicity), alkalization, and salinization. Secondary soil acidification can occur due to the long-term application of relatively high rates of N fertilizers (mostly in South Africa) or continuous cropping without organic inputs (Juo et al. 1995), leaching of nitrates, and acid rain. Some soils are naturally saline, and some are made saline by the mismanagement of soil and crops, particularly improper irrigation and drainage (i.e., changing hydrologic balance). Salinization may occur in arid and semi-arid regions of SSA where scarcity of water and high evaporation limit the leaching of salts, and in humid regions where excess irrigation or poor drainage cause the groundwater table to rise to the root zone and make soil saline. About 15 Mha have been degraded by salinization in Africa (Table 1.5). Soil degradation through salinization and alkalization is considered one of the most important threats jeopardizing sustainable irrigated rice cropping in the semi-arid regions of SSA. Alkalization can also occur when perennial vegetation is lost, or when calcareous subsoil material is incorporated into the topsoil as a result of erosion or tillage.

1.3.4 BIOLOGICAL DEGRADATION OF SOILS

The biological degradation of soils refers to the impairment or elimination of one or more populations of microorganisms that play a known ecologically significant role in soil, often resulting in changes in biogeochemical processing within the associated ecosystem. Biological degradation is closely linked to chemical degradation, where both the balance of different nutrients and their chemical forms are also important to soil fertility (Vlek et al. 2010). Soil microbial population plays a critical role of nutrient recycling that facilitates the continued production of plant- and animal-derived food. Therefore, the ability of soil microbial populations to function properly is of critical importance to the health and well-being of humans (Sims 1990). Soils are among the most biologically diverse habitats on Earth. It is estimated that 1 g of soil contains up to 1 billion bacteria cells of tens of thousands of taxa, up to 200 million fungal hyphae, and a wide range of nematodes, earthworms, and arthropods (Bardgett and van der Putten 2014). Agricultural management practices such as soil tillage, fertilization, application of pesticides, and low plant diversity adversely affect several groups of soil organisms and reduce overall soil microbial biomass (McDaniel et al. 2014). Altering soil microbial communities can impact ecosystem processes that sustain agriculture, such as nutrient cycling, especially in low-input smallholder tropical agriculture in SSA (Wood et al. 2015). Loss of biodiversity crosscuts many different types of land degradation because soil plays a vital role in several ecosystem processes and functions, underpinning food and fiber production as well as helping to regulate climate and maintain purity of water. Intensive agriculture threatens soil biodiversity, because some groups of soil biota are highly sensitive to soil management and can be severely affected by changes in land use. It can also lead to OM loss of up to 50% within several decades of cropping (Zingore et al. 2007a).

In SSA, high levels of poverty mean that many small-scale farmers cannot afford to purchase inputs such as fertilizers that are typically needed to replenish soil nutrients exported off the field through crop harvesting (Sanchez 2002). For example, an analysis of 90 samples of maize (*Zea mays L.*) from low-input smallholder farms in Siaya District, western Kenya, revealed that N inputs from fertilizers of 7 kg N ha^{-1} yr^{-1}, which the farmers could afford to provide, were much less than the amount removed in grain and crop remains, resulting in net N imbalance of 52 kg N ha^{-1} yr^{-1} (Vitousek et al. 2009). Many agriculture systems in SSA persist by drawing down the nutrient capital of what were once high-fertility soils. As a result, unlike most regions of the world, crop yields have not increased substantially in SSA, and nearly 225 million people remain chronically undernourished (FAO 2017). Population pressures in some countries have reduced or eliminated natural fallow periods, reducing nutrient and organic matter inputs (Lemenih et al. 2005) and thus causing declines in soil biological activity and soil species diversity. Introduction of pesticides and other potentially toxic materials may also affect soil biology.

Similarly, there are numerous mechanical effects of land management that have a significant impact on soil biology. Among these effects are erosion, compaction, and changes in patterns of drainage. The removal of significant amounts of surface soil material by erosion processes results in the loss of organic carbon, inorganic nutrients, and microbial biomass. Subsurface materials are generally less conducive to microbial growth, due to inadequate organic carbon, adverse chemical conditions (i.e., low or high pH, etc.), or structural properties that impede oxygen and/or water supply. Reductions in organic matter can reduce porosity (Beare et al. 1997) and infiltration, causing changes in water and nutrient cycles, plant productivity, and also affecting the energy balance of a system (Palm et al. 2007). The abundance and biodiversity of soil organisms are reduced by other management practices, including intensive grazing, biomass burning—either land clearing or crop residues, leaving soil bare, tillage and seedbed preparation, monocropping, and excess fertilizer application (Wood et al. 2015; Tully et al. 2015). The biological degradation of soil due to changes in functional diversity of soil biota affects the availability of nutrients, alters pest and disease pressure, and changes the complexity of food webs with consequences for ecosystem resilience (de Vries and Bardgett 2012; Tully et al. 2015).

1.4 SOIL DEGRADATION AND THE THREAT OF FOOD SECURITY IN SUB-SAHARAN AFRICA

Soil degradation, world food security, and the quality of environment are the challenges that assume a significant role in the 21st century since only 11% of the global land surface is considered prime cropland (Sivakumar and Stefanski 2007). Soil degradation and climate change pose an enormous risk to food production, food security, and natural resource conservation in SSA (Webb et al. 2017). Addressing these challenges is vital for building sustainable agroecosystems that can feed the fast-growing population in SSA. Soil degradation and climate change are interlinked processes that have biophysical and human drivers, impacts, and responses (Herrick et al. 2013). There is extensive knowledge about the processes and effects of land and soil degradation and climate change as separate phenomena but less is known about the links between the two challenges and their interaction under different agroecosystems and how societies can simultaneously adapt to the combined impacts (Reed and Stringer 2016). Evidence indicates that land degradation contributes to climate change, while climate change can exacerbate land degradation (Cowie et al. 2011). For example, degradation affects C sequestration and C storage in soils and vegetation, as well as C and N cycles, altering greenhouse gas (GHG) emissions and affecting the climate (Reed and Stringer 2016). Climate change affects land degradation through changes in rainfall patterns and in biomass production. However, there are knowledge gaps regarding the impact of soil degradation on overall C budget, C stability, and C stabilization mechanisms under restored soils. Therefore, research is needed to identify and quantify linkages and feedback mechanisms between climate change and soil degradation in regions affected by soil degradation, especially SSA, which is vulnerable to both degradation and climate change impact. Similarly, both climate change and soil degradation may impact biodiversity and the provision of a range of ecosystem services, which may further exacerbate land degradation and compromise capacities to adapt to climate change and maintain sustainable livelihoods in SSA (Reed and Stringer 2016).

In SSA, investments made in soil conservation are low, and food production has stagnated. Among global issues, the importance of soil degradation in SSA is enhanced because of its impact on global food security and the extent of extreme poverty in the region. It affects crop yields by influencing soil properties and microclimates, and the interactions between them. Soil degradation in SSA is taking place against a background of increasing population and deteriorating climate conditions in food-insecure parts of the world (Vlek et al. 2010). Food security is directly related to soil productivity, and soil degradation is among the principal causes of decreases in soil productivity (Tully et al. 2015). The inadequacy of the food supply in the face of a rapidly growing population is the major challenge confronting many SSA countries today. Food insecurity affects about 224 million people in SSA, and about 25% of undernourished people in 2016 lived in SSA (FAO 2017). SSA has the highest prevalence of undernourishment of all the regions in the world. Although many countries in the region have adopted policies for the development of agriculture with the expectation that agriculture can help in the eradication of poverty, increasing efforts to raise agricultural growth have led to soil degradation and land desertification. Agriculture productivity and food security in SSA are seriously threatened by soil erosion and the steady decline in soil fertility. Declining soil fertility jeopardizes the sustainability of the farming system, especially in arid and semi-arid areas of SSA that are ecologically fragile. The reduction in crop yields, as a result of land degradation, in Africa ranges from 2% to 40%, with a mean loss of 8.2% for the continent overall (Eswaran et al. 2001).

Limiting soil degradation and restoring degraded soils are integral to an adaptation plan for agriculture, because soil degradation often increases the exposure and sensitivity of agroecological systems to climate change impacts and reduces resilience to climate change and the adaptive capacity of land users (Gisladottir and Stocking 2005). Stabilizing and/or reversing soil degradation offers opportunities for increasing land productivity, increasing food production, advancing food security, and reducing poverty. Agricultural practices that help to reverse soil degradation are beneficial to long-term agricultural sustainability. Similarly, preventing soil erosion minimizes CO_2 emissions, and the restoration of degraded soils sequesters C and reduces the atmospheric concentration of GHGs. Measures for controlling soil erosion include the planting of live barriers such as grass strips, the construction of terraces and stone bunds, or applying surface mulch. Land management practices that control soil degradation promote sustainable land management, the conservation of biodiversity, the control of climate change, and the prevention of land degradation simultaneously.

Lessons from the Green Revolution in other regions point to increased fertilizer use, improved crop varieties, and irrigation infrastructure development as crucial investments to crop productivity. In SSA's situation, however, efforts to intensify crop production will require additional investment for the rehabilitation of large areas of degraded soils. Nitrogen and P are the main yield-limiting nutrients in SSA, but the optimal productivity with N and P fertilizer is only applicable to non-degraded soils. Soil degradation and soil fertility decline in Africa are deeply complex with intertwining and cyclical causes, ranging from poor inherent soil qualities to population pressure to insecure land tenure and climate change, among many other factors. In most cases, better land management practices are sacrificed for short-term needs. Soil degradation increases the complexity of soil fertility management options required to increase productivity of the land. Combining organic and inorganic fertilizer is a sound management principle for smallholder farming in the SSA tropics to improve soil fertility and enhance soil organic carbon (SOC) storage (Vanlauwe et al. 2010). SOM plays an important role in the restoration and rehabilitation of degraded soils through enhancing nutrients and water storage for plants, increasing the activity and diversity of soil biota, improving soil structure and tilth, minimizing the susceptibility of soil to erosion, and increasing the soil's resilience against the changing and uncertain climate. When the concentration of SOC falls below the threshold, key soil properties are adversely affected, inhibiting plant growth. Thus, increasing and maintaining the concentration of SOC to above the threshold level is a critical determinant of soil health and productivity. Reduced or no tillage protects vulnerable soils, improves soil structure and fertility, encourages populations of beneficial soil biota, and creates savings on labor and machinery. Likewise, the intercropping of cereals and legumes, the rotation of cereals and legumes, and the application of organic and inorganic fertilizers—either simultaneously or sequentially to the same crop—are among common nutrient management practices for increased crop yield while also stabilizing and/or reversing soil degradation in SSA. One of the challenges in restoring the productivity of severely degraded soils is a poor response to nutrients application due to the interaction of multiple constraints (physical, chemical, and biological), requiring multipurpose options that address several constraints. One such options is to increase the SOM content of such soils by the repeated application of manure and/or crop residues. For example, severely degraded sandy soil in Zimbabwe required the application of 10 Mg ha^{-1} of animal manure in combination with fertilizers for three years in order to attain a significant increase in crop productivity (Zingore et al. 2007b). Other approaches to restore severely degraded soils include the use of green manures, legume crops adapted to marginal soil conditions, deep-rooting hedgerow trees, and agro-forest practices. Soil degradation in SSA mostly affects poor smallholder farmers with limited access to fertilizers and manure, and limited resources for soil conservation practices. These farmers also have limited land to spare for rehabilitation that uses technologies that do not contribute directly to food production.

1.5 CONCLUSIONS

Soil degradation is the single largest threat to soil productivity and sustainability of crop and livestock production systems in SSA and will remain so for the near future. Soil degradation in arable land is driven by suboptimal land management practices that induce declines in the chemical, physical, and biological quality of the soil, and reduce its capacity to support production and other ecosystem and environmental functions. Deforestation, overgrazing of rangelands, mismanagement of croplands, land shortage, and socioeconomic and political factors are among the major causes of soil degradation that can be attributed directly to human activities. In many parts of SSA, these practices and the lack of inputs have led to soil erosion, salinization, acidification, loss of vegetation, and decline in crop productivity. Increasing population pressures have caused the abandonment of shifting cultivation and necessitated the cultivation of marginal lands that are prone to soil erosion and other types of degradation. It is no longer feasible to use extended fallow periods to restore soil fertility, and shorter fallow periods or continuous cultivation are not able to regenerate soil productivity. For many decades, the annual food production increase does not match the population growth in SSA, rendering food insecurity more common and more frequent in the region. In addition to low agronomic production, the low use of fertilizers (organic and inorganic) in SSA has a greater negative consequence of soil nutrient mining, leading to nutrient depletion in the croplands of SSA. The average rates of fertilizer use in SSA are just 15 kg ha^{-1}, compared to a worldwide average of about 130 kg ha^{-1}. Increasing the use of fertilizer will benefit the environment by increasing biomass production and the SOM content, which contribute to nutrient retention and restore soil quality. Controlling soil degradation and restoring degraded soils in SSA is essential for building sustainable agroecological systems that are climate resilient, while also addressing food and nutrition insecurity. Addressing soil degradation is also essential for building climate resilient systems that conserve biodiversity and meet sustainable development goals. It should be noted that soil degradation in SSA is taking place against a background of increasing population and deteriorating climate conditions, in regions that are already food insecure. SSA has the fastest growing population globally, while per capita food production is barely changing. Sustainable agriculture intensification is needed to feed the growing population in the region. But it must be implemented in a way that water and nutrient resources are used efficiently, while pressure on forests and other fragile land is also relieved. As a natural resource, soils are often overlooked. Neglected soils, over time, result in low crop productivity and food insecurity that disproportionately affect resource-poor farmers, especially in SSA. Therefore, nurturing, conserving, restoring, and enhancing this indispensable resource should be a major priority. The following steps are recommended:

- Strengthen political support for sustainable land management. As proposed in Sustainable Development Goals (SGD 15), achieving zero net soil degradation needs to be part of the development agenda for SSA. The implementation of programs such as the initiative for the Adaptation of African Agriculture (AAA), which promotes and encourages the implementation of projects to improve soil management, agricultural water control, climate risk management, and financial capability building, and the "4 per 1000" initiative, which promotes an annual growth rate of 0.4% in the soil carbon stocks, will play a significant role in restoring degraded soils and increase food security in SSA.
- Increase financial investment in soil management and commit resources dedicated to sustainable land and soil management practices.
- Secure land rights to increase care, private investment, and management of farming land.
- Implement integrated soil management that incorporates conservation agriculture, targeted use of inputs, and remedies needed to restore, conserve and enhance soil quality.
- Practice climate-smart soil research and disseminate the knowledge to farmers.

REFERENCES

Amede, T., Belachew, T., Geta, E. 2001. *Reversing the Degradation of Arable Land in the Ethiopian Highlands*. IIED, London. 23 p.

Balasubramanian, V., Sie, M., Hijmans, R.J., Otsuka, K. 2007. Increasing rice production in sub-Saharan Africa: Challenges and opportunities. *Advances in Agronomy* 94:55–133. doi: 10.1016/S0065-2113(06)94002-4.

Bardgett, R.D., van der Putten, W.H. 2014. Belowground biodiversity and ecosystem functioning. *Nature* 515 (7528):505–511. doi: 10.1038/nature13855.

Bationo, A., Hartemink, A.E., Lungo, O., Naimi, M., Okoth, P., Smaling, E.M.A., Thiombiano, L. 2006. *African Soils: Their Productivity and Profitability of Fertilizer Use: Background Paper for the African Fertilizer Summit 9–13th June 2006*. IFDC, Abuja, Nigeria, Togo. Available at: http://edepot.wur.nl/26759.

Bationo, A., Lompo, F., Koala, S. 1998. Research on nutrient flows and balances in west Africa: State-of-the-art. *Agriculture, Ecosystems and Environment* 71 (1–3):19–35. doi: 10.1016/S0167-8809(98)00129-7.

Beare, M.H., Reddy, M.V., Tian, G., Srivastava, S.C. 1997. Agricultural intensification, soil biodiversity and agroecosystem function in the tropics: The role of decomposer biota. *Applied Soil Ecology* 6 (1):87–108. doi: 10.1016/S0929-1393(96)00150-3.

Benin, S., Nin-Pratt, A., Wood, S., Guo, Z. 2011. *Trends and Spatial Patterns in Agricultural Productivity in Africa, 1961–2010. Regional Strategic Analysis and Knowledge Support System (ReSAKSS) Annual Trends and Outlook Report 2011.* International Food Policy Research Institute (IFPRI), Addis Ababa, Ethiopia. 359 p.

Borrelli, P., Robinson, D.A., Fleischer, L.R., Lugato, E., Ballabio, C., Alewell, C., Meusburger, K., Modugno, S., Schütt, B., Ferro, V., Bagarello, V., Oost, K.V., Montanarella, L., Panagos, P. 2017. An assessment of the global impact of 21st century land use change on soil erosion. *Nature Communications* 8 (1):2013. doi: 10.1038/s41467-017-02142-7.

Braimoh, A.K. 2006. Random and systematic land-cover transitions in northern Ghana. *Agriculture, Ecosystems and Environment* 113 (1–4):254–263. doi: 10.1016/j.agee.2005.10.019.

Bridges, E.M., Oldeman, L.R. 1999. Global assessment of human-induced soil degradation. *Arid Soil Research and Rehabilitation* 13 (4):319–325. doi: 10.1080/089030699263212.

Buerkert, A., Hiernaux, P. 1998. Nutrients in the West African Sudano-Sahelian zone: Losses, transfers and role of external inputs. *Zeitschrift für Pflanzenernährung und Bodenkunde* 161 (4):365–383. doi: 10.1002/jpln.1998.3581610405.

Byerlee, D., Deininger, K. 2013. Growing resource scarcity and global farmland investment. *Annual Review of Resource Economics* 5 (1):13–34. doi: 10.1146/annurev-resource-091912-151849.

Cherlet, M., Hutchinson, C., Reynolds, J., Hill, J., Sommer, S., von Maltitz, G. 2018. *World Atlas of Desertification.* Publication Office of the European Union, Luxembourg.

Cobo, J.G., Dercon, G., Cadisch, G. 2010. Nutrient balances in African land use systems across different spatial scales: A review of approaches, challenges and progress. *Agriculture, Ecosystems and Environment* 136 (1–2):1–15. doi: 10.1016/j.agee.2009.11.006.

Collins, A.L., Walling, D.E., Sichingabula, H.M., Leeks, G.J.L. 2001. Suspended sediment source fingerprinting in a small tropical catchment and some management implications. *Applied Geography* 21 (4):387–412. doi: 10.1016/S0143-6228(01)00013-3.

Cowie, A.L., Penman, T.D., Gorissen, L., Winslow, M.D., Lehmann, J., Tyrrell, T.D., Twomlow, S., Wilkes, A., Lal, R., Jones, J.W., Paulsch, A., Kellner, K., Akhtar-Schuster, M. 2011. Towards sustainable land management in the drylands: Scientific connections in monitoring and assessing dryland degradation, climate change and biodiversity. *Land Degradation and Development* 22 (2):248–260. doi: 10.1002/ldr.1086.

de Rouw, A., Rajot, J.-L. 2004. Nutrient availability and pearl millet production in Sahelian farming systems based on manuring or fallowing. *Agriculture, Ecosystems and Environment* 104 (2):249–262. doi: 10.1016/j.agee.2003.12.019.

de Vries, F.T., Bardgett, R.D. 2012. Plant–microbial linkages and ecosystem nitrogen retention: Lessons for sustainable agriculture. *Frontiers in Ecology and the Environment* 10 (8):425–432. doi: 10.1890/110162.

Dejene, A., Shishira, E.K., Yanda, P.Z., Johnsen, F.H. 1997. *Land Degradation in Tanzania: Perception from the Village.* The World Bank, Washington, D.C. 96 p.

den Biggelaar, C., Lal, R., Wiebe, K., Breneman, V. 2004. Global impact of soil erosion on productivity. I. Absolute and relative erosion-induced yield losses. *Advances in Agronomy* 81:1–48.

Dewitte, O., Jones, A., Spaargaren, O., Breuning-Madsen, H., Brossard, M., Dampha, A., Deckers, J., Gallali, T., Hallett, S., Jones, R., Kilasara, M., Le Roux, P., Michéli, E., Montanarella, L., Thiombiano, L., Van Ranst, E., Yemefack, M., Zougmore, R. 2013. Harmonisation of the soil map of Africa at the continental scale. *Geoderma* 211:138–153. doi: 10.1016/j.geoderma.2013.07.007.

DFID. 2004. Guidelines for predicting and minimizing sedimentation in small dams. Department For International Development, Report OD 152, HR Wallingford, Wallingford, UK.

Dregne, H.E. 1990. Erosion and soil productivity in Africa. *Journal of Soil and Water Conservation* 45 (4):431–436.

ECA. 2007. *Africa Review Report on Drought and Desertification.* Economic Commission for Africa, United Nations Economic and Social Council, Addis Ababa, Ethiopia. 18 p.

Eswaran, H., Almaraz, R., van den Berg, E., Reich, P. 1997. An assessment of the soil resources of Africa in relation to productivity. *Geoderma* 77 (1):1–18. doi: 10.1016/S0016-7061(97)00007-4.

Eswaran, H., Lal, R., Reich, P.F. 2001. Land degradation: An overview. In: E.M. Bridges, I.D. Hannam, L.R. Oldeman, F.W.T. Pening-de-Vries, S.J. Scherr, S. Sompatpanit (Eds.). *Responses to Land Degradation.* Proceedings of the 2nd International Conference on Land Degradation and Desertification. Oxford Press, New Delhi, Khon Kaen, Thailand. pp. 20–35.

FAO. 1995. *Land and Environmental Degradation and Desertification in Africa: Issues and Options for Sustainable Economic Development with Transformation.* Food and Agriculture Organization of the United Nations (FAO), Rome. 118 p.

FAO. 2015a. *Global Forest Resources Assessment 2015: Desk Reference.* Food and Agriculture Organization of the United Nation (FAO), Rome, Italy. 244 p. Available at: www.fao.org/forest-resources-assessment/en/.

FAO. 2015b. *Status of the World's Soil Resources: Main Report.* Food and Agriculture Organization of the United Nations (FAO), Rome, Italy. 602p. Available at: www.fao.org/3/a-i5199e.pdf.

FAO. 2017. *Regional Overview of Food Security and Nutrition in Africa 2017. The Food Security and Nutrition–Conflict Nexus: Building Resilience for Food Security, Nutrition and Peace.* Food and Agriculture Organization of the United Nations, Accra, Ghana. 92 p. Available at: www.fao.org/3/a-i7967e.pdf.

FAOSTAT. 2018. *World Food and Agriculture Statistics.* Food and Agriculture Organization of the United Nations, Statistical Division, Rome, Italy [Accessed June 2018]. Available at: http://www.fao.org/statistics/en/; http://www.fao.org/faostat/en/#home.

García-Ruiz, J.M., Beguería, S., Nadal-Romero, E., González-Hidalgo, J.C., Lana-Renault, N., Sanjuán, Y. 2015. A meta-analysis of soil erosion rates across the world. *Geomorphology* 239:160–173. doi: 10.1016/j.geomorph.2015.03.008.

Gisladottir, G., Stocking, M. 2005. Land degradation control and its global environmental benefits. *Land Degradation and Development* 16 (2):99–112. doi: 10.1002/ldr.687.

Gorse, J.E., Steeds, D.R. 1985. Desertification in the Sahelian and Sussdanian zones of west Africa. World Bank Technical Paper No. 61. The World Bank, Washington, D.C.

Haregeweyn, N., Melesse, B., Tsunekawa, A., Tsubo, M., Meshesha, D., Balana, B.B. 2012. Reservoir sedimentation and its mitigating strategies: A case study of Angereb reservoir (NW Ethiopia). *Journal of Soils and Sediments* 12 (2):291–305. doi: 10.1007/s11368-011-0447-z.

Hartemink, A.E. 2006. Assessing soil fertility decline in the tropics using soil chemical data. *Advances in Agronomy* 89:179–225.

HarvestChoice. 2010. *Updating Soil Functional Capacity Classification System.* International Food Policy Research Institute, Washington, D.C., and University of Minnesota, St. Paul, MN. Available at: http://harvestchoice.org/node/1435.

Henao, J., Baanante, C.A. 1999. *Estimating Rates of Nutrient Depletion in Agricultural Soils of Africa.* International Fertilizer Development Center (IFDC), Muscle Shoals, AL. Available at: https://pdf.usaid.gov/pdf_docs/pnacf868.pdf.

Henao, J., Baanante, C.A. 2006. *Agricultural Production and Soil Nutrient Mining in Africa Implications for Resource Conservation and Policy Development.* IFDC Technical Bulletin International Fertilizer Development Center (IFDC), Muscle Shoals, AL. Available at: www.ifdc.org.

Herrick, J.E., Sala, O.E., Karl, J.W. 2013. Land degradation and climate change: A sin of omission? *Frontiers in Ecology and the Environment* 11 (6):283–283. doi: 10.1890/1540-9295-11.6.283.

Issaka, S., Ashraf, M.A. 2017. Impact of soil erosion and degradation on water quality: A review. *Geology, Ecology, and Landscapes* 1 (1):1–11. doi: 10.1080/24749508.2017.1301053.

Jones, A., Breuning-Madsen, H., Brossard, M., Dampha, A., Deckers, J., Dewitte, O., Gallali, T., Hallett, S., Jones, R., Kilasara, M., Le Roux, P., Micheli, E., Montanarella, L., Spaargaren, O., Thiombiano, L., Van Ranst, E., Yemefack, M., Zougmore, R. 2013. *Soil Atlas of Africa.* European Commission, Publications Office of the European Union, Luxembourg. p. 176. Available at: http://esdac.jrc.ec.europa.eu/Library/Maps/Africa_Atlas/Index.html.

Juo, A.S.R., Dabiri, A., Franzluebbers, K. 1995. Acidification of a kaolinitic alfisol under continuous cropping with nitrogen fertilization in West Africa. *Plant and Soil* 171 (2):245–253. doi: 10.1007/BF00010278.

Kiage, L.M. 2013. Perspectives on the assumed causes of land degradation in the rangelands of sub-Saharan Africa. *Progress in Physical Geography: Earth and Environment* 37 (5):664–684. doi: 10.1177/0309133313492543.

Kimaro, D.N., Deckers, J.A., Poesen, J., Kilasara, M., Msanya, B.M. 2005. Short and medium term assessment of tillage erosion in the Uluguru Mountains, Tanzania. *Soil and Tillage Research* 81 (1):97–108. doi: 10.1016/j.still.2004.05.006.

Lal, R. 1976. Soil erosion on alfisols in Western Nigeria: V. The changes in physical properties and the response of crops. *Geoderma* 16 (5):419–431. doi: 10.1016/0016-7061(76)90005-7.

Lal, R. 1981. Soil erosion problems on alfisols in Western Nigeria, VI. Effects of erosion on experimental plots. *Geoderma* 25 (3–4):215–230. doi: 10.1016/0016-7061(81)90037-9.

Lal, R. 1995. Erosion-crop productivity relationships for soils of Africa. *Soil Science Society of America Journal* 59 (3):661–667. doi: 10.2136/sssaj1995.03615995005900030004x.

Lal, R. 2000. Soil management in the developing countries. *Soil Science* 165 (1):57–72.

Lal, R. 2001. Soil degradation by erosion. *Land Degradation and Development* 12 (6):519–539. doi: 10.1002/ldr.472.

Lal, R. 2003. Soil erosion and the global carbon budget. *Environment International* 29 (4):437–450. doi: 10.1016/S0160-4120(02)00192-7.

Lemenih, M., Karltun, E., Olsson, M. 2005. Assessing soil chemical and physical property responses to deforestation and subsequent cultivation in smallholders farming system in Ethiopia. *Agriculture, Ecosystems and Environment* 105 (1–2):373–386. doi: 10.1016/j.agee.2004.01.046.

Li, Z., Fang, H. 2016. Impacts of climate change on water erosion: A review. *Earth-Science Reviews* 163:94–117. doi: 10.1016/j.earscirev.2016.10.004.

Mafongoya, P.L., Bationo, A., Kihara, J., Waswa, B.S. 2006. Appropriate technologies to replenish soil fertility in southern Africa. *Nutrient Cycling in Agroecosystems* 76 (2–3):137–151. doi: 10.1007/s10705-006-9049-3.

Mbagwu, J.S.C., Lal, R., Scott, T.W. 1984. Effects of Desurfacing of alfisols and ultisols in Southern Nigeria: I. Crop Performance. *Soil Science Society of America Journal* 48 (4):828–833. doi: 10.2136/sssaj1984.0 3615995004800040026x.

McDaniel, M.D., Tiemann, L.K., Grandy, A.S. 2014. Does agricultural crop diversity enhance soil microbial biomass and organic matter dynamics? A meta-analysis. *Ecological Applications* 24 (3):560–570. doi: 10.1890/13-0616.1.

MEA. 2005. *Ecosystems and Human Well-Being: Synthesis*. Island Press, Washington, D.C. p. 137.

Montgomery, D.R. 2007. Soil erosion and agricultural sustainability. *Proceedings of the National Academy of Sciences of the United States of America* 104 (33):13268–13272. doi: 10.1073/pnas.0611508104.

Mueller, N.D., Gerber, J.S., Johnston, M., Ray, D.K., Ramankutty, N., Foley, J.A. 2012. Closing yield gaps through nutrient and water management. *Nature* 490 (7419):254–257. doi: 10.1038/nature11420.

Mullan, D. 2013. Managing soil erosion in northern Ireland: A review of past and present approaches. *Agriculture* 3 (4):684–699. doi: 10.3390/agriculture3040684.

NEPAD. 2013. *Agriculture in Africa: Transformation and Outlook*. New Partnership for African Develoment (NEPAD), Johannesburg, South Africa. 72 p.

Obalum, S.E., Buri, M.M., Nwite, J.C., Hermansah, Watanabe, Y., Igwe, C.A., Wakatsuki, T. 2012. Soil degradation-induced decline in productivity of sub-Saharan African soils: The prospects of looking downwards the lowlands with the sawah ecotechnology. *Applied and Environmental Soil Science* 2012:10. doi: 10.1155/2012/673926.

Obi, M.E., Ngwu, O.E., Mbagwu, J.S.C. 2005. Effect of desurfacing on soil properties and maize yield— Research note. *Nigerian Journal of Soil Science* 15:148–150. doi: 10.4314/njser.v6i1.28383.

OECD/FAO. 2016. *OECD-FAO Agricultural Outlook 2016–2025*. Organisation for Economic Co-operation and Development (OECD) Publishing, Paris, France. Available at: https://www.oecd-ilibrary.org.

Oldeman, L.R. 1991. Global Extent of Soil Degradation, pp. 19–36. In: *ISRIC Bi-Annual Report 1991–1992*. International Soil Reference and Information Centre (ISRIC), Wageningen, The Netherlands.

Onduru, D.D., Jager, A.D., Muchena, F.N., Gachimbi, L., Gachini, G.N. 2007. Socio-economic factors, soil fertility management and cropping practices in mixed farming systems of sub-Saharan Africa: A study in Kiambu, Central Highlands of Kenya. *International Journal of Agricultural Research* 2 (5):426–439.

Osman, K.T. 2014. *Soil Degradation, Conservation and Remediation*. Springer, New York.

Otte, M.J., Chilonda, P. 2002. *Cattle and Small Ruminant Production Systems in Sub-Saharan Africa: A Systematic Review*. Food and Agriculture Organization of the UN, Livestock Information Sector Analysis and Policy Branch, Rome, Italy. 98 p.

Oyedele, D.J., Aina, P.O. 2006. Response of soil properties and maize yield to simulated erosion by artificial topsoil removal. *Plant and Soil* 284 (1–2):375–384. doi: 10.1007/s11104-006-0041-0.

Palm, C., Sanchez, P., Ahamed, S., Awiti, A. 2007. Soils: A contemporary perspective. *Annual Review of Environment and Resources* 32 (1):99–129. doi: 10.1146/annurev.energy.31.020105.100307.

Podwojewski, P., Janeau, J.L., Grellier, S., Valentin, C., Lorentz, S., Chaplot, V. 2011. Influence of grass soil cover on water runoff and soil detachment under rainfall simulation in a sub-humid South African degraded rangeland. *Earth Surface Processes and Landforms* 36 (7):911–922. doi: 10.1002/esp.2121.

Reed, M.S., Stringer, L.C. 2016. *Land Degradation Desertification and Climate Change: Anticipating Assessing and Adapting to Future Change*. Routledge, London and New York.

Rohde, R.F., Moleele, N.M., Mphale, M., Allsopp, N., Chanda, R., Hoffman, M.T., Magole, L., Young, E. 2006. Dynamics of grazing policy and practice: Environmental and social impacts in three communal areas of southern Africa. *Environmental Science and Policy* 9 (3):302–316. doi: 10.1016/j.envsci.2005.11.009.

Salako, F.K., Dada, P.O., Adejuyigbe, C.O., Adedire, M.O., Martins, O., Akwuebu, C.A., Williams, O.E. 2007. Soil strength and maize yield after topsoil removal and application of nutrient amendments on a gravelly Alfisol toposequence. *Soil and Tillage Research*, 94 (1):21–35. doi: org: 10.1016/j.still.2006.06.005.

Sanchez, P.A. 2002. Soil fertility and hunger in Africa. *Science* 295 (5562):2019–2020. doi: 10.1126/science.1065256.

Sanchez, P.A., Palm, C.A., Buol, S.W. 2003. Fertility capability soil classification: A tool to help assess soil quality in the tropics. *Geoderma* 114 (3–4):157–185. doi: 10.1016/S0016-7061(03)00040-5.

Sanchez, P.A., Shepherd, K.D., Soule, M.J., Place, F.M., Buresh, R.J., Izac, A.-M.N., Mokwunye, A.U., Kwesiga, F.R., Ndiritu, C.G., Woomer, P.L. 1997. Soil fertility replenishment in Africa: An investment in natural resource capital, pp. 1–46. In: R.J. Buresh, P.A. Sanchez, F. Calhoun (Eds.). *Replenishing Soil Fertility in Africa.* Soil Science Society of America and American Society of Agronomy, Madison, WI.

Scherr, S.J. 1999. *Soil Degradation: A Threat to Developing-Country Food Security by 2020?* IFPRI Food, Agriculture and the Environment. Discussion paper 27. International Food Policy Research Institute (IFPRI), Washington, D.C.

Schertz, D.L. 1983. The basis for soil loss tolerances. *Journal of Soil and Water Conservation* 38 (1):10–14.

Sims, G.K. 1990. Biological degradation of soil, pp. 289–330. In: R. Lal, B.A. Stewart (Eds.). *Advances in Soil Science: Soil Degradation.* Springer, New York.

Sivakumar, M.V.K., Stefanski, R. 2007. Climate and land degradation—An overview, pp. 105–135. In: M.V.K. Sivakumar, N. Ndiang'ui (Eds.). *Climate and Land Degradation.* Springer, Berlin and Heidelberg.

Stoorvogel, J., Smaling, E. 1990. *Assessment of Soil Nutrient Depletion in Sub-Saharan Africa: 1983–2000.* The Winand Staring Centre, Wagenigen, The Netherlands. 580 p.

Stoorvogel, J.J., Smaling, E.M.A., Janssen, B.H. 1993. Calculating soil nutrient balances in Africa at different scales. *Fertilizer Research* 35 (3):227–235. doi: 10.1007/BF00750641.

Tamene, L., Le, Q.B. 2015. Estimating soil erosion in sub-Saharan Africa based on landscape similarity mapping and using the revised universal soil loss equation (RUSLE). *Nutrient Cycling in Agroecosystems* 102 (1):17–31. doi: 10.1007/s10705-015-9674-9.

Tamene, L., Park, S.J., Dikau, R., Vlek, P.L.G. 2006. Reservoir siltation in the semi-arid highlands of northern Ethiopia: Sediment yield–catchment area relationship and a semi-quantitative approach for predicting sediment yield. *Earth Surface Processes and Landforms* 31 (11):1364–1383. doi: 10.1002/esp.1338.

Tamene, L., Vlek, P.L.G. 2007. Assessing the potential of changing land use for reducing soil erosion and sediment yield of catchments: A case study in the highlands of northern Ethiopia. *Soil Use and Management* 23 (1):82–91. doi: 10.1111/j.1475-2743.2006.00066.x.

Tamene, L., Vlek, P.L.G. 2008. Soil erosion studies in northern Ethiopia, pp. 73–100. In: A.K. Braimoh, P.L.G. Vlek (Eds.). *Land Use and Soil Resources.* Springer, Dordrecht, Netherlands.

Terefe, H.A. 2012. *People in Rises: Tackling the Root Causes of Famine in the Horn of Africa.* Norwegian Agricultural Economics Research Institute, Oslo, Norway.

Thomas, D.S.G., Sporton, D., Perkins, J. 2000. The environmental impact of livestock ranches in the Kalahari, Botswana: Natural resource use, ecological change and human response in a dynamic dryland system. *Land Degradation and Development* 11 (4):327–341. doi: 10.1002/1099-145X(200007/08)11:4<327::AID-LDR395>3.0.CO;2-V.

Tully, K., Sullivan, C., Weil, R., Sanchez, P. 2015. The state of soil degradation in sub-Saharan Africa: Baselines, trajectories, and solutions. *Sustainability* 7 (6):6523–6552.

UNCCD. 2011. *Desertification: A Visual Synthesis.* United Nations Convention to Combat Desertification (UNCCD), Paris, France. 49 p.

UNDESA. 2015. *World Population Prospects: Key Findings and Advance Tables: The 2015 Revision.* United Nations Department of Economic and Social Affairs (UNDESA), New York.

UNDESA. 2017. *World Population Prospects: The 2017 Revision, Key Findings and Advance Tables.* Working Paper No. ESA/P/WP/248. United Nations, Department of Economic and Social Affairs, Population Division, New York.

UNEP. 2013. *Africa Environment Outlook 3: Our Environment, Our Health.* United Nations Environment Programme (UNEP), Nairobi, Kenya. 242 p. Available at: https://wedocs.unep.org.

UNEP. 2015. *The Economics of Land Degradation in Africa: Benefits of Action Outweigh the Costs.* United Nations Environment Program (UNEP), Economics of Land Degradation Initiative (ELDI), Nairobi, Kenya. 154 p. Available at: www.eld-initiative.org/fileadmin/pdf/ELD-unep-report_07_spec_72dpi.pdf.

UNEP. 2016. *Global Environmental Outlook 6: Regional Assessment for Africa.* United Nations Environment Programme (UNEP), Nairobi, Kenya. Available at: www.unenvironment.org/resources/report/geo-6-global-environment-outlook-regional-assessment-africa.

Ussiri, D.A.N., Lal, R. 2017. *Carbon Sequestration for Climate Change Mitigation and Adaptation.* Springer Nature, Switzerland. 545 p.

Vanlauwe, B., Bationo, A., Chianu, J., Giller, K.E., Merckx, R., Mokwunye, U., Ohiokpehai, O., Pypers, P., Tabo, R., Shepherd, K.D., Smaling, E.M.A., Woomer, P.L., Sanginga, N. 2010. Integrated soil fertility management: Operational definition and consequences for implementation and dissemination. *Outlook on Agriculture* 39 (1):17–24. doi: 10.5367/000000010791169998.

Vitousek, P.M., Naylor, R., Crews, T., David, M.B., Drinkwater, L.E., Holland, E., Johnes, P.J., Katzenberger, J., Martinelli, L.A., Matson, P.A., Nziguheba, G., Ojima, D., Palm, C.A., Robertson, G.P., Sanchez, P.A., Townsend, A.R., Zhang, F.S. 2009. Nutrient imbalances in agricultural development. *Science* 324 (5934):1519–1520. doi: 10.1126/science.1170261.

Vlek, P.L.G., Le, Q.B., Tamene, L. 2008. *Land Decline in Land-Rich Africa: A Creeping Disaster in the Making*. Consultative Group on International Agricultural Research (CGIAR), Science Council Secretariat, Rome, Italy. 55 p.

Vlek, P.L.G., Le, Q.B., Temene, L. 2010. Assessment of land degradation, its possible causes and threat to food security in sub-Saharan Africa, pp. 57–86. In: R. Lal, B.A. Stewart (Eds.). *Advances in Soil Science: Food Security and Soil Quality*. CRC Press, Heidelberg, Germany.

Vrieling, A. 2006. Satellite remote sensing for water erosion assessment: A review. *CATENA* 65 (1):2–18. doi: 10.1016/j.catena.2005.10.005.

Walling, D. 2008. The changing sediment loads of the world's rivers. Paper Read at International Symposium on Sediment Dynamics in Changing Environment, Christchurch, New Zealand.

Walling, D.E. 2009. *The Impact of Global Change on Erosion and Sediment Transport by Rivers: Current Progress and Future Challenges*. United Nations World Water Assessment Programme, Paris, France. 26 p.

Webb, N.P., Marshall, N.A., Stringer, L.C., Reed, M.S., Chappell, A., Herrick, J.E. 2017. Land degradation and climate change: Building climate resilience in agriculture. *Frontiers in Ecology and the Environment* 15 (8):450–459. doi: 10.1002/fee.1530.

Weil, R., Brady, N.C. 2016. *The Nature and Properties of Soils*. 15th Edition. Pearson Education, Columbus, OH. 1070 p.

Wood, S.A., Bradford, M.A., Gilbert, J.A., McGuire, K.L., Palm, C.A., Tully, K.L., Zhou, J., Naeem, S. 2015. Agricultural intensification and the functional capacity of soil microbes on smallholder African farms. *Journal of Applied Ecology* 52 (3):744–752. doi: 10.1111/1365-2664.12416.

Zingore, S., Murwira, H.K., Delve, R.J., Giller, K.E. 2007a. Influence of nutrient management strategies on variability of soil fertility, crop yields and nutrient balances on smallholder farms in Zimbabwe. *Agriculture, Ecosystems and Environment* 119 (1–2):112–126. doi: 10.1016/j.agee.2006.06.019.

Zingore, S., Murwira, H.K., Delve, R.J., Giller, K.E. 2007b. Soil type, management history and current resource allocation: Three dimensions regulating variability in crop productivity on African smallholder farms. *Field Crops Research* 101 (3):296–305.

2 Soil Degradation with Reference to Nutrient Mining and Soil Fertility Decline in Sub-Saharan Africa

Hamisi J. Tindwa, Ernest Semu,
Hussein B. Shelukindo, and Bal Ram Singh

CONTENTS

2.1 INTRODUCTION

The tropical zone of Earth, the region on either side of the equator between 23° S and 23° N, is a zone plagued by generally infertile soils as compared to those of the temperate zone further south and further north of these latitudes. In the Introduction to *Soil Atlas of Africa* (Jones et al. 2013), it is stated that there is a perception that tropical Africa is a region where lush rainforests grow on red soils. While the rainforests may be lush, this perception also points to the fact that most of the soils in these rainforests in tropical Africa are leached, acidic, weathered and low in plant nutrients (Jones et al. 2013). The same would be true in agricultural land adjacent to these forests. The sub-Saharan zone in Africa (SSA), most of it located to the south of the 10° N latitude line, is no exception and it is the region with perhaps the greatest extent of soil degradation, both physical and chemical, when compared to temperate zone soils.

Some studies have indicated that up to 65% of the land in SSA has been degraded, mainly due to continued cultivation coupled with inadequate nutrient replenishment, unattended soil erosion and soil acidification (Vlek et al. 2008; Zingore et al. 2015). One study in Nigeria, for example, showed that soil chemical quality, as defined by its pH, soil organic carbon (SOC), available phosphorus, total nitrogen (TN) and effective cation exchange capacity (CEC), decreased with the increase in cultivation period (Lal 1996 especially when nutrient mining was not immediately followed by deliberate replenishment efforts. Similarly, another study in central Kenya showed that losses due to soil erosion were as high as 200 Mg ha^{-1} (Mutegi et al. 2008) leading, in turn, to a direct decline in the capacity of such soils to support maize production. As a result of degradation, some soils in Zimbabwe were shown to be so low in overall fertility that they required up to 10 Mg ha^{-1} of quality animal manure for three consecutive years to be able to significantly improve crop yields (Zingore et al. 2015). Ikazaki et al. (2011) reported that soil loss that was attributed to wind erosion amounted to 55–88 Mg ha^{-1} year^{-1} in the semi-arid Sahelian region in West Africa.

Several studies have shown that soil degradation, especially all forms of soil erosion, has a direct negative effect on the productivity of the soil, as it affects important soil biochemical properties such as SOC, TN, available P, exchangeable bases and CEC (Obalum et al. 2012). Therefore, SSA is a zone where soils and their fertility have degraded considerably and agricultural productivity is relatively low. A number of factors have contributed to the low fertility of the soils of this zone: some natural ones, some man-made, as noted above, and some as a result of the interaction of the natural and the man-made. As a result of the generally low levels of soil fertility in SSA, coupled with the existing low level of soil management to replenish/raise the fertility as currently practiced by farmers, especially small-scale peasant farmers, agricultural production in this zone has for a long period of time lagged far behind that of the temperate zone. The result of this has been perpetual food insecurity, including famines, in much of the SSA region. This chapter highlights the factors that have led the SSA soils to be of such low fertility.

2.2   FACTORS THAT HAVE CONTRIBUTED TO LOW SOIL FERTILITY IN SSA

2.2.1   HUMID CONDITIONS IN SOME AREAS OF SSA

Some SSA countries, especially those in West Africa, the Congo basin and East Africa, contain large tracts of land whose climatic conditions are classified as being humid. Humid climates are characterized by high rainfall, average monthly temperatures greater than 18°C and annual precipitation greater than 1,500 mm, according to the 1936 Koppen classification system (KCC), as cited by Belda et al. (2014), and some parts of SSA qualify to be classified as such.

One of the consequences of high rainfall is that it leads to the progressive leaching of plant nutrients from the topsoil to deeper positions in the soil profile, thus impoverishing the topsoil. Plant nutrient losses are particularly intense in the highly weathered tropical soils and several studies in SSA have given quantitative data on nutrient losses in the region. Through measurements of inputs and outputs, Poss and Saragoni (1992) reported that leaching accounted for up to 85% of the total output for nitrogen (N) and practically all the outputs for magnesium (Mg) and calcium (Ca) in an Oxisol in southern Togo, which was under continuous maize cropping with mineral fertilization and crop residue return. In Zimbabwe, Nyamangara et al. (2003) reported losses of up to 56 kg N ha^{-1} yr^{-1} from a cropland of maize that was under continuous application of composted cattle manure, mineral fertilizers and their combinations. Similarly, Russo et al. (2017) reported losses of up to 81 kg N ha^{-1} year^{-1} from a maize cropland in Kenya that received up to 200 kg N ha^{-1} year^{-1}. Therefore, such nutrient losses lead to a low and declining crop productivity, especially of shallow-rooted annual crops, in the absence of fertilizer use.

Declining and low crop productivity has been the case in much of peasant farming in SSA, where fertilizer use is either minimal or non-existent. On average, farmers in SSA countries used a total of only 7.1 kg ha^{-1} yr^{-1} of fertilizers (N + P$_2$O$_5$ + K$_2$O) by the year 2008, a figure that translates to only 3% of the global fertilizer consumption rate and that had stagnated in SSA for the previous two decades (Druilhe and Barreiro-Hurle 2012). Recent estimates indicate that while global fertilizer use stands at 135 kg ha^{-1} of arable land, SSA uses only around 12–17 kg ha^{-1} year^{-1} (Harrawa and Adan 2017). This fertilizer use rate is exceptionally low, especially when compared with that in South Asia (129 kg ha^{-1}), East and South Asia (minus China and Japan) (104 kg ha^{-1}) (Druilhe and

Barreiro-Hurle 2012), or the global rate just cited. Comparing this low rate of nutrient replenishment with the combined nutrient mining rates of N, phosphorus (P) and potassium (K) of 54 kg ha^{-1} yr^{-1} in SSA (Henao and Baanante 2006; Sommer et al. 2013), it is easy to see why soil fertility in SSA continues to be low and to decline even further. Therefore, to boost crop production in SSA, deliberate and appreciable increases in investment in fertilizer use need to be instituted and sustained.

Long-term leaching of the basic nutrients (Ca, Mg, K) leads to acidification of the topsoil due to the resultant imbalance between the basic elements and those contributing to acidity. Most of the soils in the SSA humid zone are highly leached and have become extremely acidic as a result. pH levels as low as 5.0 or lower are not uncommon in many soils. For example, low-pH soils are a hindrance to crop yields and production on the western coast of Ghana (Buri et al. 2005), the highlands of Kenya (Muindi et al. 2016), in Zambia (Lungu and Dynoodt 2008) and in South Africa (Wooldridge et al. 1995), to mention just a few. Such intense soil acidity does not support the optimum growth of many crops, except in examples like tea that requires acidic soil.

In addition to the intense soil acidity condition itself not supporting plant growth per se, extreme acidity has a soil fertility implication in that it leads to the unavailability of some nutrients, especially P, due to the fixation of soluble P by iron (Fe) and aluminum (Al) in extremely acid soils (Kidd and Proctor 2001). Therefore, this presents a problem in fertilizer P management in that it requires extra effort or cost for liming to reduce the acidity. As for fertilizer use, liming is a costly undertaking, not affordable by small-scale SSA farmers. This is due, in part, to the sheer bulk of the liming material required in conventional liming practices, high transportation costs due to such bulk, poor or lack of transportation infrastructure to the fields, which are often in diverse and inaccessible locations and varied terrains, and a lack of the requisite lime application infrastructure. Therefore, some efforts need to be directed at developing effective lime application approaches that are amenable to adoption by small-scale farmers, for example, developing feasible lime-banding practices. Such approaches would reduce the volume or quantity, as well as the cost of the lime used per hectare per season, by limiting the lime application to the soil zone on which a crop plant is sown, rather than broadcasting over the entire farm area. It is worth mentioning that because of the above considerations, use of lime by small-scale farmers is almost non-existent in SSA. Efforts, therefore, need to be initiated in this direction that will reduce the bulk of the lime applied to make its use affordable to those farmers. Additionally, extreme soil acidity leads to the enhanced solubility of some nutrients, especially the micronutrients that ordinarily are required by plants only in small quantities, resulting in high concentrations that become toxic to plants. These have been observed in many SSA countries; examples include manganese (Mn) toxicity in the soils of the Mugamba region in Burundi and iron toxicity in low-pH soils of Guinea, Ivory Coast, Ghana, Nigeria and Benin (Cherrif et al. 2009; Sikirou et al. 2016). The liming of such soils is the method to reverse these toxicities and nutrient imbalances and to restore good plant growth conditions and increase crop yields.

2.2.2 NUTRIENT MINING BY CROPS

For crops to grow normally and produce high yields, they extract nutrients from the soil. Without soil nutrient replenishment via fertilizer or amendment use, the soil nutrient levels over time become depleted. This extreme extraction of nutrients, sometimes leading to their depletion, is called nutrient mining. As noted above, the fertilizer use rate in SSA is very low, and as a result, nutrient mining has been inevitable. Comprehensive studies on estimated nutrient removal per individual crop type have been reported in the literature (Henao and Baanante 2006). It has been reported, for example, that to produce 1 Mg of rice, the crop plants remove approximately 20 kg N, 11 kg P_2O_5, 30 kg K_2O, 3 kg S, 7 kg Ca, 3 kg Mg, 675 g Mn, 150 g Fe, 40 g Zn, 18 g Cu, 15 g B, 2 g Mo and 52 kg Si from soil (Roy et al. 2006a). There are similar estimates for all the major crops including those grown in SSA. When intense nutrient mining has taken place over time, crop growth will be affected, ultimately culminating in reduced crop yields. The effect on crop growth will be seen in the form of specific symptoms on crop plants that depict an insufficiency of particular nutrients in

FIGURE 2.1   Symptoms of N deficiency in maize: *left*, plants without any deficiency; *middle*, plants showing a general yellowing characteristic of N deficiency; *right*, the progression of yellowing in an individual leaf. These pictures are courtesy of Dr. Mawazo Shitindi, Department of Soil and Geological Sciences, Sokoine University of Agriculture (SUA), Tanzania.

the soil. For example, a general yellowing of plants indicates deficiency of N in the soil, an example being that for maize plants (Figure 2.1), while stunted and purple plants indicate insufficiency of P (Figure 2.2). Similarly, a lack of other nutrients will mean symptoms specific to those nutrients are displayed.

Farmers should be trained to look out for such glaring symptoms, especially those for N, so that steps to correct the deficiency and increase yields can be taken within the season by top-dressing with N fertilizer.

However, the magnitude of nutrient losses from soil due to crop removal in SSA vary from country to country, depending on several factors, including the major crop types grown, the types of land use and the general soil properties in each country. Research in Kenya has shown, for example, that after 18 years of rotational cultivation of maize and common beans without external nutrient inputs, the resultant total losses from soil were about 1 Mg of organic N ha⁻¹ and 100 kg of organic P ha⁻¹, leading to a maize yield decrease from 3 to only 1 Mg ha⁻¹ (Swift et al. 1994).

FIGURE 2.2   Purple color of maize plants characteristic of P deficiency. Courtesy of Dr. Mawazo Shitindi, SUA, Tanzania.

To reverse this trend calls for increased use of fertilizers in these countries. As the socioeconomic conditions of the majority of the small farmers in SSA are presently unfavorable, a combination of deliberate politico-economic adjustments need to be worked out and implemented by governments in order to make input use for soil management, including fertilizer use, affordable to the impoverished small-scale farmers. These politico-economic arrangements include appropriate legislation aimed at increasing fertilizer use by small farmers and deliberate, massive, fertilizer subsidies across the board for all farmers, consequently leading to low and hence affordable prices of fertilizers. Because of many socioeconomic challenges, coupled with limited financial resources, many governments in the developing world have shied away from instituting massive subsidies, especially in the agricultural sector. These governments will do well to learn the lesson from developed countries—that subsidies eventually pay back. These efforts should not be taken by individual countries independently of each other, but as coordinated and rationalized efforts within regional groupings of countries, for example, the Economic Community of West African Countries (ECOWAS), the East African Community (EAC), the South African Development Community (SADC) and the Common Market for Eastern and Southern Africa (COMESA), leading to comparable pricing across the countries within a block. In so doing, the participating countries should display a willingness to sacrifice some of their own narrow interests in order to attain the broader interests that are common to the region. These approaches, if carefully designed, will help not only to increase fertilizer usage by small farmers but will also serve as a deterrent or disincentive to illegal practices such as fertilizer smuggling from one country to another within a block, as could happen under circumstances where these rationalizations are not yet in place.

This deterrent will succeed because there will no longer be any financial benefit to be gained from moving fertilizer from one country to another when pricing across countries is pretty much the same regardless of differences in currency. The current arrangements in some countries of subsidizing some fertilizers only to small-scale farmers have just succeeded in ushering in corruption in the entire procurement-to-use chain. Some small-scale farmers would "procure" the fertilizers, only to sell them to larger farmers. In this way, their own agricultural output is not improved. In the absence of such regionally coordinated arrangements, smuggling has been reported from Ghana to other nearby countries, for example (Wortman and Sones 2017), because it must have been beneficial to do so.

To underscore the need for fertilizer use in SSA countries, requisite rates of fertilizers to use to increase crop yields have been proposed recently for these countries (Wortmann and Sones 2017). The proposals call for the use of rates higher than those sometimes used by small-scale farmers, with the result that it is even harder for the farmers to afford their purchase. That is the reason across-the-board subsidy programs should be seriously worked out, instituted and sustained. Without such a deliberate effort, SSA agricultural production will continue to decline, and the impacts and consequences of such a decline will continue to be intensified.

2.2.3   CROPPING PRACTICES THAT CONTRIBUTE TO CONTINUAL REDUCTION OF SOIL FERTILITY OVER TIME

Small-scale farmers in most of the SSA countries use little or no inorganic fertilizers, as already discussed above. Therefore, the continued cultivation of crops only leads to the perpetual removal or drain of plant nutrients from soil as the removed nutrients are not replenished. The eventual trend is continued decline in the soil's fertility. The traditional practice in much of SSA is the incorporation of a legume component in the farming system. The legume, through its capacity to fix nitrogen, replenishes the mined soil nitrogen. But, over time, cereal monocropping has taken the place of mixed cropping where a legume would be included. This non-inclusion of a legume crop component, which would replenish soil N through N_2 fixation, implies that the supply of soil N is continually decreased due to mining by the non-legume plants. Examples of reported N fixation by tropical legumes in Africa include soybean (*Glycine max*), which can fix between 159 to 227 kg N ha^{-1}, groundnuts (*Arachis hypogaea*), 17–103 kg N ha^{-1}, cowpea (*Vigna unguiculata*), 3–201 kg N ha^{-1} and the common bean (*Phaseolus vulgaris*), 8–58 kg N ha^{-1} (Peoples et al. 2009). All these are an indication that if properly included in the farming systems, and the resultant biomass returned by incorporation back into the soil, legumes have the potential to raise the N fertility of soils.

2.2.4 SOIL EROSION

Soil erosion has been perhaps the most potent agent leading to soil degradation, and in recent times, SSA has perhaps been the region with the most degraded soils. This has been due to decreased attention to soil conservation, ultimately leading to the breakdown of whatever soil conservation infrastructure that existed. An example of abandoned soil conservation structures on the slopes of Mount Kilimanjaro, Tanzania, is shown in Figure 2.3.

Examples of extremely degraded soils can be cited for most of the SSA region, including the slopes of the Ethiopian highlands, which lose soil at the rate of 290 Mg ha^{-1} year^{-1} and the Nile basin of Eastern Africa, which has an average erosion rate of 75 Mg of soil ha^{-1} year^{-1} (Tamene and Le 2015). Badly eroded land in the Kisongo area, Arusha region, Tanzania, is shown in Figure 2.4, where progressive soil erosion has led to the formation of deep gullies in large tracts of land. Such areas, and similar ones across SSA, could be classified as national disaster areas worthy of massive rehabilitation efforts to restore them to their pristine levels, both aesthetically and in terms of their soil fertility and productivity.

Reports show that the average predicted rate of soil loss from cropland under annual cropping (grain crops) in South Africa is 13 Mg ha^{-1} year^{-1} (Roux and Smith 2014). In Zimbabwe, one study reported that quantities of soil of up to 76 Mg ha^{-1} are lost every year, as compared to the rates of soil formation of only 400 kg ha^{-1} year in the same country (Makwara and Gamira 2012). A study in Uganda concluded that there are significant differences in the amounts of soil lost due to erosion under differing cropping systems. They showed that the most vulnerable is the mixed cropping

FIGURE 2.3 Unattended soil conservation terraces in Lawate village, Siha district, Kilimanjaro region, Tanzania.

N.B.: This and subsequent pictures were taken by the authors in various locations in Tanzania.

FIGURE 2.4   Extreme soil erosion in various locations of the Kisongo area, Arusha, Tanzania, that progressed from small (*left*) to very deep (*right*) gullies.

system, with losses of up to 8 Mg ha^{-1}, 19 Mg ha^{-1} and 25 Mg ha^{-1} at the lower, middle and upper slopes, respectively, as compared to sole cropping with respective loss values of 2 Mg ha^{-1}, 10 Mg ha^{-1} and 14 Mg ha^{-1} (Wambede et al. 2016).

2.2.5   INDISCRIMINATE BURNING OF VEGETATION IN LARGE TRACTS OF LAND, INCLUDING AGRICULTURAL LAND

The indiscriminate burning of vegetation is a visible feature in much of SSA where, for different reasons, vegetation in large tracts of land, including natural forests, conserved land and agricultural land, is deliberately set on fire during dry seasons. Across the region, fire is used for hunting, pasture management for livestock husbandry, pest control, food gathering, cropland fertilization via the ash, and wildfire prevention. In a majority of cases, most fires that are started for a good reason, as mentioned, are usually left unattended and they can expand into large, uncontrollable wildfires. Sometimes those fires expand into residential areas (Figure 2.5) where they can cause more damage.

FIGURE 2.5   Wildfire (foreground) that extended to a residential area in Kilosa district, Morogoro region, Tanzania.

SSA has more fires burning its vegetation than any other part of the world. Although there is lack of strong infrastructure to track, document and control wildfires in the region, there is a general agreement that a large part of SSA has its forests partly or fully burnt down every year, very much as a result of population growth (Barbosa et al. 1999). One study (JRC 2005) had estimated that up to 230 million hectares of vegetation were burnt in Africa in the year 2000. Upon this indiscriminate burning, the above-ground vegetation that is destroyed leads to the exposure of the soil and thus to increased soil erosion and the loss of the plant nutrients contained in the resultant ash. The ash from the burnt vegetation contains high quantities of K, Mg and P (Qian et al. 2009) that are lost when eroded from the sites. Perhaps the most significant long-term negative effect of the indiscriminate burning of vegetated land, including agricultural land, is the destruction by fire of the soil's organic matter, the humus, in the topsoil. Humus is generated over long periods of time upon microbial decomposition of vegetation matter in the soil, as well as by other biochemical reactions (Olk et al. 1995; Aranda and Oyonarte 2005; Abakumov 2010), and is a rich store of plant nutrients (Roy et al. 2006b), including N. Thus, its destruction and loss has led to decreased soil fertility in many countries.

Additionally, humus is a soil component that contributes to the increased aggregate stability of soil against the forces of soil erosion. Alagoz and Yilmaz (2009), for example, showed that the addition of organic matter in the form of K-humate and concentrated plant extract had a significant influence in increasing the aggregate stability of a Lithic Rhodoxeralf in Turkey. Another study found that there exists a high to very high correlation between levels of organic matter found in a given soil and the degree of aggregate stability of the soil (Chaney and Swift 1984; Wang et al. 2013). Therefore, its destruction leads to the easy erosion of the soil because the cementing agent of the soil's particles is destroyed. Destruction of humus also leads to the loss of the nutrients conserved by the humus (Rein et al. 2008; Il'icheva et al. 2011) and hence reduction in soil fertility.

2.2.6 OVERGRAZING OF PASTURELAND AND CROPLAND AFTER CROP HARVEST

In much of SSA, the mode of livestock keeping is pastoralism. Most of the pastoralists own large herds of livestock, many in the thousands. Many communities of pastoralists wander about in search of pastures, often overgrazing the land in their path, including agricultural or cultivated land. Overgrazing has been mentioned as the main contributor to land degradation in any of its four types: wind erosion, water erosion, nutrient mining and deforestation (Kirui 2016; Mirzabaev 2015). Overgrazing by pastoralist activity is a common occurrence in Tanzania where soil erosion due to overgrazing is a big problem in areas such as Babati (Pietikäinen 2006), and as the pressure grows, the pastoralists migrate to greener areas leading to farmer-pastoralist conflicts (Oladele et al. 2011). Similarly, overgrazing from three river basins in Kenya was reported to have resulted in a ten-fold increase in soil erosion between 1965 and 1986 (Sharma 1997) Grazing removes vegetation remains from the grazed site as the remains become consumed. Those remains have locked up plant nutrients that were earlier absorbed from soil when the plants were alive and growing. Therefore, overgrazing removes most of the plant remains, resulting in lower soil fertility if practiced in a given site over long periods of time. This is because the plant remains, as well as the manure containing the nutrients that were locked up in the grazed plant remains, are usually not returned back to the grazed land. Manure is not returned to the grazed location because it is lost in other areas during the wandering about of the animals. A study in the arid eastern cape of South Africa, for example, had found that heavy and light grazing reduced soil N storage by 27.5% and 22.6%, respectively, compared with the exclosure (Talore et al. 2015). Similarly, biological soil crusts (BSCs), which are an important source of SOC weathering, were shown to be seriously destroyed on overgrazed land in the Kalahari desert of Botswana, as a result of trampling and burial by livestock (Thomas 2012). Another study covering over 330 sites in South Africa had found that greater grazing pressure had a negative influence on the N availability of associated soils compared to protected areas with less grazing pressure (Craine et al. 2009).

Due to the removal of most vegetation remains, as a result of overgrazing with large herds of livestock, the soil is left bare, is loosened due to trampling by the animals, and this makes the soil more susceptible, exacerbating its erosion, both by wind and by water. One study reported that land degradation hotspots cover up to 51% and 41% of the terrestrial areas in Tanzania and Malawi, respectively (Kirui 2016). Kangalawe (2012) reported that a combination of factors that make soil erosion a serious problem in the semi-arid areas of central Tanzania included overgrazing, overcultivation and the burning of grassland and woodland. Aniah et al. (2013) concluded that overgrazing is one of many long-term factors causing soil nutrient depletion and soil erosion in Ghana's agricultural land. Efforts, education included, should be made to ensure the livestock owners appreciate and adopt practices based on the concept of the carrying capacity of their land. In extreme cases of land degradation as a result of overgrazing, rehabilitation measures should be instituted by governments. Examples of such rehabilitation are the Hifadhi Ardhi Dodoma (HADO) (Kiwsahili for "land conservation in Dodoma region")and Hifadhi Ardhi Shinyanga (HASHI) ("land conservation in Shinyanga region") initiatives taken by the Tanzanian government in the years 1973 to the early 2000s (Nshubemuki and Mugasha 1985; Barrow et al. 1992; Shrestha and Lingoja 2015) that completely rehabilitated overgrazed areas in the Dodoma and Shinyanga regions. These initiatives prohibited grazing in these areas for long periods of time, and this led to the regeneration of vegetation.

2.2.7 RESTORATION OF SOIL FERTILITY IN DEGRADED LAND: ROLE OF INTEGRATED SOIL FERTILITY MANAGEMENT

In the discussions above, the different scenarios that lead to a decline in soil fertility have been presented. One approach to restoring the fertility of degraded soils frequently mentioned above has been the use of inorganic fertilizers. But it has also been argued above that these fertilizers are expensive and that most small-scale farmers in SSA cannot afford them. Therefore, the use of these fertilizers as the only solution should not be overemphasized. Instead, inorganic fertilizers should be viewed as a component, in concert with other options, in an overall solution to reversing declining soil fertility. This leads to advocacy of the concept of integrated soil fertility management (ISFM). Thus, while inorganic fertilizers should still be recommended, increased emphasis should also be placed on ISFM components that are easily within the reach of the small-scale farmers, including taking advantage of biological nitrogen fixation (BNF), mycorrhizal crop plants, use of manures, better *in situ* management of crop-derived organic resources in the field, and direct use of rock phosphates. In this kind of integration, the rates of use of inorganic fertilizers will be reduced, thereby increasing the affordability of these inputs by the small farmers.

To illustrate the potential of the ISFM components, it is instructive to note that legume BNF systems can supply substantial quantities of fixed N. Soybeans, for example, can fix as much as 227 kg N ha^{-1} season^{-1} (Peoples et al. 2009). Likewise, common beans, cowpeas and green grams have been reported to fix as much as 200 kg N ha^{-1} season^{-1} (Peoples et al. 2009). Green manuring plants, such as *Sesbania sesban*, *Sesbania rostrata* and *Crotalaria ochreleuca* (sunn hemp), have the ability to fix similar quantities of N ha^{-1}, and can generate substantial quantities of biomass. Such biomass, if well managed and returned to the farms, would release substantial quantities of N for a subsequent crop, thereby decreasing the amounts of inorganic fertilizers that may have to be used. Agricultural extension education efforts will need to be stepped up for the farmers to correctly undertake these ISFM approaches for the benefit of increased crop productivity.

Mycorrhiza fungi scavenge P from large volumes of soil around plant rhizospheres and pass the P onto the associated higher plants. In this way, mycorrhizal plants growing in soils with low P content grow better and yield more than the same, but non-mycorrhizal, plants. Therefore, crop varieties whose plants are known to have ability to establish associations with mycorrhiza should be preferred over the same non-mycorrhizal ones.

Where crop farmers are also livestock keepers, good management and use of manures from livestock should be encouraged. In this event, farmers should be trained regarding good methods of preparation and management of livestock manures, including practices that maximize the conservation of the nutrients they contain, especially N, which is more labile.

Farm or crop residues are also an important store of the plant nutrients absorbed when the plants were growing. Good management of these residues cannot be overemphasized as they have the potential to supply substantial quantities of N, P and other nutrients locked up in the residues. Thus, destructive practices like burning the organic residues, grazing them after crop harvest, or removing them from the field for any purpose should be discouraged. Retaining such residues in the field contributes to building up SOC, or humus, whose advantages in terms of being a store of plant nutrients as well as in improving soil moisture relations are well known.

Leguminous green manuring plants also supply much N, to the benefit of crop plants so manured. Thus, the integration of green manuring plants into the cropping sequence will contribute to increasing the yields of subsequent crops. Farmers should be made aware of the importance of adopting this technology as well.

In sum, if farmers integrate as many of these technologies as is possible within their farming systems, they stand to benefit in terms of reducing the quantities of the expensive inorganic fertilizers to be purchased/used while increasing crop yields at the same time. Thus, the practice of ISFM should be advocated, and reliance should not be placed solely on inorganic fertilizers alone.

2.2.8 Land Conservation

Discussed above are some of the practices, such as forest clearing for the expansion of residential areas, burning of vegetation, overgrazing, etc., that lead to land degradation, including soil erosion. In addition, many cultivated areas in SSA are hilly, thereby exacerbating soil erosion. An example of a mountainous area in the Usambara mountains, Tanzania, that has long been cleared of vegetation is shown in Figure 2.6.

Following vegetation clearing, land/soil conservation should, therefore, form an integral part of land management to reverse land degradation and the decline of soil fertility. In so doing, this will help farmers to consolidate any gains realized in restoring soil fertility and halting/reversing land

FIGURE 2.6   A mountainous catchment area overlooking the Mlalo division in Lushoto district, northeastern Tanzania, where vegetation has been cleared for agriculture and residences.

FIGURE 2.7 Two contrasting slopes facing each other in Magamba area, Lushoto district, northeastern Tanzania. The first slope (*left*) is largely cleared of vegetation for residential purposes, while the other (*right*) is forested and well conserved.

degradation as a result of adopting the different measures discussed here. While land conservation can be a broad concept, it should at least include the following approaches. Soil conservation efforts such as the contour terraces shown in Figure 2.3 should be revived and sustained. Conservation farming should be given impetus in vulnerable areas. Reforestation of land cleared of vegetation and subsequently degraded, as exemplified by the degraded areas of Kisongo, Arusha (Figure 2.4), should be initiated. Figure 2.7 shows two contrasting slopes facing each other across a valley in the Magamba area in Lushoto district, Tanzania, whereby one slope (in the southerly side) has been cleared of vegetation for residential purposes. On the northern side is another slope in a piece of land belonging to the Sebastian Kolowa Memorial University (SEKOMU) that is well conserved, forested and aesthetic.

Examples of conserved land that was initially severely eroded and land that had not had much vegetation have already been cited above in the cases of the HADO and HASHI land conservation programs in Tanzania.

2.3 CONCLUSIONS

It has been estimated that that up to 65% of the land in sub-Saharan Africa has been degraded as a result of factors like continued cultivation with little or no plant nutrient replenishment, overgrazing, increased soil erosion and soil acidification. This has led to low soil fertility in these countries due to nutrient mining as a result of the increase in the cultivation period without concomitant land conservation. As a result, crop productivity has continued to decline. The result of this has been perpetual food insecurity, including famines, in much of the SSA region. To reverse this trend there is a need to institute politico-economic and technical correctives with the eventual aim of restoring and improving the fertility of soils for increased agricultural productivity in SSA. Governments should be increasingly involved in these efforts.

REFERENCES

Abakumov, E.V. 2010. The sources and composition of humus in some soils of West Antarctica. *Eurasian Soil Science* 43(5): 499–508.

Alagöz, Z., Yilmaz, E. 2009. Effects of different sources of organic matter on soil aggregate formation and stability: A laboratory study on a Lithic Rhodoxeralf from Turkey. *Soil and Tillage Research* 103(2): 419–424.

Aniah, P., Wedamb, E., Pukunyiemc, M., Yinimid, G. 2013. Erosion and livelihood change in north east Ghana: A look into the bowl. *International Journal of Sciences: Basic and Applied Research* 7(1): 28–35.

Aranda, V., Oyonarte, C. 2005. Effect of vegetation with different evolution degree on soil organic matter in a semi-arid environment (Cabo de Gata-Níjar Natural Park, SE Spain). *Journal of Arid Environments* 62(4): 631–647.

Barbosa, P.M., Stroppiana, D., Grégoire, J.M., Pereira, J.M.C. 1999. An assessment of vegetation fire in Africa (1981–1991): Burned areas, burned biomass, and atmospheric emissions. *Global Biogeochemical Cycles* 13(4): 933–950.

Barrow, E.G.C., Fry, P., Lugeye, S. 1992. *Hifadhi Ardhi Shinyanga (HASHI) Evaluation Report for Ministry of Tourism, Natural Resources and Environment.* United Republic of Tanzania and Norwegian Agency for International Development, Dar-es-Salaam, Tanzania.

Belda, M., Holtanová, E., Halenka, T., Kalvová, J. 2014. Climate classification revisited: From Köppen to Trewartha. *Climate Research* 59(1): 1–13. doi: 10.3354/cr01204.

Buri, M.M., Issaka, R.N., Wakatsuki, T. 2005. Extent and management of low pH soils in Ghana. *Soil Science and Plant Nutrition* 51(5): 755–759.

Chaney, K., Swift, R.S. 1984. The influence of organic matter on aggregate stability in some British soils. *Journal of Soil Science* 35(2): 223–230.

Chérif, M., Audebert, A., Fofana, M., Zouzou, M. 2009. Evaluation of iron toxicity on lowland irrigated rice in West Africa. *Tropicultura* 27(2): 88–92.

Craine, J.M., Ballantyne, F., Peel, M., Zambatis, N., Morrow, C., Stock, W.D. 2009. Grazing and landscape controls on nitrogen availability across 330 south African savanna sites. *Austral Ecology* 34(7): 731–740. doi: 10.1111/j.1442-9993.2009.01978.x.

Crawford Jr, T.W., Singh, U., Brenan, H. 2008. *Solving Agricultural Problems Related to Soil Acidity in Central African Great Lakes Region.* IFDC, Muscle Shoals, AL, 133 pp.

Druilhe, Z., Barreiro-Hurle, J. 2012. *Fertilizer Subsidies in Sub-Saharan Africa.* ESA Working Paper 12-04. FAO, Rome, 63 pp.

Harrawa, R., Adan, B.J. 2017. Foreword. In: *Fertilizer Use Optimization in Sub-Saharan Africa.* C.S. Wortmann, K. Sones (eds.). CAB International, Nairobi, Kenya.

Henao, J., Baanante, C. 2006. *Agricultural Production and Soil Nutrient Mining in Africa: Implications for Resource Conservation and Policy Development.* IFDC, Alabama, USA, 95 pp.

Il'ichev, Y.N., Ignat'ev, L.A., Artymuk, S.Y. 2011. The effect of forest fires and clearing of fire-destroyed stands on pedoecological conditions of natural forest regeneration. *Contemporary Problems of Ecology* 4(6): 634–640.

Ikazaki, K., Shinjo, H., Tanaka, U., Tobita, S., Funakawa, S., Kosaki, T. (2011). "Fallow Band System," a land management practice for controlling desertification and improving crop production in the Sahel, West Africa. 1. Effectiveness in desertification control and soil fertility improvement. *Soil Science and Plant Nutrition* 57(4): 573–586.

Jones, A., Breuning-Madsen, H., Brossard, M., Dampha, A., Deckers, J., Dewitte, O., Gallali, T., Hallett, S., Jones, R., Kilasara, M., Le Roux, P., Micheli, E., Montanarella, L., Spaargaren, O., Thiombiano, L., Van Ranst, E., Yemefack, M., Zougmore, R. (eds.). 2013. *Soil Atlas of Africa.* European Commission, Publications Office of the European Union, Luxembourg, 176 pp.

JRC-EU (Joint Research Center, European Commission). 2005. *Safari 2000 Global Burned Area Map, 1-km, Southern Africa 2000.* Joint Research Centre/European Union, Ispra, Italy.

Kangalawe, R.Y.M. 2012. Land degradation, community perceptions and environmental management implications in the drylands of central Tanzania. In: *Sustainable development-Authoritative and Leading Edge Content for Environmental Management*, S. Curkovic (ed.). IntechOpen.

Kidd, P.S., Proctor, J. 2001. Why plants grow poorly on very acid soils: Are ecologists missing the obvious? *Journal of Experimental Botany* 52(357): 791–799. doi: 10.1093/jexbot/52.357.791.

Kirui, O.K. 2016. Economics of land degradation and improvement in Tanzania and Malawi. In: *Economics of Land Degradation and Improvement—A Global Assessment for Sustainable Development,* E. Nkonya, A. Mirzabaev, J. von Braun (eds.). Springer International Publishing.

Lal, R. 1996. Deforestation and land-use effects on soil degradation and rehabilitation in western Nigeria: II. Soil chemical properties. *Land Degradation and Development* 7(2): 87–98.

Lungu, O.I.M., Dynoodt, R.F.P. 2008. Acidification from long-term use of urea and its effect on selected soil properties. *African Journal of Food, Agriculture, Nutrition and Development* 8(1): 63–76.

Makwara, E.C., Gamira, D. 2012. About to lose all the soil in Zaka's Ward 5, Zimbabwe: Rewards of unsustainable land use. *European Journal of Sustainable Development* 1(3): 457–476.

Muindi, E.M., Semu, E., Mrema, J.P., Mtakwa, P.W., Gachene, C.K. 2016. Soil acidity management by farmers in the Kenya highlands. *Journal of Agriculture and Ecology Research International* 5(3): 1–11.

Mutegi, J.K., Mugendi, D.N., Verchot, L.V., Kung'u, J.B. 2008. Combining napier grass with leguminous shrubs in contour hedgerows controls soil erosion without competing with crops. *Agroforestry Systems* 74(1): 37–49.

Nshubemuki, L., Mugasha, A.G. 1985. The Modifications to Traditional Shifting Cultivation Brought about by the Forest Development Project in HADO Area. In: *Changes in Shifting Cultivation in Africa: Seven Case Studies*. FAO Forestry Department, Kondoa, Tanzania, Rome, Italy, pp. 141–162.

Nyamangara, J., Bergström, L.F., Piha, M.I., Giller, K.E. 2003. Fertilizer use efficiency and nitrate leaching in a tropical sandy soil. *Journal of Environmental Quality* 32(2): 599–606.

Obalum, S.E., Buri, M.M., Nwite, J.C., Hermansah, H., Watanabe, Y., Igwe, C.A., Wakatsuki, T. 2012. Soil degradation-induced decline in productivity of sub-Saharan African soils: The prospects of looking downwards the lowlands with the sawah ecotechnology. *Applied and Environmental Soil Science* 2012: 1–10. doi: 10.1155/2012/673926.

Oladele O.T., Oladele O.I. (2011). Effect of pastoralist-farmers conflict on access to esources in Savanna Area. *Journal of Life Science* 8(2): 616–621.

Olk, D.C., Cassman, K.G., Fan, T.W.M. 1995. Characterization of two humic acid fractions from a calcareous vermiculitic soil: Implications for the humification process. *Geoderma* 65(3–4): 195–208.

Peoples, M.B., Brockwell, J., Herridge, D.F., Rochester, I.J., Alves, B.J.R., Urquiaga, S., Boddey, R.M., Dakora, F.D., Bhattarai, S., Maskey, S.L., Sampet, C., Rerkasem, B., Khan, D.F., Hauggaard-Nielsen, H., Jensen, E.S. 2009. The contributions of nitrogen-fixing crop legumes to the productivity of agricultural systems. *Symbiosis* 48(1–3): 1–17.

Pietikäinen, V. (2006). Measures to prevent overstocking and overgrazing in woodlands: A case study in Babati, northern Tanzania. Environment and Development Educational Program, School of Life Sciences. http://www.diva-portal.org/smash/get/diva2:16551/FULLTEXT01.pdf.

Poss, R., Saragoni, H. (1992). Leaching of nitrate, calcium and magnesium under maize cultivation on an Oxisol in Togo. *Nutrient Cycling in Agroecosystems* 33(2): 123–133.

Qian, Y., Miao, S.L., Gu, B., Li, Y.C. 2009. Effects of burn temperature on ash nutrient forms and availability from cattail (*Typha domingensis*) and sawgrass (*Cladium jamaicense*) in the Florida Everglades. *Journal of Environmental Quality* 38(2): 451–464. doi: 10.2134/jeq2008.0126.

Rein, G., Garcia, J., Simeoni, A., Tihay, V., Ferrat, L. 2008. Smouldering natural fires: comparison of burning dynamics in boreal peat and Mediterranean humus. *WIT Transactions on Ecology and the Environment* 119: 183–192. doi: 10.2495/FIVA080191.

Roux, J.L., Smith, H. 2014. Soil erosion in South Africa—Its nature and distribution. *Grain SA*. https://www.grainsa.co.za/soil-erosion-in-south-africa-its-nature-and-distribution (accessed May 20, 2019).

Roy, R.N., Finck, A., Blair, G.J., Tandon, H.L.S. 2006a. Nutrient management guidelines for some major field crops. In: *Plant Nutrition for Food Security—A Guide for Integrated Nutrient Management*. FAO, Rome, Italy, pp. 235–263.

Roy, R.N., Finck, A., Blair, G.J., Tandon, H.L.S. 2006b. *Plant Nutrition for Food Security: A Guide for Integrated Nutrient Management*. Food and Agriculture Organization of the United Nations, Rome, Italy, 368 pp.

Russo, T.A., Tully, K., Palm, C., Neill, C. 2017. Leaching losses from Kenyan maize cropland receiving different rates of nitrogen fertilizer. *Nutrient Cycling in Agroecosystems* 108(2): 195–209. doi: 10.1007/s10705-017-9852-z.

Sharma, K.D. (1997). Assessing the impact of overgrazing on soil erosion in arid regions at a range of spatial scales: Proceedings of Rabat Symposium on Human Impact on Erosion and Sedimentation 1 1Q IAHS Publ. no. 245. http://hydrologie.org/redbooks/a245/iahs_245_0119.pdf (accessed October 10, 2018).

Shrestha, R.P., Ligonja, P.J. 2015. Social perception of soil conservation benefits in Kondoa eroded area of Tanzania. *International Soil and Water Conservation Research* 3(3): 183–195.

Sikirou, M., Saito,K., Dramé, K.N., Saidou, A., Dieng, I., Ahanchédé , A., Venuprasad, R. 2016. Soil-based screening for iron toxicity tolerance in rice using pots. *Plant Production Science* 19(4): 489–496.

Sommer, R., Bossio, D., Desta, L., Dimes, J., Kihara, J., Koala, S., Mango, N., Rodriguez, D., Thierfelder, C., Winowiecki, L. 2013. *Profitable and Sustainable Nutrient Management Systems for East and Southern African Smallholder Farming Systems: Challenges and Opportunities: A Synthesis of the Eastern and Southern Africa Situation in Terms of Past Experiences, Present and Future Opportunities in Promoting Nutrients Use in Africa*. CIAT, The University of Queensland, QAAFI, CIMMYT.

Swift, M.J., Seward, P.D., Frost, P.G.H., Quershi, J.N., Muchena, F.N. 1994. Long term experiments in Africa. Developing a database for sustainable Land use under global Change. In: *Long Term Experiments in Agricultural and Ecological Sciences*. Monitoring Nutrient Flows and Ecological Performance in Tropical Farming Systems (NUTMON) Part 1. Manual for the NUTMON Toolbox, Wageningen, The Netherlands, p. 2001.

Talore, D.G., Tesfamariam, E.H., Hassen, A., Toit, J.D., Klampp, K., Jean-Francois, S. 2015. Long-term impacts of grazing intensity on soil carbon sequestration and selected soil properties in the arid Eastern Cape, South Africa. *Journal of the Science of Food and Agriculture* 96: 1945–1952.

Tamene, L., Le, Q.B. 2015. Estimating soil erosion in sub-Saharan Africa based on landscape similarity mapping and using the revised universal soil loss equation (RUSLE). *Nutrient Cycling in Agroecosystems* 102(1): 17–31. doi: 10.1007/s10705-015-9674-9.

Thomas, A.D. 2012. Impact of grazing intensity on seasonal variations in soil organic carbon and soil CO_2 efflux in two semiarid grasslands in southern Botswana. *Philosophical Transactions of the Royal Society B* 367(1606): 3076–3086.

Vlek, P.L.G., Le, Q.B., Tamene, L. 2008. CGIAR Science Council Secretariat, Rome, Italy.

Wambede, N.M., Joyfred, A., Remigio, T. 2016. Soil loss under different cropping systems in highlands of Uganda. *Natural Resources and Conservation* 4(1): 15–27.

Wang, F., Tong, Y.A., Zhang, J.S., Gao, P.C. and Coffie, J.N. 2013. Effects of various organic materials on soil aggregate stability and soil microbiological properties on the Loess Plateau of China. *Plant, Soil and Environment* 59(4): 162–168.

Wooldridge, J., Kotzé, W.A., Joubert, M.E. 1995. Acid soil management in orchard soils of the South Western Cape Province, South Africa. In: *Plant-Soil Interactions at Low pH: Principles and Management. Developments in Plant and Soil Sciences,* R.A. Date, Grundon N.J., Rayment G.E., Probert M.E. (eds.), vol. 64. Springer, Dordrecht, The Netherlands.

Wortmann, C.S., Sones, K. (eds.). 2017. *Fertilizer Use Optimization in Sub-Saharan Africa.* CAB International, Nairobi Kenya, 227 pp.

Zingore, S., Mutegi, J., Agesa, B., Desta, L., Kihara, J. 2015. Soil degradation in sub-Saharan Africa and crop production options for soil rehabilitation. *Better Crops with Plant Food* 99(1): 24–26.

3 Making Sense Out of Soil Nutrient Mining and Depletion in Sub-Saharan Africa

Giregon Olupot, Twaha Ali Ateenyi Basamba,
Peter Ebanyat, Patrick Musinguzi, Emmanuel Opolot, Alice
A. Katusabe, Mateete A. Bekunda, and Bal Ram Singh

CONTENTS

A successful revolution is where we would see soil health restored, through agroforestry techniques and organic and mineral fertilizers, among other solutions.

—Former UN Secretary General, Kofi Annan on July 5, 2004, at a meeting to
deliberate on the hunger millennium development goal (MDG) in Addis Ababa.

3.1   INTRODUCTION

Poverty, soil nutrient depletion and low agricultural productivity are interlinked (Lipper 2001) and have escalated food and nutritional insecurity (FAO 2015a; Masila et al. 2015; FAO 2017), as well as poor human health in sub-Saharan Africa (SSA) (Sanchez and Swaminathan 2005a; FAO 2015a,b; UNICEF, WHO and World Bank 2018). SSA is the only region in the world where hunger and malnutrition have remained high and are on the increase, with hunger prevalence exceeding 30%, way above the global average of 22.2% (FAO et al. 2018; UNICEF et al. 2018). Over 90% of the hungry in SSA are characterized as chronically malnourished, accounting for 80% of the world's hunger

FIGURE 3.1   (See color insert.) World hunger hot spots defined as those regions in the world where at least 20% of children under the age of five are stunted and underweight (http://www.globalhungerindex.org/result s/ accessed on October 19, 2018).

hot spots (Figure 3.1: UNICEF et al. 2018), which are defined as those regions of the world where over 20% of children under the age of five are stunted and underweight (Sanchez and Swaminathan 2005). Hunger and malnutrition cost SSA up to 10% of its gross domestic product (GDP) annually (Sanchez and Swaminathan 2005b); and at least 57% of deaths due to malaria in SSA, estimated to exceed one million, can be avoided by tackling hunger and malnutrition (Snow and Mumbo 2006), especially among those entangled by deep poverty traps beyond the reach of markets (Sachs et al. 2004).

Soil nutrient depletion is, therefore, not only a serious threat to food security and sustainable agriculture but also to the health and well-being of the people in SSA (Sanchez and Swaminathan 2005b; Baanante and Henao 2006). Soils across most countries of SSA are severely weathered, fragile and inherently of low to moderate fertility (Bationo et al. 2006; Pinheiro and Grashey-Jansen 2016). Unabated nutrient mining of such soils is the primary contributor to hunger, malnutrition and poor health, especially among households with small landholdings (Sanchez and Swaminathan 2005b). If the trend is not reversed, SSA is projected to need to import over 60 million Mg yr^{-1} of cereals, tubers and legumes by the year 2020 (Baanante and Henao 2006). Soil nutrient mining occurs when nutrients removed from the soil are not replenished, resulting in depletion of soil nutrient stocks (Majumdar et al. 2016). Soil nutrient depletion refers to all nutrient losses from soil through both natural and human-induced processes (Tan et al. 2005). Unlike natural or geological processes, anthropogenic processes of soil nutrient depletion can be controlled and their impacts minimized (Tan et al. 2005). Moreover, even natural processes of nutrient losses such as soil erosion and leaching are often exacerbated by anthropogenic activities such as overcultivation, overgrazing, insufficient or no inputs for replenishing lost nutrients, unbalanced fertilization and accelerated soil erosion (FAO 2003; Tan et al. 2005).

Continuous nutrient mining and depletion of soil nutrient stocks leads to a deterioration in socio-economic welfare, soil resource sustainability, environmental quality and reduction in crop yields (Tan 2005). Miller and Larson (1992) cautioned about the devastating impact of unchecked human-induced nutrient depletion on soil nutrient cycling and productivity. Out of the near 135 million hectares (Mha) of soils reported to be prone to nutrient exhaustion in the year 2000, about 45 Mha were in Africa alone (Osgood 2001; Tan et al. 2005). Restoration of depleted soils is therefore the first entry point toward increasing agricultural productivity and tackling hunger and malnutrition in

SSA (Sanchez and Swaminathan 2005; Bekunda et al. 2010). Calls have been made regarding the urgent need to invest in replenishing and recapitalizing soil nutrient stocks in SSA (Sanchez et al. 1997; Sanchez and Swaminathan 2005). Yet across this region, many farmers continue depleting the soil without replenishing removed nutrients and conserving soil water (Ryan and Spencer 2001), with a knock-on effect on human and environmental health (Tan et al. 2005). Moreover, much of the debate about soil nutrient mining and depletion continues to ignore the role that farmers play in shaping the processes of nutrient depletion and environmental change (FAO 2003).

This chapter focuses on soil nutrient balances as quantifiable indicators of soil nutrient mining and depletion as well as the sustainability of farming systems and practices in SSA. It then draws attention to the interpretation and application of information about soil nutrient balances as the drivers of soil nutrient mining and depletion, drawing from case studies to shed light on potential interventions to spark off the reversal of the hitherto unabated downward spiral of soil nutrient depletion in SSA. The aim is to highlight the serious situation of soil nutrient depletion in SSA and to demonstrate the important part that nutrient balances could play in assessing future prospects for tackling food and nutritional insecurity as well as ill health in SSA. Thereafter, the chapter delves into approaches to arresting soil nutrient mining and depletion to demonstrate the fact that there is no "one-size-fits-all" approach to reversing decades of unabated soil mining in SSA but a package for which organic fertilizers, biological nitrogen fixation (BNF) and mineral fertilizers are crucial, coupled with proper crop combinations, sound agronomic practices and soil water conservation. Finally, the conclusion draws attention to research imperatives to inform and guide climate-smart pathways to arresting and/or reversing soil nutrient mining and depletion in SSA.

3.2 OVERVIEW OF SOIL MACRONUTRIENT MINING AND DEPLETION IN SUB-SAHARAN AFRICA

The literature on the dynamics of nutrient stocks, flows and net balances in SSA first came to light following a continental study by Stoorvogel and Smaling (1990) based on six climate-based land water classes (cited by FAO 2003), which put average nutrient depletion rates in Africa at 22 kg N, 2.5 kg P and 15 K kg ha^{-1} yr^{-1}. Summed together, the 39.5 kg NPK ha^{-1} yr^{-1} depletion rate in this continental study was over four times the average fertilizer consumption for SSA (Henao and Baanante 1999; Africa Fertilizer Summit 2006).

3.2.1 COUNTRY-LEVEL SOIL NUTRIENT MINING AND DEPLETION IN SUB-SAHARAN AFRICA

Macronutrient balance sheets from the 44 countries considered (Table 3.1) indicated that net balances were positive for only three countries: Mauritius, Réunion and the Libyan Arab Jamahiriya. Soil macronutrient depletion rates exceeded 60 kg NPK ha^{-1} yr^{-1} in more than 52% of the countries (Table 3.1), with depletion rates ranging from −14 kg NPK ha^{-1} yr^{-1} for South Africa to −136 kg NPK ha^{-1} yr^{-1} for Rwanda (Henao and Banaante 1999; FAO 2003). This implies that even if the African Union (AU) heads of state had delivered on their commitment to raise average fertilizer consumption in SSA from 8 to 50 kg NPK ha^{-1} yr^{-1} (Africa Fertilizer Summit 2006), it would not reverse decades of unabated soil nutrient mining estimated at 60 kg NPK ha^{-1} yr^{-1} in more than 52% of the countries. The situation would even be worse in the highland areas of countries such as Rwanda, Ethiopia (Henao and Baanante 1999, 2006) and Uganda (Bekunda and Manzi 2003), with extreme levels of soil nutrient depletion rates. Yet, country-level soil nutrient depletion rates grossly underestimate actual soil nutrient depletion that occurs at farm and plot levels (see subsequent subsections).

3.2.2 COUNTRY-LEVEL SOIL NUTRIENT DEPLETION AT MACRO-, MESO- AND MICRO-SCALES

Soil nutrient depletion rates have also been quantified for countries in SSA at different scales of assessment and specified field conditions (Table 3.2). Results indicate that at the macro-scale, the

TABLE 3.1

Average Soil NPK Depletion Rates (kg ha⁻¹ yr⁻¹) in Sub-Saharan Africa, 1993–1995

Soil macronutrient depletion rates in sub-Saharan Africa (kg NPK ha⁻¹ yr⁻¹)

High (> 60)		Medium (30–60)	Moderate/low (< 30)
Burkina Faso	Mali	Benin	Algeria
Burundi	Mozambique	Cape Verde	Angola
Cameroon	Nigeria	Central African Republic	Botswana
Côte d'Ivoire	Rwanda	Chad	Egypt
DR Congo	Senegal	Congo	Morocco
Ethiopia	Somalia	Equatorial Guinea	South Africa
Gambia	Swaziland	Gabon	Tunisia
Ghana	Uganda	Lesotho	Sudan
Guinea	United Republic of Tanzania	Mauritania	Zambia
Guinea-Bissau		Niger	
Kenya		Sierra Leone	
Liberia		Togo	
Madagascar		Zimbabwe	
Malawi			

Source: FAO (2003)

soil nutrient depletion rates ranged from –12 kg N ha⁻¹ yr⁻¹ for cotton-based farming systems in Mali to –38 kg N ha⁻¹ yr⁻¹ in Kenya, and from 0 to –4 kg P ha⁻¹ yr⁻¹ in Kenya and under cocoa-based cropping systems in Ghana. The depletion rates for K were the lowest under cotton in southern Mali (–6.6 kg ha⁻¹ yr⁻¹) and the highest in Kenya (–23 kg ha⁻¹ yr⁻¹).

At the meso-scale, the nitrogen depletion rates ranged from 0 to –112 kg ha⁻¹ yr⁻¹ in southern Mali and Kisii District, Kenya, respectively. For phosphorus, the depletion rate was the lowest under cocoa-based cropping systems in Ghana (–0.5 kg P ha⁻¹ yr⁻¹) and the highest in Embu District in Kenya (–15 kg ha⁻¹ yr⁻¹). Similarly, the K depletion rate was the lowest under cocoa in Ghana (–11 kg ha⁻¹ yr⁻¹) but the highest in Kisii District in Kenya (–70 kg ha⁻¹ yr⁻¹). At the micro-scale, the N and P depletion rates were the lowest among the farmers classified as rich in southern Ethiopia (–3 kg N and –1.6 kg P ha⁻¹ yr⁻¹) but the highest in eastern Uganda and southern Ethiopia (–125 kg N and –30 kg P ha⁻¹ yr⁻¹, respectively). This wealth of information has since been coupled with advances in information science and communication technology to generate maps painting a picture of the magnitude of soil degradation engineered by soil nutrient mining and depletion in Africa (Figure 3.2). The level of soil nutrient depletion in SSA is both extreme and worrying. It lends support to the concerns already raised about the nexus between poverty, environmental degradation, food and nutrition insecurity that are destroying the lives of many in this African subregion (Figure 3.1).

3.3 INTERPRETATION OF SOIL NUTRIENT MINING AND DEPLETION IN SUB-SAHARAN AFRICA AND IMPLICATIONS

Soil nutrient balances at the continental, the country, the district and even the farm level are crucial for creating awareness among policymakers, researchers and stakeholders of the dangers to food security and to agricultural sustainability that arise from unabated soil nutrient mining and depletion. The shock waves from the Stoorvogel and Smaling (1990) study precipitated campaigns

TABLE 3.2

Soil Nutrient Balances for Selected Countries in Sub-Saharan Africa at Macro-, Meso- and Micro-Scales of Assessment[†]

			Nutrient balances (kg ha^{-1} yr^{-1})			
Scale	Site	Special	N	P	K	Source
Macro	Sub-Saharan Africa		−22	−2.5	−15	Stoorvogel and Smaling (1990)
	Africa*					Henao and Baanante (1999)
	Ghana	Spatially explicit	−27	−4	−21	FAO (2003)
	Kenya	Spatially explicit	−38	0	−23	FAO (2003)
	Mali	Spatially explicit	−12	−3	−15	FAO (2003)
	Koutiala Region, Mali	Cotton-based system	−12	1.4	−6.6	FAO (2003
Meso	Kisii District, Kenya		−112	−3	−70	Smaling et al. (1993)
	Embu District, Kenya	Tea coffee dairy system	−96	−15	−33	FAO (2003)
	Southern Mali	Optimistic and pessimistic view	−25	0	−20	Van der Pol (1992)
	Nkawie District, Ghana	Cocoa-based system	−18	−1.9	−20	FAO (2003)
	Wassa Amenfi District, Ghana	Cocoa-based system	−4	−0.5	−11	FAO (2003)
Micro	Southern Mali	Participatory approach	−8.2	8.5	7.4	Ramisch (1999)
	Southern Ethiopia	Different socioeconomic households	−55 to −6	−1.6 to −30	—	Elias et al. (1998)
	Northwest United Republic of Tanzania	Banana-based system	−76 to 80	−5 to 43	−50 to 199	Baijukya and Steenhuijsen de Baijukya et al. (1998)
	East and central Uganda		−125 to −3	−5 to −2	−11 to −9	Wortmann and Kaizzi (1998)
	United Republic of Tanzania	Sisal plantation	−13	−2.8	−38	Hartemink (2001)
	Kenya	All farms	−71	+3	−9	De Jager et al. (1998)
	East Africa		−41	−4	−31	Bekunda et al., 2002
	Southern Mali	Partial balances	−36 to −27	2.3 to 5.8	−32 to −11	Kanté (2001)
	Cameroon	Forest reserve	−6.5	−5.5	−30.8	Kanmegne et al. (2006)
	Senegal	Maize sorghum cotton groundnut rotation	−23	−2.6	−28.3	Bationo et al. (1998)

Notes: [†]Derived by updating FAO (2003).

* Nutrient balance ranged from −14 kg NPK ha^{-1} year^{-1} for South Africa to −136 kg NPK ha^{-1} year^{-1} for Rwanda.

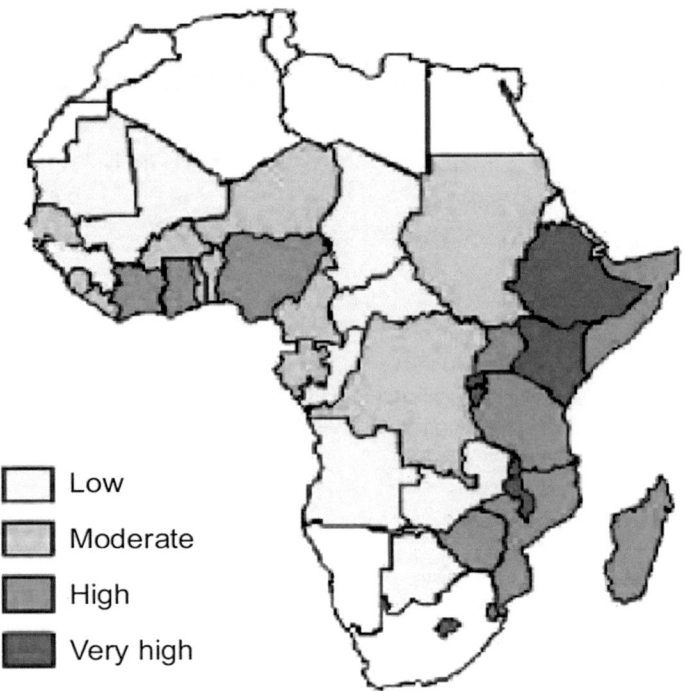

FIGURE 3.2   Soil degradation as a result of soil nutrient mining and depletion in Africa (FAO 2003).

to "recapitalize" the depleted soils of SSA, published in the 1997 special issue of the journal *Soil Science Society of America* (Sanchez et al. 1997). A year later, the 1998 special issue of the journal *Agriculture, Ecosystems and Environment* was devoted to consolidating research on soil nutrient balances in SSA (Bekunda et al. 2010).

However, soil nutrient balances at scales above the field or plot levels grossly underestimate soil nutrient depletion rates at field or plot levels because even nutrients in garbage heaps, kraals, pit latrines, abattoirs and landfills within the scale of assessment are considered as part of inflows or stocks. In reality, such nutrients only contribute to soil replenishment if they are redistributed to the actual fields where they were extracted from. Olupot et al. (2006) found net balances of +4.5 kg N, +20.2 kg P and –24.8 kg K ha^{-1} yr^{-1} at farm level, translating into a total NPK depletion rate of –0.1 kg ha^{-1} yr^{-1}. However, at plot level, depletion rates ranged from –9.4 to –138.8 kg NPK ha^{-1} yr^{-1} in the cowpea and sweet potato fields, respectively, with a soil depletion rate of –116.6 kg NPK ha^{-1} yr^{-1} in cassava fields. Similar studies across SSA report much higher soil nutrient depletion rates at plot level than those at macro-, meso- or even farm level (FAO 2003; Nkonya and Kaizzi 2004). Moreover, the soil nutrient depletion rates vary tremendously by crop, cropping system and management.

Sweet potatoes and cassava replaced cotton as the "white gold" (cash crops) in eastern Uganda from the late 1980s (Olupot et al. 2006). Their massive export to the market without replenishment of the nutrients removed by the exported crops has exacerbated soil nutrient depletion. The soybean example (Nkonya and Kaizzi 2004) in Uganda, sorghum and roselle in Sudan (El Tahir et al. 2013), tea and coffee in Kenya (Smaling et al. 1993) are all telling examples of how the market-driven export of large quantities of harvested products aggravates soil nutrient depletion. In contrast, the low nutrient depletion rates for cash crops such as cotton in southern Mali and cocoa in Ghana (FAO 2003) were attributed to the retention of biomass. Baijukya and Steenhuijsen de Piters (1998) and van den Bosch et al. (1998) demonstrated how integrating crops with livestock and proper management of both livestock and manure can lead to net positive soil nutrient balances. These peculiarities have not been considered in deriving fertilizer recommendations in SSA, pointing to a mismatch between the fertilizer recommendations and the magnitude of soil nutrient depletion. For example, the blanket average NPK application rate of 50 kg ha^{-1} yr^{-1} endorsed by the African Union heads of state in Abuja, Nigeria, in 2006 (Africa Fertilizer Summit 2006) is far below the plot-specific depletion rates of these nutrients (FAO 2003; Nkonya and Kaizzi 2004; Olupot et al. 2006). Not even the global fertilizer consumption average of 89 kg NPK ha^{-1} yr^{-1} (Henao and Banaante, 1999) might reverse decades of unabated soil nutrient mining in SSA, a concern already raised by Smaling et al. (1997).

Baijukya and Steenhuijsen de Piters (1998) evaluated soil N, P and K depletion rates of banana-based farms in three rainfall regimes: (1) high, (2) moderate and (3) low under five different banana farm management levels: (1) farm with no cattle and without brewing; (2) farm without cattle but brewing; (3) farm with indigenous cattle but with no brewing; (4) farm with indigenous cattle with brewing; and (5) farm with improved (zero grazing) cattle. The highest soil nutrient depletion rates (–76.2 kg N, –4.9 kg P and –50 kg K ha^{-1} yr^{-1}) were from banana farms in the high rainfall regime and under management level 1. In contrast, banana farms associated with zero grazing (farm management level 5) in the high rainfall regime resulted in the highest rates of soil nutrient build-up (+80.5 kg N, +42.8 kg P and +198.7 kg K ha^{-1} yr^{-1}) (see also FAO 2003). The Baijukya and Steenhuijsen de Piters (1998) study illustrates that ensuring that both the livestock and manure they generate for redistribution to the crop fields are managed well requires much more than merely integrating crop and livestock farming. The reasons for the higher quality of cattle manure under zero grazing (Nzuma et al. 1998) than under free range (Olupot et al. 2006) could include the collection of both the dung and urine as well as provision of shade and a firm floor to minimize volatilization and leaching, respectively, which can be challenging under free range-managed livestock. Positive nutrient balances under mixed farming (zero grazing and crop) have also been reported in related studies (van den Bosch et al. 1998; De Jager et al. 2004). Integrating crops and livestock, therefore, plays an effective role in sealing avoidable or wasteful nutrient losses resulting in net positive nutrient balances.

A sensitivity analysis of some input–output determinants of soil N, P and K balances for the sorghum fields in eastern Uganda (Olupot et al. 2006) indicated that raising concentrations of the initial soil N, P and K laboratory soil test results (0.08% N, 6.9 mg kg^{-1} P and 0.8 cmol(+) K kg^{-1} soil) by 0.01, 0.004 and 0.005 g kg^{-1} soil, respectively, restored net positive N, P and K balances for sorghum

TABLE 3.3
Sensitivity Analysis of NPK Input–Output Determinants for Sorghum Fields in Eastern Uganda (Olupot et al. 2006)

Determinants	Original values	Variation	Net nutrient balances (kg ha^{-1} yr^{-1}) N	P	K
Minimum rate	3%	+0.5%	–22.5	0.0	0.0
N total	0.8 g kg^{-1}	+0.01 g kg^{-1}	+0.8	0.0	0.0
P total	0.01 g kg^{-1}	+0.004 g kg^{-1}	0.0	+16.3	0.0
K total	0.02 g kg^{-1}	+0.005 g kg^{-1}	0.0	0.0	+6.5
K-factor	0.07	+0.01	–20.8	–2.7	–7.5
S-factor	4.0%	+2.0%	–21.5	–2.7	–7.1
L-factor	50m	+50m	–21.1	–2.7	–5.4
C-factor	0.26	+0.05	21.0	–2.7	–6.7
ER	1.5	+0.25	21.0	–2.7	–7.0

PLATE 3.1 Researcher (Giregon Olupot) in a farmer's field explaining to learners of Kanyum Primary School and their teacher the importance of kraal manure in replenishing nutrients removed, for example, with harvested sweet potato tubers for export to urban markets (*right*) in Kumi District, eastern Uganda in 2001.

fields (Table 3.3). Enrichment of the 2.5 Mg ha^{-1} kraal manure (field dry weight and of quality 1.1% N, 0.6% K and 1.2% K), which farmers could raise locally, with 22.5 kg N and 8.5 kg P ha^{-1} mineral fertilizers, restored net positive N, P and K balances in the sorghum fields (Table 3.3).

This same input combination, that is, 2.5 Mg kraal manure + 22.5 kg N + 8.5 kg P ha^{-1}, resulted in >2.0 Mg ha^{-1} sorghum grain yields, which was nearly three times the yield under farmers' practice during "bad" seasons (Olupot et al. 2004). However, much higher inputs of N, P and K than farmers could afford on their own were needed to restore the large negative N, P and K balances in the sweet potato and cassava fields, owing to the large quantities of nutrients exported to urban markets in tubers (Plate 3.1). The sensitivity analysis approach in sorghum plots adopted by Olupot et al. (2006) can be applied to any cropping activity if adequate information, including the laboratory soil test results and a range of locally available nutrient inputs for restoration, are available as the basis for deriving fertilizer rates to arrest and reverse soil nutrient depletion in SSA.

In a study of nutrient flows and balances in 20 low external input agroecosystems (LEIA) farms under smallholder production systems in the Campo Ma'an area in the humid forest zone of southern Cameroon, Kanmegne et al. (2006) found net soil depletion rates based on partial nutrient balances for farmer-managed flows to the tune of –65 kg N, –5.5 kg P and –30.8 kg K ha^{-1} yr^{-1}. We made calculations based on the data presented and on the five management scenarios proposed by Kanmegne et al. (2006) for restoring the negative nutrient balances in Campo Ma'an: recycling of household waste and animal manure, inclusion of legumes in the cropping systems, avoidance of bush burning, deep capture of leached nutrients (through agroforestry) and harnessing human waste. Results indicate that these could add 65.5 kg N, 5.64 kg P and 39.5 kg K ha^{-1} yr^{-1} to the soil and restore net positive balances 0.5 kg N, 0.14 kg P and 8.7 kg K ha^{-1} yr^{-1} without fertilizers.

El Tahir et al. (2013) also quantified nutrient balances as indicators of sustainability for different cropping systems and compared the 2004 nutrient stocks and balances with the 2002 baseline nutrient stocks and balances to determine the rate of soil nutrient depletion at El Demokeya Forest Reserve in the semi-arid North Kordofan in Sudan. They found higher nutrient depletion rates in the sorghum (*Sorghum bicolor* Moench.) and roselle (*Hibiscus sabdariffa* L.) treatments than in the grass treatments, particularly for the sole crops (Table 3.4). This was attributed to the export of larger quantities of above-ground biomass from crops than from grasses without replenishment of nutrients removed in the biomass. For example, between 67% of stover from sole sorghum to 80% stover from sorghum intercropped with low-density trees (266 trees ha^{-1}) was harvested. Similarly, between 99% and 100% of above-ground biomass of roselle comprising seed, calyx and stover were harvested, a common practice in most farming systems in SSA. The organic materials in the El Tahir et al. (2013) study were richer in N contents (6.8% and 3.3% for roselle and sorghum,

TABLE 3.4

Effect of Cropping and Land Management on NPK Balances and Depletion Rates at El Demokeya Forest Reserve of the Semi-Arid North Kordofan in Sudan (El Tahir et al. 2013)

	Nutrient balances in 2004			Depletion of stocks			
	(kg ha^{-1} yr^{-1})			(% of 2002 stocks)			
Land use	N	P	K	N	P	K	OC
HTD + sorghum	255	1	32	28	9	13	38
LTD + sorghum	149	−5	0	17	−42	0	5
Sorghum	−54	−13	−56	−10	−130	−21	−49
HTD + Roselle	243	−24	3	23	−218	1	39
LTD + Roselle	112	−24	−20	13	−200	−8	10
Roselle	−51	−23	−47	−8	−209	−20	
HTD + grass	280	11	74	31	100	30	78
LTD + grass	204	9	58	23	75	20	48
Grass	−3	3	13	0	25	6	7

Note: HTD and LTD denote high tree density (433 trees ha^{-1}) and low tree density (266 trees ha^{-1}).

respectively) than grass biomass (0.72%). Their removal in such large quantities therefore explained the high depletion of soil nutrient stocks under sole crops. The results highlight the importance of retaining crop residues and stubble in the field as a step toward mitigating negative soil nutrient balances and nutrient depletion.

Besides, the more positive nutrient balances for N in treatments involving intercropping sorghum and roselle with trees at high tree density (HTD) highlight the role of trees in agroforestry systems in the recapitalization of soil nutrient stocks, which should rekindle the debate among the scientific community in favor of reverting to practices that communities have traditionally deployed to restore and maintain soil productivity, such as the *Acacia senegal* bush-fallow in the semi-arid zone of the Sahel region. Pastures as effective land cover for replenishing soil nutrients can only thrive in the presence of livestock and their use highlights the crucial role of livestock in the restoration of positive soil nutrient budgets (van den Bosch et al. 1998; De Jager et al. 2001). Effective N surpluses obtained in the study of El Tahir et al. (2013) exceeded 40 kg N ha^{-1}, which is considered to be acceptable for agricultural production in most countries (Sanchez and Palm 1997), further emphasizing the importance of agroforestry-based practices in the restoration of depleted soils and sustenance of soil productivity.

3.3.1 Plot-Level Soil Nutrient Mining and Depletion in Sub-Saharan Africa

High-precision measurements of nutrient mining and depletion rates at plot or field levels that can be attributed to particular primary production units (crops or crop combinations) and their interaction with secondary production (livestock) units at farm level are beginning to emerge (Table 3.5). Staple grains and root crops to the majority of the poor in SSA are among the heaviest contributors to soil nutrient depletion. For example, soil nutrient depletion is the highest for maize (*Zea mays* L.) fields in Uganda and, indeed, SSA (−104.2 kg N, −13.6 kg P and −82.4 kg K ha^{-1} yr^{-1}), which is closely followed by sweet potatoes (*Ipomoea batatas* L.), cassava (*Manihot esculenta* L.) and yams (*Dioscorea* sp). This situation is likely to worsen with the drive to commercialize crops previously

TABLE 3.5

Crop-Specific Soil Nutrient Depletion for Selected Countries in Sub-Saharan Africa

Country	Crop	Nutrient depletion rates (kg ha^{-1} yr^{-1})		
		N	P	K
Mali[†]	Millet	−48	n/a	n/a
Uganda*	Millet	−16.7	−1.1	−22.1
Mali[†]	Sorghum	−31	n/a	n/a
Uganda*	Sorghum	−20.7	−2.7	−5.2
Mali[†]	Maize	−29	n/a	n/a
Ghana[†]	Maize	−28.1	−5.85	−16.9
Uganda[‡]	Maize	−104.2	−13.6	−82.4
Mali[†]	Rice	−43	n/a	n/a
Mali[†]	Cotton	−21	n/a	n/a
Mali[†]	Groundnuts	−40	n/a	n/a
Uganda*	Groundnuts	−33.1	−14.6	−16.9
Mali[†]	Cowpea	−21	n/a	n/a
Uganda*	Cowpea	−6.5	−0.7	−2.2
Mali[†]	Fallow	−5	n/a	n/a
Ghana[†]	Fallow	0.6	0.9	2.85
Uganda[‡]	Fallow	33.2	−1.5	−13.7
Ghana[†]	Cocoa	−2.35	−0.15	−8.85
Ghana[†]	Cassava	−60.65	−8.6	−54.65
Uganda*	Cassava	−39.7	−6.8	−70.1
Ghana[†]	Bananas	−7.45	−0.4	−35.5
Uganda[‡]	Bananas	−13.2	1.2	−35.7
Ghana[†]	Cocoyam	−42.40	−2.6	−33.0
Ghana[†]	Yam	−70.4	−4.85	−53.1
Ghana[†]	Vegetables	−57.8	−7.0	−29.3
Ghana[†]	Oil palm	−29.2	−7.2	−54.1
Uganda[‡]	Beans	−40.4	−8.8	−42.7
Uganda[‡]	Sweet potatoes	−71.3	−13.2	−78.9
Uganda*	Sweet potatoes	−64.0	−8.5	−66.3
Uganda[‡]	Soybean	−121.5	−16.4	−68.3
Uganda[‡]	Pasture	19.2	−3.3	−30.7
Uganda[‡]	Home garden	3.0	−1.8	−18.9

Notes: *Olupot et al. (2006); [†]FAO (2003); [‡]Wortmann and Kaizzi (1998); n/a: no data.

grown for subsistence or with diversification into "high-value" crops as exemplified by soybean, whose depletion rate for soils in Uganda stands at −121.5 kg N, −16.4 kg P and −68.3 kg K ha^{-1} yr^{-1}.

3.3.2 SOIL NUTRIENT DEPLETION BY FARMER TYPOLOGY AND AGROECOLOGICAL ZONE

Despite the paucity of data, a new line of research is emerging linking soil nutrient mining and depletion to the socioeconomic well-being of farming households. In southern Ethiopia, Elias et al. (1998) examined soil nutrient balances in four farms selected from two agroecological zones (high-land and lowland), representing four socioeconomic groups of farmers in terms of their resources:

TABLE 3.6

Soil Nutrient Depletion Rates Among Four Household Wealth Classes for Two Agroecological Zones in Southern Ethiopia

		Soil nutrient depletion rates (kg ha^{-1} yr^{-1}) by wealth class			
AEZ	Nutrient	Rich	Medium	Poor	Very poor
High altitude	N	−47 (−58.6)	−41 (−54.3)	−19 (−53.9)	−9
	P	11.7 (−10.7)	4.8 (−10.0)	3.6 (−8.4)	1.1
	K	(−48.3)	(−48.0)	(−53.1)	
Low altitude	N	−49 (37.7)	−41 (−35.5)	−55 (−34.0)	−20
	P	30.5 (−10.6)	17.3 (−9.3)	3.8 (−7.4)	−1.6
	K	(−32.9)	(−32.8)	(−23.3)	

Sources: FAO (2003) and Abebayehu et al. (2011) (for NPK values in parentheses).
Notes: AEZ denotes agroecological zone.

rich, medium, poor and very poor, based on ranking by the local area community. Rich farmers owned more than two oxen and a sizeable number of other livestock. Medium farmers owned two oxen and about half the number of livestock as the rich group. Poor farmers owned or shared one ox but did not own any breeding cows. Very poor farmers did not own any cattle but occasionally owned one or two goats or sheep (they borrowed animals for draught power and manure production). Abebayehu et al. (2011) also evaluated and compared nutrient balance sheets of smallholder farmers in the high and low altitudes of Gilgel Gibe catchment, Jimma zone, southwestern Ethiopia, but based on only three wealth classes: rich, medium and poor as opposed to the four by Elias et al. (1998). Generally, soil nutrient depletion rates were significantly higher for N among the rich between the high and low altitudes (Table 3.6).

For example, in the study of Abebayehu et al. (2011), N depletion was (−59 vs. −38) and K (−48 vs. −33) kg ha^{-1} yr^{-1} in the high and low altitudes, respectively, among the rich class, and similarly for N (−54 vs. −34) and K (−53 vs. −23) kg ha^{-1} yr^{-1} between the high and low altitudes, respectively, among the poor class. Two reasons could explain differences in the net nutrient balances between the high and low altitudes and between the rich and poor farmers in the studies of Elias et al. (1998) and Abebayehu et al. (2011). First, the soils in the high altitudes were inherently superior to those in the low altitudes, especially with reference to cation exchange capacity (CEC) (31.3 vs. 26.8) cmol(+) kg^{-1} soil, soil organic matter (SOM) (4.3 vs. 3.3)%, N (0.22 vs. 0.17)% and exchangeable K (2.1 vs. 1.7) cmol(+) kg^{-1} soil, respectively. Second, the rich tended to subsist on inherently richer soil than the poor, highlighting the fragility and susceptibility of soils of the poor to nutrient depletion (FAO 2003).

Nkonya and Kaizzi (2004) focused on socioeconomic determinants of soil nutrient balances with the aim of informing strategies better to address soil nutrient depletion in a study involving 58 farmers randomly selected from four villages in eastern Uganda under high and low agricultural potential zones. They applied the concept of economic nutrient depletion ratio (ENDR), coined by van der Pol (1992), meaning a share of farmers' income derived from mining soil nutrients in order to allocate a cost to soil nutrient mining, expressed as:

$$\text{ENDR} = \frac{\text{NDMV}}{\text{GM}} \times 100 \tag{3.1}$$

where NDMV denotes nutrient deficit market value, that is the value of nutrients mined ha^{-1} if such nutrients were to be replenished by applying mineral fertilizers purchased from the cheapest source, whereas GM denotes growth margin from agricultural activities per household. They observed that

95% of the farmers had negative N, P and K balances and were therefore depleting soil nutrient stocks. It would cost these farmers 20% of their total farm income estimated at US \$823 household^{-1} yr^{-1} to replenish the removed nutrients. Among the key factors that influenced nutrient balances positively were: high level of education of the household head, ownership of livestock, access to agricultural extension services, non-farm activities and incomes, and crop diversification, including perennial and annual crops, legumes and cereals. Large household size, access to poor extension services, for example, where farmers were advised to adopt piecemeal technologies, such as the use of improved varieties without the accompanying technologies to optimize their productivity, were negatively correlated with soil nutrient balances.

3.4 CLIMATE-SMART APPROACHES TO REVERSING SOIL NUTRIENT DEPLETION IN SUB-SAHARAN AFRICA

The heterogeneity of SSA agroecosystems is a function of ethnohistory and culture, climate, soils and production goals. The factors influencing the level of soil nutrient depletion are also many and complex: nutrient management, regeneration and plant protection, livestock integration, soil and water conservation, biodiversity, agricultural policies and marketing structures. All these must be factored in when designing a package of technology options aimed at arresting and reversing nutrient depletion (De Jager et al. 1998; van den Bosch et al. 1998; Bekunda et al. 2010). The case studies about nutrient mining and depletion in SSA summarized above indicate that arresting and reversing unabated nutrient mining and depletion of soil nutrient stocks spanning decades back in SSA will require complex and holistic approaches (De Jager et al. 1998; van den Bosch et al. 1998; FAO 2003; Bekunda et al. 2010). These technological options must be firmly rooted within the underlying principles (Lal et al. 2015): (i) harnessing, integrating and optimizing the use of all possible sources of nutrients within the disposal, reach and affordability of farmers; (ii) sealing all unnecessary and avoidable nutrient losses; (iii) maximizing recovery, uptake and use efficiencies of the nutrients; (iv) guaranteeing conducive conditions for optimization of nutrient recovery and uptake from the soil as well as utilization efficiencies, including soil water conservation and proper agronomic and crop husbandry practices; (v) use of the most adapted and high-yielding germplasm, (vi) integrated pest management; and (vii) following a holistic approach to building fertile grounds (De Jager et al. 1998; van Den Bosch et al. 1998; FAO 2003; Bekunda et al. 2010; Thierfeld et al. 2012; Kessler et al. 2015; Smith et al. 2016). For example, in Mali, Bationo et al. (1997) more than tripled maize grain yield by applying fertilizer P as part of a package that included planting at the right time and at the correct plant density. Arriving at an appropriate package requires a thorough understanding of its integral components.

3.4.1 HARNESSING NPK INPUTS FROM A RICH DIVERSITY OF ORGANIC FERTILIZERS (IN2)

Organic fertilizers are the most important nutrient sources for many smallholder farmers in SSA. Nutrients from organic sources are added into the soil through: (i) crop residues (*in situ* or transferred from other production areas); (ii) livestock manures (deposited directly during grazing or after collection, treatment and systematic application on land); and (iii) compost, which is a collection of a range of organic materials that have been subjected to microbial decomposition until the attainment of maturity (Bekunda et al. 2010; Adamu et al. 2014). Livestock are an integral component of many cropping systems in SSA. Nutrient inputs from livestock can originate from: poultry manure, cattle manure, pig manure, slaughter waste and feed waste (van den Bosch et al. 1998; Bekunda et al. 2010). Kitchen refuse, crop residues, agro-industrial waste and municipal solid waste are important sources of N, P and K in LEIA.

The quantities of N, P and K in organic materials, though low, can constitute a significant source of nutrients in LEIA predominant in SSA (Bekunda et al. 2010). Field trials indicate that fertilizer equivalence values of some organic fertilizers equal (Sanginga and Woomer, 2009) or even

TABLE 3.7
N, P and K Stocks in Selected Organic Materials (in kg Mg⁻¹)

Organic material	Nutrient quantities (kg of nutrient Mg⁻¹ of organic material)			Source
	N	P	K	
Poultry litter	29	18	16	Sanginga & Woomer (2009)
Poultry litter	30–50	15–35	15–33	Amanula et al. (2010)
Poultry litter	22.1	29.8	20.5	Adeniyan et al. (2011)
Cattle manure	17–20	22–16.3	10.8–20.8	Koala (2001)
Cattle manure	11	6.0	12	Olupot et al. (2006)
Cattle manure	10	2	9	Sanginga & Woomer (2009)
Cattle manure	13	5.8	21.5	Adeniyan et al. (2011)
Pig manure	15–24	4–4.4	12–32	Koala (2001)
Coffee husks	17	1.3	29	Sanginga & Woomer (2009)
Coffee husks	12.7	0.6	24.6	Dzung et al. (2013)
Coffee manure	18.3	1.1	10.2	Tewodros et al. (2010)
Rice straw	0.98	0.31	0.61	Adamu et al. (2014)
Sorghum stover	0.30	0.17	2.77	Adamu et al. (2014)
Zea mays stover	8.3	0.8	13	Sanginga & Woomer (2009)
Zea mays stover	0.20	0.04	0.68	Adamu et al. (2014)
Millet stover	0.50	0.03	0.96	Adamu et al. (2014)
Groundnut haulms	1.10	0.06	0.54	Adamu et al. (2014)
Bean stover	9.9	1.1	19	Sanginga & Woomer (2009)
Soybean prunings	27	1.9	22	Sanginga & Woomer (2009)
Cowpea prunings	24	3.1	11	Sanginga & Woomer (2009)
Cowpea stems	0.78	0.02	0.46	Adamu et al. (2014)
Cotton stalks	0.98	0.05	0.88	Adamu et al. (2014)
Crotolaria leaves	42	1.9	14	Sanginga & Woomer (2009)
Mucunapruriens	29	2.3	15	Sanginga & Woomer (2009)
Sebaniasesban leaves	35	2.1	14	Sanginga & Woomer (2009)
Tithonia leaves	38	3.8	46	Sanginga & Woomer (2009)
Municipal compost	11–13	3.5–4.1	27–45	NEMA (2011)

surpass (Jhariya and Raj 2014; Raj et al. 2014) those applied from mineral fertilizers. The organic resource database developed at the Tropical Soil Biology and Fertility Institute (Palm et al. 2001) contains information on chemical characteristics of over 300 plant species residues that can be used to optimize inputs from organic resources (Bekunda et al. 2010). One of the limitations of organic materials is that their NPK contents vary widely, even within each category (Bekunda et al. 2010). For example, nutrient stocks in poultry litter range from 22.1 to 50 kg N, 15 to 29.8 kg P and 15 to 33 kg K Mg⁻¹ of the litter. The same applies for cattle manure (Table 3.7).

The low nutrient contents in organic materials imply that they must be applied in bulk. Moreover, the quality of organic materials is higher in fertile soils than for materials derived from degraded soils (Bekunda et al. 2010), implying that even larger quantities are needed for yield gains in SSA. On field assessment of technologies for the LEIA in East Africa, De Jager et al. (2004) revealed that mulch, manure and compost amounts ranging from 8.5 to 150 Mg ha⁻¹ were needed to attain significant increases in yield and economic returns on farmers' fields. However, much smaller amounts of organic material than would be recommended for replenishing nutrients are available to the farmers (Bekunda et al. 2010), as low as 1.3 Mg ha⁻¹ of millet stover, 0.45–1.6 Mg ha⁻¹ of

PLATE 3.2   *Left–right*: Municipal solid waste (transported to a landfill) and coffee husks (*top*); chicken litter and mature municipal solid waste compost in a windrow due for sieving in Mbale Municipality, one of the 12 pilot compost plants by the World Bank in Uganda (*bottom*).

manure in the Sahel and 1–1.5 Mg per animal per year in Kenya (Palm et al., 2001). Fortunately, through initiatives such as the Clean Development Mechanism (CDM), the World Bank has aided the piloting of the conversion of biodegradable municipal solid waste (MSW) generated in cities and municipalities within SSA into high quality compost on a large scale as a strategy to combat greenhouse gas emissions from poorly managed MSW. In Uganda, for example, there are 12 municipalities selected to pilot the satellite plants, each with an installed production capacity of 50,000 Mg compost yr^{-1} (World Bank 2008). Using the windrow method, these plants convert municipal solid waste, comprised of a range of agro-wastes including garbage, coffee husks and chicken litter, into dark compost (Plate 3.2). To the World Bank initiative can be added the >33% of food produced globally that ends up as waste annually, with about 1.3 billion Mg of food waste generated in 2010 alone (HLPE 2014). Abattoirs, wastewater treatment plants and agro-industrial parks are also choking with waste, which is loaded with nutrients and waiting to be tapped into to combat soil nutrient depletion in SSA.

The contribution of organic fertilizers to farming extends beyond the provision of nutrients to include soil protection, carbon sequestration, improvement and maintenance of soil biochemical and physicochemical properties (Bekunda and Woomer 1996; Nzuma et al. 1998; Nottidge et al. 2010; Olowookere et al. 2015), boosting of populations of soil organisms and biodiversity (Jhariya and Raj, 2014; Olowookere et al. 2015; Hartati and Sudarmadji 2016); plant residues are used as animal feed or bedding materials, composted, applied as surface mulch, or ploughed into the soil and applied in combination with livestock manures or synthetic fertilizers. Retaining 30–40% crop residues in the field can reduce wind erosion by 70–90% and soil erosion exceeding 5 Mg soil ha^{-1} requires 4.9–7.4 Mg ha^{-1} of crop residues (Wortmann et al. 2008).

3.4.2 Biological Nitrogen Fixation (In4)

Biological nitrogen fixation (BNF) is one of the natural and inexpensive sources of nitrogen especially in LEIA in SSA (Ebanyat et al. 2009; Bekunda et al. 2010). It contributed to about 27.7 Tg N yr^{-1} in Africa during the 1990s (Galloway et al. 2004), 1.8 Tg yr^{-1} of which was fixed during cultivation, the equivalent of 50% of the N imported and manufactured as fertilizer in SSA (1 Tg $= 10^{12}$ g) (FAO 2015). Grain legumes in Africa seasonally fix about 15–210 kg N ha^{-1} (Dakora and Keya 1997). Besides, net soil N accrual from effectively recycled legume residue can be as much as 140 kg ha^{-1}(Giller 2001). Although this may not match the crop needs, it cuts down on the amount of N to be derived, for example, from fertilizers. As an example, soybean requires approximately 100–300 kg N ha^{-1} to achieve maximum yields (Giller 2001). As much as 30% of the N required by the crop can come from the leguminous hedgerow trees (Giller 2001). In combination with efforts to select grain legumes for BNF and an improvement in inoculant delivery systems, it is projected that inputs from BNF can increase from approximately 35 kg N ha^{-1} to over 90 kg ha^{-1}, resulting in increased total amounts of N per farm from approximately 8–30 kg N yr^{-1} across the whole area of SSA (Giller 2001; Bekunda et al. 2010). Prospects for increasing N inputs into farming systems through inoculation must be informed by research featuring need-to-inoculate trials *in situ* involving pairwise comparisons of (i) non-inoculated plots, (ii) inoculated plots, (iii) plots fertilized with substantial amounts of N (Date 1977; Hungria et al. 2005, 2006) and intra-species variability in N fixation to enable identification of varieties that can be promoted for optimization of BNF. Interest in this issue is only beginning to be picked up and research to this effect is taking place in SSA, with results underway.

3.4.3 Organic Fertilizer Fortification (In2)

3.4.3.1 Organic Fertilizer Fortification for Nitrogen

The starting point for improving the quality of organic fertilizers is to ensure that for livestock manure, both the urine and dung are collected and that all manures kept away from predisposing factors to nutrient volatilization and leaching. Protection of manure against such losses could be the reason why livestock manure from confined or zero-grazing units (Nzuma et al. 1998; Baijkya and Steenhuijsen de Piters 1998) tends to be superior to that of livestock under free range management (Olupot et al. 2006). The N content of organic fertilizers can also be increased by enriching it with biomass from legume cover crops and forage legumes. Annual N yield from five prunings of Gliricidia and Leucaena hedgerows in Nigeria was 170 $\pm$ 250 kg N ha^{-1}, as opposed to 40 $\pm$ 85 kg N ha^{-1} in the non-leguminous species *Acioa Bartic* and *Alchormea cordifolia* (Kang et al. 1990). Green manure legumes such as *Mucuna* spp can accumulate up to 100 kg N ha^{-1} in biomass and also suppress weeds (Juo and Kang, 1989; Giller and Wilson 1991). When legume residues are incorporated into the soil, they can supply N to rice and produce benefits comparable with that of 40 to 80 kg mineral fertilizer N ha^{-1} (Rahman et al. 2014). These materials could be used to fortify compost and other organic fertilizers to be rich in N, though limited progress has been made with research in this direction.

3.4.3.2 Organic Fertilizer Fortification for Phosphorus and Potassium

Africa has about 4.5 billion Mg of proven rock phosphate (RP) in deposits of igneous, sedimentary or biogenic rocks distributed across the continent (Figure 3.3). Osukuru (Sukulu) Hills, one such deposit in Uganda, contains an estimated 230 million Mg of RP stretching over 26.4 km^2 with a fertilizer grade ranging from 11 to 13% P_2O_5 (Appleton 2002). Minjingu phosphate rock in Tanzania contains 6.6 million Mg of P, with a processing capacity of 100,000 Mg yr^{-1} (Bekunda et al. 2010). In West Africa, vast phosphate rock deposits have been discovered and characterized from Tahoua (Niger), Parc-W (Niger), Kodjari (Burkina Faso), Hahotoe (Togo), Matam (Senegal) and Tilemsi

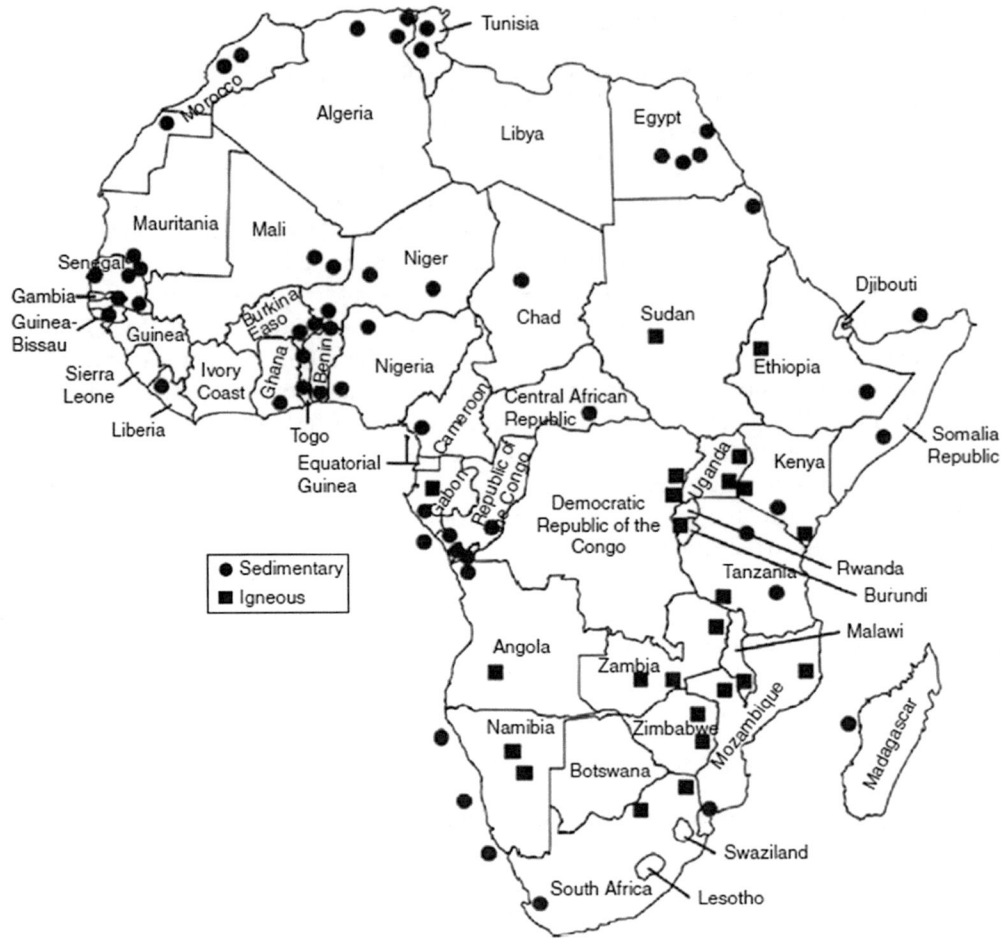

FIGURE 3.3 Deposits of sedimentary and igneous phosphate rocks in Africa (after van Kauwenbergh 2006; Bekunda et al. 2010).

(Mali) (Bekunda et al. 2010). The RP from these rocks varies widely in agronomic effectiveness (van Straaten 2002; Smalberger et al. 2006; Bekunda et al. 2010). The Tahoua, Tilemsi, Matam and Minjingu RP (MRP) are the only few known to be soluble enough for direct application. Application at a rate of 45 kg P ha^{-1} as MRP (400 kg MRP ha^{-1}) increased maize grain yield in the first year by 1.0 Mg, resulting in agronomic efficiency of 23% (Okalebo et al. 2006). Local and cheap technologies for increasing P solubilization and agronomic effectiveness of RP are needed. Khan and Sharif (2012) increased chicken litter P content by 580% through co-composting it with RP.

3.4.4 MINERAL FERTILIZERS (IN1)

Mineral fertilizers remain the most strategic option for restoring soils in SSA depleted by decades of unabated mining without replenishment of the nutrients (Stoorvogel and Smaling 1990; Smaling et al. 1997). Unfortunately, SSA, with the world's most depleted soils, is the least consumer of mineral fertilizers globally. The situation is particularly bad for Senegal, Cameroon, Zimbabwe, Sudan, Tanzania, Cote d'Ivoire and Zambia (Figure 3.4) and even worse for countries like Uganda, Ghana, Guinea and Mozambique, where per capita fertilizer consumption is not even featuring on the IFASTAT (2015) database. During the Fertilizer Summit in Abuja, Nigeria, the African Union states committed to increasing fertilizer consumption by smallholder farmers in SSA from 8 kg (the

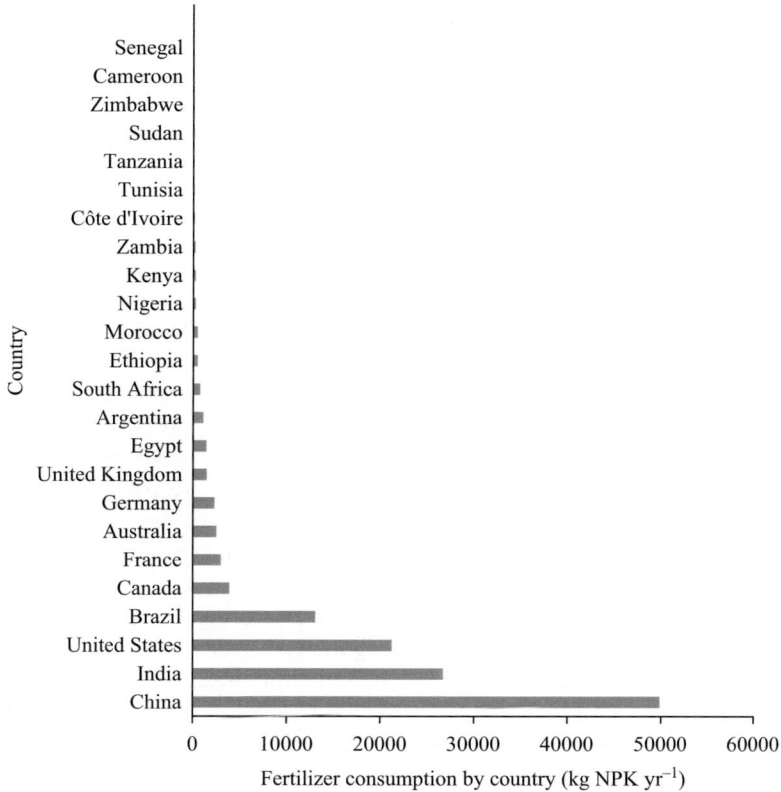

FIGURE 3.4 Fertilizer consumption in kg ha^{-1} (x-axis) in selected countries (y-axis) in 2015. (Data from IFASTAT: https://www.ifastat.org/databases/plant-nutrition accessed on October 19, 2018 to draw this figure.)

TABLE 3.8

Fertilizer N, P and K Consumption in Sub-Saharan Africa Compared with Global Consumption

Nutrient	Region	Fertilizer consumption (in thousand Mg yr^{-1}) (2014–2018)					Total
		2014	2015	2016	2017	2018	
Nitrogen	Global	147,293	151,481	155,040	158,121	161,151	**773,086**
Nitrogen	SSA	7,954	8,180	8,372	8,539	8,702	**41,747**
Phosphorus	Global	40,113	41,207	42,063	43,013	43,890	**201,286**
Phosphorus	SSA	802	824	841	860	878	**4,205**
Potassium	Global	29,058	29,807	30,568	31,407	32,287	**153,127**
Potassium	SSA	1,540	1,580	1,620	1,665	1,711	**8,116**
Total	**Global**	**216,464**	**222,495**	**227,671**	**232,541**	**237,328**	**1,136,499**
Total	**SSA**	**10,314**	**10,584**	**10,833**	**11,064**	**11,291**	**54,086**

Note: Fertilizer consumption values are derived from FAO (2015).

SSA average) to at least 50 kg ha^{-1} yr^{-1}, which is still below the global average of 89 kg ha^{-1} yr^{-1} (Henao and Banaante 1999). World fertilizer consumption in 2013 and projections for the period 2014–2018 (FAO 2015) indicated that SSA consumes <5.4% N, 2.1% P and 5.3% K of the mineral fertilizers consumed globally (Table 3.8). Stepping up mineral fertilizer use to complement organic fertilizers and BNF should be part of the broader strategy to combat soil nutrient depletion in SSA.

3.5 CONCLUSIONS AND RESEARCH IMPERATIVES

This review was aimed at highlighting the gravity of soil nutrient depletion in SSA and to demonstrate the important part that nutrient balances could play in assessing the prospects for tackling hunger, malnutrition and poor human health in the region. At the country level of assessment, soil nutrient depletion rates exceed 60 kg NPK ha^{-1} yr^{-1} in over 52% of the countries. Moreover, soil nutrient depletion rates are much higher at field or plot level than at country, district or even farm level. The commercialization of agriculture, especially among previously subsistence farming communities, without an enabling environment for replenishing the nutrients exported in harvested products, is exacerbating soil nutrient depletion. Depletion rates are particularly high for maize, sweet potatoes and cassava in Uganda; coffee and tea in Kenya; cassava, yams and vegetables in Ghana; millet, rice and sorghum in Mali; and roselle and sorghum in Sudan. Socioeconomic factors, including poverty, large household size, lack of livestock and limited access to quality extension services, are also fueling soil nutrient depletion in SSA. Restoring depleted soils is thus the entry point for increasing agricultural productivity, ending hunger, malnutrition and poor health in SSA. A systematic package of technologies is needed, building on proven indigenous initiatives already endemic to SSA.

Multiple cropping, involving intercropping and rotating cereals with legumes in agroforestry systems, has been demonstrated to abate soil nutrient depletion. Research efforts should be geared toward identifying and promoting more complementary (both below and above ground) crop-tree mixtures in order to optimize: (1) interception and utilization efficiency for growth resources; (2) soil protection against agents of erosion; and (3) symbiotic relationships such as biological nitrogen fixation and mycorrhizae as well as use of iron-mining bacteria (siderophores) to free the phosphorus that could be locked up by iron in red soils (Gerard 2016) in much of SSA. There is need to harness urine and dung as well as conserve animal manure from free range-managed livestock to minimize gaseous losses and leaching of nutrients. Nontraditional sources of manure, including food waste, municipal solid waste and biosolids polluting many cities, municipalities and urban centers as well as human waste in pit latrines and sanitary toilets, should be tapped into to offset the shortage of organic fertilizers for large-scale field operations. The sensitivity analysis approach should be applied to inform the least expensive pathway to restoring neutral-to-positive soil nutrient balances per plot or field of interest. This should then be followed by joint identification of input combinations with the farmer(s) in question and on-farm evaluation(s) of responses to the inputs by particular crop(s) of interest. Research into the biofortification of these organic fertilizers with abundantly available substrates such as rock phosphates, slaughter waste and agro-industrial processing waste to improve their quality could reduce fertilizer N, P and K demand from mineral fertilizers even to the 50 kg ha^{-1} yr^{-1} that the African Union heads of state committed to achieving for SSA by the year 2015. However, research in this area is in its nascent stages. It will require innovative communication techniques to translate the figures and numbers about nutrient balances, soil nutrient depletion and prospects for a bright future in a manner that makes sense to all stakeholders, especially decision makers. Part of the broader strategy could involve overlaying soil degradation and environmental degradation maps with maps about poverty, hunger and malnutrition as well as disease prevalence in order to highlight the linkage among all these with soil nutrient depletion. This should be part of the broader strategy to keep alive the vision of former United Nations Secretary General Kofi Annan (RIP) that "A successful revolution is where we would see soil health restored, through agroforestry techniques and organic and mineral fertilizers, among other solutions."

REFERENCES

Adamu, U.K., Almu, H., Adam, I.A., Sani, S.A. 2014. Evaluation of nutrient composition of some cereals and legumes crops residues as compost materials. *Bayero Journal of Pure and Applied Sciences* 7(2): 52–54.

Adeniyan, O.N., Ojo, A.O., Adediran, J.A. 2011. Comparative study of different organic manures and NPK fertilizer for improvement of soil chemical properties and dry matter yield of maize in two different soils. *Journal of Soil Science and Environmental Management* 2(1): 9–13.

Africa Fertilizer Summit. 2006. *Africa Fertilizer Summit Proceedings*. IFDC, Muscle Shoals, AL.

Amanullah, M.M., Sekar, S., Muthukrishnan, P. 2010. Prospects and potential of poultry manure. *Asian Journal of Plant Sciences* 9: 172–182.

Appleton, J.D. 2002. Local phosphate resources for sustainable development in sub-Saharan Africa. British Geological Survey Research Report CR/02/121/N. National Environment Research Council Publication, Keyworth, Nottingham, UK.

Aticho, A., Elias, E., Diels, J. 2011. Comparative analysis of soil nutrient balance at farm level: A case study in Jimma Zone, Ethiopia. *International Journal of Soil Science* 6: 259–266.

Baijukya, F.P., Steenhuijsen de Piters, B. 1998. Nutrient balances and their consequences in the banana-based land use systems of Bukoba district, northwest Tanzania. *Agriculture, Ecosystems and Environmentalist* 71: 147–158.

Bationo, A., Ayuk, E., Ballo, D., Kon'e, M. 1997. Agronomic and economic evaluationof Tilemsi phosphate rock in different agroecological zones of Mali. *Nutrient Cycling in Agroecosystems* 48: 179–189.

Bationo, A., Hartemink, A., Lungu, O., Naimi, M., Okoth, P., Smaling, E., Thiombiano, L. 2006. *African Soils: Their Productivity and Profitability of Fertilizer Use*. Background Paper prepared for the African Fertiliser Summit. IFDC, Muscle Shoals, AL.

Bationo, A., Lompo, F., Koala, S. 1998. *Research on Nutrient Flows and Balances in West Africa: State-of-the-Art.*

Bekunda, M., Manzi, G. 2003. Use of partial nutrient budget as an indicator of nutrient depletion in the highlands of south-western Uganda. *Agroecosystems* 67: 187–195.

Bekunda, M., Sanginga, N., Woomer, P.L. 2010. Restoring soil fertility in sub-Sahara Africa. *Advances in Agronomy* 108: 183–236.

Bekunda, M.A., Nkonya, E., Mugendi, D., Msaky, J.J. 2002. Soil fertility status, management, and research in east Africa. *East African Journal of Rural Development* 20: 94–112.

Bekunda, M.A., Woomer, P.L. 1996. Organic resource management in banana-basedcropping systems of the Lake Victoria Basin, Uganda. *Agriculture, Ecosystems and Environmentalist* 59: 171–180.

Dakora, F.D., Keya, S.O. 1997. Contribution of legume nitrogen fixation tosustainable agriculture in sub-Saharan Africa. *Soil Biology and Biochemistry* 29: 809–817.

Date, R.A. 1977. Inoculation of tropical pasture legumes. In *"Exploiting the Legume-Rhizobium Symbiosis in Tropical Agriculture,"* Miscellaneous Publication number 145, J.M. Vincent, Whitney, A.S., Bose, L.J. (Eds.), pp. 293–311. College of Tropical Agriculture, University of Hawaii, Honolulu, HI.

De Jager, A., Nandwa, S.M., Okoth, P.F. 1998. Monitoring nutrient flows and economicperformance in African farming systems (NUTMON). I. Concepts and methodologies. *Agriculture, Ecosystems and Environment* 71: 37–48.

De Jager, A., Onduru, D., Van Wijk, M.S., Vlaming, J., Gachini, G.N. 2001. Assessing sustainability of low-external-input farm management systems with the nutrient monitoring approach: A case study in Kenya. *Agricultural Systems* 69: 99–118.

De Jager, A., Onduru, D.D., Walaga, C. 2004. Facilitated learning in soil fertilitymanagement: Assessing potentials of low-external-input technologies. *Agricultural Systems* 79: 205–223.

Dzung, N.A., Dzung, T.T., Khanh, V.T.P. 2013. Evaluation of coffee husk compost for improving soil fertility and sustainable coffee production in rural central highland of Vietnam. *Resources and Environment* 3: 77–82.

Ebanyat, P. 2009. *A Road to Food? Efficacy of Nutrient Management Options Targeted to Heterogeneous Soilscapes in the Teso Farming System, Uganda,* 1st edition. Wageningen, The Netherlands.

El Tahir, B.A., Daldoum, M.A., Ardö, J. 2013. Nutrient Balances as Indicators of Sustainability in *acacia senegal*Land use Systems in the Semi-arid Zone of North Kordofan, Sudan. Standard. *Scientific Research and Essays* 1: 93–112.

Elias, E., Morse, S., Belshaw, D.G.R. 1998. Nitrogen and phosphorus balances of KindoKoisha farms in Southern Ethiopia. *Agriculture, Ecosystems and Environmentalist* 71: 93–113.

FAO. 2003. *Assessment of Soil Nutrient Balance: Approaches and Methodologies*. Food and Agriculture Organization Fertilizer and Plant Nutrition Bulletin 14, Rome, Italy, 87p.

FAO. 2015a. *The State of Food Insecurity in the World 2015: Meeting the 2015 International Hunger Targets: Taking Stock of Uneven Progress.* FAO, Rome, Italy. http://www.fao.org/3/a-i4646e.pdf.

FAO. 2015b. *World Fertilizer Trends and Outlook to 2018.* Food and Agriculture Organization of the United Nations, Rome, Italy.

FAO, IFAD, UNICEF, WFP and WHO. 2018. *The State of Food Security and Nutrition in the World 2018. Building Climate Resilience for Food Security and Nutrition.* Food and Agriculture Organization of the United Nations, Rome, Italy. Licence: CC BY-NC-SA 3.0 IGO.

FAO, IFAD, WFP, and WHO. 2017. *The State of Food Security and Nutrition in the World. Building Resilience for Peace and Food Security.* Food and Agriculture Organization of the United Nations, Rome, Italy.

Galloway, J.N., et al. 2004. Nitrogen cycles: Past, present and future. *Biogeochemistry* 70: 153–226.

Gérard, F. 2016. Clay minerals, iron/aluminum oxides, and their contribution to phosphate sorption in soils - A myth revisited. *Geoderma* 262: 213–226 doi: 10.1016/j.geoderma.2015.08.036.

Giller, K.E. 2001. *Nitrogen Fixation in Tropical Cropping Systems*, 2nd edition. CABI, Wallingford, UK.

Giller, K.E., Wilson, K.J. 1991. *Nitrogen Fi Xation in Tropical Cropping Systems.* CAB International, Wallingford, UK.

Hartati, W., Sudarmadji, T. 2016. Relationship between soil texture and soil organic matter content on mined-out lands in Berau, East Kalimantan, Indonesia. *Nusantara Bioscience* 8(1): 83–88. doi: 10.13057/nusbiosci/n080115.

Hartemink, A.E. 2001. *Soil Fertility in the Tropics with Case Studies on Plantation Crops.* Ph.D. thesis. The University of Reading, Reading, UK.

Henao, J., Baanante, C. 1999. *Estimating Rates of Nutrient Depletion in Soils of Agriculturallands of Africa.* International Fertilizer Development Center (IFDC), Muscle Shoals, Alabama. 76 pp.

Henao, J., Baanante, C. 2006. *Agricultural Production and Soil Nutrient Mining in Africa Implications for Resource Conservation and Policy Development: Summary an International Center for Soil Fertility and Agricultural Development.* IFDC.

HLPE. 2014. *Food Loss and Waste in the Context of Sustainable Food Systems. A Report by the High Level Panel of Experts on Food Security and Nutrition of the Committee on World Food Security*, Rome.

Hungria, M., Campo, R.J., Mendes, I.C., Graham, P.H. 2006. Contribution of biological nitrogen fixation to the nutrition of grain crops in the tropics: The success of soybean (Glycine max L. Merr.) in South America. In: *Nitrogen Nutrition in Plant Productivity*, R.P. Singh, Shankar, N., Jaiwal, P.K. (Eds.), pp. 43–93. Studium Press, Houston, TX.

Hungria, M., Franchin, J.C., Campo, R.J., Graham, P.H. 2005. The importance of nitrogen fixation to soybean cropping in South America. In: *Nitrogen Fixation in Agriculture, Forestry, Ecology and Environment*, D. Werner, Newton, W.E. (Eds.), pp. 25–42. Springer, Dordrecht, The Netherlands.

IFASTAT. 2015. *International Fertilizer Association Database.* www.ifastat.org/databases/plant-nutritio naccessedon (Accessed on October 19, 2018).

Jhariya, M.K., Raj, A. 2014. Human welfare from biodiversity. *Agrobios Newsletter* 12: 89–91.

Juo, A.S.R., Kang, B.T. 1989. Nutrient effects of modification of shifting cultivation in west Africa. In: *Mineral Nutrients in Tropical Forest and Savanna Ecosystems*, Protor, J. (Ed.), pp. 289–300. Blackwell, Oxford, UK.

Kang, B.T., Reynolds, L., Atta-Krah, A.N. 1990. Alley farming. *Advances in Agronomy* 43: 315–359.

Kanmegne, J., Smaling, E.M.A., Brussaard, L., Gansop-Kouomegne, A., Boukong, A. 2007. Nutrient flows in smallholder production systems in the humid forest zone of southern Cameroon. *Nutrient Cycling in Agroecosystems* 76: 233–248. doi: 10.1007/s10705-005-8312-3.

Kanté, S. 2001. *Gestion de la fertilité des sols par classe d'exploitation au Mali-Sud.* Ph.D. thesis. Wageningen University, Wageningen, The Netherlands.

Kaola, S. 2001. *Integrated Soil Fertility Management for Central Kenya Highlands.* ISFM Extension Module, AFNET.

Kessler, C.A., van Duivenbooden, N., Nsabimana, F., van Beek, C.L. 2015. Bringing ISFM to scale through an integrated farm planning approach: A case study from Burundi. *Nutrient Cycling in Agro-Ecosystems* 105: 249–261. doi: 10.1007/s10705-015-9708-3.

Khan, M., Sharif, M. 2012. Solubility enhancement of phosphorus from rock phosphate through composting with poultry litter. *Sarhad Journal of Agriculture* 28(3): 415–420.

Lal, R. 2015. Sustainable intensification for adaptation and mitigation of climate change and advancement of food security in Africa. In: *Sustainable Intensification to Advance Food Security and Enhance Climate Resilience in Africa*, R. Lal, Singh, B.R., Mwaseba, D.L. et al. (Eds.), pp. 3–17. Springer, Berlin.

Lipper, L. 2001. *Dirt poor: Poverty, farmers and soil resources investment.* FAO Economic and Social Development Paper 149. www.fao.org/DOCREP/.

Majumdar, K., Sanyal, S.K., Dutta, S.K., Satyanarayana, T., Singh, V.K. 2016. Nutrient mining: Addressing the challenges to soil resources and food security. In: *Biofortification of Food Crops*, pp. 177–198. Springer, India.

Masila, T., Udonto, M.O. 2015. The influence of soil and water conservation technologies on household food security among small-scale farmers in Kyuso sub-county, Kitui County, Kenya. *Academia Journal of Agricultural Research* 3: 23–28. doi: 10.15413/ajar.2015.0101.

Miller, F.P., Larson, W.E. 1992. Lower input effects on soil productivity and nutrient cycling. In: *Sustainable Agricultural Systems,* C.A. Edwards, Lal, R., Madden, P., Miller, R.H., House, G. (Eds.), pp. 549–568. Soil Conservation Society of America, Ankeny, IA.

NEMA. 2011. *Clean Development Mechanism (CDM).* National Environment Management Authority (NEMA), Uganda, Kampala.

Nkonya, E., Pender, J., Kaizzi, C. 2004. Determinants of soil nutrient balances and implications for addressing land degradation and poverty in Uganda. Paper presented at the American Agricultural Economics Association Annual Meetings, August 1–4, Denver, CO.

Nottidge, D.O., Ojeniyi, S.O., Nottidge, C.C. 2010. Grain legumes residues effects on soil physical conditions, growth and grain yield of maize in an ultisol. *Nigerian Journal of Soil Science* 20(1): 150–153.

Nzuma, J.K., Murirwa, H.K., Mpepereki, S. 1998. Cattle manure management options for reducing nutrient losses. In: Proceedings of the Soil Fertility Network Results and Planning Workshop, R. Waddington, Murwira, H.K., Kumwenda, J.D.T., Hikwa, D., Tagwira, F. (Eds.), pp. 183–189. Africa University, Mutare, Zimbabwe.

Okalebo, J.R., Othieno, C.O., Woomer, P.L., Karanja, N.K., Semoka, J.R.M., Bekunda, M.A., Mugendi, D.N., Muasya, R.M., Bationo, A., Mukhwana, E.J. 2007. Available technologies to replenish soil fertility in east Africa. *Nutrient Cycling in Agroecosystems* 76: 153–170.

Olowookere, B.T., Oyerinde, A.A., Diamond, A.U. 2015. Assessment of impacts of varying rates of cow dung (organic manure) on soil physicochemical properties and production efficiency of okra (*Abelmoschusesculentus L. Moench.*) in the Federal Capital Territory (FCT) Abuja Nigeria. *International Journal of Agriculture and Biosciences* 4: 69–74.

Olupot, G., Etiang, J., Aniku, J., Ssali, H., Nabasirye, M. 2004. Sorghum yield response to kraal manure combined with mineral fertilisers in eastern Uganda. *MUARIK Bulletin* 7: 30–37.

Olupot, G., Etiang, J., Aniku, J., Ssali, H., Nabasirye, M. 2006. Nutrient inflow and outflow at plot and farm level in Kumi District, eastern Uganda. *Makerere University Research Journal* 1: 63–72.

Osgood, D. 2001. Dirt poor: Poverty, farmers and soil resources investment. FAO Economic and Social Development Paper 149. www.fao.org/DOCREP/ .

Palm, C.A., Gachengo, C.N., Delve, R.J., Cadisch, G., Giller, K.E. 2001. Organic inputs for soil fertility management: Some rules and tools. *Agricola Ecosys. Environmentalica* 83: 27–42.

Pinheiro, R., Rehm, M., Grashey-Jansen, S. 2016. Soil sequence-studies on the tropical Buganda- Catena (Masaka District, Uganda). *African Journal of Soil Science* 4: 295–304.

Rahman, M.M., Islam, A.M., Azirun, S.M., Tropical Boyce, A.N. 2014. Tropical legume crop rotation and nitrogen fertilizer effects on agronomic and nitrogen efficiency of rice. *The Scientific World Journal*, 490841. doi: 10.1155/2014/490841.

Raj, A., Jhariya, M.K., Toppo, P. 2014. Cow dung for eco-friendly and sustainable productive farming. *Environmental Sciences* 3: 201–202.

Ramisch, J. 1999. *In the Balance? Evaluating Soil Nutrient Budgets for an Agro-Pastoral Village of Southern Mali.* Managing Africa's Soils No. 9. IIED Drylands Programme, London.

Ryan, J.G., Spencer, D.C. 2001. *Future Challenges and Opportunities for Agricultural R and D in the Semi-Arid Tropics.* ICRISAT, Patancheru, India.

Sachs, J., McArthur, J.W., Schmidt-Traub, G., Kruk, M., Bahadur, C., Faye, M., McCord, G. 2004. Ending Africa's poverty trap. *Brookings Papers on Economic Activity* 2004(1): 117–240.

Sanchez, P.A., Shepherd, K.D., Soule, M.J., Place, F.M., Buresh, R.J., Izac, A.N., Mokwunye, A.U., Kwesiga, F.R., Ndiritu, C.G., Woomer, P.L. 1997. *Soil Fertility Replenishment in Africa: An Investment in Natural Resource Capital.* SSSA Special Publication Number 51, pp. 1–46. SSSA, Madison, MI.

Sanchez, P.A., Swaminathan, M.S. 2005a. Hunger in Africa: The link between unhealthy people and unhealthy soils. *Lancet* 365(9457): 442–444.

Sanchez, P.A., Swaminathan, M.S. 2005b. Cutting world hunger in half: Policy forum on health. *Science* 307: 357–359.

Sanginga, N., Woomer, P.L., (Eds.) 2009. *Integrated Soil Fertility Management in Africa: Principles, Practices and Developmental Process.* Tropical Soil Biology and Fertility Institute of the International Centre for Tropical Agriculture, Nairobi, Kenya.

Smalberger, S.A., Singh, U., Chien, S.H., Henao, J., Wilkens, P.W. 2006. Development and validation of a phosphate rock decision support system. *Agronomy Journal* 98: 471–483.

Smaling, E.M.A., Nandwa, S.M., Janssen, B.H. 1997. Soil fertility is at stake! In *Replenishing Soil Fertility in Africa: An Investment in Natural Resource Capital*, SSSA Special Publication Number 51, pp. 47–61. SSSA, Madison, MI.

Smaling, E.M.A., Stoorvogel, J.J., Windmeijer, P.N. 1993. Calculating soil nutrientbalances at different scales: II. District scale. *Fertility Resources* 35: 227–235.

Smith, A., Snapp, S., Dimes, J., Gwenambira, C., Chikowo, R. 2016. Doubled-up legume rotations improve soil fertility and maintain productivity under variable conditions in maize-based cropping systems in Malawi. *Agricultural Systems* 145(2016): 139–149.

Snow, R.W., Omumbo, J.A. 2006. Malaria. In: *Disease and Mortality in Sub-Saharan Africa*, 2nd edition, D.T. Jamison, Feachem, R.G., Makgoba, M.W., Bos, E.R., Baingana, F.K., Hofman, K.J., Rogo, K.O. (Eds.), pp. 194–213.

Stoorvogel, J.J., Smaling, E.M.A. 1990. *Assessment of soil nutrient depletion insub-Sahara Africa: 1983–2000*, 4 Volumes. Report 28. The Winand Staring Centrefor Integrated Land, Soil and Water Research, Wageningen, The Netherlands.

Tan, Z.X., Lal, R., Wiebe, K.D. 2005. Global soil nutrient depletion and yield reduction. *Journal of Sustainable Agriculture* 26: 123–146.

Tewodros, T., Giorgis, M.M. 2010. Determinants of coffee husk manure adoption: A case study from southern Ethiopia. *Indian Journal of Agricultural Economics* 65: 159–172.

Thierfelder, C., Cheesman, S., Rusinamhodzi, L. 2012. A comparative analysis of conservation agriculture systems: Benefits and challenges of rotations and intercropping in Zimbabwe. *Field Crops Research* 137: 237–250.

UNDP. 2012. *The Nutrition Challenge in Sub-Saharan Africa*. Working Paper WP 2012-012, January 2012.

UNICEF, WHO, International Bank for Reconstruction and Development/World Bank. 2018. *Levels and Trends in Child Malnutrition: Key Findings of the 2018 Edition of the Joint Child Malnutrition Estimates*. World Health Organization, Geneva, Switzerland.

UNSCN. 2009. Landscape analysis on countries' readiness to accelerate action in nutrition. In: *SCN News*. UNSCN, UK.

Van den Bosch, H., Gitari, J.N., Ogaro, V.N., Maobe, S., Vlaming, J. 1998. Monitoring nutrient flows and economic performance in African farming systems (NUTMON): III. Monitoring nutrient flows and balances in three districts in Kenya. *Agriculture, Ecosystems and Environment* 71: 63–80.

Van der Pol, F. 1992. *Soil Mining: An Unseen Contributor to Farm Income in Southern Mali*. Bulletin 35. The Royal Tropical Institute (RTI), Amsterdam, The Netherlands.

Van Kauwenbergh, S.J. 2006. *Fertiliser Raw Material Resources of Africa*. Reference Manual 16. IFDC, Muscle Shoals, AL.

Van Straaten, P. 2002. *Rocks for Crops: Agrominerals of Sub-Saharan Africa*. International Centre for Research in Agroforestry (ICRAF), Nairobi, Kenya.

World Bank. 2008. *Uganda Municipal Waste Compost Programme*. The World Bank. CDM Executive Board, Washington D.C.

Wortmann, C.S., Kaizzi, C.K. 1998. Nutrient balances and expected effects of alternative practices in farming systems of Uganda. *Agriculture, Ecosystems and Environment* 71: 115–129.

Wortmann, C.S., Klein, R.N., Wilhelm, W.W., Shapiro, C. 2008. *Harvesting Crop Residues*. NebGuide G184b (revised 2012). University of Nebraska–Lincoln Extension, Lincoln, NE.

4 Land-Use Impacts on Soil Physical Properties of an Alfisol in Western Nigeria

Rattan Lal

CONTENTS

4.1 INTRODUCTION

The average cereal yield in Africa is merely 1.6 Mg/ha with global average yield of 3.9 Mg/ha. Low and stagnating crop yields are attributed to low soil moisture and nutrient reserves and poor soil structure (Tadele 2017), as well as the overall low fertility of soils in sub-Saharan Africa (SSA) (Vanlaue et al. 2017). Similarly, the yield of yam tubers is also low (~10 Mg/ha) because of soil degradation (Frossard et al. 2017). Soil degradation is a serious problem in SSA (Hartemink and van Keulen 2005). The Montpellier report indicated that neglecting the health of Africa's soil will lock the continent into a cycle of food insecurity for generations to come (Kinver, 2015), with serious land degradation affecting 65% of available land, 30% of grazing land, and 20% of forests (Montpellier Panel 2014; Zingore et al. 2015). The soil organic matter (SOM) content, with strong effects on soil properties and processes and the human dimensions or social, cultural and economic conditions (Ayuk 2001), has been depleted in soils of arable land use in West Africa because of extractive farming practices. Tully et al. (2015) observed clear indications of degradation across multiple indicators that have different trajectories. Whereas pH and cation exchange capacity (CEC) may decline linearly, soil organic carbon (SOC) and yield decline non-linearly. These degradation trends lead to severe economic loss (Bojo 1996; Nkonya et al. 2016). Alfisols of the sub-humid and semi-arid regions of West Africa predominantly contain kaolinite (1:1 non-expanding) minerals, have low charge density, weak soil structure (aggregation), and low plant available nutrients and water reserves (Vanlaue et al. 2017). In the context of resource-poor farmers and small landholders, SOM is an important source of plant nutrients (especially nitrogen, N), the formation and stabilization of aggregates, water retention and transmission, soil erodibility, and CEC (Jones 1973; Ouedraogo et al. 2007; Nakamura et al. 2011; Party et al. 2018). It is also essential to enhancing soil biodiversity and keeping drylands alive (Laban et al. 2018). The low CEC often, 1 cmol (+)/kg of soil (Bationo and Buerkert 2001), is the cause of low nutrient reserves and high losses due to leaching, surface runoff, and erosion. Furthermore, SOM content in the soils of West Africa is low because of inappropriate land use, deforestation, and overexploitation of natural resources (Batjes 2001). Low aggregation and weak stability of aggregates render soils prone to crusting, surface runoff, and accelerated erosion (Valentin et al. 2004). Traditionally, the low inherent soil fertility has been addressed by addition of ash/coal and household waste to enhance nutrient reserves (Solomon et al. 2016). Therefore, nutrient management and cover cropping through organic inputs can be a useful strategy to enhance soil fertility (Manyong et al. 2001; Bekunda et al. 2010) and improve soil health and functionality. In this context, fallow management and cover cropping can also play an important role (Koutika et al. 2002). The soil properties responsible for decline in crop yield through erosion-induced degradation include physical (e.g., crusting, compaction, low plant available water capacity, PAWC, high soil temperature), chemical (acidification and nutrient/elemental imbalance), and biological (decline in activity and species diversity of soil biota, including reduction in the microbial biomass carbon) (Oyedele and Aina 2006). Therefore, the incorporation of legumes in the rotation cycle can restore the productivity of eroded soils (Salako et al. 2006). Residue retention and conservation agriculture (CA) can improve soil fertility by recycling nutrients (Naab et al. 2015), improving soil physical properties (Lal 1976), and moderating soil temperature (Lal 1973).

Nutrient cycling and improved fallow management (planted fallows) can enhance soil fertility and also improve soil physical properties through integrated soil fertility management (Bellwood-Howard 2014). The strategy of sustainable intensification (with residue retention and managed fallowing) can improve agronomic productivity (Cafer and Qin 2017). Building a resilient and sustainable agriculture in SSA (Shimless et al. 2018) is essential to advancing nutritional and food security (Devereux et al. 2001). Therefore, the objective of this article is to deliberate the impact of land use and residue management on soil physical and hydrological properties of some Alfisols in western Nigeria.

4.2   CHALLENGES OF CONTINUOUS CULTIVATION OF FRAGILE SOILS OF THE TROPICS

Low and stagnating crop yields in SSA are partly because of weak soil physical, chemical, and biological properties, and consequently high vulnerability to a range of degradation processes. Soils of the humid, sub-humid, and semi-arid regions of West Africa are characterized by low aggregation and weak structural stability, rapid decline in SOM content, acidification, salinization, decline in microbial biomass carbon (MBC) and soil biodiversity, vulnerability to crusting and compaction (hard setting), and accelerated erosion by water and wind. Furthermore, the soils of West Africa are subjected to a negative nutrient budget (more nutrients are removed in crop/animal harvest than added through fertilizers and amendments) because of the widespread use of extractive farming practices whereby more nutrients are removed when the produce is harvested than are added through meagre, if any, inputs. The adverse impacts of soil degradation trends on agronomic productivity and the environment are exacerbated by the changing and uncertain climate, especially the droughts and heatwaves. Decline in the PAWC of the root zone, coupled with an increase in surface runoff and soil evaporation, reduces the green-water supply and aggravates the frequency and intensity of drought. Increases in the maxima and the diurnal fluctuations in soil temperature (especially in the 0–5 cm layer, where the meristem or the growing point of cereals remains for 4 to 6 weeks) affect the agronomic yield potential. Thus, maintenance of a protective vegetative cover on the soil surface (CA, cover cropping, and residue mulching) is critical to restoring SOC concentration and stock, improving the soil structure, and enhancing soil biodiversity. The maintenance of SOC concentrations in the root zone to above the critical level (1.5–2.0% by weight) supports ecosystem services including: (i) supporting services such as biodiversity, water cycling, nutrient cycling, and soil formation; (ii) provisioning services such as habitat for biodiversity, clean water, food production, climate control, and biological control; (iii) regulating services such as cognitive services, recycling waste, denaturing and filtering contaminants, and control of hydrological processes; and (iv) cultural services such as spiritual, recreational, and aesthetic (Safriel et al. 2005).

4.3   TECHNOLOGICAL OPTIONS FOR ALLEVIATING SOIL-RELATED CONSTRAINTS

The outline in Table 4.1 lists some basic concepts of soil management options to minimize the risks of degradation, to alleviate soil-related constraints, and also to reduce anthropogenic-related environmental degradation.

Most of the management options outlined in Table 4.1 are bio-based systems of soil, crop, water, or animal management. The strategy is to enhance soil structure (% aggregation and aggregate stability), and improve pore continuity and stability. The management of soil physical properties and processes is essential to improving soil chemical fertility and biological attributes (e.g., earthworm activity or MBC).

4.4   MANAGED FALLOWING

The rapid decline in soil quality with continuous cropping of soils of the humid and sub-humid tropics can be addressed by several options, such as CA, cover cropping, recycling of biomass, agroforestry systems, and the incorporation of managed fallows within the rotation cycle (Table 4.1). A long-term experiment (Juo et al. 1995, 1996) was established at the International Institute of Tropical Agriculture (IITA) to assess the effects of planted versus natural fallowing (leaving land fallow for regrowth of natural vegetation cover), crop residue management, and rotation on soil properties.

The experiment was coordinated by the late Dr. A.S.R. Juo. In addition to monitoring soil chemical and nutritional properties (Juo et al. 1995; 1996), soil physical properties were monitored, following the standard procedures (Dane and Topp 2002), during the dry season (December–February). Soil physical properties monitored were bulk density by the core method (Grossman and Reinsch 2002), particle size distribution by the hydrometer method (Gee and Or 2002), aggregate stability and the size distribution including the mean weight diameter (MWD) by the wet sieving technique (Nimmo and Perkins 2002), penetrometer resistance by a handheld pocket penetrometer (Lowery and Morrison Jr. 2002), soil water retention at different tensions by using a tension table and pressure plate extractors (Dane and Hopmans 2002), and water infiltration by a double-ring technique (Reynold et al. 2002). The infiltration data were fitted to Philip's (1957) model to compute soil water "sorptivity" (S) and transmissivity (A). All data were statistically analyzed to compute the ANOVA according to the *completely randomized* block design.

TABLE 4.1

Management Option to Alleviate Soil-Related Constraints for Agricultural Land Use

Constraint	Management option
Erosion	Mulching, cover cropping, conservation agriculture, contour hedges
Compaction	Precision/guided traffic, deep-rooted cover crops, mulch farming, enhancing activity and species diversity of soil biota (e.g., earthworms)
Drought	Mulch farming, conservation agriculture, water harvesting and recycling, moderating soil temperature
Nutrient depletion	Biological N fixation, inoculation with rhizobium, recycling crop, tree, or animal waste as compost, mycorrhizal waste, and supplemental input of fertilizers
Acidification	Judicious inputs of fertilizers, minimizing use of ammonium sulfate etc.
Salinization	Improving drainage, both internal and surface, salt-tolerant species, irrigation with a good quality water

4.5 RESULTS AND DISCUSSION

There were significant differences among the fallowing and continuous cultivation treatments on several soil structure, water retention, and water infiltration properties.

4.5.1 Soil Structure and Mean Weight Diameter

Land use and management strongly affected the relative proportion of macro- and micro-aggregates in soil. Expectedly, the highest proportion of macro-aggregates was observed under guinea grass and bush fallow systems, and the reverse trends were observed for micro-aggregates (Table 4.2). The proportion of macro-aggregates decreased significantly in soil that was under crop cultivation, and especially so with the removal of crop residues. Regardless of the land use, macro-aggregation also decreased with the increase in soil depth, probably because of the decrease in SOC concentration. Similar to the effects on macro-aggregation, land use and management also affected the MWD (Table 4.3). The MWD was significantly lower in soil under cultivation than that under planted fallowing. Similar to the effect on macro-aggregates, the MWD also decreased with the increase in soil depth (Table 4.3). The decline in structural aggregation and the MWD in cultivated soils may partly be attributed to the reduction in SOC concentration (Table 4.4), and the latter was exacerbated by the

TABLE 4.2
Land-Use Effect on Macro- and Micro-Aggregates and Structural Properties

| Treatment | Macro-aggregates (%) | | | | Micro-aggregates (%) | |
	0–5 cm	5–10 cm	10–15 cm	15–20 cm	0–5 cm	5–10 cm
Guinea grass	93.3 a	79.4 a	83.2 a	73.1 a	6.7	20.6 b
Bush fallow	90.9 a	88.4 a	62.7 ab	70.5 a	9.2 c	29.6 b
Cassava + maize	89.3 ab	63.0 bc	47.8 b	47.2 b	10.6 c	37.0 ab
Leucaena	80.9 abc	64.2 bc	58.0 ab	53.6 ab	13.5 bc	35.8 ab
Maize + residue	73.1 bcd	56.5 cd	34.0 b	70.6 a	26.9 ab	44.3 b
Pigeon pea	71.2 d	70.2 bc	53.2 ab	52.9 ab	28.8 ab	29.8 b
Maize – soybean	57.0 d	38.6 d	43.5 b	41.5 b	43.0 a	61.4 a
Maize – residue	56.5 d	43.0 d	48.7 b	53.1 ab	43.5 a	57.0 a

Note: Figure within a column followed by the same letter are statistically similar.

TABLE 4.3
Land-Use Effect on Mean Weight Diameter of Aggregates

| Treatment | MWD (mm) | | | |
	0–5 cm	5–10 cm	10–15 cm	15–20 cm
Guinea grass	3.90 a	2.81 ab	2.51 a	1.99 a
Bush fallow	4.22 a	3.87 a	1.52 b	1.31 ab
Cassava + maize	3.60 a	1.74 bc	0.85 b	0.92 ab
Leucaena	2.73 ab	1.66 bc	1.50 b	1.32 ab
Maize + residue	1.55 b	1.37 c	0.87 b	1.98 a
Pigeon pea	2.47 ab	1.32 c	0.89 b	0.75 ab
Maize – soybean	1.53 b	0.79 c	0.95 b	1.23 ab
Maize – residue	1.34 b	0.60 c	1.02 b	1.40 ab

Note: Figure within a column followed by the same letter are statistically similar.

TABLE 4.4

Stover Removal Effects on the Physical Properties of an Alfisol in Western Nigeria (15 years after, 0–5 cm)

Parameter	Leucaena	+ Residue	– Residue
Macro-aggregates (%)	93.3 a	73.1	56.5
Micro-aggregates (%)	6.7	26.9	43.5
Mean weight diameter (mm)	3.9	1.6	1.3
Transmissivity (cm/min)	35.5	8.6	–

Source: Adapted from Lal (2018, this chapter).

Note: Figure within a column followed by the same letter are statistically similar.

TABLE 4.5

Land-Use, Rotation, and Residue Management Effects on Soil Moisture Retention for 0–5 cm Depth of an Alfisol in Western Nigeria in 1979 (7 years after starting the experiment)

Treatment	Moisture retention (g/g, %) at different potentials (bar)							
	0.1	0.3	0.5	1	2	3	15	AWC
Guinea grass	30.0 a	24.5 a	22.9 a	21.6 a	18.8 a	17.1 a	16.1 a	8.4
Bush fallow	28.2 ab	21.9 ab	20.1 ab	17.8 ab	15.7 ab	10.6 ab	9.6 ab	12.3
Leucaena	19.5 bc	14.9 bc	12.3 abc	11.9 bc	11.4abc	9.8 ab	9.3 ab	5.6
Pigeon pea	14.9 c	12.3 bc	9.9 bc	7.3 c	6.9 c	6.1 b	5.4 b	6.9
Maize + cowpea	17.7 c	13.0 bc	10.6 bc	10.4 bc	10.1 bc	9.5 ab	8.8 ab	4.2
Maize + residue	16.3 c	12.5 bc	10.2 bc	9.7 bc	9.1 bc	8.5 b	8.0 b	4.5
Maize – residue	11.3 c	8.5 c	6.8 c	6.2 c	5.3 c	4.9 ab	4.1 b	4.4
Cassava	12.3 c	9.6 c	8.0 c	6.5 c	5.6 c	5.2 b	4.2 b	5.4
LSD (0.05)	9.6	9.7	10.3	9.6	8.4	7.8	7.8	1.9

Notes: AWC = available water capacity, LSD = Least Significant Difference. Figure within a column followed by the same letter are statistically similar.

removal of crop residues (Table 4.5). The adverse effects of cultivation on structural properties are also reflected in the data from soil mechanical properties, such as the penetration resistance (PR) and soil bulk density. Expectedly, the PR was high in cultivated soils, and especially in treatments from where the crop residues were removed (Table 4.6). In accord with the trends in PR, soil bulk density also increased in soil under cultivation. The increase in these mechanical properties is attributed to the reduction in biotic activity, raindrop impact on bare soil surface, and the reduction in SOC concentration. These adverse changes in soil mechanical properties affect root growth and thus uptake of water and nutrients. Adverse effects on crop growth and yield were observed on plots with residues removal.

4.5.2 Hydrological Properties

The decline in aggregation and the increase in soil bulk density and PR also affect water retention and transmission pores. Land-use and management effects on soil moisture retention at different suctions (matric potential) for 0–5cm depth are shown in Table 4.7. Regardless of the matric potential, soil moisture retention was lower in soil from cultivated than from soil under planted and

TABLE 4.6
Land-Use Rotation and Management on Soil Bulk Density of the Whole Fraction, Gravel-Free Material, and the Penetration Resistance of 0–5 cm Larger of an Alfisol (Western Nigeria, 1979)

Treatment	Penetration resistance (kg/cm²)	Soil bulk density Whole soil (Mg/m³)	Soil bulk density Fine fraction (Mg/m³)
Guinea grass	0.62 cd	1.04 ab	1.00 ab
Bush fallow	0.42 d	0.83 b	0.82 b
Leucaena	0.88 cd	1.09 ab	0.98 ab
Pigeon pea	2.22 a	1.19 ab	1.16 ab
Maize + cowpea	1.14 bcd	1.10 ab	1.06 ab
Maize + residue	1.53 ab	1.26 ab	1.19 ab
Maize – residue	1.91 ab	1.35 a	1.29 a
Cassava	1.21 bcd	1.34 a	1.24 ab
LSD (0.05)	0.96	0.46	0.44

Note: LSD = Least Significant Difference.

TABLE 4.7
Land-Use, Rotation, and Residue Management Effects on Modeling of Water Infiltration Characteristics of an Alfisol in 1985 (13 years after starting the experiment)

$$I = St^{1/2} + At$$

Treatment	Soil water sorptivity (S) [cm/(min)$^{1/2}$]	Soil water transmissivity (A) [cm/min]	R²
Guinea grass	8.70	1.12	0.89
Bush fallow	14.50	1.31	0.96
Leucaena	27.92	3.11	0.69
Pigeon pea	11.20	2.55	0.76
Maize + cowpea	29.68	2.02	0.98
Maize + residue	26.87	0.96	0.94
Maize – residue	20.91	0.69	0.98
Cassava	13.70	2.47	0.80

Note: t = time (minutes after start of water infiltration).

natural fallowing. The PAWC, the difference between the moisture retention at 0.3 bar and 15 bar, was also affected by fallowing and cropping system treatments. In addition to differences in pore size distribution and aggregation, differences in moisture retention may also be attributed to the changes in SOC concentration and stock, especially in the surface 0–5cm layer.

There were also differences in water infiltration amount (I) and rates (i) as affected by land use and management (Figure 4.1, Tables 4.8 and 4.9). The magnitude of soil water sorptivity (S) and transmissivity (A) differed between years (1985 and 1987) because of differences in the antecedent soil moisture content. The infiltration data strongly fitted the experimental model of Philip (1957).

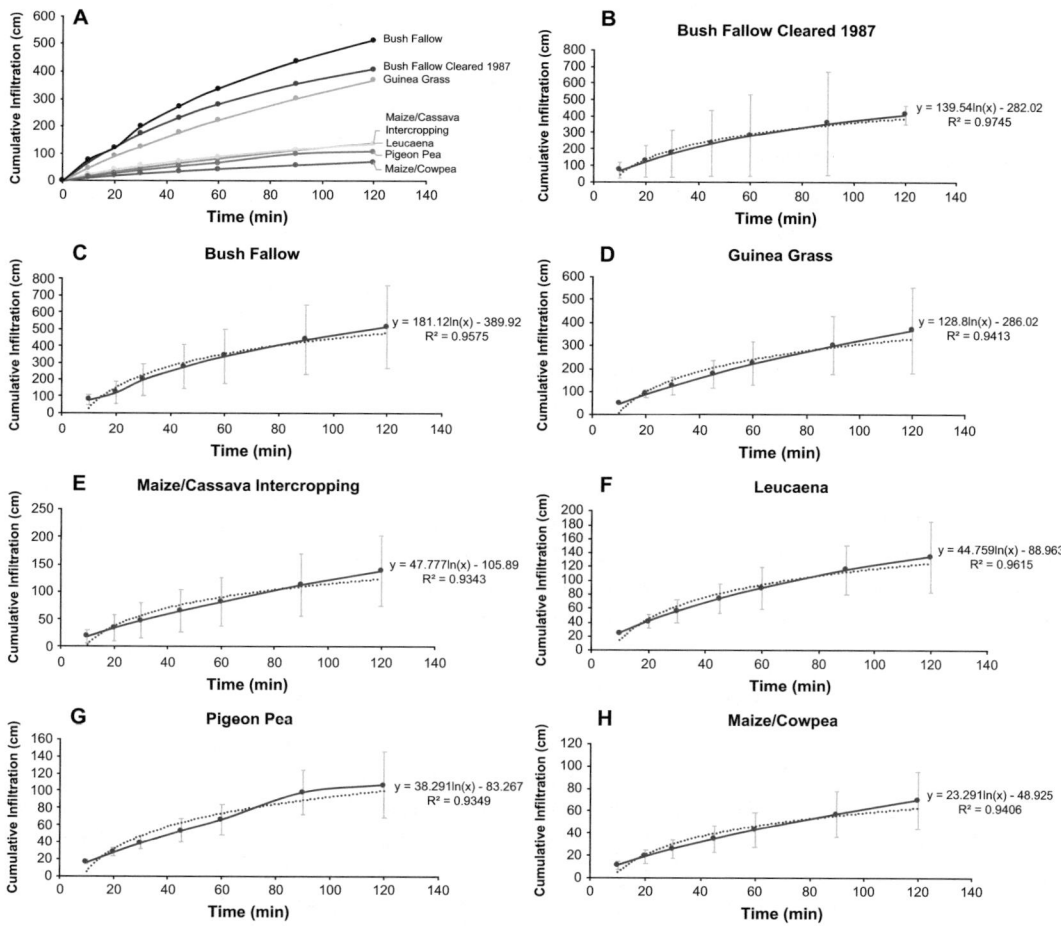

FIGURE 4.1 (See color insert.) Land-use effects on cumulative infiltration into Alfisols in western Nigeria. Different land-use and cropping system treatments are marked A–H.

TABLE 4.8

Land-Use, Rotation, and Residue Management Effects on Modeling of Water Infiltration Characteristics of an Alfisol in 1987 (15 years after starting the experiment)

$$I = St^{1/2} + At$$

Treatment	Soil water sorptivity (S) $[cm/(min)^{1/2}]$	Soil water transmissivity (A) [cm/min]
Guinea grass	35.50	1.41
Bush fallow	54.12	2.54
Leucaena	19.40	0.96
Pigeon pea	12.24	0.54
Maize + cowpea	8.58	0.40
Maize + residue	—	—
Maize – residue	—	—
Cassava	13.57	0.55

TABLE 4.9

Stover Removal Effects on Properties of an Alfisol in Western Nigeria (15 years after, 0–10 cm)

Parameter	Leucaena	+ Residues	– Residue
Soil organic carbon (g/kg)	21.5	14.5	12.5
Bulk density (Mg/m³)	1.3	1.56	1.62
pH	6.1	5.0	4.5
Ca^{+2}	6.0	3.5	1.1
Grain yield (maize)	—	4.1	2.3

Source: Adapted from Juo et al., 1995.

The graphic plots of cumulative infiltration (I) versus time (t) indicate strong differences between treatments. The highest cumulative water infiltration was measured for the fallowing treatments under bush fallow, then under guinea grass, followed by that under pigeon pea and leucaena, and the least under maize cultivation with residue removal. A higher cumulative infiltration and rate are indicative of the relative proportion and continuity of macropores.

4.6   GENERAL CONCLUSIONS AND RECOMMENDATIONS

The management of soil structure, along with that of soil mechanical and hydrological properties, is essential to several soil processes including water retention and susceptibility to drought, water infiltration, surface runoff and erosion, root growth, aeration and gaseous exchange, water and nutrient uptake, and the incidence and severity of drought and heatwave. The process of soil physical degradation is set in motion by the decline in soil structure and the proportion, size, and strength of aggregates. The decline in soil structure, exacerbated by plowing and residue removal, leads to crusting, compaction, hard setting, and a decline in water infiltration rate and amount. The last is the principal cause of accelerated runoff and erosion by interrill or rill processes, loss of water and of the topsoil, and regressive decline in productivity. Therefore, the periodic incorporation of planted fallows (guinea grass or pigeon pea) in the rotation cycle and the retention of crop residues as mulch in conjunction with no-till (NT) and fallowing is the recommended management practice applicable under wide physiographic and climatic conditions. A system-based conservation tillage (Lal 2013; 2015) is an appropriate option of sustainable management of these fragile soils under the harsh climate of SSA.

The sustainable management of soil physical, mechanical, and hydrological properties and processes, through fallowing with site-specific species (grasses, legumes, and shrubs) and system-based CA with cover cropping during the off-season (e.g., mucuna), would reduce or minimize both the on-site and off-site damage, while enhancing productivity and minimizing the need for bringing any new land under agriculture through deforestation. Important among on-site benefits are high and sustained production, with improvements in use efficiency of inherent and applied inputs. Notable off-site benefits include a decreased risk of non-point source pollution and eutrophication along with the reduction in algal bloom and the related concerns. Furthermore, the decrease in the rate and magnitude of erosion would also reduce the erosion-induced transport or redistribution of SOC and the attendant emissions of greenhouse gases (e.g., CO_2, CH_4, N_2O) (Lal 2003). Thus, an effective soil erosion control through fallowing and CA, also reduces the emission of greenhouse gases and the depletion of the SOC pool. The use of planted fallows and conservation agriculture, with residue retention as mulch, are critical to developing climate-smart agriculture and facing climate volatility in SSA (Partey et al. 2018).

A judicious combination of fallowing and CA can be a viable alternative to the traditional shifting cultivation and bush fallow systems of extractive farming. These land-saving options are important strategies of eco-intensification and of producing more from less.

ACKNOWLEDGMENTS

These data were obtained from the measurement of soil properties in a cooperative experiment conducted by the late Professor A.S.R. Juo.

REFERENCES

Ayuk, E.T. 2001. Social, economic, and policy dimensions of soil organic matter management in sub-Sahara Africa: Challenges and opportunities. *Nutrient Cycling in Agroecosystems* 61:183–195.

Bationo, A., Buerkert, A. 2001. Soil organic carbon management for sustainable land use in Sudano-Sahelian West Africa. *Nutrient Cycling in Agroecosystems* 61:131–142.

Batjes, N.H. 2001. Options for increasing carbon sequestration in West African soils: An exploratory study with special focus on Senegal. *Land Degradation and Development* 12(2):131–142.

Bekunda, M., Sanginga, N., Woomer, P.L. 2010. Restoring soil fertility in sub-Saharan Africa. *Advances in Agronomy* 108:183–236.

Bellwood-Howard, I.R.V. 2014. Smallholder perspectives on soil fertility management and markets in the African Green Revolution. *Agroecology and Sustainable Food Systems* 38:660–685.

Bojö, J. 1996. The costs of land degradation in sub-Saharan Africa. *Ecological Economics* 16:161–173.

Cafer, A.M., Qin, H. 2017. Sustainable intensification, community, and the Montpellier Panel: A meta-analysis of rhetoric in practice in sub-Saharan Africa. *Journal of Agriculture Food Systems and Community Development* 7(3):123–137.

Dane, J.H., Hopmans, J.W. 2002. Pressure plate extractor. In: *Methods of Soil Analysis: Part 4 Physical Methods*, SSSA Book: 5, J.H. Dane, G.C. Topp (eds.). Soil Science Society of America, Madison, WI, pp. 688–689.

Dane, J.H., Topp, G.C. (eds.). 2002. *Methods of Soil Analysis: Part 4 Physical Methods*. SSSA Book Series: 5, Soil Science Society of America, Madison, WI, 1692pp.

Devereux, S., Maxwell, S. (eds.). 2001. *Food Security in Sub-Saharan Africa*. Practical Action, Rugby, UK. 280pp.

Frossard, E., Aighewi, B.A., Ake, S., Barjolle, D., Baumann, P., Bernet, T., Dao, D.D. Diby, L.N., Floquet, A., Hgaza, V.K., Ilboudo, L.J. 2017. The challenge of improving soil fertility in yam cropping systems of West Africa. *Frontiers in Plant Science* 8:1953.

Gee, G.W., Or, D. 2002. Particle size analysis. In: *Methods of Soil Analysis: Part 4 Physical Methods*. SSSA Book Series: 5, J.H. Dane, G.C. Topp (eds.). Soil Science Society of America, Madison, WI, pp. 255–294.

Grossman, R.B., Reinsch, T.G. 2002. Bulk density and linear extensibility. In: *Methods of Soil Analysis: Part 4 Physical Methods*. SSSA Book Series, J.H. Dane, G.C. Topp (eds.). Soil Science Society of America, Madison, WI, pp. 201–228.

Hartemink, A., van Keulen, H. 2005. Soil degradation in Sub-Saharan Africa. *Land Use Policy* 22(1):1–74. doi: 10.1016/landusepol.2004.01.001.

Jones, M.J. 1973. The organic matter content of the savanna soils of west Africa. *Soil Science* 24:42–53.

Juo, A.S.R., Franzluebbers, K., Dabiri, A., Ikhile, B. 1996. Soil properties and crop performance on a kaolinitic Alfisol after 15 year of fallow and continuous cultivation. *Plant and Soil* 180:209–217.

Juo, A.S.R., Franzluebbers, K., Dabiri, A., Ikhile, B.I. 1995. Changes in soil properties during long-term fallow and continuous cultivation after forest clearing in Nigeria. *Agriculture, Ecosystems and Environment* 56:9–18.

Kinver, M. 2014. African soil crisis threatens food security, says study. *Science and Environment. BBC News*, 4 December, 2014. https://www.bbc.com/news/science-environment-30277514.

Koutika, L.S., Sanginga, N., Vanlauwe, B., Weise, S. 2002. Chemical properties and soil organic matter assessment under fallow systems in the forest margins benchmark. *Soil Biology and Biochemistry* 34:757–765.

Laban, P., Metternicht, G., Davies, J. 2018. *Soil Biodiversity and Soil Organic Carbon: Keeping Drylands Alive*, IUCN, Global Dryland Initiative, 24 pp.

Lal, R. 1973. Effects of methods of seedbed preparation and time of planting of maize in western Nigeria. *Experimental Agriculture* 9:303–313.

Lal, R. 1976. No-tillage effects on soil properties under different crops in western Nigeria. *Soil Science Society of America Journal* 40:762–768.

Lal, R. 2003. Soil erosion and the global carbon budget. *Environment International* 29:437–450.

Lal, R. 2013. Enhancing ecosystem services with no-till. *Renewable Agriculture and Food Systems* 28(2):102–114.

Lal, R. 2015. A system approach to conservation agriculture. *Journal of Soil and Water Conservation* 70(4):82A–88A.

Lowery, B., Morrison, J.E., Jr. 2002. Soil penetrometers and penetrability. In: *Methods of Soil Analysis: Part 4 Physical Methods.* SSSA Book Series: 5, J.H. Dane, G.C. Topp (eds.). Soil Science Society of America, Madison, WI, pp. 363–388.

Manyong, V.M., Makinde, K.O., Sanginga, N., Vanlauwe, B., Diels, J. 2001. Fertilizer use and definition of farmer domains for impact-oriented research in the northern Guinea savanna of Nigeria. *Nutrient Cycling in Agroecosystems* 59(2):129–141.

Montpellier Panel. 2014. No ordinary matter: Conserving, restoring and enhancing Africa's soils. A Montpellier Panel Report. International Fund for Agriculture Development (IFAD), Rome, Italy, 40 pp.

Naab, J.B., Mahama, G.Y., Koo, J., Jones, J.W., Boote, K.J. 2015. Nitrogen and phosphorus with crop residue retention enhances crop productivity, soil organic carbon, and total soil nitrogen concentrations in sandy-loam soils of Ghana. *Nutrient Cycling in Agroecosystems* 102:33–43.

Nakamura, S., Hayashi, K., Omae, H., Ramadjita, T., Dougbedji, F., Shinjo, H., Saidou, A.K., Tobita, S. 2011. Validation of Soil Organic Carbon Dynamics Model in the Semi-Arid Tropics in Niger, West Africa, *Nutrient Cycling in Agroecosystems* 89(3):375–385.

Nimmo, J.R., Perkins, K.S. 2002. Aggregate stability and size distribution. In: *Methods of Soil Analysis: Part 4 Physical Methods.* SSSA Book Series: 5, J.H. Dane, G.C. Topp (eds.). Soil Science Society of America, Madison, WI, pp. 317–328.

Nkonya, E., Johnson, T., Kwon, H.Y., Kato, E. 2016. Economics of land degradation in sub-Saharan Africa. In: *Economics of Land Degradation and Improvement—A Global Assessment for Sustainable Development*, E. Nkonya, A. Mirzabaev, J. von Braun (eds.). IFPRI/University of Bonn/Springer, Cham, Bonn, Germany, pp. 215–259.

Ouédraogo, E., Mando, A., Brussaard, L., Stroosnijder, L. 2007. Tillage and fertility management effects on soil organic matter and sorghum yield in semi-arid West Africa. *Soil and Tillage Research* 94:64–74.

Oyedele, D.J., Aina, P.O. 2006. Response of soil properties and maize yield to simulated erosion by artificial topsoil removal. *Plant and Soil* 284:375–384.

Partey, S.T., Zougmoré, R.B., Ouédraogo, M., Campbell, B.M. 2018. Developing climate-smart agriculture to face climate variability in West Africa: Challenges and lessons learnt. *Journal of Cleaner Production* 187:285–295.

Philip, J.R. 1957. The theory of infiltration: 4. Sorptivity and algebraic infiltration equations. *Soil Science* 84(3):257–264.

Reynold, W.D., Elrick, D.E., Youngs, E.G., Amoozegar, A. 2002. Ring or cyclinder infiltrometers. In: *Methods of Soil Analysis: Part 4 Physical Methods.* SSSA Book Series: 5, J.H. Dane, G.C. Topp (eds.). Soil Science Society of America, Madison, WI, pp. 817–843.

Safriel, U., Adeel, Z., Niemeijier, D., Puigdefabregas, J., White, R., Lal, R., Winslow, M., Ziedler, J., Prince, S., Archer, E., King, C. 2005. Dryland systems. In: *Ecosytem Human Wellbing. Findings of the Conditions Trends Working Group of the Millennium Ecosystem Assessment*, vol. 1, R. Hassan, R.J. Scholes, N. Ash (eds.), Osland Press, Washington, D.C., pp. 623–662.

Salako, F.K., Kirchhof, G., Tian, G. 2006. Management of previously eroded tropical Alifsol with herbaceous legumes: Soil loss and physical properties under mound tillage. *Soil and Tillage Research* 89:185–195.

Shimless, A., Verdier-Chouchane, A., Boly, A.A. (eds.). 2018. *Building a Resilient and Sustainable Agriculture in Sub-Saharan Africa*, 302 pp.

Solomon, D., Lehmann, J., Fraser, J.A., Leach, M., Amanor, K., Frausin, V., Kristiansen, S.M., Millimouno, D., Fairhead, J. 2016. Indigenous African soil enrichment as a climate-smart sustainable agriculture alternative. *Frontiers in Ecology and the Environment* 14(2):71–76.

Tadele, Z. 2017. Raising crop productivity in Africa through intensification. *Agronomy* 7(1):22.

Tully, K., Sullivan, C., Weil, R., Sanchez, P. 2015. The state of soil degradation in Sub-Saharan Africa: Baselines, trajectories and solutions. *Sustainability* 7:6523–6552.

Valentin, C., Rajot, J.-L., Mitja, D. 2004. Responses of soil crusting, runoff and erosion to fallowing in the sub-humid and semi-arid regions of West Africa. *Agriculture, Ecosystems and Environment* 104(2):287–302.

Vanlauwe, B., AbdelGadir, A.H., Adewopo, J., Adjei-Nsiah, S., Ampadu-Boakye, T., Asare, R., Baijukya, F., Baars, E., Bekunda, M., Coyne, D., Dianda, M. 2017. Looking back and moving forward: 50 years of soil and soil fertility management research in sub-Saharan Africa. *International Journal of Agricultural Sustainability* 15(6):613–631.

Zingore, S., Mutegi, J., Agesa, B., Desta, L.T., Kihara, J. 2015. Soil degradation in Sub-Saharan Africa and crop production options for soil rehabilitation. *Better Crops with Plant Food* 99:24–26.

5 Soil Degradation in the Senegal Lower Valley

Laurent Barbiero and Claude Hammecker

CONTENTS

The Senegal River is regulated by two dams, the Diama anti-salt dam constructed in 1986, close to the river mouth (Figure 5.1), and the Mantanali reservoir dam that was impounded two years later in Mali. Since then, there has been a spectacular development of irrigated agriculture in the valley. The initial plan for the development of the Senegal river valley envisaged a potential of at least 250,000 ha, which could be increased to 375,000 ha, mainly in the two countries, Senegal and Mauritania. This intensification of agriculture and the increase in irrigated areas has disrupted the existing environmental balance. It is therefore important to determine whether these changes have caused soil degradation, or simply offered a new framework for the development and exploration of the Senegal river middle valley. Impact studies prior to development are generally insufficient to understand the environment and the complexity of its functioning. As will be shown in this study, recent pedogenetical processes in the Senegal river valley have led to soil diversity and salinity distribution. The first objective is to illustrate the mechanisms that dominated this environment up until the commissioning of the two dams. In a second step, we will present the evolution of some

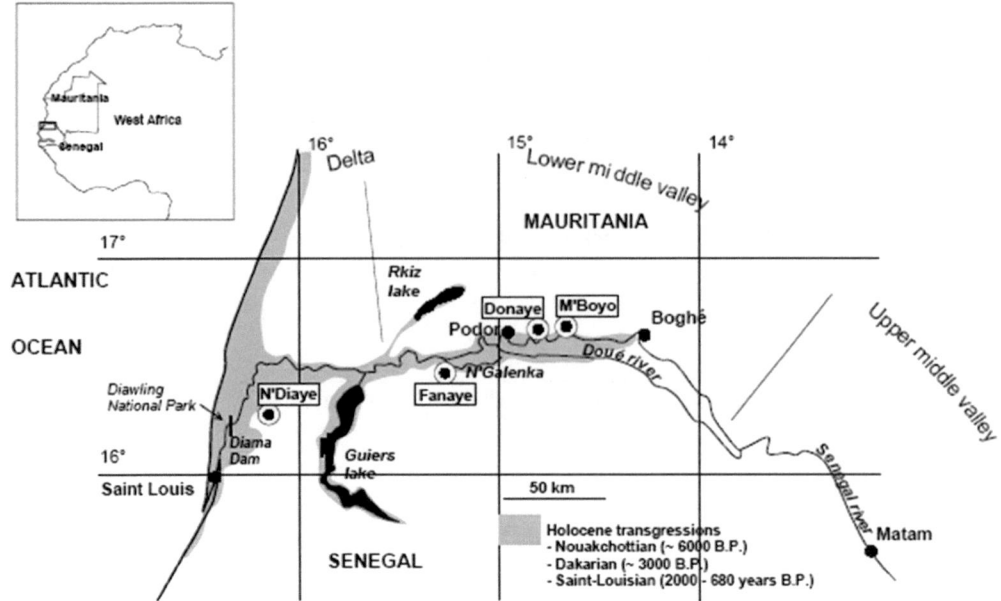

FIGURE 5.1 The Senegal valley, extension of the Holocene transgressions and location of the monitored sites.

sites being studied at the beginning of irrigated agriculture, focusing on soil salinization and alkalinization hazards.

5.1 GENERALITIES AND DEFINITIONS

5.1.1 SALINE SOILS, SALINIZATION PROCESS, SODIZATION, ALKALIZATION

A soil is described as saline when the morphological, physicochemical or agronomic properties of the soil are modified by the presence of significant quantities of soluble salts, whether in the soil solution or crystallized. Generally, when the electrical conductivity of the soil solution is greater than 0.4 dS m^{-1} (or mS cm^{-1}), the soil is considered saline. Except in some special cases (Hammecker et al. 2012), saline soils are formed generally in areas with highly evaporative conditions when the water balance has a leaching deficit. During the drying of the soil, the water vaporizes and the mineral salts remain *in situ* and accumulate, generally in the upper part of the soil profile. Saline soils are therefore found mainly in arid zones around the world and particularly in intertropical areas where soil conservation is paramount. The management of this type of soil is a major environmental problem for the development of irrigated agriculture in these arid or semi-arid zones. All types of salinization processes are not equivalent in so far as their effects on soil quality are different and where the restoration of initial soil properties is more or less difficult. There are generally three types of processes:

- The salinization process is an excessive accumulation of soluble salts (chlorides, sulphates, carbonates etc.) in the soil profile, which reduces the availability of water to plants by subjecting them to osmotic stress. Depending on the nature of the salts, that is, they are neutral (chlorides, sulphates etc.) or alkaline (bicarbonates, carbonates etc.), we can distinguish different evolutions, which will have particular physicochemical consequences.
- Sodization corresponds to an accumulation of sodium on the cationic soil exchange complex. It is manifested by a decrease in the porosity as well as the structural stability of the soils.
- Alkalization is an accumulation of weak acid bases such as bicarbonates, which leads to an increase in soil pH. Increasing the pH can become problematic for plant growth as it

can block the uptake of some nutrients. Moreover, the phenomenon of alkalization also accelerates sodization. Indeed, precipitation of calcite ($CaCO_3$) enriches the solution of the soil correspondingly in sodium, which therefore preferentially adsorbs onto the exchangeable complex.

5.1.2 Quality of Irrigation Water

Evolution of soil physicochemical characteristics and their suitability for cultivation depend directly on the quality of irrigation water. Several indicators have been defined to assess the intrinsic quality of water:

- electrical conductivity (EC) of water is a very easy to measure field indicator, expressed in mS cm^{-1} or dS m^{-1}. It is directly related to the total concentration of salts (CTS) in solution that can be determined by the following relationship: $CTS\_(mg/l) = 0.64 \times 10^{-3} \times EC\_(ds/m)$
- The sodium adsorption ratio (SAR) is an indicator that attempts to assess the risk of sodium adsorption onto the exchange complex. It is defined as: where the concentrations of the different cations are expressed in centi-mole-charge per liter of solution. Thus, the higher the SAR of water, the greater the risk of soil sodization.

Based on these two basic indicators, the Riverside laboratory team (1954) defined a diagram to assess the risks for the soil, when water of a given composition is used for irrigation. The risk of salinization and sodization can thus be determined from the electrical conductivity and the SAR.

However, this representation only gives an instantaneous view of the risks associated with the composition of the solution but does not anticipate its evolution in the event of concentration. As shown in Figure 5.2, the water composition of Doué is apparently of very good quality for irrigation as it has a low concentration, low SAR and is positioned in box C1-S1, indicating a very low risk of soil degradation. However, when concentrated, the composition of Doué water becomes sodic (the SAR increases strongly) and the risk of sodization becomes very high. However, as irrigation is practiced in arid zones with high evaporation demand, it is essential to be able to anticipate the composition of irrigation water when it is concentrated.

Therefore, other indicators are used to qualitatively predict the evolution of the chemical facies of the solution in case of concentration. One of the main indicators being calcite residual alkalinity

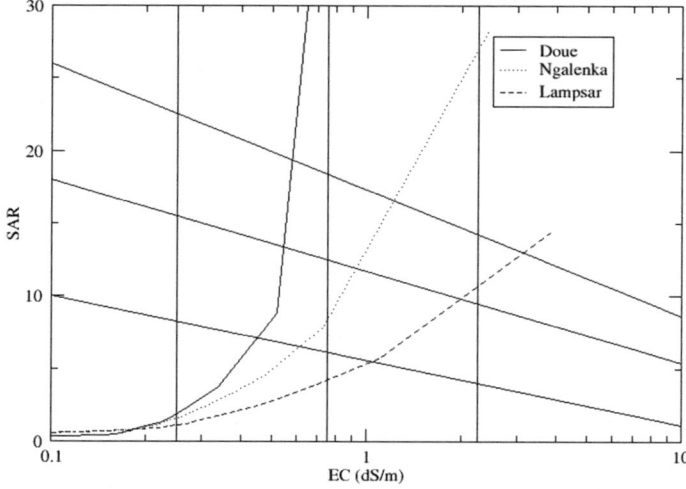

FIGURE 5.2 Chemical evolution of Senegal river irrigation water in Doué, N'Galenka and Lampsar during evaporation.

($RA_{calcite}$), as, in natural surface water, alkalinity generally corresponds to the content of bicarbonates (HCO_3^-) and / or carbonates (CO_3^{2-}). By concentration calcite ($CaCO_3$) is the first mineral crystallizing. Calcite residual alkalinity ($RA_{calcite}$), which predicts whether the solution evolves towards a neutral or carbonated saline facies, is defined as follows: $RA_{calcite}$ = Alk. − Ca^{2+} where the concentrations of alkalinity and calcium are expressed in $mmol_c\ l^{-1}$.

- if $RA_{calcite}$ < 0 the solution contains more Ca than alkalinity and during the concentration process, and crystallization of calcite, the solution will deplete in alkalinity and concentrate in calcium. There is therefore no risk of sodization because calcium will be the dominant cation in solution and on the exchangeable complex. This condition is sometimes referred to as a "neutral pathway"
- if $RA_{calcite}$ = 0 calcite will precipitate at equilibrium and the pH of the solution will not exceed 8.5
- if $RA_{calcite}$ > 0 the solution contains more alkalinity than calcium and crystallization of calcite will deplete the stock of calcium. While concentrating, the solution will become more and more alkaline, the pH will increase beyond 8.5 and above all the sodium will become the major cation in solution and on the exchange complex. Such an evolution is referred to as an "alkaline pathway."

Nevertheless, the concentration of magnesium also contributes in controlling the geochemical evolution of the solutions. In fact, magnesium can enter the composition of magnesium silicates (sepiolites, stevensite) or magnesium carbonates (magnesite, magnesian calcite)

$$2Mg^{2+} + 3H_4SiO_4 \rightleftarrows Mg_2Si_3O_{7.5}OH \cdot 3H_2O + 4H^+ + 0.5H_2O \qquad (5.1)$$

$$Mg^{2+} + CO_3^{2-} \rightleftarrows MgCO_3 \qquad (5.2)$$

In the crystallization reaction of sepiolite, 4 protons (H^+) are released, and they react with two carbonates to form water and carbon dioxide: $4H^+ + 2CO_3^{2-} \rightleftarrows 2CO_2 + 2H_2O$.

Whereas during the precipitation of magnesite, each Mg^{2+} uses a carbonate CO_3^{2-}; therefore, it is useful to define an additional indicator, called the residual sodium carbonate (RSC) sometimes referred to as calcite+magnesite residual alkalinity ($RA_{calcite+magnesite}$):

$$RSC = Alk - (Ca + Mg)(mmol_c\ l^{-1}) \qquad (5.3)$$

The evolution of the soil solution can be summarized by the diagram in Figure 5.3 where, depending on the values of the different indicators, it is possible to predict the type of salinity likely to affect the irrigated soil.

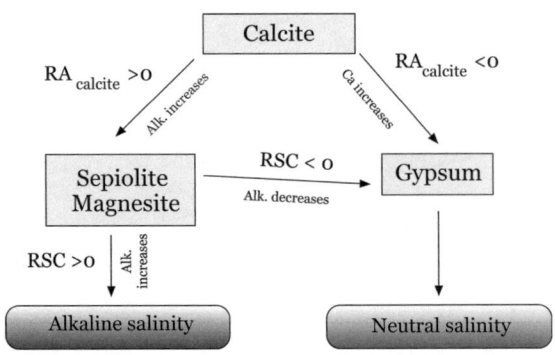

FIGURE 5.3   Main chemical evolution processes for irrigation water.

5.1.3 Consequences for Soils

Whether by dilution or by adsorption of monovalent cations (particularly sodium), when clays are dispersed, the aggregates burst and the soil structure collapses, showing some adverse properties:

- soils without structure behave like mud when wet;
- when dry they become extremely hard and practically impossible to cultivate.

Moreover, dispersed clays move, rearrange and eventually block the access to the macropores, and thus drastically reduce the soil hydraulic conductivity. The conditions of oxygenation of the environment are then strongly disturbed and the development of the roots, and therefore plants, is affected. On the other hand, imbibition of water in the soil during irrigation or during rain events becomes extremely slow or almost impossible. The practice of agriculture becomes very difficult, if not impossible, especially in areas with low technical and financial resources to rehabilitate these soils.

It is therefore important to monitor the parameters that can lead to this catastrophic situation. For cultivated soils, sodium is usually responsible for these types of degradation. The exchangeable sodium percentage (ESP) is therefore a particularly important indicator to follow. If ESP values are close or superior to 15%, then the risk of clay dispersion is high.

5.2 A COMPLEX PHYSICAL FRAMEWORK

5.2.1 Late Holocene history

The lower Senegal valley is characterized by a very low altitude, since the level at low water period reaches only 20 m in Kayes, a city currently located at 900 km from the river mouth. As a result, before the river regulation of 1986, at the end of the dry season, saline water reached as far as the town of Podor, 215 km from the ocean. Such a topographical configuration motivated the construction of an anti-salt dam for river management, but it has above all allowed the sea to transgress several times into the valley during the Quaternary (Faure et al. 1980). The presence of saline soils observed in the middle valley has long been attributed to residual lenses of salt deposits from these transgressions, but we will see that this assumption does not match with the chemical profile of salinity, or with its distribution in the landscape, and that the origin of the salt and its distribution are much more complex. The most documented transgression is the Nouakchottian transgression that occurred in the middle Holocene, soon after 6000 BP (before present), with a shoreline that reached about 1 to 3 m above the present sea level (Barusseau et al. 2007; Sarr, Sow and Sarr 2008). Another short transgression up to 2 m is referred to as the Dakarian event at around 3000 BP. The last marine incursion started about 2000 years BP and ended about 680 BP. Mangrove vegetation grew in the fluviomarine environment. Then, during the last 600 years, the sea and the mangrove regressed from Boghé, located 380 km from the present coast, progressing downriver to the present Senegal delta, where remnants of the mangrove vegetation are still present along small watercourses. Soil development from muddy sediments accompanied this regression. Thus, upriver soils, being the first to have been drained, are those with the most advanced development, whereas the soils in the delta show only a beginning of pedogenesis. In this context, soils of the lower valley of the Senegal River must be regarded as a chronosequence, which makes it possible to trace not only the gradual evolution of the soils from an unripe mud to acid sulphate soils and finally to vertisols, but also to follow the various processes of salt accumulations.

5.2.2 Major Processes of Soil Formation

The mangrove vegetation that colonized the floodplain is a key point to understand the soil formation and evolution. In the waterlogged environment around the mangrove roots, sulfur-fixing bacteria accumulated pyrite derived from inputs of seawater sulphate at each flood tide. Pyrite contents are usually between 1 and 2%, although values up to 5% of the soil mass have been reported in the delta (Seiny Boukar 1983; Deckers et al. 1997). This mineral can be considered as storing potential acidity as long as the reduced state was maintained. The last regression, which started about 600 years ago, favored oxidation, releasing strong acidity in the soil profile. The initial pyritic sediment is then transformed into acid sulphate soil through two major processes, on the one hand an irreversible loss of water, and on the other hand, iron oxidation and the appearance of jarosite, and then hydrolysis into hematite and goethite minerals from the bottom to the top of the profile (Barbiero et al. 2005). Then a series of chemical reactions was induced, related to acid neutralization by the soil acid neutralizing capacity (ANC). The first neutralization of acidity is by the dissolution of the shells and other carbonates according to the reaction:

$$CaCO_3 + 4H^+ + SO_4^{2-} \rightarrow CaSO_4 \cdot 2H_2O + CO_2 \tag{5.4}$$

This reaction can be observed today in the delta, where shell layers (*Anadara senilis* fragments and *Tympanotonus fuscatus* shells) abruptly and laterally transform into gypsum, which appears as centimeter scale nodules. Parallel to the dissolution of carbonates, the protonation of the clay minerals (a mixture of kaolinite, smectite and illite) contributed to the acid neutralization. The protonated clays are then destabilized by the penetration of protons into the octahedral layers by substituting for the aluminum ions, which are released from the network and adsorb to the clay surface (van Breemen, Mulder and Driscoll 1983; van Breemen 1988). Consequently, the Al no longer binds the tetrahedral layers and the constituent ions (Si, Al, Fe, Mg, K etc.) are released into the sulphate-rich solution. The influence on the soil profiles is twofold. Various types of aluminum sulphates precipitate on the topsoil (Le Brusq et al. 1987) and a sandy horizon resulting from the residual accumulation of quartz appears around 20 cm deep, progressively reaching down the soil profile (Furian et al. 2011). The above described reactions transform a soil profile initially consisting of an unripe mud with pyrite and carbonate shell beds into a sandy soil strongly colored by iron oxides, with gypsum horizons about 60 to 90 cm deep, more or less rich depending on the initial amount of shells. These reactions can be grouped into a first soil transformation front (Figure 5.4), which progressed downriver into alluvial deposits, accompanying the last marine regression from the area around Boghé to the present Senegal delta.

   A second transformation front induced by the overflows of the Senegal River impacted the soils in turn (Figure 5.5). The river water has a typically continental chemical profile, with low mineral charge and neutral to slightly alkaline pH. During floods, the waters infiltrated the soils and came into contact with acid solutions rich in iron, aluminum and silica. On contact between these two chemically contrasted environments, kaolinite precipitated, initially as clay laminae in the sandy

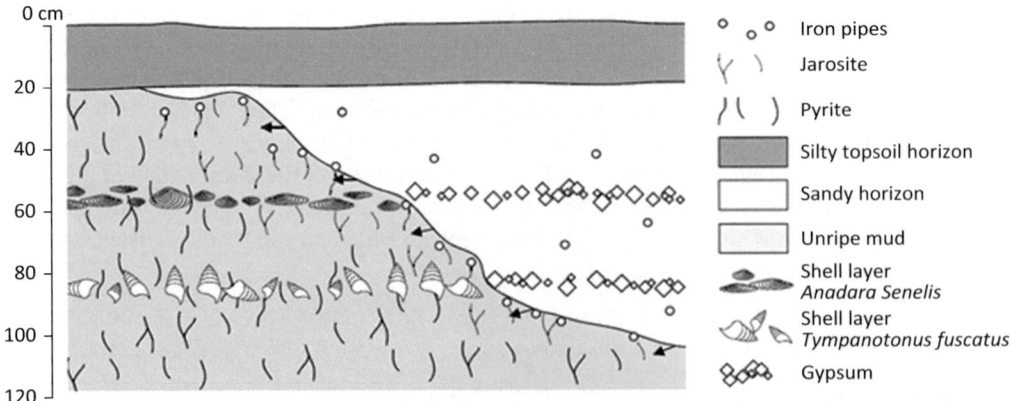

FIGURE 5.4   Schematic description of the first soil transformation front in the lower Senegal valley.

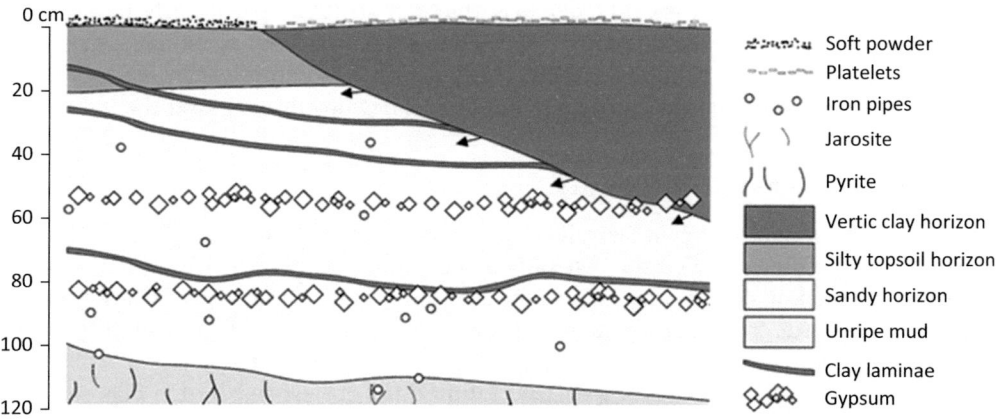

FIGURE 5.5 Schematic description of the second soil transformation front in the lower Senegal valley.

horizon (Barbiero et al. 2005). Progressively, the acid conditions were sufficiently neutralized by the large amounts of base cations present in the solution to allow the progressive appearance of a smectite of iron-beidellite type. Eventually, the clay precipitation resulted in the formation of a superficial clay horizon, which thickened laterally and also acquired vertic properties. The final soil profile consists of three main horizons of variable thickness, with, from top to bottom, a clay horizon with vertic properties, a sandy horizon with some clay laminae and an unripe mud with pyrite and jarosite. Such contrasting soil horizons with abrupt transition look like alluvial deposits and have therefore been interpreted stratigraphically by many authors as indicating changes in the sedimentation regime of the river (Michel 1973), whereas they actually result from two successive soil transformation processes.

5.2.3 Aeolian Salt Deposits

The phenomenon of soft powder formation, consisting of silt and sand-size particles, is caused by the rapid crystallization of salts, and is currently occurring in the delta. However, two cases should be distinguished: when wind deflation is active before the first transformation front, the precipitation sequence is gypsum > halite (Mohamedou et al. 1999). The particles are therefore mainly enriched with these two minerals. On the other hand, when the wind deflation is active after the first transformation front, pyrite oxidation has occurred and enriched the soil solution with sulphates. The precipitation sequence was changed to gypsum > thénardite/mirabilite > halite. The wind accumulation can then be relatively enriched by sodium sulphates. The deflated particles deposit behind any wind-breaking obstacle, such as saline lunettes or clay dunes. Because watercourses, still occupied by the remnants of mangrove vegetation, are the main obstacles to prevailing winds from NE to NW, aeolian saline accumulation mainly occurs on their southern bank, forming elongated deposits (Barbiero et al. 2001; Barbiero et al. 2004). The second transformation front matches with a change in the topsoil texture from 15% to 30% clay. It abruptly modifies the aspect of the soil surface from soft powder, vulnerable to aeolian deflation, to platelets less vulnerable to wind erosion. A laboratory outflow experiment according to Wind (1968) was carried out on two topsoil blocks, collected a few meters apart on each side of the limit between the soft powder and the platelets. The results emphasized that before the passage of this second transformation front, the solutes reached more easily the surface of the soil where they precipitate, causing a deflatable soft, powdery structure. On the other hand, after the passage of the front, the change of texture in the topsoil horizon modifies the transfers in unsaturated conditions, favoring the crystallization of the salts at a depth of about 3 cm below the surface, causing the formation of platelets (Furian et al. 2011). As a conclusion, the second transformation front stops the wind deflation. Therefore, the deflation of saline particles is now observed only in some parts of the delta (Barbiero et al. 1998), particularly in the Diawling National Park (Mauritania) downstream of the Diama dam. However, in the past, deflation and accumulation in the form of saline clay dunes have occurred more upriver in the lower Senegal valley, representing the second transformation front the valley downstream over the last 600 years.

5.2.4 Saline Areas in the Lower Valley

To summarize, the soil formation in the valley has implemented two processes of salt accumulation. The first one is of sedimentary origin, generally at a depth of about 0.6 to 1.2 m, and corresponds to the transformation of shell layers. The chemical profile of this salinity is of Ca-SO$_4$ type. The quantity of salt depends on the initial quantity of carbonate shells. As a result this salinity is referred to as "deep gypsum."

The second one is of wind origin, a surface salinity that corresponds to the deposits of wind deflated saline particles. The chemical profile is predominantly of Na-Cl type, secondarily Ca-SO$_4$ and Na-SO$_4$. The thickness of the aeolian accumulations is variable, from a few centimeters to 1 m thick. It will be thereafter referred to as "aeolian salinity." In the Senegal middle valley, since the passage of these two above described transformation fronts, seasonal rains and river floods have contributed to the elimination of these two kinds of salt stocks, although these have interacted to control the presence and distribution of saline soil in the landscape.

A database collected from survey reports and other grey literature (mainly in French) shows that the general trend is a decrease in soil salinity towards upriver, in accordance with the chronosequence mentioned in Section 5.2.1 (Barbiero et al. 2004). The aeolian salinity, initially at the top-soil, has been translocated about 0.3 to 0.6 m down in the soil profile. Therefore, although the salts are not immediately detectable, farmers are locally confronted with the presence of saline soils, which are evident from the first crops after irrigation is started by remobilization of the salts up to the top of the soil. Many farmers have deserted their plots whose crops have been burnt by salinity. A key aspect for a successful implementation of irrigated agriculture is to detect where the saline areas are located in the landscape.

The distribution of landscape units has changed considerably since the saline accumulations have ceased. The anastomosed meanders of the river have moved, some channels have been degraded, filled, and some depressions have formed. Thus, it is not possible to relate the occurrence of salt to the present geomorphology in the valley. There are saline and non-saline river banks, saline and non-saline depressions, and so on. On the other hand, saline areas are well correlated with a former geomorphology, which can usually be detected from aerial or satellite imagery. From these tools, transects can be positioned for electromagnetic salinity surveys (Barbiero et al. 2001). Figure 5.6

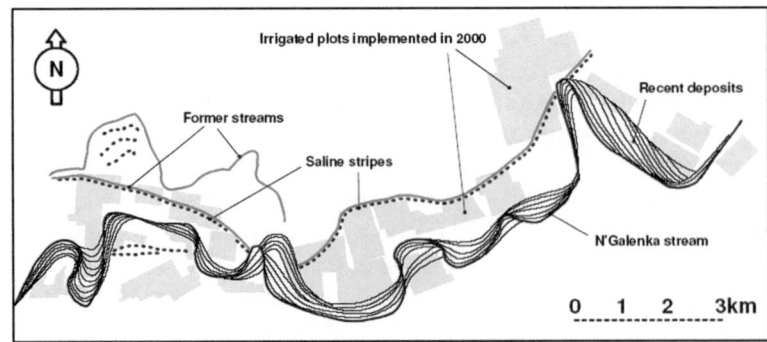

FIGURE 5.6 Results of an electromagnetic survey in the N'Galenka area showing the distribution of the salinity as stripes accompanying the former streams. Note that the salinity is intersected by the current N'Galenka stream and its recent deposits, attesting an ancient salt accumulation.

[N'Galenka] shows such a survey in the N'Galenka area together with the simplified geomorphology in the Podor region, where irrigated schemes were implemented in the early 2000s.

The N'Galenka flows south of the area and its meanders intersect the bed of an ancient watercourse. The electromagnetic survey reveals that salinity is distributed as strips about 100–200 m wide and several kilometers long, mainly on the southern margin of the former channel. This distribution seems in agreement with the presence of former wind accumulations of salt. The study of the chemical profile of salinity, from concentration diagrams based on chloride ions, reveals that water was not the vector of this salt accumulation (Barbiero et al. 2004). Moreover, the chemical profile is predominantly of $Ca-SO_4$, $Na-SO_4$ and $Na-Cl$ type, although various intermediate chemical compositions between these three end-members have been detected, within the same saline area, and sometimes within the same soil profiles over short distances of about 0.2 to 0.3 m.

The distribution, morphology and chemical profile diversity of saline areas can be explained by the interactions between both, deep gypsum and aeolian salinity. Soil desalinization by rains and stream overflows has occurred before the river regulation, and is still occurring in the valley, although with less intensity. In this framework, former clay dunes remained saline for a longer time for two reasons: first, they are on slightly higher ground and less vulnerable to desalinization during floods, and second, the initial amount of salt is usually higher than outside the deposit. In these clay dunes, by dissolving some superficial gypsum, soil water reaches saturation with respect to this mineral and therefore becomes non-aggressive when percolating through the deep gypsum that has remained in the soil profile. The deep gypsum began to dissolve only after almost complete elimination of the superficial aeolian salinity. In this context, former clay dunes are detected nowadays in the Senegal middle valley as saline areas with salts from two sources, and with amounts that depend on (i) the thickness of the aeolian deposit, (ii) the quantity of initial carbonate shells, and (iii) the progress of desalinization.

5.2.5 Conclusion

The model of soil formation in the lower Senegal valley, which is summarized in Figure 5.7, involved very powerful processes, initially supplied by extreme acidic geochemical conditions resulting from the oxidation of reduced iron minerals accumulated in the form of pyrite by tidal cycle in a waterlogged environment. The proposed model explains salt distribution, and the intensity and chemical variability of salinity (Furian et al. 2011). This diversity will be illustrated in the following examples, which correspond to sites monitored under cultivated conditions for several years. Two main aspects will be developed: on the one hand, the dynamics of salts after cultivation, and on the other hand, the secondary risk of soil degradation by alkalization under irrigated conditions with weakly mineralized Senegal river water, having a positive calcite residual alkalinity ($RA_{calcite} > 0$), and thus with an alkalizing potential.

5.3 EVOLUTION OF SOIL AND GROUNDWATER WITH IRRIGATION

5.3.1 Study Sites

In this section, evolution of the quality of both groundwater and soils in irrigated conditions will be discussed using several examples selected along the soil chronosequence. The selected experimental sites are N'Diaye, Fanaye, Donaye and M'Boyo, located at 35, 165, 215 and 230 km west from the city of St Louis, respectively (Figures 5.1 and 5.7). In the Senegal valley, the climate is qualified as Cape Verdean up to 30 km from the Atlantic coast because of maritime influences, and then as Sahelian in the more continental part. It is semi-arid, with an average yearly rainfall of about 220 mm concentrated during the rainy season from July to September. The dry season is composed of a cool dry season from October to February and a hot dry season from March to June, during which dry warm winds from the Sahara desert (Harmattan) enhance evaporation. Average Class A pan

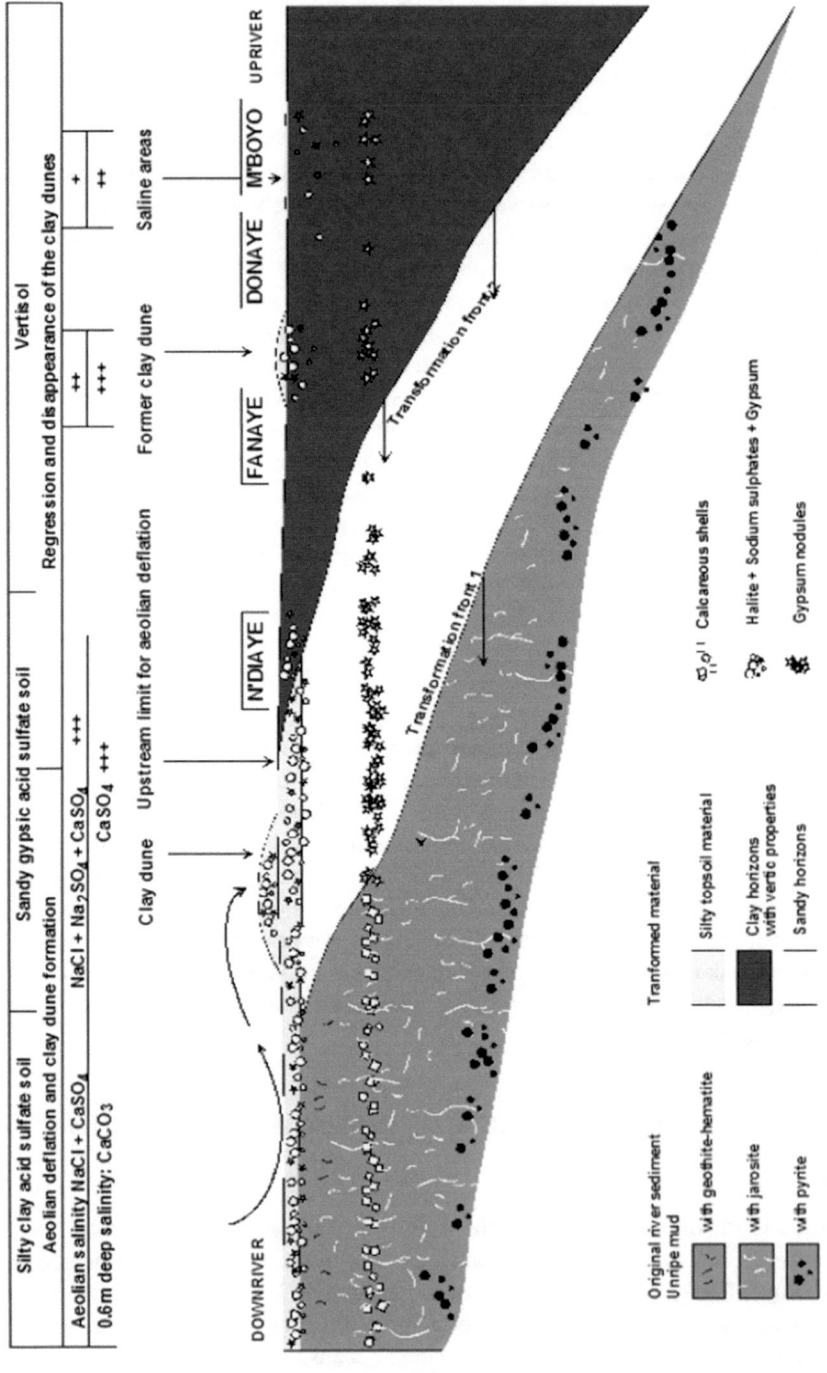

FIGURE 5.7   Schematic evolution of the Senegal floodplain sediment and saline accumulation from the delta to the middle valley.

evaporation exceeds 2,000 mm annually. The monitoring of the water chemistry was carried out in the irrigation water, in the water table from piezometers installed in the plots, and in the soil solution from porous cups installed at 0.1, 0.2, 0.3 and 0.6 m deep.

5.3.1.1 The Site of N'Diaye

N'Diaye is an experimental site of AfricaRice (previous WARD, West Africa Rice Development Association) in the Senegal delta, 35 km west of St Louis. It is located in a transition zone between the Cape Verdean and Sahelian climate. The overall topography is flat, and before the river regulation, the area used to get submersed by the floods. The typical soil profile of the N'Diaye site shows two very distinct superimposed materials. In the upper 0.6 m, the soil profiles shows a clayey, compact material with coarse prismatic to vertic structure, and a light brown color. Numerous iron pipes were found from 0.5 cm downwards associated with coarse (1 or 2 cm) gypsum nodules. Underneath, the soil texture is sandy to sandy loam with marked hydromorphic features, illustrated by a dominant grey color. Like most of the soils in the Senegal delta, the soil at N'Diaye is saline and generally acidic, due to the remains of pyrite or by-products, witnessing the ancient mangrove vegetation. In particular, many pale yellow, jarosite mottles can be observed below a depth of 0.8 m.

The saline water table is shallow, It can be observed at about 1 m in depth in the absence of irrigation during the dry season, and reaches 0.4 m and 0.2 m during the irrigated cropping season in the off-season and rainy period, respectively. The irrigation water is pumped from the Lampsar, a distributary watercourse of the Senegal River. The electrical conductivity fluctuates from 0.1 to 0.7 dS m^{-1} throughout the year. The chemical profile is HCO$_3$/Ca-Mg with a positive RA$_{calcite}$, but with a negative RSC.

5.3.2.2 The Site of Fanaye

On Fanaye site, paddy fields are irrigated with water pumped from the N'Galenka stream, another distributary from the Senegal River. N'Galenka water has a low mineral charge (EC of about 0.1 dS m^{-1}) with a Ca-Mg-HCO$_3$ chemical profile, a positive RA$_{calcite}$ and also a positive RSC throughout the year. A risk of soil degradation by alkalization must therefore be considered in irrigated plots. The monitoring was carried out within a slight depression of approximately 30 ha, and such a situation favored the development of the clayey horizons, down from the soil surface (second transformation front described above). The soils are vertisols with 48% to 55% of clay content, characterized by a gilgai microrelief on the surface inherited from continuous wetting-drying cycles. The clay horizons constitute a 2 m thick homogeneous material overlying a sandy material, with an abrupt interface. The groundwater was not reached with auger prospection to a depth of 6 m.

5.3.1.3 The Site of Donaye

For several years, the plots on this site were generally exploited with two crops of rice per year, sown during the hot dry season, from March to June, and the wet season from July to October. Paddy fields are flooded for more than 100 days during each growth cycle. However, during the monitoring there were only rainy season crops, and no hot dry season crops. The soil is a vertisol with a clay texture (>65% clay consisting of 70% smectite and 30% kaolinite, Boivin et al. 2002), a very marked vertic structure, abruptly overlaying a sandy material at about 1.2 m deep. The water table fluctuates between 1.2 and 2.5 m deep throughout the year and shows a SO$_4$-Cl/Na chemical profile. The irrigation is supplied with water from the Senegal River. It has a low mineral charge with EC of about 0.06 dS m^{-1}. RA$_{calcite}$ is positive, whereas RSC is usually slightly positive or close to zero.

5.3.2.4 The Site of M'Boyo

The site of M'Boyo is on a 15 ha irrigated area developed by EDF (European Development Fund) and located on an island between the Senegal River and the Gayo, a distributary-tributary stream from the Senegal. The irrigation water arises from the Senegal River, just like on the site of Donaye, and shows the same chemical characteristics. The monitored plot was cultivated with rice crops during the wet season from August to November without any crops during the hot dry season. The soil profile of the M'Boyo site is very similar to that of Donaye, with 1.9 m of clay material (58% to 66% clay), abruptly overlaying a sandy material (about 1% clay), strongly colored by iron oxides. The sandy horizons are 1.5 to 2 m thick. Below, there is a grey loamy sand material with some pyrite and jarosite minerals. In the monitored plot, the soils are saline and clearly show two saline horizons at 0.6 m and 1–1.2 m deep, with Cl-SO$_4$/Na and SO$_4$/Ca chemical profile, respectively. These two saline levels are in agreement with the model of salt accumulation presented in Section 5.2.4. The first level may correspond to a migration towards a salinity of aeolian origin to a depth of 0.6 m, whereas the second level presents the characteristics of a deep gypsum accumulation from the transformation of a shell layer. The water table is very saline, with a mineral charge close to the seawater, and a chloride sodium chemical profile.

5.3.1.5   Irrigation Water

The chemical composition of the irrigation water used in the four sites, does not represent a threat for soil quality when plotted in the Riverside diagram. However, when simulating the concentration with PHREEQC (Parkhurst and Appelo 1999), the solution becomes extremely sodic, especially for the Senegal river water used in Mboyo and Donaye (Figure 5.2).

5.3.2   Soil and Groundwater Table Evolution Under Irrigated Crops

5.3.2.1   N'Diaye

Evolution of soil solution chemistry in plots follows clear trends for all major elements, except for alkalinity, which shows a more erratic behavior over time. During the rainy season, Na, Ca, Mg, K, SO$_4$ and Cl contents in soil solution increase at all depths. However, the soil is more saline at a depth than on the surface. Consequently, this saline profile is not due to a concentration process driven by evaporation, but is supplied by the rise of the saline water table under the influence of the Senegal River flood, which is confirmed by the piezometric measurements. During the off-season crops, the salinity gradient remains unchanged, but soil solution concentration declines at all depths. Salts are evacuated downwards, during the drawdown of the water table. The chemistry of irrigation water is dominated by carbonates, and this chemical facies dominates in surface water, and in the most superficial soil solution, up to 10 cm deep. Underneath, alkalinity is very low in soil solution. Alkalinity concentration oscillates during the year, namely with noticeable local temporary decreases during the rainy season. This behavior of the alkalinity is attributed to the upraise of the water table with a more acidic composition during the rainy season. Indeed, at this site, potential soil acidity inherited from the mangrove deposits has not yet been completely evacuated: the oxidation pyrite and hydrolysis of jarosite during the intercropping periods when the water table is low release H$^+$ ions, which are transferred to the root zone during the upraise of the groundwater during the rainy season (Figure 5.8).

5.3.2.2   Fanaye

At the site of Fanaye, located further upstream than N'Diaye, the natural soil desalination process is more advanced. Although located close to the meanders of the Senegal River (less than 2 km), the groundwater could not be reached at a depth of 6m (Figure 5.9). The absence of a shallow water table allows irrigation water to infiltrate deeper into the soil, avoiding excessive accumulation of carbonate salts brought by irrigation during intercultural periods. The soil solutions have moderate electrical conductivity (total concentration) but are in equilibrium with calcite, and pH is maintained at around 8.5. As 90% of the cation exchange capacity is saturated with Ca and Mg, these elements buffer the soil solution by limiting the increase in carbonate alkalinity and pH. No current salinization nor alkalinization is affecting the soil quality on this site. Moreover, considering the thickness of the vertic clayey horizons and the buffer capacity of soil, at short- to mid-term, irrigation should not jeopardize soil quality.

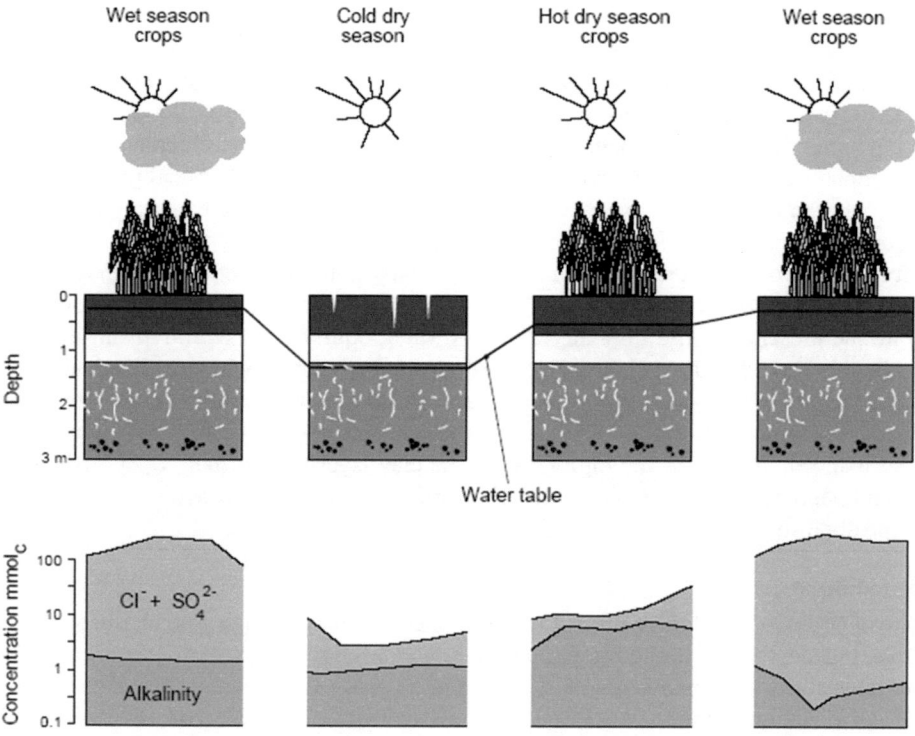

FIGURE 5.8 Evolution of groundwater level and soil solution chemistry at 0.3 m depth during the different seasons at the N'Diaye site. Note the synchronous movements of salts with the water table.

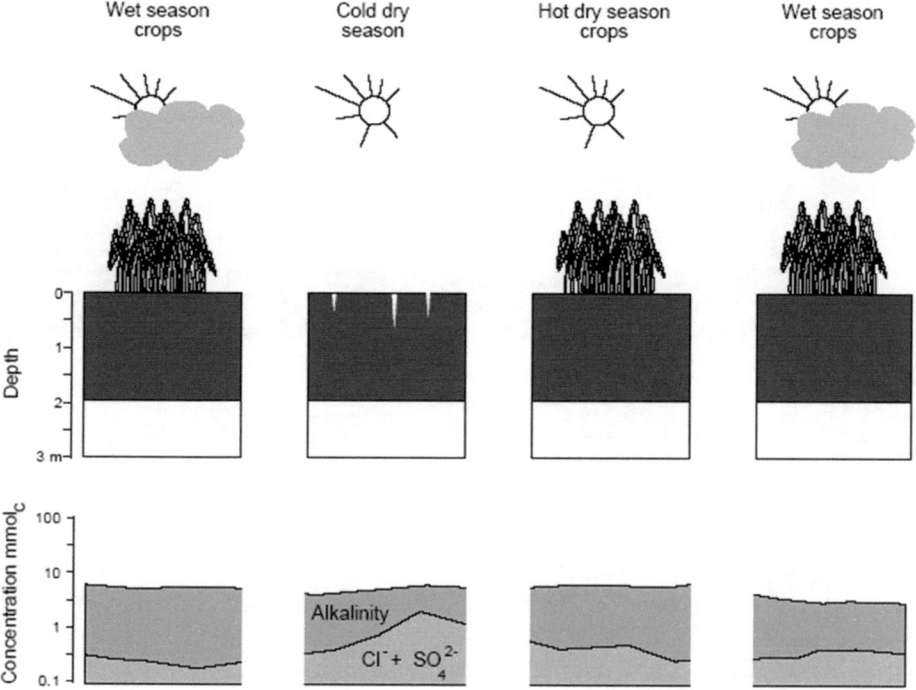

FIGURE 5.9 Evolution of soil solution chemistry at a depth of 0.3 m during the different seasons at the Fanaye site. On this site, the desalination process is the most advanced.

5.3.2.3   Donaye

As the site is located several meters above the river level, the groundwater is partly supplied by the irrigation canal and partly by the river, the amplitude of the water table level variation is about 3 m (Figure 5.10). Two distinctive chemical profiles are represented in the soil solution. Near the surface, it has a calcium–magnesium carbonate profile, whereas, at depth, the chemical profile is clearly chloride–sulphate sodium. In the upper part of the soil, the alkaline chemical profile imposes during the cropping season, especially since the water infiltration is blocked by entrapped air caused by submersion irrigation (Hammecker et al. 2003).

Calcium is controlled by precipitation as calcium carbonate that accumulates in the 0.6 m upper part of the soil after the cropping cycles. The soil solutions are saturated with respect to calcite, but despite the presence of the sulphate in groundwater, equilibrium with gypsum is not reached. Locally, the remains of pyrite and jarosite contribute to acidification when oxidized and hydrolyzed and control the pH of the soil solution. Fe-mineral oxidation and acidification of the soil solution at depth appears to be more pronounced during the years without off-season crops, while the water table level is lower. Moreover, the high CEC of the clay layer (>35 mmol$_c$ kg^{-1}), mainly saturated with Ca and Mg, represents a huge buffer controlling sodization and avoids a further evolution towards the alkaline pole.

5.3.2.4   M'Boyo

At the site of M'Boyo, the flow regime of the river seems to play a major role on the evolution of the water table. Indeed, the rise of the water table starts with high river flow, before irrigation. Similarly, the water table starts to draw down with the flood recession, while the plots are still immersed. The amplitude of variation of the water table level is about 1 to 1.5 m. The river flooding also impacts the chemistry of the water table, causing a decrease of the total concentration but an increase in

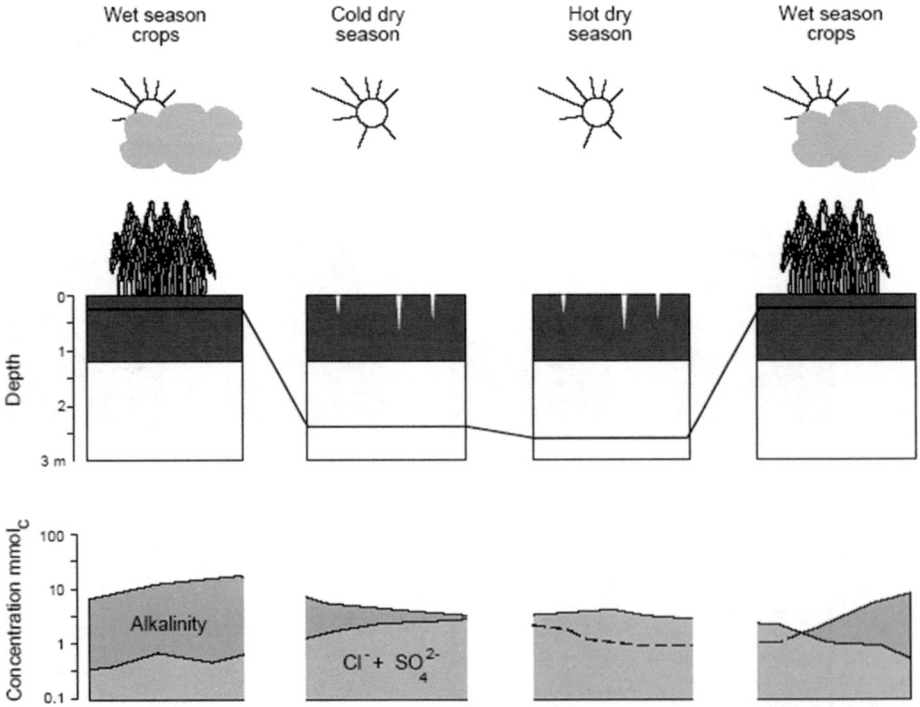

FIGURE 5.10   Evolution of groundwater level and soil solution chemistry at 0.3 m depth during the season at the Donaye site.

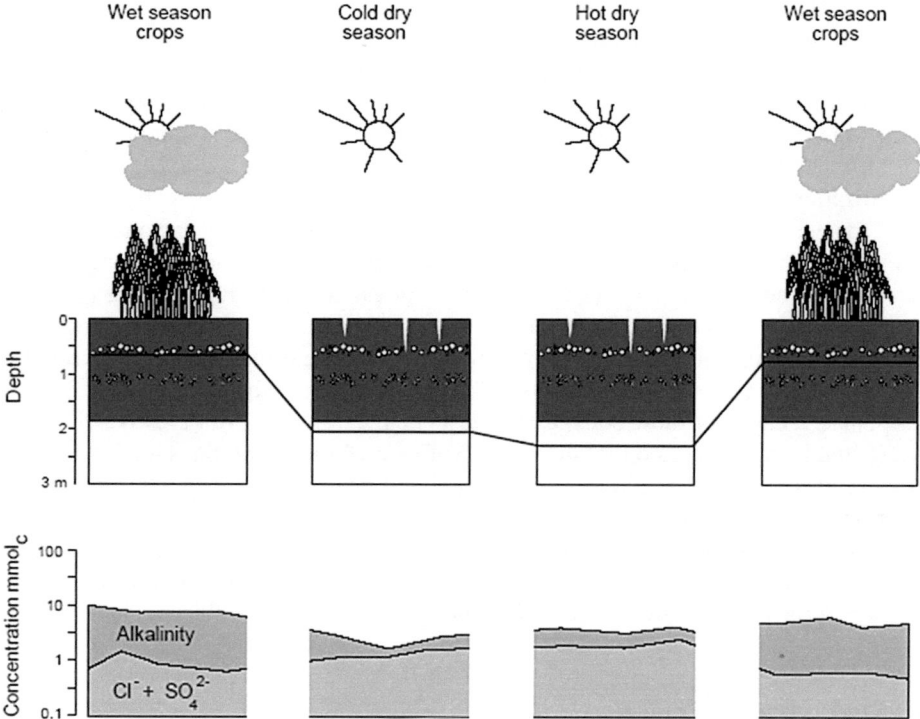

FIGURE 5.11 Evolution of groundwater level and soil solution chemistry at 0.3 m depth during the season at the M'Boyo site. Note that despite the presence of salt accumulation at a depth of 0.6 m, the soil solution concentration at 0.3 m remains moderate.

alkalinity and pH. This effect decreases with the increasing distance to the river. The aquifer is saline and at equilibrium with gypsum and therefore not very aggressive with respect to the saline accumulations at about 1.2 m of depth. In the part of the irrigated perimeter closest to the river, the dilution causes a temporary undersaturation with this mineral, associated with a rise in the water table. It is therefore in these plots closest to the river that the elimination of deep gypsum is possible. As in the case of the Donaye site, the influence of irrigation is negligible on the chemistry of aquifers. Though the soil solution at M'boyo is very concentrated below 60 cm, it remains much less mineralized at 30 cm (Figure 5.11) and soluble salts don't precipitate at the soil surface even during the dry season. In fact, self-mulching (Grant and Blackmore 1991), very common in clay soil, impedes the evaporation front from reaching the soil surface and therefore the soil solution from concentrating sufficiently to allow the saturation of gypsum or halite. On the other hand, flooding during the rainy season, or irrigation during the dry season, drains these soluble salts downwards, in a generalized desalinization dynamics.

5.3.3 Conclusion

The four sites under study have distinct hydrodynamic and chemical conditions. In the delta, the N'Diaye site is characterized by the presence of a saline water table very close to the surface and here the chemical facies is strongly influenced by the various processes of neutral salt accumulation (Ca-Na/Cl-SO$_4$). In the short- to medium-term, the Lampsar irrigation water, whose alkalinizing potential is limited, has no effect on the chemical evolution of the irrigated profiles. Irrigation simply mechanically provokes the raising of the water table level. Neutral salts migrate towards the top of the soil profile during the rainy season when the water table rises, but are evacuated downwards during the drawdown of the groundwater in the dry season. Alkalinity provided by irrigation water is counteracted by the potential acidity liberated during the dry season when pyrite oxidizes and/or jarosite hydrolyses, and rises with the water table during the rainy season.

On the site of Fanaye, further upriver and on the left bank of the Senegal valley, despite the use of an alkaline water for irrigation, the absence of the water table allows the solutions to percolate down the soil profiles, and the strong soil exchange capacity, mainly saturated with Mg and Ca, avoids the risk of alkalization in the short term.

On the Donaye site, unlike in the previous other sites, the evolution of the chemistry of the soil solution in the topsoil is disconnected from the groundwater. The presence of trapped air in the soil separates two geochemical pools physically. A stock of calcite forms in the upper part of the soils under the influence of both the alkalinity of irrigation water and the calcium from remnants of deep gypsum and/or aeolian salinity. This stock of calcium carbonate is the result of the irrigation cycles that the plots have undergone. Both the rise of the water table with neutral saline facies and the strong exchange capacity of these soils, acting as a chemical buffer, limit the alkalization process. However, it should be mentioned that the reserve of neutral saline solutions tends to be evacuated during the rainy season floods, and the soil solution will evolve towards an alkaline chemical profile, with a risk of alkalization in the long term.

In the case of M'Boyo, salinization is essentially driven by the mobilization of significant ancient salt deposits (a mixture of aeolian salinity and deep gypsum accumulation) during the water table movements governed by the Senegal River. Irrigation has very little influence on saline dynamics on this site.

5.4   GENERAL CONCLUSION

The recent history of the Senegal river valley is complex and marked by several marine intrusions during the Late Holocene period. The development of mangrove vegetation and the oxidation of soils following the last regression have favored the emergence of acidic conditions that are the driving force not only for soil formation from unripe muddy sediments, but also for two different processes of salt accumulation. One process generated salinity from the transformation of shell levels into gypsum, and another kind of salinity was brought by the wind and accumulated as clay dunes. The river regime was regulated about 30 years ago by the construction of two dams, one upstream in Mali used as a water reservoir (Manantali), the other in the delta aimed at blocking marine intrusions (Diama). This development increased salinity in the Diawling National Park in Mauritania downstream of the Diama dam, but also favored the development of irrigated agriculture in the middle valley. Initially, the presence of these salts, at least in appearance, randomly distributed in the landscape of the middle valley worried the farmers during the first years of irrigation because of the remobilization of these soluble salts towards the topsoil, mainly in vegetable crops, which do not use enough water to drain the salts towards the soil profile. Data from the downstream-upstream soil distribution as a chronosequence, show that the natural dynamic before the regulation of the river was a gradual elimination of the salts, mainly during the floods. Moreover, monitoring of the few sites presented shows that this dynamic continues under irrigated conditions, mainly in submerged paddy fields. Water in the Senegal River is very little mineralized, but presents a potential risk for soil alkalization. The composition of this water is very similar to the Niger River, which led to soil alkalization in the interior delta of Mali and on the alluvial terraces in Niger. However, in the case of the Senegal valley, this risk remains minimal mainly for three reasons: (i) a reserve of potential acidity in soils, which is still available in the most downstream part of the valley and in the delta; (ii) the presence of neutral salts inherited from former marine deposits and related accumulation processes; and (iii) clayey soils with high CEC and saturated with Ca and Mg, a consequent reserve to restore equilibrium with the carbonates in the soil solution, that drastically slows down any alkalization process, especially if the water management in the plots is performed adequately (Hammecker et al. 2009).

The presence of clayey soils in the Senegal river valley contributed physically to limit the superficial evaporation by self-mulching. Therefore, soluble salts (gypsum or halite) did not tend to concentrate at the soil surface during the dry season and show typical efflorescences, unlike in more silty soils.

REFERENCES

Barbiéro, L., Cunnac, S., Mané, L., Laperrousaz, C., Hammecker, C., and Maeght, J.L. 2001. Salt distribution in the Senegal middle valley: Analysis of a saline structure on planned irrigation schemes from N'Galenka creek. *Agricultural Water Management* 46(3): 201–213. doi: 10.1016/S0378-3774(00)00088-3.

Barbiero, L., Mohamedou, A.O., and Caruba, R. 1998. Influence de la maturation des sols de mangrove sur la déflation éolienne et la formation de dunes argileuses dans le delta du fleuve Sénégal. *Comptes rendus de l'Académie des Sciences*, t. 327, Série IIa, 115–120. doi: 10.1016/S1251-8050(98)80041-4.

Barbiéro, L., Mohamedou, A.O., Laperrousaz, C., Furian, S., and Cunnac, S. 2004. Polyphasic origin of salinity in the senegal delta and middle valley. *Catena* 58(2): 101–124. doi: 10.1016/j.catena.2004.03.003.

Barbiéro, L., Mohamedou, A.O., Roger, L., Furian, S., Aventurier, A., Rémy, J.C., and Marlet, S. 2005. The origin of vertisols and their relationship to acid sulfate soils in the Senegal valley. *Catena* 59(1): 93–116. doi: 10.1016/j.catena.2004.05.007.

Barusseau, J.P., Vernet, R., Saliège, J.F., and Descamps, C. 2007. Late holocene sedimentary forcing and human settlements in the Jerf El Oustani - Ras El Sass region (Banc d'Arguin , Mauritania). *Géomorphologie: Relief, Processus, Environnement* 13(1) doi: 10.4000/geomorphologie.634.

Deckers, J., Dondeyne, S., Vandekerckhoven, L., and Raes, D. 1997. Major soils and their formation in the West African Sahel. In: *Irrigated Rice in the Sahel: Prospects for Sustainable Development*, M. Miézan, M.C.S. Woperies, M. Dingkhun, J. Deckers, and T.R. Randokph (eds.), pp. 23–35. Saint Louis, Senagal: West Africa Rice Development Association (WARDA/ADRAO).

Faure, H., Fontes, J.C., Hebrard, L., Monteillet, J., and Pirazzoli, P.A. 1980. Geoidal change and shore-level tilt along holocene estuaries: Sénégal River area, West Africa. *Science* 210(4468): 421–423.

Furian, S., Mohamedou, A.O., Hammecker, C., Maeght, J.-L., and Barbiero, L. 2011. Soil cover and landscape evolution in the Senegal floodplain: A review and synthesis of processes and interactions during the late Holocene. *European Journal of Soil Science* 62(6): 902–912. doi: 10.1111/j.1365-2389.2011.01398.x.

Grant, C., and Blackmore, A. 1991. Self-mulching behavior in clay soils—Its definition and measurement. *Australian Journal of Soil Research* 29(2): 155–173.

Hammecker, C., Antonino, A.C.D., Maeght, J.L., and Boivin, P. 2003. Experimental and numerical study of water flow in soil under irrigation in northern Senegal: Evidence of air entrapment. *European Journal of Soil Science* 54(3): 491–503.

Hammecker, C., Maeght, J.L., Grünberger, O., Siltacho, S., Srisruk, K., and Noble, A. 2012. Quantification and modelling of water flow in rain-fed paddy fields in NE Thailand: Evidence of soil salinization under submerged conditions by artesian groundwater. *Journal of Hydrology* 456: 68–78.

Hammecker, C., van Asten, P., Marlet, S., Maeght, J.L., and Poss, R. 2009. Simulating the evolution of soil solutions in irrigated rice soils in the Sahel. *Geoderma* 150(1–2): 129–140.

Le Brusq, J.Y., Loyer, J.Y., Mougenot, B., and Carn, M. 1987. Nouvelles paragenèses à sulfates d'aluminium, de fer, et de magnésium, et leur distribution dans les sols sulfatés acides du Sénégal. *Science du Sol* 25: 173–184.

Michel, P. 1973. Les bassins des fleuves Sénégal et Gambie: Étude géomorphologique. *Mémoires ORSTOM* 63, 752 p.

Mohamedou, A.O., Aventurier, A., Barbiero, L., Caruba, R., and Valles, V. 1999. Geochemistry of clay dunes and associated pan in the Senegal delta (Mauritania). *Arid Soil Research and Rehabilitation* 13(3): 265–280. doi: 10.1080/089030699263302.

Parkhurst, D.L., and Appelo, C.A.J. 1999. User's Guide to PHREEQC (Version 2): A Computer Program for Speciation, Batch-Reaction, One-Dimensional Transport, and Inverse Geochemical Calculations. Water-Resources Investigations Report 99-4256. Denver, CO: U.S. Geological Survey : Earth Science Information Center.

Sarr, R., Sow, E.H., and Sarr, B. 2008. Holocene marine intrusions in Retba and Mbawane lakes (Senegal) evidenced by ostracod faunas. *Revue de Micropaléontologie* 51(4): 327–338. doi: 10.1016/j.revmic.2007.03.002.

Seiny Boukar, L. 1983. *Etude Pédologique de la Cuvette de N'Der, Lac de Guiers (Région du Fleuve Sénégal)*. Rapport d'élève ORSTOM, Centre ORSTOM de Dakar, Senegal.

Van Breemen, N. 1988. Effects of seasonal redox processes involving iron on the chemistry of periodically reduced soils. In: *Iron in Soils and Clay Minerals*, J.W. Stucki, B.A. Goodman, and U. Schwertmann (eds.), pp. 797–809. Dordrecht, Netherlands: Springer.

Van Breemen, N., Mulder, J., and Driscoll, C.T. 1983. Acidification and alkalinization of soils. *Plant and Soil* 75(3): 283–308. doi: 10.1007/BF02369968.

6 Within-Field Monitoring of Secondary Salinity in Irrigated Areas of South Africa

Sybrand Jacobus Muller and Adriaan van Niekerk

CONTENTS

6.1 INTRODUCTION

Soil salinity is defined as the accumulation of soluble salts in the topsoil. Salinity can be classified as primary (i.e., when soil salinity occurs naturally), or secondary (when induced by human interference). Secondary salinity mainly occurs due to a lack of proper planning or maintenance of irrigation infrastructures and is a major concern, especially in irrigated areas. Extreme salt accumulation leads to the death of vegetation and consequently to a reduction in crop yield and general degradation of fertile soil (Ghassemi et al. 1995). Previous estimates of global salinity effects indicate that around 77 million hectares (Mha) of the Earth's surface is affected by secondary salinity and that more than half of this area is made up of irrigated soil (Metternicht and Zinck 2003). The negative economic impact of salt-affected soils on the agricultural sectors' production capacity is immense, with the added pressure of maintaining sustainable food production in a world with a growing population. With quality crop and soil management, good irrigation and drainage systems design, as well as political and social support, irrigation can be a thriving, sustainable practice (Hillel and Vlek 2005; Van Rensburg et al. 2012). Quality management practice includes tracking and identifying salinity changes to anticipate further degradation and to undertake suitable reclamation and rehabilitation practices (Metternicht and Zinck 2003).

Waterlogging, a process very closely related to soil salinity, is the saturation of the soil as a result of the water table being located at or near the surface (Barrett-Lennard 2003; McDonagh and Bunning 2009). Waterlogging can also be caused by high sodicity levels in soils, which limit the infiltration of surface water as a result of compaction and crusting (Ghassemi et al. 1995; Qureshi and Barrett-Lennard 1998). Soil sodicity occurs when clay particles in the soil become dominantly charged with sodium (Na) ions, rather than calcium (Ca) ions (Hillel 2000; Qureshi and Barrett-Lennard 1998).

It has been estimated that 18% of South Africa's irrigated land is either salt-affected or water-logged (Backeberg et al. 1996). Although this proportion is relatively small compared to the proportion of salt-affected areas in other countries – for example, Argentina 34%, Egypt 33%, Iran 30%, Pakistan 26% and the United States 23% (Ghassemi et al. 1995) – one should bear in mind that only 13.7% of South Africa's land area is suitable for irrigation (Department of Agriculture, Forestry and Fisheries 2012). Irrigation is critical for food security, specifically, given that large parts of the country receive insufficient rainfall for dryland crop production (Backeberg 2003; Backeberg et al. 1996). Therefore, proactive measures need to be taken to prevent further loss of this limited resource. However, conventional methods for monitoring salt accumulation within irrigation schemes involve regular field visits to collect soil samples followed by laboratory analyses. Many field visits are often required to monitor large irrigation schemes effectively. The effort and cost involved in monitoring the status of salt accumulation thus inhibit regular and comprehensive assessments. In fact, most assessments are limited to specific irrigation schemes and most are outdated. There is consequently little information available about the extent of secondary salt accumulation in South Africa.

Remote sensing has been proposed as a less time-consuming and more cost-effective method for monitoring salt accumulation, as satellite images cover large areas on a regular, timely basis (Abbas et al. 2013). Earth observation can capture data on both a temporal and spatial scale and, if successfully applied, can serve as a monitoring solution that would not be as costly and labor intensive as regular field surveys and laboratory analyses.

Various techniques and applications of remote sensing for the identification of salt-affected areas have been published. Salt accumulation can be detected either directly or indirectly using remotely sensed images (Bastiaanssen et al. 1998; Mougenot et al. 1993). The direct detection of the accumulation of salts involves identifying salt encrustations on the bare ground, while the indirect method focuses mainly on vegetation responses to salt accumulation.

Several authors have successfully applied the direct approach for identifying salt accumulation (Abood et al. 2011; Dwivedi and Sreenivas 1998; Elnaggar and Noller 2010; Iqbal 2011; Khan et al. 2005; Metternicht and Zinck 2003; Rao et al. 1995; Setia et al. 2013; Sidike et al. 2014). Most of these studies reported that salt crusts generally have high reflectance in the visible and near- to mid-infrared regions of the electromagnetic spectrum, depending on the chemical composition of the salts (Metternicht and Zinck 2003). Specific spectral ranges for direct salinity detection (identified by laboratory analyses) are in the visible region (550–770 nm), near-infrared region (900–1,030 nm, 1,270–1,520 nm) and middle-infrared region (1,940–2,150 nm, 2,150–2,310 nm, 2,330–2,400 nm) (Csillag et al. 1993). Metternicht and Zinck (2009) observed that applying laboratory techniques to optical remote sensing data to detect salt accumulation is complicated by the variations in reflectance that cannot be attributed to a single soil property and salt type. Spectral variation in salt crusts can be attributed to the difference in the quantity of salts, the mineralogy of the salts (e.g., carbonates, sulphates or chlorides), soil moisture content, color of the salt crust (white to dark) and surface roughness of the salt crust (smooth to rough), which all vary among chemical structures. The direct approach also fails to take into account salt accumulation that occurs in the subsurface, since it is limited to monitoring surface conditions.

The indirect approach, which focuses mainly on vegetation monitoring, has also been successfully applied to remotely identify salt-affected areas. One approach is to identify halophytic vegetation types that commonly occur in salt-affected areas (Dehaan and Taylor 2002, 2003; Dutkiewicz et al. 2009; García Rodríguez et al. 2007). However, this method is less suitable for application in irrigated areas where natural vegetation is removed during field preparations. A more common approach in irrigated areas is to monitor vegetation (crop) vigor. For this purpose, vegetation indices (VIs) are primarily used to distinguish between healthy vegetation and stressed vegetation. Two of the most used VIs are the Normalized Difference Vegetation Index (NDVI) (Abbas et al. 2013; Aldakheel 2011; Bouaziz et al. 2011; Elnaggar and Noller 2010; Hamzeh et al. 2012a; Platonov et al. 2013; Sidike et al. 2014; Zhang et al. 2011) and the Soil-Adjusted Vegetation Index (SAVI) (Abood et al. 2011; Allbed et al. 2014; Bouaziz et al. 2011; Elnaggar and Noller 2010; Hamzeh et al. 2012b; Koshal 2010; Zhang et al. 2011). However, using vegetation response for identifying salinity conditions should be viewed with caution because many factors besides soil salinity (e.g., farming practices) can contribute to loss of vegetation vigor (Metternicht and Zinck 2009). Different crops in different growing phases also have different tolerances to soil salinity, which further complicates the implementation of the indirect approach.

Despite the advances in the use of remote sensing for salt accumulation detection, its use in South Africa has been limited, mainly because salt-affected areas in the region are relatively small in extent (often as narrow as 1–2 m) (Figure 6.1). Most of the examples described above were carried out in regions where salt-affected areas are large in extent, which allowed for the use of high-resolution (e.g., 15–250 m) imagery such as those captured by the ASTER, LANDSAT and MODIS satellites. Such imagery will have little value within a South African context. However, the increasing availability of very high resolution (VHR) satellite imagery offers new possibilities for detecting small patches of saline soils and provides a solution to quantifying the extent of salt accumulation in South African irrigation schemes. Multitemporal analyses of such imagery can also be carried out to gain a better understanding of the dynamic nature of salt accumulation. In addition, early detection of affected areas (i.e., while they are still small) will facilitate the implementation of mitigating measures and may prevent losses of fertile agricultural land.

This chapter overviews a new remote sensing technique – called within-field anomaly detection (WFAD) – that was specifically developed for regions such as South Africa, where the extent of salt-affected areas is small. The WFAD technique was implemented in nine irrigation schemes throughout South Africa to determine (quantify) the status of salt accumulation at national scale. The results of these estimations are presented and discussed within the context of setting up a system whereby salt accumulation monitoring can be operationalized.

FIGURE 6.1   (See color insert.) Examples of typical salt-affected patches in the Vaalharts irrigation scheme.

6.2 SELECTED IRRIGATION SCHEMES

To represent the true variability of South African irrigation schemes, nine schemes across the country, namely Vaalharts, Loskop, Makhatini, Olifants River, Tugela River, Breede River, Sundays River, Pondrift and Douglas, were selected for analysis. Figure 6.2 shows the location of each scheme while Table 6.1 provides a summary of key characteristics for each area. A short description of each scheme follows.

6.2.1 Vaalharts Irrigation Scheme

The Vaalharts irrigation scheme is situated on the borders of the Northern Cape, North West province and Free State province, near the towns of Jan Kempdorp, Hartswater and Pampierstad. Fields considered in the Vaalharts cover 26,384 ha and the irrigation scheme is one of the largest in South Africa (Van Rensburg et al. 2012). The area borders two plateaus on the east and west of the Harts river valley. The valley slopes towards the south with very little topographical change due to the low gradient of the non-incising Harts River (Gombar and Erasmus 1976; Liebenberg 1977). Vaalharts is known for its sandy soils, which are prone to waterlogging and salt accumulation. This is exacerbated by insufficient natural drainage and soil compaction (Maisela 2007). Typically the soils consist of 8% clay, 2% silt, 68% fine sand and 22% medium and coarse sand (Streutker 1997). The scheme is located at an altitude of 1,175 m above sea level and is known for its cold winters and long warm summers (Maisela 2007). The area receives a mean annual rainfall of 400 mm and has a mean annual temperature of 19°C (Schulze 2006).

   Due to its semi-arid climate, irrigation is required for crop production (Van Rensburg et al. 2012). Vaalharts receives its irrigation water from the Vaal and Harts rivers. Approximately 70% of the irrigation infrastructure comprises of flood irrigation, while pivot irrigation contributes to the

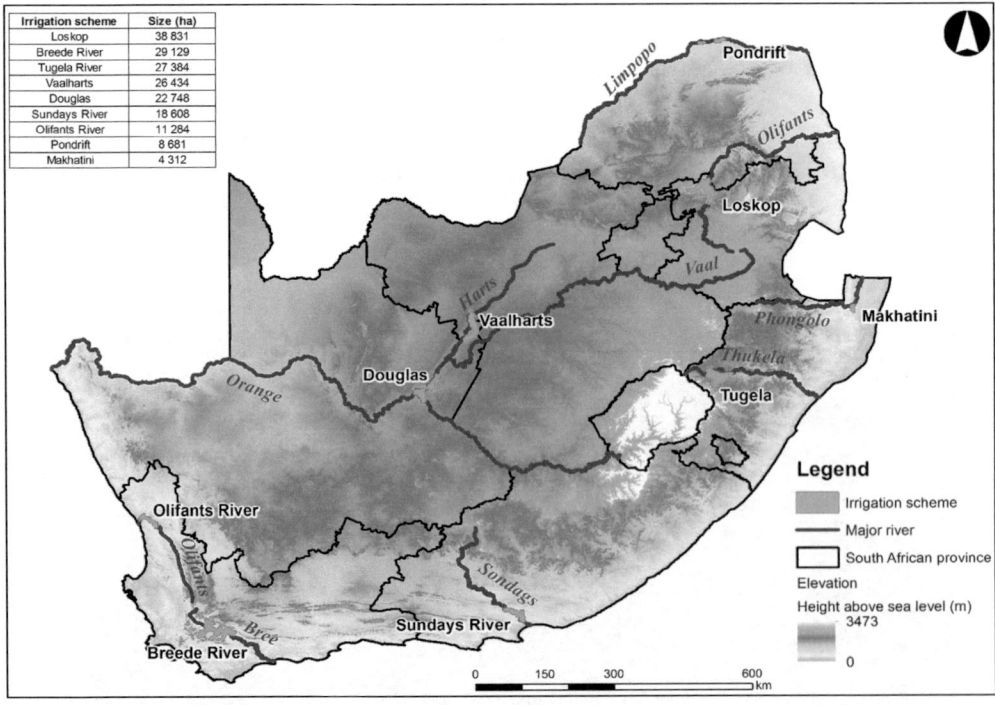

Irrigation scheme	Size (ha)
Loskop	38 831
Breede River	29 129
Tugela River	27 384
Vaalharts	26 434
Douglas	22 748
Sundays River	18 608
Olifants River	11 284
Pondrift	8 681
Makhatini	4 312

FIGURE 6.2 (See color insert.) Geographical distribution of the nine irrigation schemes across South Africa, with elevation data as backdrop.

TABLE 6.1

Summary of Descriptive Characteristics of Selected Irrigation Schemes

Irrigation scheme	Total area (ha)	Mean annual temperature (°C) (Schulze 2006)	Mean annual rainfall (mm/year) (Schulze 2006)	Mean elevation (MAMSL) (Farr et al. 2007)	Mean slope (degrees) (Farr et al. 2007)	WRB soil classification (top 3) (Hengl et al. 2017)	Surface water quality (TDS concentrations) (mg/l) (Bailey and Pitman 2015)
Vaalharts	26,384	19	400	1,175	0.67	Haplic Luvisols – 63% Haplic Lixisols – 27% Calcic Luvisols – 5%	1,000–1,500 – 93% 0–500 – 7%
Breede River	29,129	17	290	233	2.64	Haplic Luvisols – 74% Haplic Cambisols – 18% Haplic Fluvisols – 4%	>2,000 – 46% 0–500 – 40% 1,500–2,000 – 14%
Loskop	38,831	20	552	916	1.47	Haplic Luvisols – 93% Haplic Lixisols – 3% Calcic Luvisols – 1%	0–500 – 60% 1,500–2,000 – 27% 500–1,000 – 10%
Makhatini	4,312	22	577	92	0.86	Luvic Phaeozems – 30% Haplic Lixisols – 22% Albic Arenosols – 18%	0–500 – 75% 15,00–2,000 – 25%
Tugela	27,384	18	745	1,104	2.67	Haplic Acrisols – 71% Haplic Lixisols – 26% Haplic Luvisols – 3%	0–500 – 100%
Olifants River	11,284	18.5	134	31	2.07	Haplic Cambisols – 66% Haplic Arenosols – 13% Albic Arenosols – 6%	0–500 – 57% >2000 – 42%
Sundays River	18,608	18.5	380	83	1.87	Haplic Luvisols – 85% Haplic Luvisols (Ferric) – 6% Haplic Cambisols – 3%	500–1,000 – 78% 0–500 – 22%
Pondrift	8,681	22	314	506	0.62	Haplic Luvisols – 95% Calcic Luvisols – 2% Haplic Arenosols – 0.4%	0–500 – 100%
Douglas	22,748	19	293	1,014	0.98	Haplic Cambisols – 83% Haplic Arenosols – 7% Haplic Luvisols – 5%	500–1,000 – 51% 0–500 – 35% 150,00–2,000 – 14%

Note: MAMSL = metres above mean sea level, WRB = world reference base and TDS = total dissolved solids.

remainder (Maisela 2007). The principal crops planted are maize (*Zea mays*), wheat (*Triticum aestivum*), barley (*Hordeum vulgare*), lucerne (*Medicago sativa*), groundnuts (*Arachis hypogaea*) and, increasingly, pecan nuts (*Carya illinoinensis*) (Kruger et al. 2009; Barnard 2013).

6.2.2   BREEDE RIVER IRRIGATION SCHEME

The Breede River irrigation scheme is mainly sourced from the Breede River, which originates in the Ceres valley, ±100 km northwest of Cape Town, and flows 320 km in a southeasterly direction, where it reaches the Indian Ocean at Witsand (Kirchner 1995). The river drains a catchment of about 1.26 million ha. Fields covering 29,129 ha in the middle parts of the greater catchment, between Worcester and Robertson, was the focus in this catchment. It has a gentle, hilly relief defined by high mountain ranges to the north (Langeberg mountain) and south (Riviersonderend mountain) parallel to the river. These mountain ranges are formed by resistant quartzite and sandstone (Ghassemi et al. 1995), while the valleys are mainly underlined with shale and siltstone from the Malmesbury and Bokkeveld basements (Beuster et al. 2003). The focus area is located at a mean altitude of 233 m above sea level and has a Mediterranean climate with dry, hot summers and moderately warm, wet winters (Flügel and Kienzle 1989). It receives an annual rainfall of 290 mm, mainly during the winter rainfall season (May to October), and is thus classified as semi-arid (Schulze 2006). The annual mean temperature for the area is 17°C (Schulze 2006). The majority of the farm irrigation infrastructure is made up of drip and microjet systems (Beuster et al. 2003). The crop mix in the Breede River irrigation scheme is less diverse compared to the Vaalharts scheme, with wine grape (*Vitis*) variations being the primary (65%) crop. Other crops include peaches (*Prunus persica*) and apricots (*Prunus armeniaca*) (13%), vegetables, mainly tomatoes (*Solanum lycopersicum*) (3%) and irrigated pastures (7%) (Moolman et al. 1999).

6.2.3   LOSKOP IRRIGATION SCHEME

The Loskop irrigation scheme is located in the Limpopo province, intersecting the towns of Marblehall and Goblersdal. The scheme, with 38,831 ha fields, forms part of the Olifants river basin, where salinization has been identified as the main agricultural pollution problem (Aihoon et al. 1997). The area is situated at 916 m above sea level and the geography ranges from mountainous bushveld to undulating terrain with thorn trees (Tren and Schur 2000). The region comprises a mix of the central sandy bushveld region and the Springbokvlakte and Loskop thornveld regions (Mucina and Rutherford 2006). Loskop has a mean annual temperature of 20°C and a mean annual rainfall of 552 mm (Schulze 2006). The main crops grown in the scheme are citrus (*Citrus*), table grapes, maize, wheat, soya bean (*Glycine max*), cotton (*Gossypium*), tobacco (*Nicotiana*) and groundnuts.

6.2.4   MAKHATINI IRRIGATION SCHEME

The Makhatini irrigation scheme is situated in the KwaZulu-Natal (KZN) province, east of the town Jozini and south of the Phongolo River. The scheme is situated 92 m above sea level in the Makhatini and western Maputaland clay bushveld region (Mucina and Rutherford 2006), has a humid subtropical climate with a mean annual temperature of 22°C and a mean annual rainfall of 577 mm (Schulze 2006). The main crop is sugar cane (*Saccharum*), while cotton and a variety of vegetables are also grown on the 4,312 ha represented fields.

6.2.5   TUGELA RIVER IRRIGATION SCHEME

Located 1,104 m above sea level in the KZN province, the Tugela River irrigation scheme includes the towns Bergville and Winterton and has a total of 27,384 ha fields represented in the study. Drained by the upper part of the Tugela River, the area falls within the moist grassland bioclimatic region of KZN (Mucina and Rutherford 2006). Over the past 30 years, the scheme has changed from furrow and overhead sprinkler irrigation to mainly center pivot irrigation systems. The scheme has a mean annual temperature of 18°C and a mean annual rainfall of 745 mm (Schulze 2006). Crops grown in this area are mostly commodity and industrial crops, mainly maize and winter wheat with soybeans as a rotational crop (Phipson 2012).

6.2.6   Olifants River Irrigation Scheme

The Olifants River irrigation scheme is located in the Western Cape province and includes the towns of Lutzville and Vredendal. Irrigation water is provided by the Olifants River, which has the second largest catchment in South Africa. Bordering the Atlantic Ocean, the scheme has a mean elevation of 31 m above sea level and consists mainly of the Namaqualand Riviera region (Mucina and Rutherford 2006). The mean annual temperature of the area is 18.5°C, while the mean annual rainfall is 134 mm (Schulze 2006). Crops in this region are mostly grapes and citrus of which 11,284 ha are represented.

6.2.7   Sundays River Irrigation Scheme

The Sundays River irrigation scheme is located at 83 m above sea level at the foot of the Eastern Cape, bordering the Indian Ocean. Major settlements in the vicinity of the irrigation scheme are Addo and Kirkwood. The scheme is situated predominantly between two series of alluvial terraces, and according to Mucina & Rutherford (2006), is mostly made up of part of the Albany alluvial vegetation region. Sundays River has a mean annual temperature of 18.5°C and a mean annual rainfall of 380 mm (Schulze 2006). Citrus is the dominant crop in this irrigation scheme and fields that are included in this study cover an area of 18,608 ha.

6.2.8   Pondrift Irrigation Scheme

The Pondrift irrigation scheme receives its water supply from the Limpopo River, which forms the boundary between the Limpopo province, Zimbabwe and Botswana. The scheme, located in the northern-most water management area, near the town of Pondrift, has a mean height of 506 m above sea level. The area consists of alluvial deposits from the Quaternary system and forms part of the subtropical, alluvial vegetation region (Mucina and Rutherford 2006). The mean annual temperature in the Pondrift irrigation scheme is 22°C, and the mean annual rainfall is 314 mm (Schulze 2006). Cotton, grain sorghum (*Sorghum bicolor*) and tobacco are the main crops in this region, which only has 8,681 ha fields represented.

6.2.9   Douglas Irrigation Scheme

The Douglas irrigation scheme receives water from the Orange River and the Vaal River, and is located in the eastern part of the Northern Cape near the towns of Douglas and Salt Lake. The irrigation scheme is located 1,014 m above sea level with 22,748 ha of fields. The area has a mean annual temperature of 19°C and a mean annual rainfall of 293 mm (Schulze 2006). According to Armour (2002), 28% of the area is flood-irrigated while 70% is sprinkler-irrigated. The trend is, however, towards the conversion to center pivots. The region comprises a combination of the upper Gariep alluvial vegetation and the Kimberly thornveld regions (Mucina and Rutherford 2006). The main crops grown in this irrigation scheme are wheat, maize, lucerne, potatoes (*Solanum tuberosum*), cotton and groundnuts (Armour 2002).

6.3   WITHIN-FIELD ANOMALY DETECTION (WFAD)

Vegetation response, as an indirect indicator of salt accumulation, is ambiguous because of varying tolerances of plants to saline or waterlogged conditions. This is exacerbated by crops being in different growing phases, as it confounds inter-field response comparisons (Zhang et al. 2011; Hanson et al. 2006). Direct indicators are similarly ambiguous as they can vary from white salt crusts to greasy black surfaces and surface ponding (McGhie and Ryan 2005). To effectively quantify salt accumulation and waterlogging at a scheme level, each of these indicators needs to be considered. However, developing a methodology to identify each unique indicator separately will result in a highly complex system. Such an approach will likely not be robust as the indicators may vary between regions.

The indicators of salt accumulation and waterlogging share a common characteristic that can be exploited for their detection and delimitation. If it can be assumed that a salt accumulation or waterlogging indicator only affects a part of a field (i.e., it is spatially smaller than the entire field in which it occurs), each indicator will appear as an area of abnormality (anomaly) in an otherwise spectrally homogenous field. When such an anomaly is compared to the average spectral response of the field in which it occurs, it becomes more apparent. Conceptually, this approach – called WFAD – would be ideal for the characteristically small and patchy nature of salt-affected and waterlogged areas in South African irrigation schemes and would be applicable to both fallow and vegetated fields. Potentially, the WFAD should also accommodate the limitations posed by the spectral ambiguity of direct and indirect indicators of salt accumulation and waterlogging. For example, using this approach in a wheat field that has a relatively high root zone salinity tolerance, the anomaly would be compared to the wheat field itself and not to a more sensitive maize field.

6.4   IMAGE AND FIELD BOUNDARY DATA

According to Lobell et al. (2010), the risk of producing poor results is high when single date remotely sensed imagery is used to identify salinity. When using multitemporal remote sensing data, certain characteristic trends become more apparent, and if poor vegetation conditions persist throughout multiple growing seasons, they are more likely to be caused by salt accumulation (Furby et al. 1995; Lenney et al. 1996). Two-year multitemporal SPOT-5 scenes covering most of the nine study areas were consequently acquired from the South African National Space Agency (SANSA). In some areas, multiple scenes were required as a single SPOT-5 scene did not cover the full extent of the study area (Table 6.2). With a relatively high temporal resolution and the capability of pan-fusing multispectral bands to 2.5 m, SPOT-5 imagery provided a cost-effective alternative to other VHR alternatives such as WorldView-2, QuickBird or IKONOS imagery.

Geometric and radiometric corrections of all images were done using the software package PCI Geomatica (v. 2013 SP2), while the nearest neighbor method was used to do all necessary resampling during preprocessing in order to preserve the original digital numbers (DN) (Campbell 2007; Lillesand et al. 2004). A north-oriented implementation of the Gauss conform coordinate system (also known as the LO coordinate system), with the central meridian adjusted for each scheme, was used. The mathematical model ATCOR-2 was employed to convert the DNs into percentage reflectance, while the Pansharp algorithm was used to increase the resolution of all the 10 m and 20 m multispectral bands to 2.5 m.

To compare each field to itself, detailed field boundaries were needed for each study area. Field boundaries in GIS vector format were obtained from the Department of Agriculture, Forestry and Fisheries (DAFF) and refined and improved for each area using manual editing.

6.5   FIELD SURVEYS

Field surveys were conducted from June 2012 to October 2014 to collect suitable reference data. Accessibility was often restricted by canal systems and fencing, but an attempt was made to include sites that represented great variation in terms of salt accumulation, waterlogging and unaffected areas. Surveyed points varied from being clustered to being randomly spaced across each scheme.

TABLE 6.2
SPOT-5 Scenes Acquired for the Study Areas

Irrigation scheme	Scene date
Vaalharts	27 April 2012
	20 February 2011
Loskop	30 September 2012
	3 June 2012
	18 May 2012
	15 June 2011
	5 May 2011
	17 August 2011
Makhatini	26 July 2012
	2 August 2011
Tugela	22 March 2012
	1 August 2011
Olifants	26 January 2013
	10 April 2012
Breede	16 January 2013
	26 February 2013
	26 March 2012
	11 December 2012
Sundays	18 February 2013
	12 April 2012
Pondrift	23 February 2013
	12 March 2012
Douglas	13 August 2013
	16 April 2013

The location of the surveyed points was guided by an initial analysis of each area and expert knowledge of the scheme (i.e., areas known to be affected).

Soil samples were collected by means of a soil auger and analyzed for their electroconductivity (EC) value using laboratory analyses (saturated paste technique). Soil samples with EC values of 4.0 dS/m or higher were regarded as being salt-affected. Some waterlogging was caused by shallow or rising water tables and does not necessarily result in surface ponding (Dwivedi et al. 1999). Waterlogging status in such areas can only be established by analyzing the vertical soil profile, therefore this analysis was carried out up until 1 m below the soil surface at each sample site. Saturation of the top soil was determined *in situ* to identify waterlogging status. All survey points were accompanied by a GPS coordinate, notes on the visual appearance of the immediate area, and in some cases, a photograph was also recorded.

6.6 OBJECT-BASED IMAGE ANALYSES

Sophisticated and well-established pixel-based techniques that successfully classify lower resolution images exist (Blaschke et al. 2014). However, high resolution images often lead to a high within-class spectral variability that can decrease the accuracy of pixel-based approaches (Hay et al. 1996). In this study, the extent of salt-affected and waterlogged areas is greater than that of the image pixels of the pan-fused SPOT-5 imagery (2.5 m), making it susceptible to within-class variabilities. Geographical object-based image analysis (GEOBIA) mimics higher order logic, similar to human interpretation, for identifying useful shapes, sizes and textures from image data (Campbell 2007).

This new paradigm in remote sensing emerged from the realization that pixels are no longer the optimal spatial unit for mapping landscape elements (Addink et al. 2012; Blaschke 2010). GEOBIA starts with a segmentation process that divides a high resolution image into segments of spatially continuous and homogenous regions, called objects. Image segmentation algorithms are commonly based on one of the two basic properties of the pixel's grey level values – discontinuity and similarity. The discontinuity (abrupt changes in pixel levels) principle is used to partition the image into several non-overlapping objects, while the similarity principle is used for methods like region growing and merging (Addink et al. 2012; Haralick and Shapiro 1985). Objects hold key advantages over pixels, in particular, an increase in the number of available spectral variables (e.g., mean, median, maximum, minimum and variance) and spatial features (e.g., distances, neighborhood, topologies and hierarchical properties) (Blaschke and Strobl 2001; Flanders et al. 2003; Hay and Castilla 2008). The use of spatial information in object-based methods also allows for the integration of vector and raster data, enabling a GIS-like functionality for classification (Blaschke et al. 2014).

The WFAD technique was implemented in eCognition Developer 8 and 9. The first step in the WFAD process is to perform image segmentation. The implementation of the segmentation for each of the nine irrigation schemes was based on three SPOT-5 spectral bands, as they have higher spatial resolutions (10 m) compared to the SWIR band (20 m). Segmentation was done on each SPOT-5 image for each year. The scale factor of the multi-resolution segmentation (MRS) algorithm, which determines the size of the objects, was adjusted individually (using visual interpretation) to produce meaningful objects that best represent the levels of homogeneity within a field for each irrigation scheme.

Hierarchical segmentation, one of the spatial features of GEOBIA, enables more than one level of segmentation sharing inherent properties (Campbell 2007). For each individual scheme, the vector layer delineating the fields was used for the first (parent) segmentation level. Below the parent-level objects, MRS was used to populate the parent objects with smaller child-level objects (Figure 6.3). Each child object inherits all the properties (e.g., mean, median, mean ratio and variance per band) of its relative parent object.

The second step in the WFAD process was to classify the image objects. Although a wide range of classification approaches and procedures to assign classes to the objects is available, a rule-based

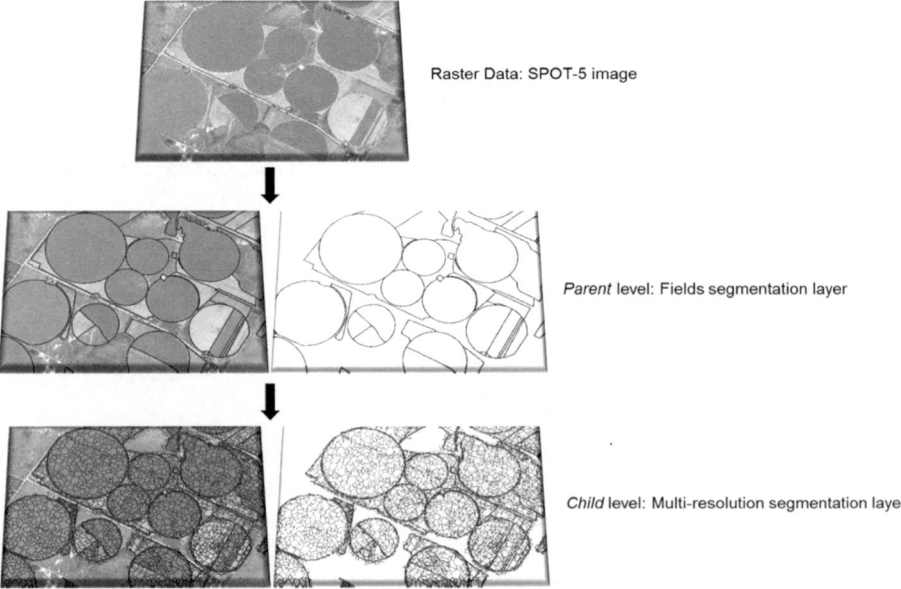

Raster Data: SPOT-5 image

Parent level: Fields segmentation layer

Child level: Multi-resolution segmentation layer

FIGURE 6.3   Hierarchical WFAD segmentation process.

(expert system) classification approach was preferred as it has the ability to accommodate observable differences and changes within the data. A rule-set approach also does not require training data and the rules can progressively be applied and refined while maintaining full control of the classification process (Lucas et al. 2007).

Fields of each irrigation scheme were first classified as *vegetated* or *bare* at the parent level using the NDVI. Second, for the identification of anomalies, the spectral response of each child object was compared to the average spectral response of its relative parent object. If a substantial difference occurred between a child object and the relative parent object, the child object was identified as an anomaly. This mean difference (MeaD) process is illustrated in Figure 6.4.

A MeaD equation was used to identify threshold values that highlight anomalies with a substantially different spectral response. The equation is described as:

$$MeaD = SR_{Child\ object} - SR_{Relative\ Parent\ object} \qquad (6.1)$$

where

$SR_{Child\ object}$ is the mean spectral response of a child object; and

$SR_{Relative\ Parent\ object}$ is the mean spectral response of the relative parent object.

A positive MeaD threshold identifies a child object with a substantially higher spectral response compared to the relative parent object, while a negative MeaD threshold identifies a child object with a substantially lower spectral response compared to the relative parent object.

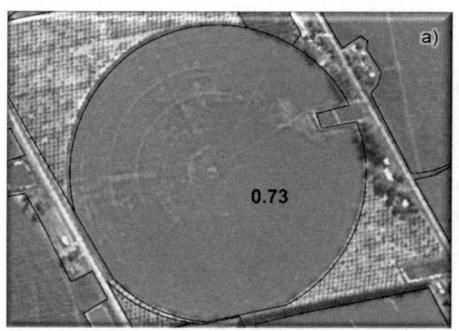

Mean *NDVI* response for the whole field

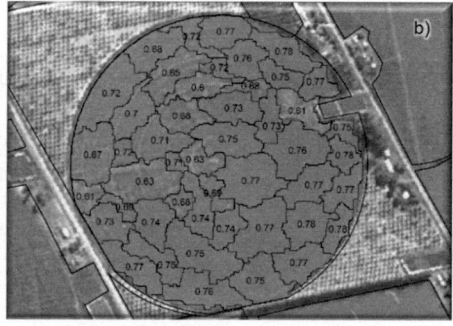

Mean *NDVI* response for each object

(*Child* layer)

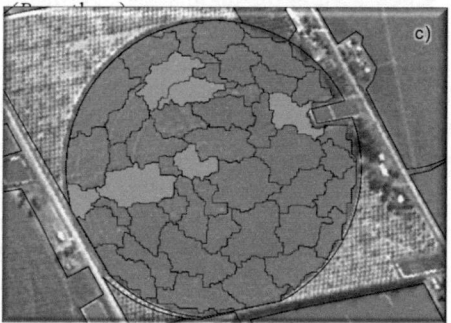

Objects identified if mean difference threshold of -0.07

is applied

FIGURE 6.4 Mean difference threshold process. (a) Parent-level segmentation with mean NDVI value; (b) child-level segmentation with individual NDVI values; (c) identified anomalies with MeaD equation.

The spectral response used for comparing the child and parent objects differed for vegetated (indirect indicators) and fallow (direct indicators) fields to accommodate the different indicators for soil salinity and waterlogging.

The NDVI was used to identify anomalies in vegetated fields (Figure 6.4). Given that the main indicator of salt accumulation and waterlogging in vegetated fields is physiological stress (Wiegand et al. 1994; Abood et al. 2011; Lenney et al. 1996; Koshal 2010; Fernández-Buces et al. 2006; Lobell et al. 2010; Peñuelas et al. 1997; Zhang et al. 2011), only a negative MeaD threshold was implemented. Although halophytic vegetation can occur within vegetated fields, it was assumed that its vegetation response would be less than that of a commercially grown crop.

For fallow fields, the NDVI and a brightness band ratio (Br) (Equation 6.2) were used. A positive MeaD threshold was implemented for the NDVI on bare fields to identify potential halophytic plants (Dehaan and Taylor 2002, 2003; García Rodríguez et al. 2007; Dutkiewicz et al. 2009; Elnaggar and Noller 2010). To detect salt encrustations, with generally high reflectance values in the visible and near-infrared regions, a positive MeaD threshold was used for the Br (Rao et al. 1995; Elnaggar and Noller 2010; Metternicht and Zinck 2003; Abood et al. 2011; Iqbal 2011; Khan et al. 2005; Setia et al. 2013; Sidike et al. 2014; Dwivedi and Sreenivas 1998). A negative MeaD threshold was used to accommodate the generally low reflectance values associated with waterlogging (ponding) (Dwivedi and Sreenivas 1998). The Br (brightness) can be defined as:

$$Br = (G + R + N)/3 \tag{6.2}$$

where

    G is the reflectance in the green band;

    R is the reflectance in the red band; and

    N is the reflectance in the near-infrared band.

Because of its lower (20 m) resolution, the SWIR band was not included in the brightness ratio. The positive and negative MeaD threshold values were manually determined for each SPOT-5 image during analysis. To improve the automation of WFAD, further research should be directed towards achieving a higher level of image standardization in order to use single threshold values for all areas.

6.7 MULTITEMPORAL ANALYSIS AND QUANTIFICATION

A multitemporal analysis was carried out for each scheme using the anomalies identified for both years. ArcGIS software was used for this purpose. Anomalies occurring in both years were exclusively considered as potential salt-affected or waterlogged areas. The multitemporal analysis would potentially eliminate anomalies caused by factors unrelated to salt accumulation or waterlogging (Lobell et al. 2010; Furby et al. 1995; Lenney et al. 1996).

An anomaly ratio (Ar), describing the relationship between anomalies caused by salt accumulation and waterlogging and other non-related anomalies, was introduced to account for the likelihood of failing to completely eliminate all anomalies unrelated to salt accumulation or waterlogging. The Ar (Equation 6.3) was calculated using categories (A_{sw} and A_o) derived from the ground surveyed data of each scheme.

$$Ar = A_{sw}/(A_{sw} + A_o) \tag{6.3}$$

where

    A_{sw} represents the anomalies related to salt accumulation or waterlogging; and

    A_o represents anomalies related to other factors.

Final estimations of affected areas were calculated by using the *Ar* to alter the total percentage covered by the anomalies (Equation 6.4). The most recent extents of the anomalies were used for area calculations.

$$\text{Quantification}(\%) = \left(\left(\text{Area}_{\text{anomalies}} / \text{Area}_{\text{total}} \right) \times 100 \right) \times \text{Ar} \tag{6.4}$$

where

Area$_{\text{anomalies}}$ represents the total area of anomalies in hectares; and

Area$_{\text{total}}$ represents the total area of fields in the relative irrigation scheme in hectares.

Because no training data was needed for the rule-based classification, the field surveyed points were exclusively used for testing the accuracy of the WFAD method. An accuracy assessment for each scheme was carried out to produce error matrices and the final quantification percentages.

Error matrices (also called confusion matrices) show the accuracies achieved by the WFAD method for each irrigation scheme (Tables 6.3 to 6.11). The field verification columns represent three categories, namely *Salt-affected/Waterlogged*, *Stressed* and *Unaffected* areas. Anomalies that appear to be caused by salt accumulation and waterlogging (e.g., indicators such as plant stress and the occurrence of halophytic vegetation), but were found to show no sign thereof during the field surveys, are represented by the Stressed category. This category essentially represents anomalies unrelated to salt accumulation and waterlogging that the multitemporal imagery was unable to eliminate. These observations were used to calculate *Ar*. Anomalies that were found to have been

TABLE 6.3
Vaalharts Irrigation Scheme Error Matrix

		Field verification data				
		Affected	Stressed	Unaffected	Total	UA (%)
Predicted data	Anomaly1	**18**	0	6	24	81.3
	Anomaly2	0	**8**	0	8	
	Not Anomaly	11	3	**24**	38	63.2
	Totals	29	11	30	**70**	
	PA (%)	62.1	72.73	80		
	Overall accuracy	**71.64**				
	Kappa	**0.58**				

TABLE 6.4
Loskop Irrigation Scheme Error Matrix

		Field verification data				
		Affected	Stressed	Unaffected	Total	UA (%)
Predicted data	Anomaly1	**35**	0	5	40	90.7
	Anomaly2	0	**14**	0	14	
	Not Anomaly	8	3	**25**	36	69.5
	Totals	43	17	30	**90**	
	PA (%)	81.4	82.4	83.3		
	Overall accuracy	**82.20**				
	Kappa	**0.72**				

TABLE 6.5

Makhatini Irrigation Scheme Error Matrix

		Field verification data				
		Affected	Stressed	Unaffected	Total	UA (%)
Predicted data	Anomaly1	25	0	6	31	86.04
	Anomaly2	0	12	0	12	
	Not Anomaly	5	0	8	13	61.54
	Totals	30	12	14	56	
	PA (%)	83.3	100	57.14		
	Overall accuracy	80.4				
	Kappa	0.672				

TABLE 6.6

Tugela Irrigation Scheme Error Matrix

		Field verification data				
		Affected	Stressed	Unaffected	Total	UA (%)
Predicted data	Anomaly1	43	0	10	53	85.3
	Anomaly2	0	15	0	15	
	Not Anomaly	5	3	7	15	46.67
	Totals	48	18	17	83	
	PA (%)	89.58	83.33	41.18		
	Overall accuracy	78.3				
	Kappa	0.6				

TABLE 6.7

Olifants River Irrigation Scheme Error Matrix

		Field verification data				
		Affected	Stressed	Unaffected	Total	UA (%)
Predicted data	Anomaly1	34	0	13	47	79.7
	Anomaly2	0	17	0	17	
	Not Anomaly	6	0	12	18	66.67
	Totals	40	17	25	82	
	PA (%)	85	100	48		
	Overall accuracy	76.8				
	Kappa	0.6				

caused by salt accumulation and/or waterlogging were arbitrarily labelled "Anomaly1" in the error matrices, while the predicted observations that were verified to be caused by other factors were labelled as "Anomaly2." Given that the WFAD method does not differentiate between anomalies caused by salt accumulation and/or waterlogging, the user's accuracy (UA) percentages for the two types of anomalies were combined for accuracy assessment purposes. Accurate delineation of salt-affected or waterlogged areas is vital for quantification purposes. Field survey points were used to

TABLE 6.8
Breede River Irrigation Scheme Error Matrix

		Field verification data				
		Affected	**Stressed**	**Unaffected**	**Total**	**UA (%)**
Predicted Data	Anomaly1	**26**	0	5	31	85.7
	Anomaly2	0	**4**	0	4	
	Not Anomaly	8	7	**35**	50	70
	Totals	34	11	40	**85**	
	PA (%)	76.47	36.36	87.5		
	Overall accuracy	**76.5**				
	Kappa	**0.58**				

TABLE 6.9
Sundays River Irrigation Scheme Error Matrix

		Field verification data				
		Affected	**Stressed**	**Unaffected**	**Total**	**UA (%)**
Predicted data	Anomaly1	**15**	0	3	18	84.2
	Anomaly2	0	**1**	0	1	
	Not Anomaly	7	0	**25**	32	78.13
	Totals	22	1	28	**51**	
	PA (%)	68.18	100	89.29		
	Overall accuracy	**80.4**				
	Kappa	**0.6**				

TABLE 6.10
Pondrift Irrigation Scheme Error Matrix

		Field verification data				
		Affected	**Stressed**	**Unaffected**	**Total**	**UA (%)**
Predicted data	Anomaly1	**14**	0	11	25	65.6
	Anomaly2	0	**7**	0	7	
	Not Anomaly	17	2	**26**	45	57.78
	Totals	31	9	37	**77**	
	PA (%)	45.16	77.78	70.27		
	Overall accuracy	**61.0**				
	Kappa	**0.326**				

visually verify the accuracy of the delineated areas (Figure 6.5). Hanson et al. (2006) showed that crop types have a unique rate at which relative yield is reduced relative to soil salinity increases. The transition between affected and unaffected areas will thus vary among different crop types and can have a significant effect on the delineated extent of an anomaly. Figures 6.5a and b indicate a highly accurate delineation of the extent of the affected areas, while Figures 6.5c and d show some inconsistencies.

TABLE 6.11

Douglas Irrigation Scheme Error Matrix

		Field verification data				
		Affected	Stressed	Unaffected	Total	UA (%)
Predicted data	Anomaly1	**21**	0	9	30	72.7
	Anomaly2	0	**3**	0	3	
	Not Anomaly	13	0	**20**	33	60.61
	Totals	34	3	29	**66**	
	PA (%)	61.76	100	68.97		
	Overall accuracy	**66.7**				
	Kappa	**0.387**				

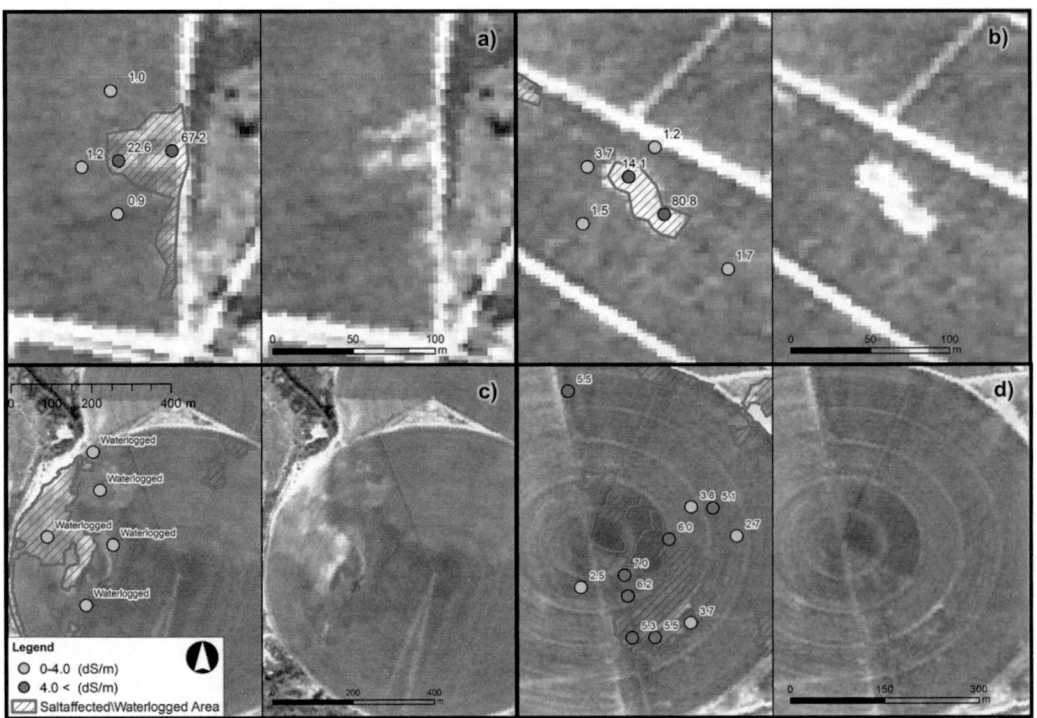

FIGURE 6.5 Indication of the extent of anomaly delineation for four areas, with (a) and (b) representing examples of good delineations, and (c) and (d) highlighting some inconsistencies.

The mean overall accuracy (OA) for the nine schemes is 75% with a mean kappa of 0.6. The error matrices (Tables 6.3 to 6.11) aided in the interpretation of the salt-affected and waterlogging quantification percentage for each scheme in Table 6.12.

6.8   DISCUSSION

The WFAD achieved a mean accuracy of 75% across the nine study areas, demonstrating its robustness and potential for monitoring salt-affected and waterlogged areas in South Africa. The producer's accuracies (PAs) of salt-affected and waterlogged areas ranged from high (Tugela 89.6%; Olifants 85%; Makhatini 83.3%; Loskop 81.4%; Breede 76.5%), to medium (Sundays 68.2%;

TABLE 6.12

Salt-Affected and Waterlogging Quantification

Irrigation scheme	Total area considered (ha)	Anomalies (ha)	Ar	Estimated affected area (ha)	Estimated affected area (%)
Vaalharts	26,434	572	0.725	414.7	1.57
Loskop	38,831	1,239	0.716	887.124	2.28
Makhatini	4,312	194	0.714	138.516	3.21
Tugela River	27,384	2.032	0.727	1,477.264	5.39
Olifants River	11,284	320	0.702	224.64	1.99
Breede River	29,129	1,807	0.773	1,396.811	4.80
Sundays River	18,608	556	0.95	528.2	2.84
Pondrift	8,681	604	0.775	468.1	5.39
Douglas	22,748	1,614	0.919	1,483.3	6.52

Vaalharts 62%, Douglas 61.8%) to low (Pondrift 45.16%). A likely explanation for the poor PA in the Pondrift and Douglas irrigation schemes is the occurrence of false positives due to recent flood damage. Flood damage leads to heterogeneous patches in fields unrelated to salt accumulation or waterlogging. During a flood, salts are temporarily flushed from the soil, only to return at a later stage. The rehabilitation period after a flood also varies. All of these factors highlight the importance of synchronizing the capture of satellite and field data. In some cases, the area of salt accumulation or waterlogging is greater than the area of normal growth in a field (often also related to flood damage). As a result, the MeaD algorithm highlights the areas of normal growth as anomalies and these were consequently recorded as false positives.

Field digitizing errors also contributed to false positives. A common cause is when vector boundaries do not accurately delineate the true field boundary, especially when areas slightly outside a field is included. The spectral properties of these areas do not match the associated field and are consequently identified as anomalies by the MeaD equation. Using object relationships, the rule-set was modified to eliminate most of these errors, but some errors could not be avoided.

False positives may also occur when a fixed threshold value (e.g., 4.0 dS/m (EC) for salt accumulation) is used to verify if an area is affected or not. Even though the WFAD accommodates various crop sensitivities to salt accumulation and waterlogging, the accuracy of the method can be compromised if a salt-tolerant crop displays no physiological stress at an EC level of 4.0 dS/m.

The two-year multitemporal approach failed to effectively eliminate all anomalies not associated with salt accumulation or waterlogging. Anomalies caused by features such as ground compaction, drought stress and poor farming practices frequently occurred in both images. Increasing the period of analysis to three or four years may improve the elimination of non-related anomalies. The multitemporal imagery did, however, eliminate most anomalies unrelated to salt accumulation and waterlogging. Harvesting remnants such as symmetrical tracks caused by harvesters and stacking of bales all lead to areas of high heterogeneity within a field, but were successfully eliminated by the multitemporal approach.

Considering the false positives and varying anomaly extents, the quantification of salt-affected and waterlogged areas presented in this study should be considered as an estimation and not absolute figures (Table 6.12). The quantifications (Table 6.12) only represent the current active fields in each scheme and do not take account of fields that were abandoned due to salt accumulation and/or waterlogging. If abandoned fields were to be included, the percentages would be higher. Some fields have been abandoned for an extended period of time and a much larger temporal analysis would be needed to credibly identify such cases. Nevertheless, the estimations based on the WFAD method is a good representation of the current status of each scheme and serves as a starting point for further monitoring. Apart from the abandoned areas, the percentages indicate that salt accumulation and waterlogging is under control, especially when compared to the percentages of other areas such as Egypt, Iran, and Argentina that have affected areas of 30% or more (Ghassemi et al. 1995).

6.9 CONCLUSIONS

Excessive accumulation of salt in the plant root zone has a deteriorating effect on vegetation growth, resulting in reduced crop yield and rendering fertile soil barren, which ultimately leads to a decrease in food production. There is a critical need for active salinity monitoring so that rehabilitation and preventive measures can be implemented. Conventional salinity monitoring methods, such as regular field visits and laboratory analyses of soil samples, are ineffective for frequent salt accumulation monitoring over large areas. Earth observation techniques can complement conventional methods and potentially improve the cost- and time-efficiency of regular salt accumulation monitoring.

A number of direct and indirect Earth observation approaches for detecting accumulated salts exists. The direct approach is based on identifying spectral characteristics directly related to salt encrustations on the soil surface, while the indirect approach detects salt accumulation indirectly through the identification of vegetation stress or the classification of specific halophytic vegetation. This chapter presents a methodology that considers both direct and indirect salinity indicators for individual fields, called within-field anomaly detection (WFAD). WFAD was explicitly developed for regions such as South Africa where the extent of salt-affected areas are relatively small in extent. The methodology is implemented through a rule-based classification scheme in a geographical object-based image analysis (GEOBIA) environment and makes use of a multitemporal image analysis approach. The WFAD was applied to nine different irrigation schemes in South Africa representing sufficient climatic, environmental and agricultural variability to test for model robustness. The WFAD achieved an average classification accuracy of 75% (0.6 kappa) over the nine irrigation schemes, which suggests robustness and transferability when identifying salt accumulation under various climatic and cultivation conditions.

> This chapter overviews the WFAD and the quantitative experiments carried out to test it. The results will be instrumental in the establishment of a salinity monitoring system for South Africa. The aim of such a system is to proactively identify areas where salt accumulation are likely occurring so that they can be targeted for mitigation and rehabilitation, with the ultimate goal of maximizing agricultural production and sustainable food production.

A robust WFAD approach to quantify salinity and waterlogging was applied to nine different irrigation systems in South Africa. The great variety of crop types, field size and different indirect and direct salinity indicators across the irrigation schemes posed a unique challenge. The WFAD method uses a GEOBIA approach with a rule-based classification scheme. MeaD thresholds were applied separately for vegetated and bare fields to accommodate the various salt accumulation and waterlogging indicators. The WFAD has various strengths and weaknesses. The weaknesses include:

- The manual delineation of field boundaries for a large irrigation scheme is a time-consuming process. Inaccuracies in delineations led to an increase in false positives.
- The multitemporal approach does not eliminate all anomalies unrelated to salt accumulation and waterlogging. The addition of more seasons is suggested to remedy this.
- Currently, the approach cannot distinguish between salt-affected or waterlogged areas.

The strengths of the WFAD are:

- Unlike with supervised classification, training data is not necessary and ground truth data (where available) can thus be used exclusively for accuracy assessment.

- The technique is not restricted to only vegetated or only bare areas.
- Most of the known salinity and waterlogging indicators are accounted for by the various MeaD thresholds.
- The approach accounts for (at least to a certain extent) the different crop growing phases and sensitivities to salt accumulation and waterlogging by evaluating each field relative to itself.
- The approach is easily transferable across different irrigation schemes.
- Seasonal factors unrelated to salt accumulation and waterlogging are accounted for by using multitemporal imagery.
- The approach produces a quantifiable result.

The monitoring of salt-affected and waterlogged areas is necessary for the mitigation of potential damage to agricultural areas (De Villiers et al. 2003). With a mean OA and kappa of 75% and 0.6 respectively, the WFAD presents a robust, systematic approach for keeping track of these areas. Recent satellite imagery can easily be incorporated to identify salt-affected and waterlogged areas and to monitor spatial and temporal changes.

More research is needed to automate the field boundary delineation, as the manual digitizing of fields is very time-consuming and error prone. Further analyses of the spectral properties of the anomalies could also help in distinguishing between salt-affected or waterlogged areas.

6.10 ACKNOWLEDGMENTS

This chapter formed part of the first author's MSc research, carried out at Stellenbosch University. The authors would like to thank the Water Research Commission (WRC) for initiating and funding the project titled "Methodology to monitor the status of waterlogging and salt-affected soils on selected irrigation schemes in South Africa" (contract number K5/1880//4), of which this work forms part. More information about this project is available in the 2015 WRC Report (No. TT 648/15; ISBN 978-1-4312-0739-8), available at www.wrc.org.za. We also acknowledge and thank the project leader, Dr Piet Nell of the Agricultural Research Council, for providing the soil data used in this study and for his invaluable guidance and insight, particularly during the field surveys.

REFERENCES

Abbas, A., Khan, S., Hussain, N., Hanjra, M.A., and Akbar, S. 2013. Characterizing soil salinity in irrigated agriculture using a remote sensing approach. *Physics and Chemistry of the Earth, Parts A/B/C* 55–7: 43–52.

Abood, S., Maclean, A., and Falkowski, M. 2011. *Soil Salinity Detection in the Mesopotamian Agricultural Plain Utilizing WorldView-2 Imagery.* Digital Globe Corporation. Houghton: Michigan Technological University.

Addink, E.A., Van Coillie, F.M.B., and de Jong, S.M. 2012. Introduction to the GEOBIA 2010 special issue: From pixels to geographic objects in remote sensing image analysis. *International Journal of Applied Earth Observation and Geoinformation* 15(1): 1–6. doi: 10.1016/j.jag.2011.12.001.

Aihoon, J.K., Groenewald, J.A., and Sartorius von Bach, H.J. 1997. Agricultural salinization in the Olifants River at Loskop Valley, Mpumalanga/Landbou-geïnduseerde versouting in die Olifantsrivier by die Loskopvallei, Mpumalanga. *Agrekon* 36(3): 268–83. doi: 10.1080/03031853.1997.9523465.

Aldakheel, Y.Y. 2011. Assessing NDVI spatial pattern as related to irrigation and soil salinity management in Al-Hassa Oasis, Saudi Arabia. *Journal of the Indian Society of Remote Sensing* 39(2): 171–80. doi: 10.1007/s12524-010-0057-z.

Allbed, A., Kumar, L., and Aldakheel, Y.Y. 2014. Assessing soil salinity using soil salinity and vegetation indices derived from IKONOS high-spatial resolution imageries: Applications in a date palm dominated region. *Geoderma* 230–1: 1–8. doi: 10.1016/j.geoderma.2014.03.025.

Armour, R.J. 2002. *The Economic Effects of Poor and Fluctuating Irrigation Water Salinity Levels in the Lower Vaal and Riet Rivers.* Bloemfontein, South Africa: University of the Free State.

Backeberg, G.R. 2003. Water usage and irrigation policy. In: *The Challenge of Change: Agriculture, Land and the South African Economy.* Pietermaritzburg, South Africa: University of Natal Press. pp. 149–170.

Backeberg, G.R., Bembridge, T.J., Bennie, A.T.P., Groenewald, J.A., Hammers, P.S., Pullen, R.A., and Thompson, H. 1996. *Policy Proposal for Irrigated Agriculture in South Africa*. Pretoria, South Africa: Water Research Commission (WRC).

Bailey, A.K., and Pitman, W.V. 2015. *Water Resources of South Africa 2012 Study (WR2012)*. Pretoria, South Africa: Water Research Commission (WRC).

Barnard, J.H. 2013. *On-Farm Management of Salinity Associated with Irrigation for the Orange-Riet and Vaalharts Schemes*. Bloemfontein, South Africa: University of the Free State.

Barrett-Lennard, E.G. 2003. The interaction between waterlogging and salinity in higher plants: Causes, consequences and implications. *Plant and Soil* 253(1): 35–54. doi: 10.1023/A:1024574622669.

Bastiaanssen, W.G.M., Meneti, M., Feddes, R.A., and Holtslag, A.A.M. 1998. A remote sensing surface energy balance algorithm for land (SEBAL) 1. Formulation. *Journal of Hydrology* 212–13: 198–212.

Beuster, H., Shand, M.J., and Carter, C.A. 2003. *Breede River Basin Study*. Pretoria, South Africa: Department of Water Affairs and Forestry.

Blaschke, T. 2010. Object based image analysis for remote sensing. *ISPRS Journal of Photogrammetry and Remote Sensing* 65(1): 2–16. doi: 10.1016/j.isprsjprs.2009.06.004.

Blaschke, T., Hay, G.J., Kelly, M., Lang, S., Hofmann, P., Addink, E., Feitosa, Q., Van der Meer, F., Van der Werff, H., Van Coillie, F., and Tiede, D. 2014. Geographic object-based image analysis–towards a new paradigm. *ISPRS Journal of Photogrammetry and Remote Sensing* 87: 180–91. doi: 10.1016/j.isprsjprs.2013.09.014.

Blaschke, T., and Strobl, J. 2001. Was ist mit den Pixeln los? Neue Entwicklungen zur Integration von Fernerkundung und GIS. *Geo-Informations-Systeme* 14(6): 12–7.

Bouaziz, M., Matschullat, J., and Gloaguen, R. 2011. Improved remote sensing detection of soil salinity from a semi-arid climate in Northeast Brazil. *Comptes Rendus Geoscience* 343(11–12): 795–803. doi: 10.1016/j.crte.2011.09.003.

Campbell, J.B. 2007. *Introduction to Remote Sensing*, 4th ed. London: Taylor & Francis.

Csillag, F., Pasztor, L., and Biehl, L. 1993. Spectral band selection for the characterization of salinity status of soils. *Remote Sensing of Environment* 43(3): 231–42. doi: 10.1016/0034-4257(93)90068-9.

De Villiers, M.C., Nell, J.P., Barnard, R.O., and Henning, A. 2003. *Salt-Affected Soils: South Africa*. Pretoria, South Africa: Food and Agricultural Organization.

Dehaan, R.L., and Taylor, G.R. 2002. Field-derived spectra of salinized soils and vegetation as indicators of irrigation-induced soil salinization. *Remote Sensing of Environment* 80(3): 406–17. doi: 10.1016/S0034-4257(01)00321-2.

Dehaan, R.L., and Taylor, G.R. 2003. Image-derived spectral endmembers as indicators of salinisation. *International Journal of Remote Sensing* 24(4): 775–94. doi: 10.1080/01431160110107635.

Department of Agriculture, Forestry and Fisheries. 2012. Abstract of agricultural statistics. Pretoria, South Africa: Department of Agriculture, Forestry and Fisheries. https://www.nda.agric.za/docs/statsinfo/Ab2012.pdf.

Dutkiewicz, A., Lewis, M., and Ostendorf, B. 2009. Evaluation and comparison of hyperspectral imagery for mapping surface symptoms of dryland salinity. *International Journal of Remote Sensing* 30(3): 693–719. doi: 10.1080/01431160802392612.

Dwivedi, R.S., and Sreenivas, K. 1998. Delineation of salt-affected soils and waterlogged areas in the Indo-Gangetic plains using IRS-1C LISS-III data. *International Journal of Remote Sensing* 19(14): 2739–51. doi: 10.1080/014311698214488.

Dwivedi, R.S., Sreenivas, K., and Ramana, K.V. 1999. Inventory of salt-affected soils and waterlogged areas: A remote sensing approach. *International Journal of Remote Sensing* 20(8): 1589–99. doi: 10.1080/014311699212623.

Elnaggar, A.A., and Noller, J.S. 2010. Application of remote-sensing data and decision-tree analysis to mapping salt-affected soils over large areas. *Remote Sensing* 2(1): 151–65. doi: 10.3390/rs2010151.

Farr, T.G., Rosen, P.A., Caro, E., Crippen, R., Duren, R., Hensley, S., Kobrick, M., Paller, M., Rodriguez, E., Roth, L., and Seal, D. 2007. The shuttle radar topography mission. *Reviews of Geophysics* 45(2): RG2004. doi: 10.1029/2005RG000183.

Fernández-Buces, N., Siebe, C., Cram, S., and Palacio, J.L. 2006. Mapping soil salinity using a combined spectral response index for bare soil and vegetation: A case study in the former lake Texcoco, Mexico. *Journal of Arid Environments* 65(4): 644–67. doi: 10.1016/j.jaridenv.2005.08.005.

Flanders, D., Hall-Beyer, M., and Pereverzoff, J. 2003. Preliminary evaluation of eCognition object-based software for cut block delineation and feature extraction. *Canadian Journal of Remote Sensing* 29(4): 441–52. doi: 10.5589/m03-006.

Flügel, W.A., and Kienzle, S. 1989. Hydrology and salinity dynamics of the Breede River, Western Cape Province, Republic of South Africa. In: *Regional Characterization of Water Quality*. Proceedings of the Baltimore Symposium, May 1989). Baltimore, MD: IAHS Publications, pp. 221–8

Furby, S.L., Wallace, J.F., Caccetta, P.A., and Wheaton, G.A. 1995. *Detecting and Monitoring Salt-Affected Land*. Canberra, Australia: Land and Water Resources Research and Development Corporation, Canberra ACT.

García Rodríguez, P., Pérez González, M.E., and Guerra Zaballos, A. 2007. Mapping of salt-affected soils using TM images. *International Journal of Remote Sensing* 28(12): 2713–22. doi: 10.1080/01431160600928658.

Ghassemi, F., Jakeman, A.J., and Nix, H.A. 1995. *Salinisation of Land and Water Resources: Human Causes, Extent, Management and Case Studies*. Wallingford, UK: CAB International.

Gombar, O., and Erasmus, C.J.H. 1976. Vaalharts Ontwateringsprojek. Technical Report No. GH2897. Pretoria, South Africa: Department of Water Affairs.

Hamzeh, S., Naseri, A.A., AlaviPanah, S.K., Mojaradi, B., Bartholomeus, H.M., Clevers, J.G.P.W., and Behzad, M. 2012a. Estimating salinity stress in sugarcane fields with spaceborne hyperspectral: Vegetation indices. *International Journal of Applied Earth Observation and Geoinformation* 21(1): 282–90. doi: 10.1016/j.jag.2012.07.002.

Hamzeh, S., Naseri, A.A., AlaviPanah, S.K., Mojaradi, B., Bartholomeus, H.M., and Herlod, M. 2012b. Mapping salinity stress in sugarcane fields with hyperspecteral satellite imagery. In: *Remote Sensing for Agriculture, Ecosystems, and Hydrology*, vol. XIV. Edinburgh, UK: SPIE.

Hanson, B.R., Grattan, S.R., and Fulton, A. 2006. *Agricultural Salinity and Drainage*, 2nd ed. Oakland, CA: Department of Land, Air and Water Resources, University of California.

Haralick, R.M., and Shapiro, L.G. 1985. Image segmentation techniques. *Computer Vision, Graphics, and Image Processing* 29(1): 100–32. doi: 10.1016/S0734-189X(85)90153-7.

Hay, G.J., and Castilla, G. 2008. Geographic object-based image analysis (GEOBIA): A new name for a new discipline. In: *Object-Based Image Analysis*, T. Blaschke, S. Lang, and G. Hay. Berlin/Heidelberg: Springer, pp. 75–89.

Hay, G.J., Niemann, K.O., and McLean, G.F. 1996. An object-specific image-texture analysis of H-resolution forest imagery. *Remote Sensing of Environment* 55(2): 108–22. doi: 10.1016/0034-4257(95)00189-1.

Hengl, T., de Jesus, J.M., Heuvelink, G.B.M., Gonzalez, M.R., Kilibarda, M., Blagotić, A., Shangguan, W., et al. 2017. SoilGrids250m: Global gridded soil information based on machine learning. *PLOS ONE* 12(2): e0169748. doi: 10.1371/journal.pone.0169748.

Hillel, D. 2000. *Salinity Management for Sustainable Irrigation*. Washington, D.C.: The International Bank for Reconstruction.

Hillel, D., and Vlek, P. 2005. The sustainability of irrigation. *Advances in Agronomy* 87(5): 55–84. doi: 10.1016/S0065-2113(05)87002-6.

Iqbal, F. 2011. Detection of salt affected soil in rice-wheat area using satellite image. *African Journal of Agriculture Research* 6(21): 4973–82. doi: 10.5897/AJAR11.634.

Khan, N.M., Rastoskuev, V.V., Sato, Y., and Shiozawa, S. 2005. Assessment of hydrosaline land degradation by using a simple approach of remote sensing indicators. *Agricultural Water Management* 77(1–3): 96–109. doi: 10.1016/j.agwat.2004.09.038.

Kirchner, J. 1995. *Investigation into the Contribution of Groundwater to the Salt Load of the Breede River, Using Isotopes and Chemical Tracers*, Report No 344/1/95. Pretoria, South Africa: Water research Commission.

Koshal, A.K. 2010. Indices based salinity areas detection through remote sensing & GIS in parts of South West Punjab. In: *13th Annual International Conference and Exhibition on Geospatial Information Technology and Applications*. Gurgaon, India, pp. 1–11.

Kruger, M., Van Rensburg, J.B.J., and Van den Berg, J. 2009. Perspective on the development of stem borer resistance to Bt maize and refuge compliance at the Vaalharts irrigation scheme in South Africa. *Crop Protection* 28(8): 684–9. doi: 10.1016/j.cropro.2009.04.001.

Lenney, M.P., Woodcock, C.E., Collins, J.B., and Hamdi, H. 1996. The status of agricultural lands in Egypt: The use of multitemporal NDVI features derived from Landsat TM. *Remote Sensing of Environment* 56(1): 8–20. doi: 10.1016/0034-4257(95)00152-2.

Liebenberg, L. 1977. *Die Geologie van die Gebied 2724D (Andalusia)*. Bloemfontein, South Africa: University of the Free State.

Lillesand, T.M., Kiefer, R.W., and Chipman, J.W. 2004. *Remote Sensing and Image Interpretation*, 5th ed. New York: John Wiley & Sons.

Lobell, D.B., Lesch, S.M., Corwin, D.L., Ulmer, M.G., Anderson, K.A., Potts, D.J., Doolittle, J.A., Matos, M.R., and Baltes, M.J. 2010. Regional-scale assessment of soil salinity in the red river valley using multi-year MODIS EVI and NDVI. *Journal of Environmental Quality* 39(1): 35–41. doi: 10.2134/jeq2009.0140.

Lucas, R., Rowlands, A., Brown, A., Keyworth, S., and Bunting, P. 2007. Rule-based classification of multi-temporal satellite imagery for habitat and agricultural land cover mapping. *ISPRS Journal of Photogrammetry and Remote Sensing* 62(3): 165–85. doi: 10.1016/j.isprsjprs.2007.03.003.

Maisela, R.J. 2007. *Realizing Agricultural Potential in Land Reform: The Case of Vaalharts Irrigation Scheme in the Northern Cape Province.* Cape Town, South Africa: University of the Western Cape.

McDonagh, J., and Bunning, S. 2009. *Field Manual for Local Level Land Degradation Assessment in Drylands.* Norwich: Food and Agricultural Organization.

McGhie, S., and Ryan, M. 2005. *Salinity Indicator Plants. Australia.* New South Wales, Australia: Department of Infrastructure, Planning and Natural Resources.

Metternicht, G.I., and Zinck, J.A. 2003. Remote sensing of soil salinity: Potentials and constraints. *Remote Sensing of Environment* 85(1): 1–20. doi: 10.1016/S0034-4257(02)00188-8.

Metternicht, G.I., and Zinck, J.A. 2009. *Remote Sensing of Soil Salinization: Impact on Land Management.* Boca Raton, FL: CRC Press Taylor and Francis.

Moolman, J.H., De Clercq, W.P., Wessels, W.P.J., Meiring, A., and Molman, C.G. 1999. *The Use of Saline Water for Irrigation of Grapevines and the Development of Crop Salt Tolerance Indices,* 303/1/1999, Pretoria, South Africa: Water Research Commission.

Mougenot, B., Pouget, M., and Epema, G.F. 1993. Remote sensing of salt affected soils. *Remote Sensing Reviews* 7(3–4): 241–59.

Mucina, L., and Rutherford, M.C. 2006. *The Vegetation of South Africa, Lesotho and Swaziland.* Pretoria, South Africa: South African National Biodiversity Institute.

Peñuelas, J., Isla, R., Filella, I., and Araus, J.L. 1997. Visible and near-infrared reflectance assessment of salinity effects on barley. *Crop Science* 37(1). doi: 10.2135/cropsci1997.0011183X003700010033x.

Phipson, J.S. 2012. *Agricultural and Agribusiness Status Quo Assessment: Uthukela District Municipality: KwaZulu-Natal.* Mtunzini, South Africa: Uthukela District Municipality.

Platonov, A., Noble, A., and Kuziev, R. 2013. Soil salinity mapping using multi-temporal satellite images in agricultural fields of Syrdarya province of Uzbekistan. In: *Developments in Soil Salinity Assessment and Reclamation: Innovative Thinking and Use of Marginal Soil and Water Resources in Irrigated Agriculture,* S.A Shahid, M.A Abdelfattah, and F.K Taha (eds.). Dordrecht, The Netherlands: Springer. pp. 87–98

Qureshi, R.H., and Barrett-Lennard, E.G. 1998. *Agriculture for Irrigated Land in Pakistan: A Handbook.* Queanbeyan, Australia: Australian Centre for International Agricultural Research.

Rao, B.R.M., Sharma, R.C., Ravi Sankar, T., Das, S.N., Dwivedi, R.S., Thammappa, S.S., and Venkataratnam, L. 1995. Spectral behaviour of salt-affected soils. *International Journal of Remote Sensing* 16(12): 2125–36.

Schulze, R.E. 2006. *South African Atlas of Climatology and Agrohydrology.* Pretoria, South Africa: Water Research Commission.

Setia, R., Lewis, M., Marschner, P., Raja Segaran, R., Summers, D., and Chittleborough, D. 2013. Severity of salinity accurately detected and classified on a paddock scale with high resolution multispectral satellite imagery. *Land Degradation and Development* 24(4): 375–84. doi: 10.1002/ldr.1134.

Sidike, A., Zhao, S., and Wen, Y. 2014. Estimating soil salinity in Pingluo County of China using QuickBird data and soil reflectance spectra. *International Journal of Applied Earth Observation and Geoinformation* 26: 156–75. doi: 10.1016/j.jag.2013.06.002.

Streutker, A. 1997. The dependence of permanent crop production on efficient irrigation and drainage at the Vaalharts government water scheme. *Water SA* 3(2): 90–102.

Tren, R., and Schur, M. 2000. *Olifants River Irrigation Schemes, Reports 1 & 2.* Colombo, Sri Lanka: IWMI.

Van Rensburg, L.D., Barnard, J.H., Bennie, A.T.P., Sparrow, B., and du Preez, C.C. 2012. *Managing Salinity Associated with Irrigation at Orange-Riet and Vaalharts Irrigation Schemes,* 1647/1/12, Pretoria, South Africa: Water Research Commission.

Wiegand, C.L., Rhoades, J.D., Escobar, D.E., and Everitt, J.H. 1994. Photographic and videographic observations for determining and mapping the response of cotton to soil salinity. *Remote Sensing of Environment* 49(3): 212–23. doi: 10.1016/0034-4257(94)90017-5.

Zhang, T.T., Zeng, S.L., Gao, Y., Ouyang, Z.T., Li, B., Fang, C.M., and Zhao, B. 2011. Using hyperspectral vegetation indices as a proxy to monitor soil salinity. *Ecological Indicators* 11(6): 1552–62. doi: 10.1016/j.ecolind.2011.03.025.

7 Changes in Soil Organic Matter Content and Quality in South African Arable Land

Chris C. Du Preez, Cornie W. van Huyssteen and Wulf Amelung

CONTENTS

7.1 INTRODUCTION

In South Africa, as in the rest of the world, a renewed appreciation of soil organic matter (SOM) emerged toward the end of the last century and continues to this day. This can be attributed inter alia to the paradigm of industrial agriculture, which is premised on the view that the best way to deal with individual problems is to resolve each one separately. The remedies, according to this view, lie, for example, in the application of specific purchased inputs like fertilizers and pesticides, which may in turn degrade soil quality. Soil quality is usually defined as the ability to function efficiently in the long term within the boundaries of an ecosystem, so that biological productivity is sustained, environmental quality remains intact, and the health of plants, animals and humans is promoted (Weil and Magdoff 2004).

In this regard, soil mediates many ecosystem processes, which may be grouped into five fundamental, though somewhat overlapping, functions: promotion of plant growth; biogeochemical cycling of carbon (C) and nutrients; provision of habitat for soil organisms; partitioning, storage, translocation and decontamination of water; and support and protection of human structures and artifacts. With the exception of the latter, SOM usually exerts a positive impact on a soil's capacity to perform the other functions. This is ascribed to the integrating effects of SOM on numerous soil properties and processes (Brady and Weil 2008).

In this context, a meta-analysis of research was done on SOM and its indicator elements (C, N, P and S) in some cultivated soils in southern Africa (Swanepoel et al. 2016). The results of this study showed that 70% of the research was conducted in South Africa. This provides an opportunity to take cognizance of the change in SOM content and quality in some arable lands of South Africa subjected to three diverse crop production systems. First, the physiographical features that determine the development and distribution of soil types are addressed, followed by a discussion of the spatial variability of SOM in undisturbed soils under natural ecosystems. The focus then shifts to agricultural land use and the impact that the production of field crops, sugarcane (*Saccharum officinarum*) and pasture had on SOM content and quality. Lastly, research needs are identified and recommendations are made with respect to management of SOM in arable lands.

7.2 PHYSIOGRAPHICAL FEATURES

It is generally assumed that SOM reaches the maximum equilibrium level in natural land under specific environmental conditions where inputs equal losses. This level of equilibrium is influenced in order of importance by the following factors: climate > vegetation > topography ~ parent material > time, with all five soil-forming factors being partially interactive. Human intervention changes this equilibrium through the effects of agriculture, which can be either positive or negative towards SOM content and quality (Stevenson and Cole 1999). In understanding these changes, some knowledge of the physiographical features of the land is therefore required.

South Africa is bordered by the Indian Ocean, with its warm Agulhas current on the east coast, and the Atlantic Ocean, with its cold Benquela current, on the west coast (Figure 7.1). These surrounding oceans have a great influence on the climate of the country, due to its relative narrowness, even at the widest east–west transect. The climate, with the exception of the extreme southern

FIGURE 7.1   Orientation map, indicating the research sites in South Africa.

coastal region, which has a Mediterranean climate with winter rains, is typical subtropical bimodal with a summer rain season from October to March. During this period, the incidence of annual rainfall is more than 75%, which ranges from <100 mm near the Atlantic seaboard to >1,000 mm near the Indian seaboard (Figure 7.2). Conversely, annual evaporation varies from <1,800 mm in the east to >3,000 mm in the west. In addition to the difference in annual rainfall across the country, total rainfall is erratic and highly variable in both space and time, especially at the beginning and middle of the wet season, when it tends to occur in the form of intense storms of short duration. The daytime annual temperature varies from 20°C to 25°C, but drops to around 15°C at night. At higher elevations, however, it may even freeze during the night in June and July. Annual rainfall under the Mediterranean climate of the southern coastal region varies from 200 mm to >1500 mm in the mountains, with an 80% to 85% incidence in April to September. This rain results from cold fronts moving from the southwest over the interior during winter. The region therefore has hot, dry summer months from October to March (De Jager 1993).

A detailed examination of long-term meteorological data of South Africa suggests that there is an 18 (±2)-year cycle of rainfall. This cycle comprises several good years followed by several poor years. Evidence exists that this cycle is strongly related to the southern oscillation because of either of the El Niño or El Niña phenomena. El Niña dominates during the wetter period, while El Niño dominates during the drier period (Twomlow et al. 2006).

The land surface of South Africa is divided into several physiographical regions. These regions, however, combine into two broad natural groups. One is the marginal zone, ranging in altitude between 0 m at the coast and 2,300 m in the great escarpment. The other is the vast interior plateau, ranging in altitude between 900 m and 3,400 m. Topography is generally smooth to undulating on the elevated tablelands, except for isolated mountain ranges such as the Drakensberg mountains.

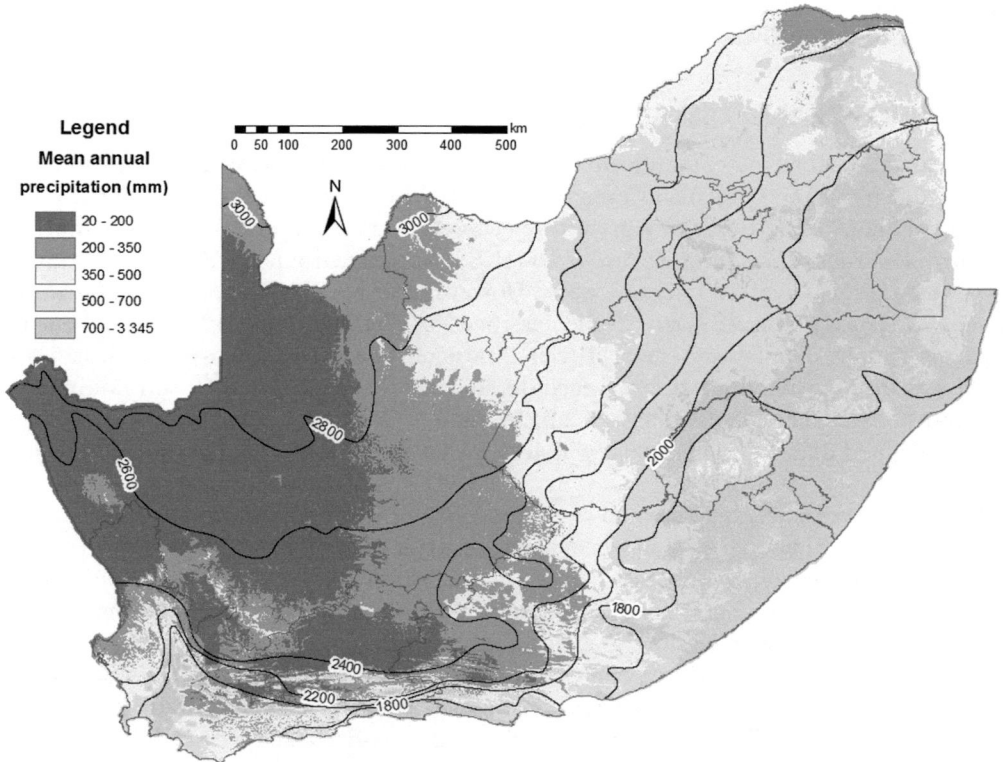

FIGURE 7.2   (See color insert.) Average annual precipitation and average annual evaporation isohyets (mm) in South Africa (interpolated from Schulze et al. 2001).

The occurrence of dolerite or granite kopjes scattered in the landscape is fairly typical. A progressive diminution in altitude occurs from the east to the west (Jones et al. 2013).

In its native state, the interior plateau mainly comprises the flat scrubland of the Karoo and extensive grassland plains of the central and northeastern Highveld. However, sands predominate in the northwest, in which the thorn trees of the southern Kalahari thrive. Thorn trees, bush and savannah characterize the Bushveld basin to the north. Further to the northeast, baobab trees (*Adansonia digitata*) and thick bush become more prolific. Vast differences in the vegetation are found in the marginal zone, where the great escarpment forms a divide between the interior and the coast. These differences are mainly due to the changing altitude, surface form and climate. The vast Great Karoo in the interior and Little Karoo of the south and southwest, constitute a large proportion of the marginal zone and are characterized by sparse Karoo shrubs. Fynbos vegetation covers the major area of the marginal zone. Much of the marginal zone is taken up by the barren peaks of the Cape Fold mountains (Low and Rebelo 1996).

Soils in South Africa are in some instances derived from the oldest rocks on Earth, ranging in age between 1 million and 3,500 million years. These igneous rocks are comprised inter alia of basalt, dolerite and granite. Sedimentary rocks such as conglomerate, sandstone and shale serve in other instances as parent material. Some soils also developed from metamorphic rocks, which are altered versions of either the igneous or sedimentary rocks. Another important source of soil formation is the recent aeolian sediments, known as Kalahari Sand, covering an area of 2.5 million ha (Mha), and considered to be the largest terrestrial body of sand on Earth. Land to the northwest is dominated by the Kalahari basin, a large lowland area. The rocks of the geological stable Karoo basin and Kaap Vaal Craton cover most of the land to the east and south (Jones et al. 2013).

On account of these disparate physiographical features, a large range of soils with divergent properties is found in South Africa (Fey 2010). The content and quality of organic matter in the soils therefore vary accordingly.

7.3 SOIL TYPES AND DISTRIBUTION

Soil classification aims to communicate differences in soils that develop due to the influence of climate, topography and organisms on parent material over time (Fanning and Fanning 1989). The United States Department of Agriculture (USDA) *Soil Taxonomy* (Soil Survey Staff 2014) and *World Reference Base for Soil Resources* (IUSS Working Group WRB 2014) soil classification systems are widely used internationally. However, in South Africa, *Soil Classification – A Taxonomic System for South Africa* (Soil Classification Working Group 1991) is used due to the unique soils in the country and the user-friendliness of the system. This system classifies soils into 74 soil forms at the highest level, each divided into between 2 and 12 soil families at the next, lowest, level. The identification of soil forms is based on a unique sequence of morphologically defined diagnostic horizons, which is, in turn, based on the identification of master horizons.

The potential for arable land use and crop production is determined by soil, climate and topography. It is therefore imperative to understand the country's soil types and distribution in relation to climate and topography in any discussion on the current status and future challenges of agricultural production. Land use for cropping is, in general, only possible where the average annual rainfall exceeds 400 mm and where the slope gradient is less than 12%.

The land type survey (Land Type Survey Staff 2004) provides soil information for the whole of South Africa at a scale of 1:250,000. The survey was conducted by mapping terrain types and pedosystems (with uniform soil patterns) within each terrain type. Each land type therefore has marked uniformity in terms of terrain and soil pattern. The land type map was subsequently generalized to yield broad soil pattern groups (Figure 7.3), briefly described in Table 7.1. The resultant map is consequently only applicable at a scale of 1:10,000,000.

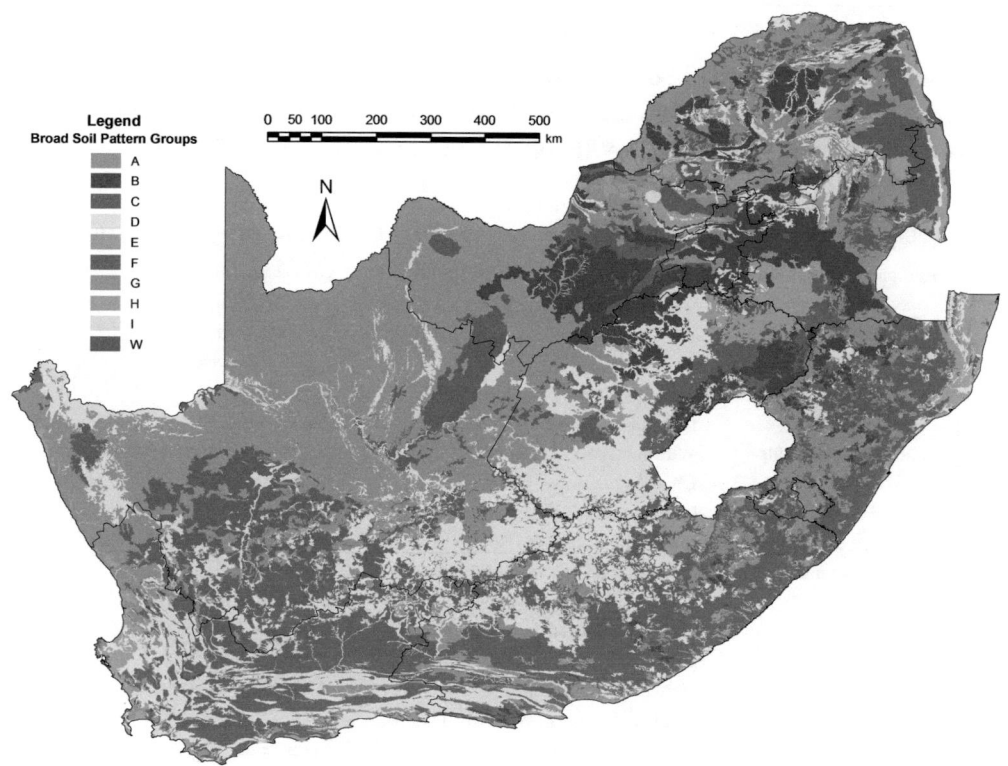

FIGURE 7.3   (See color insert.) Generalized broad soil pattern groups map (adapted from Land Type Survey Staff 2004).

A. The red and yellow, apedal, freely drained soils mapping unit is the largest (33.8%). Shallow apedal soils (>40% of the mapping unit) overlying lime and duripans dominate in the arid central and northwestern parts, while deeper apedal soils occupy the wetter areas of the mapping unit. Soils in this mapping unit vary from dystrophic to mesotrophic and may or may not have illuvial clay accumulation. Arable lands in this mapping unit are restricted to the wetter (>400 mm annual rainfall) parts of the country.

B. Plinthic catena, with rare upland duplex and marginal soils, occupy a large area (8.8%) of the South African interior. Plinthic soils cover more than 10% of the mapping unit, while gley soils occur in lowland positions. Deeper variants of the plinthic soils in this mapping unit are well suited to arable land use, especially in the drier regions (350–550 mm average annual rainfall) of the country, while the marginal soils are not suitable for arable agriculture.

C. Plinthic catena, with common upland duplex and marginal soils, also occur in the South African interior, but to a much lesser extent (2.7%). Plinthic soils cover more than 10% of the mapping unit, but here gley and duplex soils also occur in upland positions. Deeper variants of the plinthic soils in this mapping unit are well suited to arable land use, especially in the drier regions (350–550 mm average annual rainfall) of the country, while the marginal soils are not suitable for crop production.

D. Duplex soils (9.7%) have marked clay accumulation in the subsoil, as evidenced by an abrupt clay increase and strongly developed soil structure in more than half of the soils of this mapping unit. These soils are therefore not suitable for crop-based agriculture.

E. Structured soils (vertic, melanic, red structured) occupy 4.1% of South Africa. These soils are characterized by clayey, high-base status, dark and/or red colored soils, typically

TABLE 7.1

Brief Description of the Broad Soil Pattern Groups (Land Type Survey Staff 2004), the Dominant Soil Forms (Soil Classification Working Group 1991), the Equivalent Reference Soil Groups (IUSS Working Group WRB 2014) and the Area of Each Group

Map symbol	Brief description	Dominant soil forms	Equivalent reference soil groups	Area (10^6 ha)	(%)
A	Red and yellow, apedal, freely drained soils	Inanda, Kranskop, Magwa, Hutton, Griffin, Clovelly	Acrisols, Lixisols, Alisols, Luvisols, Ferralsols (Durisols, Gypsisols, Calcisols)	41.4	33.8
B	Plinthic catena, with rare upland duplex and margalitic soils	Hutton, Bainsvlei, Avalon, Longlands (Rensburg, Willowbrook, Katspruit, Champagne in lowlands)	Plinthosols (with Vertisols, Stagnosols, Histosols in lowlands)	10.8	8.8
C	Plinthic catena, with common upland duplex and margalitic soils	Hutton, Bainsvlei, Avalon, Longlands (Arcadia, Bonheim, Tambankulu, Mayo, Milkwood, Estcourt, Sterkspruit, Swartland, Valsrivier, Kroonstad in uplands)	Plinthosols (with Vertisols, Leptosols, Planosols, Solonetz, Stagnosols, Histosols in uplands)	3.3	2.7
D	Duplex soils	Estcourt, Sterkspruit, Swartland, Valsrivier, Kroonstad	Planosols, Solonetz	11.9	9.7
E	Structured soils (vertic, melanic, red structured)	Arcadia, Rensburg, Willowbrook, Inhoek, Mayo, Immerpan, Milkwood, Steendal, Bonheim, Shortlands	Vertisols, Chernozems, Kastanozems, Phaeozems, Umbrisols, Nitisols	5.0	4.1
F	Rocky soils	Glenrosa, Mispah	Leptosols	35.4	28.9
G	Podzolic soils	Lamotte, Houwhoek	Podzols	0.2	0.2
H	Grey regic sands	Fernwood, Constantia, Shepstone, Vilafontes	Arenosols, Podzols	1.6	1.3
I	Youthful, rocky, and miscellaneous soils	Dundee, Oakleaf, rock, stones, boulders	Cambisols, Fluvisols, Regosols, rock	12.5	10.2
W	Water	Dams, rivers		0.2	0.2
	Total			122.4	100.0

associated with basic parent material in more than half of the soils of the mapping unit. These soils may or may not be suitable for arable agriculture. Vertic soils are extremely resilient, but also severely prone to drought. Melanic and red structured soils are particularly suitable to arable land use, provided that they are deep enough and the rainfall is high enough.

F. Rocky soils cover about one-third (28.9%) of South Africa. This mapping unit largely has shallow soils overlying rock, weathering rock, hardpan carbonate or duripan subsoils.

G. Podzolic soils (0.2%) occupy the smallest part of South Africa. More than 10% of the mapping unit has podzolic subsoils. These soils are not suitable for arable land use, primarily due to the slope, but also due to the sandy nature of them.

H. Grey regic sands (1.3%) refer to areas where deep grey aeolian sands predominate (20% to >80%) the mapping unit. These soils normally have minimal soil development, but may have undergone clay eluviation. They are common along the coastal regions and are not particularly suitable for crop production, primarily due to their sandy nature. These soils do, however, form the basis for production of root crops (vegetables) in some localized areas.

I. Youthful, rocky and miscellaneous soils cover 10.2% of South Africa. Soils with weak profile development, stratified soils along riverbanks, and rock outcrops predominate in this mapping unit. These soils are normally quite shallow and therefore not suitable for arable land use. They are, however, important irrigable soils in localized areas, typically along rivers where water is available.

The information presented here only aims to provide an overview of the soil distribution and the arability thereof, with the tentative WRB translation given in Table 7.1. The scale of representation excludes localized and small incidences of exceptional soils. Managerial input can also impact on soil suitability for crop production and should therefore not be omitted when evaluating soil suitability.

7.4 SPATIAL VARIABILITY OF SOIL ORGANIC MATTER

No systematic study has yet been done to determine the status and spatial distribution of SOM in South Africa. However, the land type survey (Land Type Survey Staff 2004), which commenced in 1970, described and analyzed 2,235 modal profiles of undisturbed soils under natural vegetation across the country. Morphological and chemical data from these profiles were augmented by various other soil surveys, resulting in the data of 7,079 virgin soil profiles being available in a database.

These data were used by Barnard et al. (2000) to produce a generalized map of organic C content for virgin topsoils (<300 mm) in South Africa. Using this map as a frame of reference, although it does not appear here, the soil organic carbon (SOC) content of topsoils ranged from <5 g kg^{-1} to >40 g kg^{-1}. Only 4% of the country is covered with topsoils that contain >20 g kg^{-1} SOC, whilst 58% of the country is covered with topsoils that contain less than 5 g kg^{-1} SOC. The remaining 38% of the country has topsoils that contain between 5 and 20 g kg^{-1} SOC content.

Rantoa et al. (2015) used the data of the 7,079 virgin soil profiles to statistically analyze the SOC contents in the master horizons (Table 7.2). The results are in accord with what can be expected from the master horizons' definitions in either the binomial (Macvicar et al. 1977) or taxonomical (Soil Classification Working Group 1991) soil classification systems, which were used in the surveys. The SOC content was the highest in the O horizon, followed by that in the A horizon, which contains well-mixed humified SOM. The E horizon is defined as a horizon with a lower SOC content than the overlying or underlying horizons, and normally underlies A horizons and overlies B and G horizons. The SOC content was larger in the G horizon than that in the E, B and C horizons,

TABLE 7.2
Statistical Summary of SOC Contents Data for South African Master Horizons

Horizon	Count (*n*)	Average	SE	25th percentile	Median	75th percentile
				----------g kg^{-1}----------		
O	11	166.4	24.5	104.0	133.0	215.0
A	7,173	16.3	0.4	6.0	11.2	22.0
E	423	4.8	0.2	2.0	4.0	6.6
B	4,232	6.2	0.4	2.1	4.0	7.0
G	102	8.0	2.5	2.0	4.0	6.0
C	657	2.7	0.1	1.0	2.0	3.0
Total	12,598	11.9	0.3	3.2	7.0	14.8

Source: Adapted from Rantoa et al. 2015.

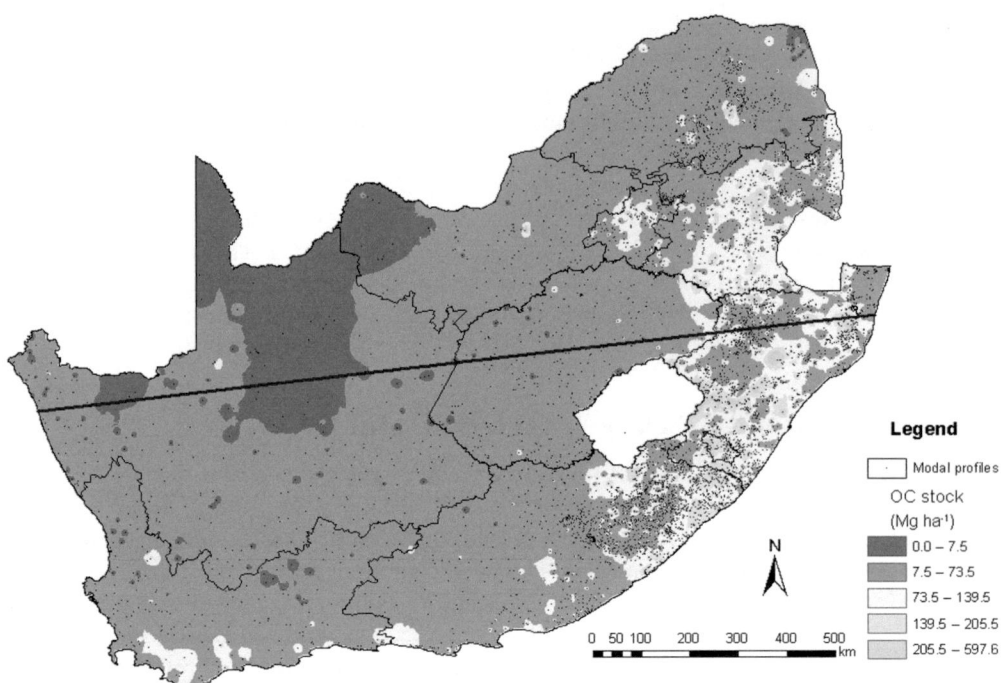

FIGURE 7.4 **(See color insert.)** Map showing soil organic carbon stocks to 300 mm depth, estimated with a bulk density of 1.5 Mg m^{-3} (adapted from Rantoa 2009). The cross section is given in Figure 7.5.

probably due to more water and clay contents. The C horizon, being a mineral horizon that is located deep in the soil and that includes weathered rock, exhibited, as usual, the lowest SOC content.

The SOC contents of the master horizons were poorly correlated with the climate variables: annual rainfall, evaporation and the aridity index. For all master horizons, the influence of the topographical variables (viz., slope angle, shape and aspect) on their SOC content was also not very clear. Clay content provided a better explanation for the variation in SOC content of the master horizons, although the algorithms cannot be used with any predictive confidence. The low coefficient of determinations is attributed by Rantoa et al. (2015) to the large variability in the physiographical features of South Africa. These trends suggest that disaggregation of the country into more homogeneous ecoregions might result in better correlations and should be considered in future studies.

The SOC stocks, to a 300 mm soil depth, were estimated for South Africa by Rantoa (2009), using the SOC content of master A horizons and a bulk density of 1.5 Mg m^{-3} (Figure 7.4). This bulk density was based on measured values of various studies over the country. The horizontal cross section of this map showed that the SOC stock increased from <10 Mg ha^{-1} at the western Atlantic seaboard to >80 Mg ha^{-1} at the eastern Indian seaboard (Figure 7.5), revealing the influence of increasing rainfall and decreasing temperature.

7.5   AGRICULTURAL LAND USE

South Africa covers an area of 122 Mha, of which 101 Mha is used for various agricultural production systems (Figure 7.6). Most of this land, viz. 84.2 Mha, is used for grazing of veld by livestock. The remaining 16.8 Mha of land are potentially arable and hence suitable for either commercial or communal crop production. Currently 12.9 Mha are used for crop production, comprising 11.5 Mha under dryland and 1.4 Mha under irrigation (Table 7.3).

The major farming activity on arable land is the production of field crops, especially grains, on 9.5 Mha (Figure 7.6). In decreasing order, based on annual production over the last five years,

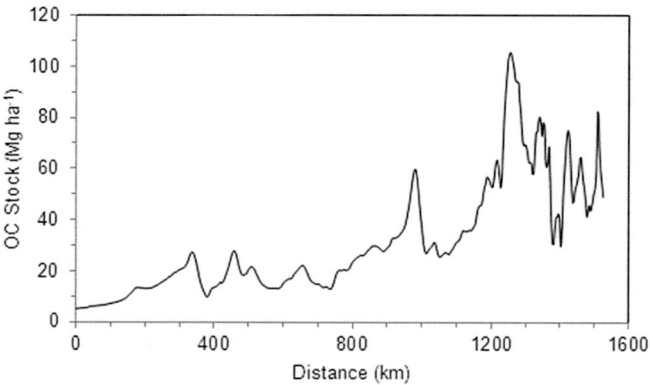

FIGURE 7.5   A cross section graph of the soil organic carbon stock map from west to east, as depicted in Figure 7.4 (adapted from Rantoa 2009).

the crops comprise maize (*Zea mays*), wheat (*Triticum aestivum*), soybean (*Glycine max*), sunflower (*Helianthus annuus*), barley (*Hordeum vulgare*), grain sorghum (*Sorghum bicolor*), canola (*Brassica napus*), dry bean (*Phaseoulus vulgaris*) and groundnut (*Arachis hypogaea*). Canola is produced exclusively in the winter rainfall region, and soybean, sunflower, grain sorghum, dry-bean, and groundnut in the summer rainfall region (DAFF 2016). In both regions, tillage is done by conventional moldboard and disc ploughing despite a strong trend toward mulch or no tillage to conserve soil resources and reduce production costs (Beukes et al. 2004).

Sugarcane is produced on 0.45 Mha in parts of KwaZulu-Natal, Mpumalanga and Limpopo, where climate and soil are suitable (Figure 7.6). This perennial crop lasts for several ratoons before yield drops

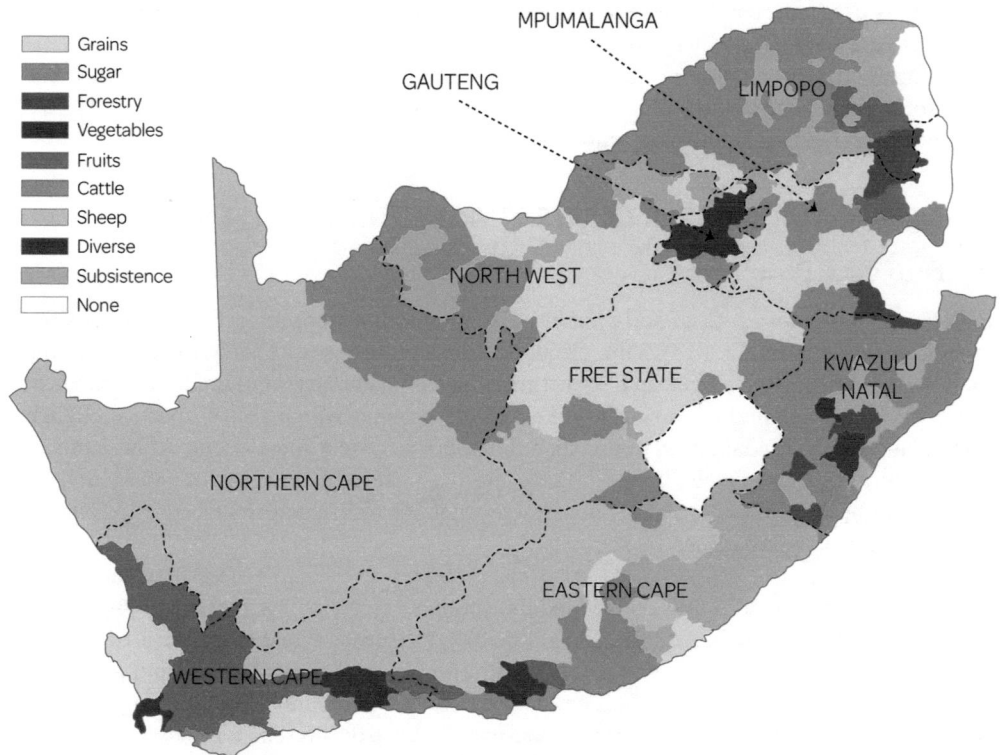

FIGURE 7.6   (See color insert.) Agricultural production areas in South Africa (adapted from FAO 2005).

TABLE 7.3

Synopsis of Arable Land Utilization in South Africa and Its Provinces, 1991 (DAFF 2016)

		----------------------------------- Farmland -----------------------------------				
		Potentially arable land			Arable land utilized	
Country and provinces	Total area	Total	Commercial agriculture	Communal agriculture[1]	Dryland	Irrigation
			--------------------------------------- million ha ---------------------------------------			
South Africa	122.32	100.67	14.19	2.55	11.54	1.36
Eastern Cape	17.06	14.82	0.64	0.53	0.41	0.19
Free State	12.94	11.79	4.19	0.04	3.86	0.14
Gauteng	1.89	0.83	0.44	–	0.38	0.03
Kwazulu-Natal	9.15	6.53	0.84	0.36	0.70	0.13
Limpopo	11.96	10.55	1.17	0.53	0.50	0.16
Mpumalanga	8.18	4.98	1.60	0.14	1.61	0.13
Northern Cape	36.33	29.54	2.41	–	0.03	0.19
North West	11.87	10.10	0.45	0.95	2.21	0.10
Western Cape	12.94	11.56	2.45	–	1.84	0.29

Note: [1] In former homelands.

to an uneconomical level that justifies replanting. In some instances, replanting of sugarcane coincides with the cultivation of an annual field crop for a year or two, while in others the opposite applies.

Pasture crops are planted in the southern coastal region on 0.6 Mha for dairy production (Figure 7.6). Kikuyu is used as a base pasture and either ryegrass or clover species are incorporated annually to improve both production and quality of the herbage.

These three crop production systems are practiced under diverse climate and soil conditions and therefore provide the opportunity to achieve better insight into human-induced changes of soil organic matter content and quality in arable land.

7.6   SOIL ORGANIC MATTER CONTENT AND QUALITY CHANGES

7.6.1   Field Crop Production

On a global scale, almost 24% of soils potentially suitable for cropping are already considered as being degraded (Jones et al. 2013). This problem is particularly severe in Africa, where land degradation and desertification affect up to two-thirds of the productive land area. To combat or even reverse such trends, cognizance must be taken of the rates at which soil properties change with prolonged land use, and adaptation to the specific land use system must occur. At best this is done within long-term research trials, because these are least affected by spatial heterogeneity in soil properties, and because there are also undesirable variations in management by different farmers. Some of these results are outlined in Section 7.6.3.

Running long-term field experiments over decades has however, hardly been practicable, owing to years of war and crisis in both the country and the continent (see also Swift et al. 1994). Even if concentrating on SOM as the main indicator for soil quality, the turnover rate of different SOM pools and constituents extends from years to centuries (e.g., Buyanovski et al. 1994; Derrien and Amelung 2011; Paul 2016). In the Free State, however, farmers started to break the savannah for cropping more than 100 years ago, and with farm expansion, have continued to do so up to the present. It is thus possible to sample the so-called pseudo chronosequences, that is, sites under different durations of land use. The next section deals with related studies on the vulnerability and resilience of sandy arable soils to land-use change.

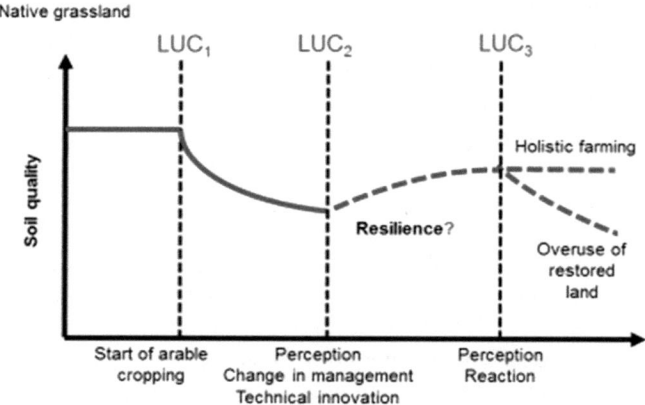

FIGURE 7.7   Conceptual figure of the response of soil quality to land-use change (LUC)

7.6.2   Chronosequence Studies in Free State Province

In general, conversion of natural to agroecosystems and ploughing leads to a rapid loss in soil quality, whereas, after the re-establishment of perennial cultures, soils may recover. Figure 7.7 is a conceptual indication that after breaking the savannah sod (LUC1 = first land-use change), soils continue to degrade until the point when farmers no longer accept the declining crop yields and lack of economic revenues. At this stage, farmers might be willing to invest in new technologies, such as irrigation, soil conservation measures or soil conditioners, or might simply change land use again (LUC2), for example, by converting formerly degraded cropland back into pastureland. If this is done, soils may restore – at best they are even resilient to former land-use change; in other words, their properties re-adapt to the earlier state of an undisturbed savannah. Finally, land use is frequently altered again (LUC3), for example, by novel holistic farming approaches that maintain soil properties, or by cash-crop production that might repeat the risk of a temporal decline in soil properties. Similar processes occurred in the Free State Province of South Africa: The savannah soils of the grassland biome were converted to cropland for the production of wheat, maize and sunflower (LUC1). About 50 years ago, some farmers started to reseed grasses for the re-use of formerly degraded soils as secondary pastures (LUC2). The system nowadays offers options of holistic farming practices (LUC3) in which arable sites are rotated with secondary grassland for a more sustainable use of land.

7.6.2.1   Conceptual Chronosequences of Land Degradation and Restoration

Several studies on temporal changes in soil properties with land-use change have been performed in three agroecosystems close to Harrismith, Kroonstad and Tweespruit in the Free State Province of South Africa. The characteristics of the sites are given by Lobe et al. (2001). The altitude of the sites ranged between 1,350 and 1,830 m above sea level, dominated by Plinthustalfs (Soil Survey Staff 2014), which are equivalent to Westleigh or Avalon soil forms in the South African soil classification system (Soil Classification Working Group 1991). Variations in soil texture were confined to a narrow range. All three agroecosystems are in the summer rainfall region (six months from October to March, with >50 mm rainfall), with 516 to 625 mm mean annual rainfall, and 13.8–16.6°C mean annual temperature. The three agroecosystems belong to the grassland biome – Harrismith: moist cold highveld grassland; Kroonstad: dry sandy highveld grassland; Tweespruit: moist cool highveld grassland (Bredenkamp et al. 1996). The native grassland was used as primary pasture for grazing cattle or sheep. The botanical composition was maintained by: narrow-leaved turpentine (*Cymbopogon plurinodis*), red (*Themeda triandra)*, creeping bristle (*Setaria sphacelata*), wire (*Elionurus muticus*) and weeping love (*Eragrostis curvula*) grasses in Harrismith; Lehmann's love

(*Eragrostis lehmanniana*), dew (*Eragrostis obtusa*), small buffalo (*Panicum coloratum*) and silky bushman (*Stipagrostis uniplumis*) grasses in Kroonstad; and red grass in Tweespruit. The stocking density on the primary grassland was on average 0.4 large stock units per hectare at Harrismith, 0.5 at Kroonstad and 0.6 at Tweespruit; and the grazing period was between 1 and 3 months in total per year. The arable land had been ploughed to a depth of 200–300 mm. Inorganic fertilizer was applied annually (maize: 50–70 kg N, 10–25 kg P, 0–10 kg K ha^{-1}; wheat: 10–40 kg N, 10–25 kg P, 0–15 kg K ha^{-1}; sunflower: 20–50 kg N, 10–20 kg P, 2–6 kg K ha^{-1}). Average grain yields varied from 2.2–3.75 Mg ha^{-1} year^{-1} for maize, 1.2–2.75 Mg ha^{-1} year^{-1} for wheat, and 1.0–1.25 Mg ha^{-1} year^{-1} for sunflower. Within a rotation cycle of one to two years, the soils were kept clear of vegetation for up to six months in the dry season so that the soil retained its stored water. This technique might enhance the potential loss of SOM because little, if any, organic matter is added back. One of the main risks is enhanced erosion during the period without protective plant cover to stabilize the soil.

In each agroecosystem, nine individual sampling sites with varied lengths of time under cultivation were used for analysis. All sites were adjacent to the primary pasture. The sites thus comprised a gradient with increasing soil degradation after LUC1. In each of the three agroecosystems, additional grassland sites could be selected. These sites had formerly been cropped for >20 years and were now continuously used as pastures for a known number of years (1.5 to 31 years). The native primary grassland served as additional control. These secondary pastures contained both commercially seeded (pasture) species (e.g., common finger (*Digitaria smutsii*) and weeping love grasses), as well as native grasses that invaded secondary pastures (e.g., love (*Eragrostis* sp.), red and common thatching (*Hyparrhenia hirta*) grasses). After 20 years of cropping, the SOC contents of the degraded fields deviated less than 10% from the steady state SOC content reached after prolonged cropping (derived from Lobe et al. 2001). Both secondary pastures and primary grassland sites were grazed by cattle and, in some places, by sheep. The stocking density for secondary pastures in the three agroecosystems ranged between 0.25 and 0.65 livestock units ha^{-1} year^{-1}. At a few sites, the secondary pastures had also been used occasionally for hay production (~2 t DM (dry matter) ha^{-1} year^{-1}, information provided by farmers). The sites were never irrigated or fertilized with organic fertilizers, as was the case with the primary grassland sites. The seeded secondary pastures received either a fertilizer mixture or single fertilizers, such as super phosphate and limestone ammonium nitrate, to promote pasture establishment, amounting to 0–56 kg N ha^{-1} and 0–33 kg P ha^{-1} administered once at the beginning of pasture establishment. At most younger sites, the sown grass species still dominated the vegetation cover at the time of sampling. At older secondary pastures, the diversity of plant species tended to be higher. The sites thus comprised a land restoration sequence, with increasing time for soil restoration and secondary pasture establishment (LUC2).

7.6.2.2 Soil Degradation Rates with Prolonged Arable Cropping

With increasing duration of cultivation, the yields declined exponentially, both for wheat and maize (Figure 7.8; Lobe et al. 2005). The results for sunflower are not shown as they are only grown on selected fields. Two phases of land changes could be distinguished: the yields declined rapidly within 1–2 decades (Phase I) until a new steady-state equilibrium was maintained (Phase II). The final yields, however, only ascertained a 5–10% revenue of investment costs. It has to be noted that during the period of observation, the development of new crop cultivars by breeding improved. As a result, the apparent stability in yields could not be taken as an indicator that soil quality remained unaffected, because better adapted cultivars might have sustained yield at worsened soil property levels.

As cropping proceeded, SOM and available N were lost (Du Toit et al. 1994; Du Preez and Du Toit 1995; Lobe et al. 2001), because little if any organic fertilizer was available to replenish SOC losses. With declining yields, the annual C input rates declined rapidly after conversion of the grassland to cropland, resulting in rapid annual C output rates that approached steady-state equilibrium after ca. 20 years of cropping (Lobe et al. 2005). With declining yields, the abundance of arbuscular mycorrhizal fungi was also reduced, as indicated by losses of glomalin-related soil proteins, the

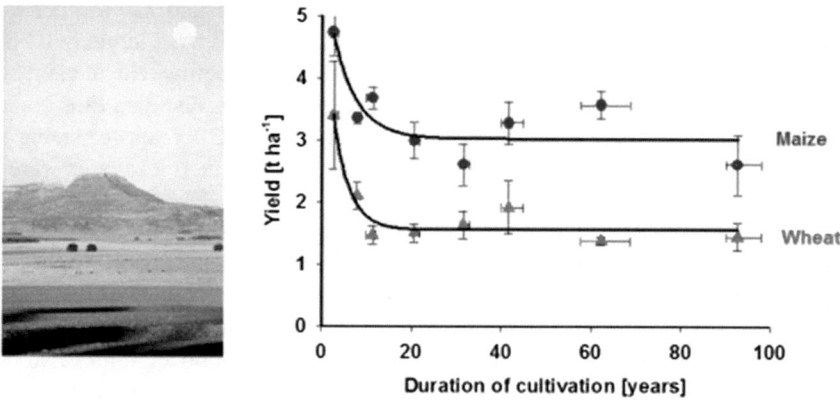

FIGURE 7.8 Yield development with prolonged arable cropping of sandy soils in the South African Highveld (from Lobe et al. 2005).

contents of which reached steady-state conditions after 11–92 years on a level of 39–69% of the initial contents (Preger et al. 2007). Nevertheless, the overall SOC stocks for the top 200 mm of soil, did not reach steady-state conditions but continued to decline (Lobe et al. 2001), thus reflecting continued soil degradation that was not immediately reflected in the yields. Hence, two phases of SOM loss could be distinguished: a phase of rapid SOM loss within 1–2 decades (Phase I), and a phase of slow continued SOM loss, when yields already were stabilized with progress in cultivar development (Phase II) (Figure 7.9). Prolonged arable cropping thus resulted in an overall loss of 60% SOC stored at the corresponding depth of the native primary pasture, at an absolute SOC loss of about 2 kg SOC m^{-2}, that is, 20 t ha^{-1} over 90 years.

The SOC losses could be described by a bi-exponential equation (Equation 7.1), assuming a labile and a stable pool (Gregorich et al. 1996; Lobe et al. 2001):

$$Xt = X1\exp(-k1\,t) + X2\exp(-k2\,t) \tag{7.1}$$

where Xt is the concentration of C or N in the soil at cultivation time t, $X1$ is the concentration of C or N of the labile pool, $X2$ is the concentration of C or N of the stable pool, and $X2 = X0 - X1$, $X0$ is the initial concentration of C or N in the soil ($t = 0$, concentration of the grassland), $k1$ is the rate constant of the labile pool (year^{-1}), and $k2$ is the rate constant of the stable pool (year^{-1}). Overall,

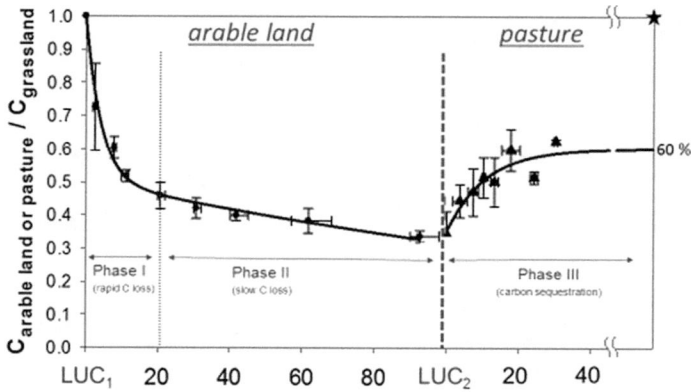

FIGURE 7.9 Losses and gains of SOC after converting native savannah to cropland (LUC1) and back to secondary pastures (LUC2) in sandy South African Highveld (after Lobe et al. 2001; Preger et al. 2010; redrawn by Kösters 2018).

about 50% of the SOC was assigned to the labile and stable pool X1 and X2, respectively (Lobe et al. 2001). The respective rate constants were $k1 = 0.23$ year^{-1} for the first, labile SOC pool, and $k2 = 0.0046$ years^{-1} for the second (Lobe et al. 2001). The inverse of k defines the mean residence time of SOC within a given SOC pool (e.g., Derrien and Amelung 2011). The data thus correspond to a mean residence time of 4.3 years for SOC in the labile pool, but 220 years for stable SOC. After about 17 years, the labile SOC pool was largely lost, and the continued loss rates were dominated by stable SOC pool dynamics.

The identification of different C pools also allowed mathematical models that conceptualize the SOM into pools of different stability, such as the RothC carbon model (Jenkinson and Rayner 1977; Skjemstadt et al. 2004). When applying the Roth C carbon model in combination with stable isotope-tracing of C3-derived crop carbon within this C3 (wheat, sunflower) and C4 (grasses, maize) mixed agroecosystem, the data indicated that only minor portions of novel humus had been formed from the crop itself (Figure 7.10; Lobe et al. 2005); that is, the systems had lost their function to efficiently re-form the SOM after prolonged land use.

When soils are ploughed, aggregates are broken down and oxidative decomposition of the SOM accelerates with increased aeration and loss of physical protection. Fractionating soil into aggregates of different sizes thus mimicked the fast loss rates of the SOC. As shown in Figure 7.11, the rapid loss of the SOC (Phase I in Figure 7.9) coincided with a rapid loss of soil macroaggregates >2,800 µm. After 17 years, a new steady-state equilibrium was reached, which corresponded to the time at which stable SOC kinetics controlled the overall SOM loss rates (the aggregate fraction of 2,000 to 2,800 µm showed similar dissipation kinetics as the 2,800–8,000 µm fraction, with a slight delay in reaching equilibrium; Lobe et al. 2011). As a result, it seems reasonable to assume that rapid disintegration of large macroaggregates (peds) accounted for both loss of labile SOC pools with prolonged arable cropping (Phase I of SOC losses; Figure 7.9) as well as the functional lack of the soil to build up novel SOM efficiently from the crop C input (Figure 7.10).

The similarity of the figures suggests that labile SOC might, at least in part, comprise SOC that was formerly protected within aggregates, even if SOC losses occurred from all aggregate fractions, not only the largest (Lobe et al. 2011). The macroaggregates that are broken down during initial

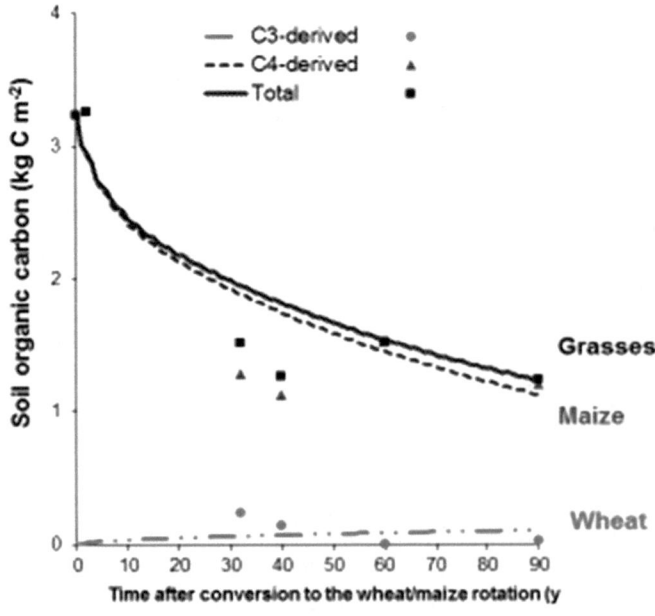

FIGURE 7.10   Modelled total, C3- and C4-derived soil C stock using the Roth-C model for Tweespruit agroecosystem (from Lobe et al. 2005).

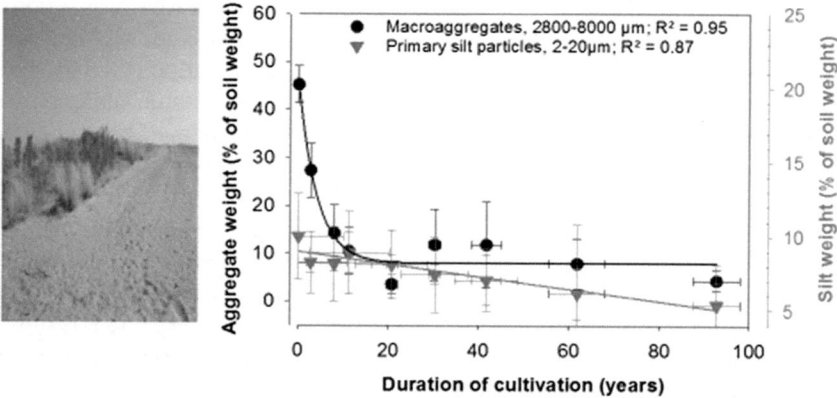

FIGURE 7.11   Soil macroaggregation and content of silt-sized particles with prolonged arable cropping of sandy soils in the South African Highveld (redrawn with data from Lobe et al. 2001, 2011).

phases of soil management are known to stabilize particulate plant materials rich in lignin (Golchin et al. 1994; Amelung and Zech 1996; Kögel-Knabner and Amelung 2014). Indeed, with the rapid loss of macroaggregates, this plant-derived C was lost as rapidly as the macroaggregates: the total contents of VSC-lignin declined with prolonged arable cropping until a new steady-state equilibrium was reached after 24.9 years. This is slightly slower than the loss of bulk SOC, confirming that lignin might be more resistant to degradation than other SOC constituents; however, since most of the lignin was associated with the labile SOC pool, the overall loss rate was not significantly different from that of bulk SOC (Lobe et al. 2002). The process was accompanied by a rapid increase in phenolic acid-to-aldehyde ratios, due to the rapid oxidative alteration of the side-chains in the remaining lignin macromolecule (Lobe et al. 2002; data not shown here). However, it is noteworthy that Dungait et al. (2012) stated in their review that the major C stabilization mechanisms in soils are now recognized to be biological, non-preferred soil spaces where SOM is physically protected from microbial activity, regardless of its initial chemical structure.

The steady-state equilibrium of both macroaggregate disintegration and lignin losses coincides with the stabilization of yields (Figure 7.8) but cannot explain the ongoing SOC losses (Figure 7.9). Hence, additional processes must account for this loss. The bare soil between harvest and reseeding remains prone to wind erosion. Mainly silt-associated C has been blown away, thus contributing to the slow, long-term loss of total SOC (Figure 7.11; Lobe et al. 2001). It therefore seems reasonable to conclude that the ongoing loss of SOC (Phase II of SOC loss with prolonged cropping; Figure 7.9) was caused by wind erosion. This finding agrees with the observations of the farmers who contend that soil is lost from the sites in small but significant quantities due to wind erosion. Much of this blown-off silt likely re-accumulates at the edges of the old arable fields, as evidenced by the formation of small dunes with some grass species near the main cropping sites (Figure 7.11; slide). Yet, since silt also contains old C and N, this loss of humus forms cannot be replaced easily; that is, SOM properties underwent irreversible change upon management (Brodowski et al. 2004).

When plant materials are decomposed, microorganisms may recycle part of this plant-derived C in their biomass. Hence, although the total contents of plant residues and microbial biomass residues declined, the contribution of microbial residues to the remaining SOC tended to increase with prolonged duration of continuous cropping (Amelung et al. 2002). The increased microbial transformation also left a fingerprint: increasing D-contents of alanine pointed to a higher degree of bacterial transformation of the remaining SOM (Brodowski et al. 2004), whereas elevated ratios of glucosamine to muramic acid reflected that even greater quantities of the remaining microbial residues originated from fungi (see also Guggenberger et al. 1999; Amelung et al. 2002).

The overall composition of SOM was not strikingly different from what was reported for other soils. With microbial conversion of plant remains however, it shifted after breaking the savannah. As a result, the soils under cropland showed: (i) a continued degradation of O-alkyl and acetal-C structures found in carbohydrate and holocellulose biomolecules, and some labile aliphatic-C functionalities; (ii) an increased side-chain oxidation of phenylpropane units of lignin; and (iii) a selective accumulation of recalcitrant H and C substituted aryl-C and aliphatic-C components relative to the native grassland (Solomon et al. 2007).

With changing SOM stocks and quality, the cropland soils also exhibited different bonding forms of its major organic nutrients, such as P and S. While total P contents were maintained due to adapted fertilization, Von Sperber et al. (2017) reported that the contributions of inorganic P (Pi) to the total P of bulk soil increased from 37% to 63%, whereas those of organic P (Po) were reduced respectively. After approximately 60 years of cultivation, that is, about 40 years later than for SOC, a new steady-state equilibrium was approached, which was characterized by overall smaller Po and larger Pi concentrations. In contrast to P, total S contents were not maintained with prolonged arable land use, but reduced by about 40% relative to the native savannah. The remaining S forms were depleted in reduced S structures (thiols, monosulphides, disulphides and polysulphides) and enriched in oxidized S forms (ester sulphates) (Solomon et al. 2005).

It has to be recalled that all current data described in Section 7.6.2.2 refer to Plinthustalfs, with loamy sand as the major particle size class. For the different soils occurring in South Africa (see Figure 7.3), as well as for the different climatic regimes, the SOC loss rates will differ. Du Toit et al. (1994) as well as Du Preez and Du Toit (1995) sampled similar chronosequences in other regions of the Free State. The sites differed slightly in aridity, but mainly in soil order and texture. As shown in Figure 7.12, the SOC loss rates vary significantly across soil orders and particle size classes. The largest SOC stocks were assessed for the clayey Vertisols (Haplusterts according to *Soil Taxonomy*), the smallest ones for the sandy Arenosols (Psamments), while those of the loamy Plinthosols (Plinthustalfs) were intermediate. The mean residence time (MRT) for the SOC (MRT = $1/k$; k derived from mono-exponential equations fitted to curves in Figure 7.12) thus declined in the order of the clayey Vertisols (MRT = 33 years) > loamy Plinthosols (MRT = 16 years) > sandy Arenosols (MRT = 4 years). The degree of the above-described land-use effects on the SOM stocks and quality is thus very likely site-specific.

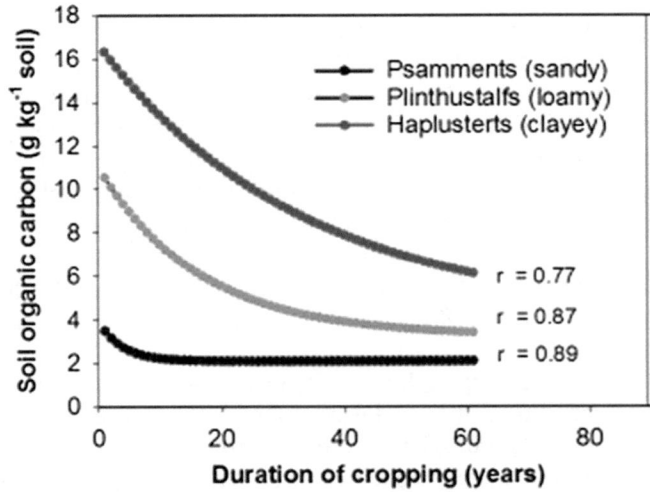

FIGURE 7.12 Range of SOC losses with prolonged arable cropping in South African agroecosystems dominated by different soil orders (data from Du Toit et al. 1994; the *r* value represents the correlation coefficient for a mono-exponential relationship).

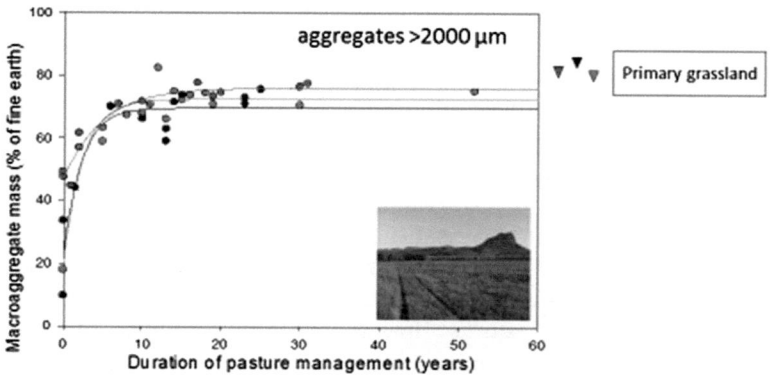

FIGURE 7.13 Regeneration of soil macroaggregation after reseeding of grasses into formerly degraded arable land of three agroecosystems within the South African Highveld (redrawn from Kösters et al. 2013).

7.6.2.3 Soil Regeneration Rates Under Prolonged Secondary Pasture

When the sites were reconverted to grassland, the soil structure was restored almost completely. This spanned a time period of 9.5–18 years until aggregate-size distribution in the secondary pasture soils resembled that of the primary pastures in the native savannah (Figure 7.13; Kösters et al. 2013). Hence, the soils more or less restored their main physical soil properties, which gave rise to the question whether the potential to sequester C by physical stabilization processes was also restored.

Although the surface soils recovered SOC, the overall C sequestration was however incomplete (Figure 7.9). The C sequestration rates showed significant differences among ecosystems (Preger et al. 2010). Moreover, C sequestration was pronounced only for the very surface soil (0–50 mm), in part also for the 50–100 mm depth interval; yet, there was little, if any, gain in SOC in the 100–200 mm depth interval and thus likely also not below this depth (Preger et al. 2010). Overall, between 9.0 and 15.3 Mg ha^{-1} the SOC was sequestered within 25–30 years in the 0–200 mm of surface soil following this land conversion. Finally, the SOC stocks of secondary pastures reached 60% of the SOC stocks of the primary pastures (Figure 7.9).

Lignin-derived phenols showed similar trends to those observed with SOC when degraded cropland has been converted to secondary pasture (Kösters et al. 2018). Overall it took 8–19 years until crop lignin was replaced by grass lignin. The degree of lignin side-chain oxidation was less pronounced in the secondary pasture than in the cropland. The accrual of OM in the 0–50 mm soil therefore went along with a rejuvenation of its lignin.

The SOC sequestered in soil under secondary pasture was comprised of different humic substances, as outlined by Kotzé et al. (2016). This recovery of SOC amounted on average to 49% in crude humic substances, 34% in extractable humic substances, 26% in humic acids and 44% in fulvic acids. Although humic substances are recalcitrant in nature, management practices that prohibit soil disturbance are needed to complement the refractory nature of humic substances against biological oxidation. In addition, the humic substances might be essential in the re-aggregation process of cultivated soils.

The stocks of SOM bound to large macroaggregates reached only 50% of the primary grassland. As the SOC concentration in the aggregates did not change significantly with prolonged pasture use, the increase in SOC stocks in the secondary pasture soils was mainly due to a rebuilding of the large macroaggregates. They contained more SOC than the smaller-sized aggregate classes, but not more SOC than the respective aggregates in the degraded cropland. Kösters et al. (2013) suggested that only the amount of large macroaggregates regenerated upon land conversion, while their protective capacity was obviously not restored, and was thus not strong enough to account for a full sequestration of SOM.

It must be noted that the secondary pasture soils exhibited a slightly finer texture than the primary grassland. On the one hand, this texture offset might be due to the fact that the soils had lost silt through wind erosion (Lobe et al. 2001). On the other hand, these data might also indicate that the farmers primarily withdrew the finer soil from arable land use, because finer textures enhance evaporation, whereas infiltration is facilitated for sandy soils, resulting in larger amounts of plant-available soil water above the plinthite layer. Nevertheless, the slow changes in specific soil properties indicate that previous losses of SOM cannot easily be reversed, suggesting that the native primary grassland soils are only partially resilient to land-use change.

The restored SOM also changed its quality. Using amino sugars as markers for microbial residues in soil (Amelung et al. 2008), Lauer et al. (2011) found that the concentration of microbial residues also increased with increasing SOC stocks. However, the final proportion of amino sugars in the SOM corresponded to only 90% of the amino sugar level in the primary grassland. The amino sugar concentration in the clay-sized fraction recovered to a higher end-level than in the bulk soil, and also at a faster annual rate. This confirms that the finer particles specifically contained elevated amounts of amino sugars and were thus responsible for the restoration of microbially derived C and N. The incomplete recovery of amino sugars in bulk soil is in line with results from the US Crop Reserve Program (Amelung et al. 2001), and can only in part be attributed to a slightly coarser texture of secondary grassland that had lost silt through wind erosion. The soils particularly had also lost the ability to restore microbial residues below a soil depth of 50 mm. Overall, the ratios of glucosamine to muramic acid also increased with increasing duration of pasture usage, suggesting that fungi dominated the microbial sequestration of C and N, whereas the re-accumulation of bacterial cell wall residues was less pronounced. However, the glucosamine-to-muramic acid ratios exceeded those of the primary grassland, indicating that some irreversible changes of the soil microbial residue composition persisted, and were thus a likely result of microbial nutrient cycling by former intensive crop management. Future establishment of secondary pastures might want to address this issue, for example, by N, S fertilization, decompaction, as well as the introduction of legumes as perennial crops into improved pasture management. Some of these options have already been put into practice by a few farmers.

Similar to the United States tallgrass prairies (Bach et al. 2010), microbial phospholipid fatty acid (PLFA) richness, total PLFA biomass, mean-weighted aggregate diameter, and proportion of C and N in aggregate fractions increased exponentially to a maximum across the Harrismith chronosequence in the South African Highveld, with the conversion of cropland to grassland (Baer et al. 2015). However, the time frame for restoration of total soil C and N stocks to steady-state conditions in the Highveld was shorter than in the tallgrass prairies. This more rapid soil recovery from cultivation through prolonged secondary pasture was attributed to greater C saturation deficit at the onset of restoration, coupled with development of microbial biomass and aggregate structure, and climate limitation in the Highveld relative to the tallgrass prairies.

7.6.3 Long-Term Field Trial Studies

In comparison with the summer rainfall region, more field trial studies were completed or still continue in the winter rainfall region to examine inter alia the long-term effects of crop cultivation management practices on SOM content and quality. The results of selected trials in the two regions are presented and concisely discussed here.

7.6.3.1 Winter Rainfall Region

The first trial of this particular nature commenced in 1976 on the Langgewens research farm (33°16′34″S, 18°45′51″E; altitude 260 m) near Malmesbury. This farm receives a mean annual rainfall of 395 mm and has a mean annual temperature of 18.2°C. The trial was dominated by a Swartland soil form with 20–25% clay to 400 mm depth. From 1976 until 1990 tillage practices were tested with wheat in monoculture, and from 1990 onward with an additional wheat-lupine-wheat-canola rotation. The tillage practices included conventional clean tillage, conventional mulch tillage, minimum tillage and no tillage. In addition to annual minimum tillage, this practice was also conducted on a second, third and fourth yearly basis, with no tillage in between (Smit 2004).

In 2001, the SOC content to 100 mm depth tended to be larger under monocropping than under crop rotation, especially in the 50–100 mm layer although not to a significant degree. Conversely, crop rotation resulted in a larger SOC content than was the case with monocropping at the 100–400 mm depth, with a significant difference in the 300–400 mm layer (Smit 2004).

The SOC content in the 0–50 mm layer varied from 4.6 g kg^{-1} with conventional clean tillage to 11.5 g kg^{-1} with no tillage. An intermediate SOC content of about 8.2 g kg^{-1} was found with conventional mulch tillage and minimum tillage. This trend was repeated in the 50–100 mm layer, although not significantly. However, SOC content for the 100–200 mm layer was significantly higher with conventional clean tillage than with conventional mulch, minimum and no tillage. These tillage practices had no significant effect on SOC content below 200 mm depth (Smit 2004). These results confirm those of Agenbag and Maree (1989) to a large extent, implying that the treatment effects on SOC content manifested relatively early in this trial.

In a follow-up study, the effects of tillage practice and crop rotation on SOC were investigated by Cooper (2016) in two other trials at Langgewens research farm. One of the trials is laid out on a sandy loam Swartland soil form. The SOC stocks to 400 mm depth ranged in the nineteenth year of this trial from 13 Mg ha^{-1} (conventional tillage combined with wheat-canola-wheat-lupine rotation) to 31 Mg ha^{-1} (no tillage combined with wheat monoculture) (Table 7.4). However, with either wheat monoculture or wheat-medic-wheat-medic rotation, no tillage resulted in the highest SOC stocks, followed by minimum and no tillage. In the wheat-clover-wheat-lupine rotation, minimum tillage had the largest SOC stocks followed by no and conventional tillage.

In the other trial, on a sandy loam Glenrosa soil form, no-tillage crop rotations were examined. Stocks of SOC to 400 mm depth in the eighth year of this trial amounted to 22.1 Mg ha^{-1} for wheat-medic-wheat-medic rotation, 24.9 Mg ha^{-1} for wheat-medic/clover-wheat-medic/clover rotation, 25.2 Mg ha^{-1} for natural vegetated soil and 32.5 Mg ha^{-1} for wheat monoculture (Cooper 2016).

From Cooper's (2016) results, it can be concluded that tillage practices had a stronger influence on SOC stocks than crop rotation. The best combination for the sequestration of SOM was no-tillage with wheat monoculture.

Smith (2014) also investigated the effects of no-tillage crop rotations on SOC at Tygerhoek research farm (34°29′32″S, 19°54′30″E; altitude 158 m) near Riviersonderend. A mean annual rainfall of 450 mm and mean annual temperature of 17.5°C are noted for this farm. The trial is located

TABLE 7.4

SOC Stocks (Mg ha^{-1}) to 400 mm Depth under Various Tillage Practice and Crop Rotation Combinations

	Crop rotation[1]		
Tillage practice	WWWW	WMWM	WCML
No	31[a]	30[a]	22[c]
Minimum	28[b]	27[b]	28[b]
Conventional	22[c]	21[c]	13[d]

Source: Compiled from Cooper 2016.

Notes: [1] WWWW = wheat monoculture

Lower case letters indicate significant differences at P<0.05.

WMWM = wheat-medic-wheat-medic

WCML = wheat-canola-medic-lupine

on a Lithosol (locally a Glenrosa soil form) with 20% to 25% clay, predominantly kaolinite and illite. In this trial, which still continues, five no-tillage crop rotation systems are compared: lucerne (100% pasture), medic-medic-wheat (67% pasture and 33% crop), medic-medic-wheat-wheat (50% pasture and 50% crop) and wheat-barley-canola-wheat-barley-lupine (100% crop) in two different phases. Soil with natural fynbos vegetation serves as reference.

After 11 years, medic-wheat rotations had the largest SOC contents (15.2–18.6 g kg^{-1}) and stocks (70.2–74.9 Mg ha^{-1}) to 300 mm depth, compared to continuous cropping (13.3–14.1 g kg^{-1} and 54.7–58.9 Mg ha^{-1}), lucerne (15.0 g kg^{-1} and 63.4 Mg ha^{-1}) or natural vegetation soil (13.2 g kg^{-1} and 60.5 Mg ha^{-1}). This was attributed to larger below-ground inputs through roots and less soil disturbance in the upper 100 mm (Smith 2014).

Organic matter fractions were determined (Smith 2014) in the 50–100 mm depth only. In this layer the contribution of free particulate organic matter (fPOM: labile C) to SOC in the cultivated treatments (6–9%) was lower than in the natural vegetated soil (13%). This fraction is the most sensitive SOC pool to detect changes due to management practices, which include quantity and quality of OM inputs, extent of physical disturbance and fertilization. The medic-wheat rotations had the largest C contents (1.37–1.74 g kg^{-1}) and stocks (1.06–1.14 Mg ha^{-1}) of the cultivated treatments, followed by lucerne (1.30 g kg^{-1} and 0.88 Mg ha^{-1}) and continuous cropping (0.9–1.0 g kg^{-1} and 0.62–0.66 Mg ha^{-1}). These cultivated treatments, however, had lower contents and stocks in the occluded particulate organic matter (oPOM: moderately stabile C) fraction that contributed on average only 0.4–1.0% to SOC. The major part (85–93%) of SOC was associated with the mineral fraction (stabile C). In this fraction the C contents were 11.0, 15.5, 17.3 and 18.2 g kg^{-1} and stocks 8.4, 9.8, 11.2 and 11.9 Mg ha^{-1} for the natural vegetated soil, continuous cropping, lucerne and medic-wheat rotations, respectively (Smith 2014).

7.6.3.2 Summer Rainfall Region

The data of one trial only are worth dealing with here. This trial is on the ARC-Small Grain Institute research farm (28°09′00″S, 28°17′00″E; altitude 1,680 m) near Bethlehem. For this farm, a mean annual rainfall of 743 mm and mean annual temperature of 13.7°C were recorded. The trial was laid out on a Plinthosol (locally an Avalon soil form) with 18% clay in the topsoil. Monocropped wheat has been planted since the trial started in 1979 and subjected to the following treatments in factorial arrangement: two straw management treatments (burned and unburned), three tillage methods (ploughing, stubble mulch and no tillage), two weed control methods (mechanical and chemical) and three levels of nitrogen fertilization (20, 30 and 40 kg ha^{-1} until 2002, and thereafter 20, 40 and 60 kg ha^{-1}).

The change of SOC content was studied in the intermediate N level plots after periods of 10 years (Wiltshire and Du Preez 1993), 20 years (Kotzé and Du Preez 2007) and 31 years (Loke et al. 2012) had elapsed. Results of the three studies indicated that the effects, especially of straw burning and to a lesser extent of weeding, were small compared to tillage practice. The impact of tillage practice manifested significantly in the 0–50 mm layer and showed similar trends in the 50–100 mm layer. On a relative basis, the mulched and no-tilled plots respectively contained 10% and 22% more SOC than the ploughed plots in the 0–50 mm layer after 10 years, and 20% and 39% more after 20 years, respectively. At 20 years the SOC content of the ploughed, mulched and no-tilled plots were 6.0, 7.2 and 8.4 g kg^{-1}, respectively. Over the next 11 years, the SOC content increased by 12% in the ploughed plots and decreased by 7% in the mulched and 11% in the no-tilled plots. Thus, after 32 years, the no-tilled plots contained, on a relative basis only, 12% more SOC in the 0–50 mm layer than either the mulched or ploughed plots. The presumption was that after 20 years, the amount of recycled wheat residues of this trial was inadequate to sustain SOC in either the mulched or no-tilled plots, when compared to the ploughed plots.

It is noteworthy that although conservation tillage practices seemed to improve SOM, especially in the surface soil, wheat grain yield remained significantly lower in these management systems compared to that in the conventional tillage practices. For example, from 1979 to 2010, an average grain yield of 2.11, 2.19 and 2.28 Mg ha^{-1} was realized for the no-tilled, mulched and ploughed treatments, respectively. This discrepancy is discussed elsewhere by Loke et al. (2012) in some detail.

7.7   SUGARCANE PRODUCTION

The production of sugarcane is of some importance in parts of KwaZulu-Natal, Mpumalanga and Limpopo, where climate and soil are suitable for the cultivation of this perennial crop (Figure 7.6). Most of the 0.45 Mha land on which sugarcane is cultivated has been converted from indigenous forest and grassland. Traditionally, sugarcane is monocropped, replanted after an average of eight ratoons and burnt prior to each harvest for removal of leafy, non-sucrose-containing biomass. The residue after harvesting is left mostly on the soil surface. This practice may result in the depletion of SOC and much research was done. especially in KwaZulu-Natal on this topic.

The SOM content and stock of virgin and adjoining fields at 29 sites in northern KwaZulu-Natal (15 sites originating from dryland and 14 sites from irrigated sugarcane areas) were compared by Van Antwerpen and Meyer (1996). These sites, with an annual rainfall of 900 to 1,200 mm, represent 16 soil forms (Bonheim, Fernwood, Glenrosa, Hutton, Inanda, Katspruit, Kroonstad, Mayo, Millwood, Mispah, Nomanci, Oakleaf, Shortlands, Swartland, Westleigh and Willowbrook), which are accommodated in all the broad soil pattern groups, except the podzolic group (Table 7.1). The production of sugarcane, ranging from 2 to more than 50 years, depleted SOM content and stock significantly under dryland to 150 mm depth, and under irrigation to 300 mm depth (Table 7.5). On a relative basis the depletion of both the content and stock of SOM was more pronounced in the irrigated than in the dryland areas.

In another study, Qongqo and Van Antwerpen (2000) determined the impact of dryland sugarcane production on SOM in the midlands (918 mm annual rainfall) and on the south coast (1,025 mm annual rainfall) of Kwazulu-Natal (Table 7.6). On the clayey dolerite-derived Hutton soils of the midlands, organic matter content decreased in the 0–100 mm layer and increased in the 100–200 mm layer, though not significantly, over 30 years of sugarcane cultivation. The SOM stock, however, tended to increase in the two layers. The cultivation of sugarcane over a period of 50 years decreased SOM content and stock significantly to 200 mm depth in the sandy loam, granite-derived soils on the south coast. This discrepancy between the Glenrosa and Hutton soils was attributed to clay protection of SOM.

The above-mentioned studies prompted Dominy et al. (2001, 2002) to establish the effects on SOM and related properties of increasing the period under sugarcane in a Glenrosa soil (18% clay,

TABLE 7.5

SOM Content and Stock under Dryland and Irrigated Sugarcane Production in Northern KwaZulu-Natal

Soil depth mm	Dryland (n = 15)			Irrigated (n = 14)		
	Virgin	Cultivated	P-value	Virgin	Cultivated	P-value
Content (g kg^{-1})						
0–150	38.7	33.1	<0.05	24.0	18.8	<0.05
150–300	33.3	31.9	>0.05	20.8	16.9	<0.05
300–450	31.6	30.4	>0.05	14.6	13.9	>0.05
Stock (Mg ha^{-1})						
0–150	70.1	66.5	<0.05	50.3	40.8	<0.05
150–300	70.8	69.9	>0.05	46.9	40.6	<0.05
300–450	67.1	69.6	>0.05	32.9	33.4	>0.05

Source: Compiled from Van Antwerpen and Meyer 1996.

TABLE 7.6
**Change in SOM Content and Stock under Dryland Sugarcane
Production in the Midlands and on the South Coast of KwaZulu-Natal**

Soil depth mm	Midlands (n = 4)			South coast (n = 7)		
	Virgin	Cultivated	P-value	Virgin	Cultivated	P-value
Content (g kg^{-1})						
0–100	60.6	58.0	>0.05	46.8	29.4	<0.05
100–200	53.6	56.4	>0.05	33.5	25.3	<0.05
Stock (Mg ha^{-1})						
0–100	71.8	76.9	>0.05	73.1	52.1	<0.05
100–200	64.6	72.1	>0.05	54.3	44.1	<0.05

Sorce: Compiled from Qongqo and Van Antwerpen 2000)

predominately kaolinite) on the south coast (mean annual rainfall 965 mm and temperature 19.9°C) and in a Hutton soil (61% clay, dominated by kaolinite and halloysite) in the midlands (mean annual rainfall 821 mm and temperature 18.3°C) of Kwazulu-Natal. The cultivation of sugarcane ranged from 0 to 50 years on the Hutton soil and from 0 to 70 years on the Glenrosa soil.

The SOC content in the upper 100 mm soil layer at both sites under undisturbed vegetation was 46 g kg^{-1}. The SOC content declined exponentially with increasing years under sugarcane. It declined rapidly in the Glenrosa soil over the first 10 years and reached a new equilibrium of 13 g kg^{-1} after 30 years. In the Hutton soil the decline of SOC was rapid over the first 20 years, reaching a new equilibrium of 34 g kg^{-1} after 50 years. The larger SOM content maintained at the Hutton soil rather than the Glenrosa soil, viz. 74% versus 28%, was attributed to clay protection of SOM. The loss of SOM resulted in a concomitant decline in soil microbial biomass C, basal respiration, microbial quotient arylsulfatase, acid phosphatase and aggregate stability. After a period of sugarcane cultivation of 50 years and beyond, only 39% of the SOC in the upper 100 mm of the Glenrosa soil was still forest-derived, with ^{13}C studies showing that 61% was sugarcane-derived. At the Hutton soil sites, the native vegetation was dominated by C_4 grasses, and as a consequence the contribution of sugarcane-derived C to SOC was not measured (Dominy et al. 2001, 2002).

The reasons for a decline in SOM under sugarcane cultivation are well documented (Hartemink and Wood 1998; Haynes and Hamilton 1999) and can be summarized as follows:

- A much lower allocation of organic residues to the soil on account of relatively wide spacing of crop plants, removal of harvested cane and burning of crop residues.
- Enhanced aggregate disruption and exposure of physically protected organic matter to microbial action following cultivation.
- Increased rates of SOM decomposition due to more favorable conditions for microbial activity.

The following cultivation practices should therefore be considered as methods of protecting SOM under sugarcane production:

- Green cane harvesting with residue retention to replace the traditional system of burning before harvesting.
- Planting green manure crops, which will return greater amounts of biomass to the soil.
- Implementation of controlled traffic with minimum tillage that can reduce physical damage to aggregates.

The impacts of some of these cultivation practices on SOM content and quality were studied occasionally in a long-term trial on the research farm (29°04′20″S, 31°04′20″E; altitude 52 m) of the South African Sugarcane Research Institute at Mount Edgecombe. The mean annual rainfall is 950 mm and the mean annual temperature 20.4°C. This trial, which is still in progress, commenced in 1935 and is therefore the oldest in the world where sugarcane cultivation practices, with or without fertilizer applications, are tested (Mthinkhulu et al. 2016). The cultivation practices involve green cane harvesting with all residues mulched over the area (Gm), cane burning prior to harvest with cane tops left scattered evenly, covering two-thirds of the surfaces area (Bt), and cane burning prior to harvest with all residues removed (Bto). Plots associated with these cultivation practices were split into no fertilization (Fo) and fertilization (F). For the latter treatment, 140 kg N ha^{-1}, 28 kg P ha^{-1} and 140 kg K ha^{-1} as a 5:1:5(46) fertilizer mixture at 670 kg ha^{-1} was applied annually about 40 days after harvesting (Van Antwerpen et al. 2001). The trial is located on a southwest facing slope with a 13.5% upper slope and an 18.5% lower slope. A Nitisol (locally a Bonheim form) dominates the lower slope and a Cambisol (locally a Mayo form) dominates the upper slope (Table 7.1). Both soils have a similar particle size distribution of 23% sand, 34% silt and 43% clay to 200 mm depth, with the latter fraction predominantly kaolinite (Mthinkhulu et al. 2016).

Green cane harvesting with complete residue mulching induced higher SOM content than that when the cane was burned prior to harvest, although not at significant levels in all instances (Table 7.7). However, in Brazil significantly higher SOC stocks were noted in unburned than burned sugarcane (Galdos et al. 2009). In general fertilization made an important contribution to the maintenance of organic matter in this clayey soil. Conversely, the removal of cane tops after harvesting burned cane impacted negatively on SOM contents. Differences in SOM content between the various treatments were greatest in the 0–50 mm layer, followed by those in the 100–200 mm layer, and the 50–100 mm layer. The only exception occurred with the comparison of the BtF and BtoF treatments, where the differences declined with depth, viz. 11.3, 9.6 and 6.7 g kg^{-1} in the 0–50, 50–100 and 100–200 mm layers, respectively.

In a follow-up study by Graham et al. (2002a, 2002b) it was confirmed that accumulation of SOC was most pronounced in the upper 50 mm layer, although this was not the case with labile C fractions. Labile C fractions had increased markedly to 300 mm depth by residue retention, reflecting downward leaching of soluble C and/or deposition of particulate C at depths below the residue blanket. Hence, microbial biomass C to 300 mm depth was greater under green cane harvesting than under burning (Graham et al. 2001). Below 100 mm, there were no significant treatment effects on organic carbon. Fertilization also induced the accumulation of SOC, and the effect was evident to 100 mm depth. Compared with the grass plots in this long-term trial, a net loss of SOC from the clayey soil under sugarcane in all treatments was recorded. This loss was more pronounced in the unfertilized plots (Graham et al. 1999).

From these long-term trial results it can be concluded that the best management practice with respect to the conservation of SOM is the production of properly fertilized sugarcane harvesting green, with all residues being mulched.

The production of green manures as cover or break crops was introduced in the 1930s to reduce the loss of SOM and achieve rejuvenation of SOM under sugarcane production. This strategy was slowly phased out due to the advent of inexpensive commercial fertilizers and improved sugarcane varieties, which shifted the economics of sugarcane production away from fallow periods and green manuring. The awareness of soil quality has lately inspired several studies of the utilization of green manure crops to develop strategies for improving, inter alia, SOM in sugarcane fields (Schumann et al. 2000).

The use of filter cake as an organic amendment for sugarcane soils is actively advocated nowadays. Applications of this waste product from sugar mills increased the organic matter and microbial activity in soils (Dee et al. 2002).

TABLE 7.7

Mean SOM Contents after 58 Years of Sugarcane Monocropping under Various Cultivation Treatments at Mount Edgecombe in KwaZulu-Natal

Treatment comparison	Soil depth mm	SOC $g\ kg^{-1}$		P-value
Gm vs. B	0–50	60.0	53.6	<0.05
n = 16	50–100	54.1	53.0	>0.05
	100–200	52.6	49.8	>0.05
Gmf vs. BFo	0–50	62.2	48.3	<0.05
n = 8	50–100	54.7	51.6	>0.05
	100–200	54.4	48.0	<0.05
Gmf vs. BF	0–50	62.2	56.8	>0.05
n = 8	50–100	54.7	53.3	>0.05
	100–200	54.4	51.7	>0.05
Gmfo vs. BFo	0–50	57.7	48.3	<0.05
n = 8	50–100	53.4	51.6	>0.05
	100–200	50.8	48.0	>0.05
Gmfo vs. BF	0–50	57.7	56.8	>0.05
n = 8	50–100	53.4	53.4	>0.05
	100–200	50.8	50.8	>0.05
F vs. Fo	0–50	59.5	54.0	<0.05
n = 12	50–100	54.0	53.0	>0.05
	100–200	53.0	49.4	<0.05
GmF vs. GmFo	0–50	62.2	57.7	>0.05
n = 8	50–100	54.7	53.4	>0.05
	100–200	54.4	50.8	>0.05
BF vs. BFo	0–50	56.8	50.3	<0.05
n = 8	50–100	53.3	52.7	>0.05
	100–200	51.6	48.0	>0.05
BtF vs. BtoF	0–50	62.4	51.1	<0.05
n = 4	50–100	58.1	48.5	<0.05
	100–200	55.0	48.3	<0.05

Source: Compiled from Van Antwerpen and Meyer 1998.

Notes: Gm = green cane harvesting and mulched;

B = cane burned prior to harvest;

T = cane tops scattered after burning;

to = cane tops removed after burning;

F = cane fertilized;

Fo = cane not fertilized.

7.8 PASTURE CROP PRODUCTION

The southern Cape coastal region is documented as the highest milk producing region in South Africa (Figure 7.6). Large areas of the natural fynbos vegetation were therefore cleared and converted to pasture. This conversion involved the conventional tillage of soil, the amendment of soil fertility with lime and fertilizer, the installation of irrigation systems and the establishment of highly productive pasture crops. Kikuyu is the dominant pasture species but milk production from kikuyu pastures is restricted by the seasonability and low nutritional value of the herbage. This is overcome by using kikuyu as a pasture base and incorporating either ryegrass or clover species, which creates a challenge for the introduction of conservational practices such as minimum tillage (Milne 2002; Swanepoel 2014; Phohlo 2016).

The region, which extends from Stormsvlei (34°05′05″S, 20°05′08″E) in the Western Cape Province to Van Stadens River (33°54′33″S, 25°11′50″E) in the Eastern Cape Province, covers about 600,000 ha. Almost half of the arable land in the region is managed as cultivated pastureland in dairy production systems. The dominant soils of the region comprise well-sorted sands or sandy loams (Land Type Survey Staff 2004) and form part of the podzol or duplex soil groups (Figure 7.3; Soil Classification Working Group 1991; IUSS Working Group WRB 2014), otherwise known as Spodosols and Alfisols (Soil Survey Staff 2014). Cultivated pastures are usually established on slopes generally ranging between 1% and 3%, but also possibly ranging from zero to 10%.

The region has a temperate climate with rainfall distributed throughout the year. Annual rainfall varies between 700 and 1,000 mm, depending on the distance from the Indian Ocean and Langeberg, Outeniqua or Tsitsikamma mountain ranges. Ambient temperatures become cooler from west to east, with average daily maximum temperatures that vary between 18°C and 25°C, and minimum temperatures between 7°C and 15°C in winter and summer, respectively (ARC-ISCW 2014). The region can be subdivided into three distinct sections according to these climate differences: western, central and eastern.

A comprehensive survey by Swanepoel et al. (2015c) at a large number of dairy production systems in the region showed that these pasture mixtures impact significantly on SOM stocks when virgin soils served as reference (Table 7.8). In the 0–100 mm layer the SOM stocks of cultivated pasture soil in all three sections were significantly larger than those of virgin soil. This trend was reversed with depth since the virgin soil had larger SOM stocks in the 200–300 mm layer than the cultivated pasture soil.

Based on information given by farmers, these dairy production systems can be grouped into the following tillage practices (Swanepoel et al. 2015c):

1. No disturbance (ND): pure kikuyu grass pasture without any soil disturbance or over sowing practice ($n = 20$).
2. Kikuyu-ryegrass pasture (KR): kikuyu grass-based pasture oversown annually with annual or perennial ryegrass, using a minimum-till seed drill after the kikuyu base was mulched to ground level ($n = 62$).

TABLE 7.8

SOM Stocks of Cultivated Pasture Soil of the Western, Central and Eastern Sections of the Southern Cape Coastal Region of South Africa with Virgin Soil as Reference

Soil depth mm	Cultivated pasture soil			Virgin soil	P-value
	Western	Central	Eastern		
	------------------Mg ha^{-1}------------------				
0–100	129.4[a]	95.8[b]	95.8[b]	74.2[c]	≤0.001
100–200	86.2[a]	62.1[c]	62.1[c]	70.9[bc]	≤0.001
200–300	51.1[b]	38.8[cd]	38.2[d]	68.5[a]	≤0.001

Source: Compiled from Swanepoel et al. 2015c.

Note: Lower case letters indicate significant differences at P<0.05.

3. Herbicide-treated pasture (HT): pure, annual ryegrass established with minimum tillage methods, following eradication of the summer self-sown crops or weeds by herbicide treatment ($n = 29$).
4. Shallow tillage (ST): a kikuyu-based pasture reinforced with annual or perennial ryegrass by means of tillage less than 150 mm deep ($n = 16$).
5. Deep tillage (DT): conventionally tilled kikuyu-ryegrass pasture soil using an offset disc plough ($n = 15$).

The number of assessments per tillage practice represents the approximate total area covered under those practices. In the study, SOM ranged from 39 to 86 g kg^{-1}, SOC from 23 to 63 g kg^{-1} and active C from 2.3 to 2.7 g kg^{-1} (Table 7.9). Weil et al. (2003) claimed that microbial biomass carbon (MBC) or active C is a better indicator of soil quality than either SOC or SOM. Predictably, the highest values of the three parameters were recorded for the ND treatment, because pure kikuyu builds up a matt in the top 100 mm soil through time, and could therefore accumulate large amounts of SOM. The lowest values were recorded for the DT treatment, since deep tillage enhances decomposition of SOM (Haynes et al. 2003).

The above-mentioned results are largely supported by those of Phohlo (2016), who studied ten pasture-based dairy farms in the eastern section, better known as the Tsitsikamma. Owing to a change in climate and soil from east to west, the eastern section was subdivided into the upper Tsitsikamma, represented by five farms and the lower Tsitsikamma represented by five farms. The SOC contents increased after the introduction of kikuyu pasture mixtures, with surprisingly little difference between the upper and lower Tsitsikamma. The total SOC stocks to 600 mm soil depth measured 11.9 Mg ha^{-1} for the upper Tsitsikamma and 11.5 Mg ha^{-1} for the lower Tsitsikamma. In both the lower and upper Tsitsikamma, organic C stocks decreased significantly from the 0–150 mm layer to the 150–300 mm layer, and from this latter layer to the 300–450 mm layer. The SOC stocks in the 300–450 mm and 450–600 mm layers were almost similar. However, the farmers' management practices had significant influence on organic C stocks of cultivated soils, especially in the lower Tsitsikamma where a greater diversity of either inorganic or organic fertilizers were used. The use of organic fertilizers offer greater benefit regarding the build-up of SOM than inorganic fertilizers. Stocks of active C mirrored that of SOC, probably because of its proportional quantities rise to 16% and 19% of SOC (Phohlo 2016).

TABLE 7.9

Mean SOM, SOC and Active C Contents to 100 mm Depth under Different Tillage Practices for Cultivated Pastures in the Southern Cape Coastal Region of South Africa

Tillage practice	SOM	SOC	Active C
	g kg^{-1}		
ND: no disturbance ($n = 20$)	86[a]	63[a]	2.7[a]
KR: kikuyu-ryegrass ($n = 62$)	49[b]	44[bc]	2.5[a]
HT: herbicide treated ($n = 29$)	44[b]	32[d]	2.4[b]
ST: shallow tillage ($n = 16$)	50[ab]	42[cd]	2.4[ab]
DT: deep tillage ($n = 15$)	39[b]	23[e]	2.3[b]
P-value	0.048	0.001	0.009

Source: Compiled from Swanepoel et al. 2015b.
Note: Lower case letters indicate significant differences at P<0.05.

In another study on four commercial dairy farms in the Tsitsikamma, Milne (2002) reported that in comparison with undisturbed native vegetation, soils under annual cultivation and permanent pasture both gained and lost SOM in certain instances. This phenomenon was attributed inter alia to differences in natural vegetation and, hence, soil types. In spite of this, SOC was lower under annual ryegrass than permanent kikuyu pasture at all the sites, reflecting the degrading effect of annual cultivation on SOM. As a consequence, labile, extractable and microbial biomass carbon were all lower under annual ryegrass than under permanent kikuyu pasture at all sites.

On the Outeniqua research farm (33°58′38″S, 22°25′16″E; altitude 204 m), situated near George in the central section, Swanepoel et al. (2014c) assessed whether 19 years of irrigated, minimum-till, kikuyu-ryegrass pasture on a podzolic soil altered stocks of SOM and its indicators: SOC, active C and MBC, formerly developed under undisturbed natural fynbos rangeland (Table 7.10). A mean annual rainfall of 728 mm and mean annual temperature of 16.2°C were recorded for this farm. The stocks of these parameters in the cultivated pasture soil were significantly larger in the upper two layers than in the virgin soil, with the exceptions of MBC in the 0–100 mm and SOM and active C in the 100–200 mm layer stock values for the parameters, which were mostly similar. In the 200–300 mm layer, stock values for the parameters were mostly similar, except MBC, which was higher in cultivated pasture soil than in virgin soil.

Conversion of virgin soil to cultivated pasture soil improved conditions supporting the build-up of SOM levels and, therefore, the soils' microbial community. Kikuyu, with its high density of rhizomes and stolons, is a pasture crop that is particularly good in enhancing SOM levels (Skjemstadt et al. 1990). Yet, neither SOM nor a specific SOC fraction alone are suitable to estimate biological activity, since concurrent changes to adopted management affect SOM likely much slower than microbial activity parameters.

The active C (MBC) stock in the cultivated pasture soil was ~530 times higher than that of the virgin soil at 0–100 mm depth, reflecting the importance of the soil surface as a biologically active interface and entry point for additions of readily available organic material (Lopez-Garrido et al. 2011). In these intensively grazed dairy pastures, high volumes of labile organic matter are added in the forms of manure, moribund forage material and forage waste. Active C provided additional

TABLE 7.10

Mean Stocks of SOM, SOC, Active C and MBC after 19 Years of Irrigated Minimum-Tillage Kikuyu-Ryegrass Pasture in the Southern Cape Coastal Region of South Africa

Parameter	Depth mm	Cultivated pasture	Virgin soil	P-value
		kg m^{-3}		
SOM	0–100	119.4	71.9	0.001
	100–200	78.0	68.7	0.359
	200–300	42.2	65.2	0.276
SOC	0–100	59.2	35.6	0.001
	100–200	43.2	25.9	0.001
	200–300	19.7	24.0	0.102
Active C	0–100	4.19	0.09	0.001
	100–200	1.79	0.11	0.791
	200–300	0.24	0.09	0.290
MBC	0–100	0.24	0.19	0.579
	100–200	0.44	0.18	0.002
	200–300	0.08	0.16	0.026

Source: Compiled from Swanepoel et al. 2014c.

TABLE 7.11

Microbial Quotient for Cultivated Pasture Soil and Virgin Soil in the Southern Cape Coastal Region of South Africa

Soil depth mm	Cultivated pasture soil	Virgin soil	P-value
	%		
0–100	42.1	53.7	0.046
100–200	101.7	70.6	0.085
200–300	36.7	66.1	0.045

Source: Compiled from Swanepoel et al. 2014c.

information to that relating to SOC, by proving that cultivated pasture soil improved the system by introducing high volumes of vital energy substrates for microbial metabolism in the surface layer of the soil. These microbes play an important role in SOM decomposition and nutrient cycling (Granatstein et al. 1987). Mean MBC comprised 0.41% and 0.26% of SOM in the cultivated pasture soil and virgin soil, respectively. The larger MBC stock in the cultivated pasture soil, when compared to the virgin soil, could be ascribed to the vigorous and large root system of the kikuyu-ryegrass pasture with improved external environmental conditions such as water supply by irrigation, and nutrient supply by fertilization, liming and manure from grazing animals (Carter 1986). These conditions create an environment that supports microbes by providing active C by means of root exudates and plant nutrients from manure.

The microbial quotient, shown in Table 7.11, indicates the substrate-use efficiency of the microbial community and its importance in regulating SOM transformations (Moore et al. 2000). Insam and Domsch (1988) state that the microbial quotient serves as indicator of C accumulation or release. The microbial quotients of the 0–100 mm and 200–300 mm depths were higher in the cultivated pasture soil, but did not differ between sites at the 100–200 mm depth. The upshot of this is that more microorganisms were sustained in the cultivated pasture soil than in the virgin soil.

Martens (1995) concluded that the microbial quotient should therefore be used as a reference point during steady-state conditions. In this regard the assumption could be made that the virgin soil is in equilibrium and sustainable for its relevant land use. Sudden deviations from this level, as with the cultivated pasture soil, indicate that the system is changing and C is being released or accumulated. Changes in the microbial quotient might therefore help to forecast C sequestration actions.

The establishment of kikuyu pasture mixtures on what was formerly rangeland dominated by fynbos vegetation impacts positively on SOM content and quality to 200 mm depth. Minimum-till kikuyu, oversown annually with either ryegrass or clover, gave the best results in terms of herbage production and quality on the one hand (Swanepoel et al. 2014a), and sustenance of soil quality on the other (Swanepoel et al. 2013; Swanepoel et al. 2017). For more details on the challenges of and opportunities for managing cultivated pastures to improve soil quality characteristics, we here refer to the work of Swanepoel et al. (2014b, 2015a).

7.9 SUMMARY

South Africa is a diverse country in terms of its climate, soil and topography and, therefore, its agriculture potential. Approximately 14% of the country's 122 Mha land is regarded as arable. Currently, 77% of this arable land is used for crop production.

In their virgin state, only 4% of the country's topsoils contain >20 g C kg^{-1}, 38% between 5 and 20 g C kg^{-1} and 58% even <5 g C kg^{-1}. This SOC is subject to depletion under prolonged, conventional crop production systems, amounting to 70% of the SOC loss relative to the virgin SOC levels in some instances. The implementation of conservation crop production systems restore SOC at different degrees. Here three diverse crop production systems were selected to illustrate their differential impacts on SOC. They included the production of field crops, sugarcane and pasture.

Results from either chronosequence studies or long-term field trial studies confirmed that the conventional cultivation of field crops, sugarcane and pasture depleted SOM when naturally vegetated soil served as reference. However, the establishment of minimum-tillage kikuyu pasture mixtures impact positively on SOM. Conservation of SOM is promoted best by properly fertilized sugarcane that is harvested green, with all residues being mulched. The restoration of SOM under field crop production is a greater challenge since the best combination of tillage practice and crop rotation must be established. There is unanimity regarding the introduction of no-tillage and avoidance of monocropping. The latter is often hampered because crop rotation is usually site-specific, an aspect that warrants further research, with concomitant long-term field trials.

Changes in SOM quality were also addressed where the data justified investigation. This provides some clarification on the mechanisms involved in the depletion and restoration of SOM in arable soils. A better understanding of these mechanisms could contribute to the development of proper strategies for SOM management, particularly when disentangling the different reactions that might be involved in SOC losses. This entails aspects such as erosion and a lowering of SOC input rates, or SOC degradation with the physical breakdown of aggregates and/or going along with microbial adaptation strategies to changes in nutrients supply, for instance.

REFERENCES

Agenbag, G.A., and Maree, P.C.J. 1989. The effect of tillage on soil carbon, nitrogen and soil strength of simulated surface crusts in two cropping systems of wheat. *Soil Till. Res.* 14: 53–65.

Amelung, W., Brodowski, S., Sandhage-Hofmann, A., and Bol, R. 2008. Combining biomarker with stable isotope analyses for assessing the transformation and turnover of soil organic matter. *Adv. Agron.* 100: 155–250.

Amelung, W., Kimble, J.M., Samson-Liebig, S., and Follett, R.F. 2001. Restoration of microbial residues in soils of the Conservation Reserve Program. *Soil Sci. Soc. Am. J.* 65: 1704–1709.

Amelung, W., Lobe, I., and Du Preez, C.C. 2002. Fate of microbial residues in sandy soils of the South African Highveld as influenced by prolonged arable cropping. *Eur. J. Soil Sci.* 53: 29–35.

Amelung, W., and Zech, W. 1996. Organic species in ped surface and core fractions along a climosequence in the prairie, North America. *Geoderma* 74: 193–206.

ARC-ISCW. 2014. *Agro-Climatology Database.* Agricultural Research Council - Institute for Soil, Climate and Water, Pretoria, South Africa.

Bach, E.M., Baer, S.G., Meyer, C.K., and Six, J. 2010. Soil texture affects soil microbial and structural recovery during grassland restoration. *Soil Biol. Biochem.* 42: 2182–2191.

Baer, S.G., Bach, E.M., Meyer, C.K., Du Preez, C.C., and Six, J. 2015. Below ecosystem recovery during grass restoration: South African Highveld compared to US tallgrass prairies. *Ecosystems* 18: 390–403.

Barnard, R.O. 2000. Carbon sequestration in South Africa soils. ARC-ISCW Report No. GW/A/2000/48. Agricultural Research Council-Institute for Soil, Climate and Water, Pretoria, South Africa.

Beukes, D.J., Bennie, A.T.P., and Hensley, M. 2004. Optimizing soil water balance components for sustainable crop production in dry areas of South Africa. In: *Challenges and Strategies of Dryland Agriculture,* eds. S.C. Rao and J. Ryan, pp. 291–313. Special Publication No. 32. Crop Science Society of America, Madison, WI.

Brady, N.C., and Weil, R.R. 2008. *The Nature and Properties of Soils,* 14th edition. Prentice-Hall, Upper Saddle River, NJ.

Bredenkamp, G., Van Rooyen, N., and Jubke, R.A. 1996. Moist cold Highveld grassland (40), Dry sandy Highveld grassland (37) and moist cool Highveld grassland (39). In: *Vegetation of South Africa, Lesotho and Swaziland,* eds. A.B. Low and A.G. Rebelo, pp. 41–44. Department of Environmental Affairs and Tourism, Pretoria, South Africa.

Brodowski, S., Amelung, W., Lobe, I., and Du Preez, C.C. 2004. Losses and biogeochemical cycling of soil organic nitrogen with prolonged arable cropping in the South African Highveld – Evidence from D- and L-amino acids. *Biogeochemistry* 71: 17–42.

Buyanovsky, G.A., Aslam, M., and Wagner, G.H. 1994. Carbon turnover in soil physical fractions. *Soil Sci. Soc. Am. J.* 58: 1167–1173.

Carter, M.R. 1986. Microbial biomass as an index for tillage-induced changes in soil biological properties. *Soil Till. Res.* 7: 29–40.

Cooper, G.D. 2016. Long-term effect of tillage and crop rotation practices on soil C and N in the Swartland, Western Cape, South Africa. M.Sc. Agric. Dissertation. University of Stellenbosch, Stellenbosch, South Africa.

DAFF 2016. Abstract of agricultural statistics. Department of Agriculture, Forestry and Fisheries, Pretoria, South Africa.

De Jager, J.M. 1993. Geographical and agro-ecological features of southern Africa. In: *Livestock Production Systems: Principles and Practices*, eds. C. Maree and N.H. Casey, pp. 14–31. Agric-Development Foundation, Brooklyn, Pretoria, South Africa.

Dee, B.M., Haynes, R.J., and Meyer, J.H. 2002. Sugar mill wastes can be important soil amendments. *Proc. S. Afr. Sugar Technol. Assoc.* 76: 51–60.

Derrien, D., and Amelung, W. 2011. Computing mean residence time of soil carbon fractions by the use of stable isotopes: Impacts of model framework. *Eur. J. Soil Sci.* 62: 237–252.

Dominy, C.S., Haynes, R.J., and Van Antwerpen, R. 2001. Long-term effects of sugarcane production on soil quality in the South Coast and Midlands areas of KwaZulu-Natal. *Proc. S. Afr. Sugar Technol. Assoc.* 75: 222–227.

Dominy, C.S., Haynes, R.J., and Van Antwerpen, R. 2002. Loss of soil organic matter and related soil properties under long-term sugarcane production on two contrasting soils. *Biol. Fert. Soils* 36: 350–356.

Du Preez, C.C., and Du Toit, M.E. 1995. Effect of cultivation on the nitrogen fertility of selected agroecosystems in South Africa. *Fert. Res.* 42: 27–32.

Du Toit, M.E., Du Preez, C.C., Hensley, M., and Bennie, A.T.P. 1994. Effect of cultivation on the organic matter content of selected dryland soils in South Africa. *S. Afr. J. Plant Soil* 11: 71–79.

Dungait, J.A.J., Hopkins, D.W., Gregory, A.S., and Whitmore, A.P. 2012. Soil organic matter turnover is governed by accessibility not recalcitrance. *Glob. Change Biol.* 18: 1781–1796.

Fanning, D.S., and Fanning, M.C.B. 1989. *Soil Morphology, Genesis and Classification*. John Wiley & Sons, New York.

FAO 2005. *Fertilizer Use by Crops in South Africa. Land and Plant Nutrition Service.* Food and Agriculture Organization of the United Nations (FAO), Rome, Italy.

Fey, M.V. 2010. *Soils of South Africa: Their Distribution, Properties, Classification, Genesis, Use and Environmental Significance.* Cambridge University Press, New York.

Galdos, M.V., Cerri, C.C., and Cerri, C.E.P. 2009. Soil carbon stocks under burned and unburned sugarcane in Brazil. *Geoderma* 153: 347–352.

Golchin, A., Oades, J., Skjemstad, J., and Clarke, P. 1994. Study of free and occluded particulate organic matter in soils by solid state C-13 CP/MAS NMR-spectroscopy and scanning electron-microscopy. *Soil Res.* 32: 285–309.

Graham, M.H., Haynes, R.J., and Meyer, J.H. 1999. Green cane harvesting promotes accumulation of soil organic matter and an improvement in soil health. *Proc. S. Afr. Sugar Technol. Assoc.* 73: 53–57.

Graham, M.H., Haynes, R.J., and Meyer, J.H. 2002a. Soil organic matter content and quality. Effects of fertilizer applications, burning and trash retention on a long-term sugarcane experiment in South Africa. *Soil Biol. Biochem.* 34: 93–102.

Graham, M.H., Haynes, R.J., and Meyer, J.H. 2002b. Changes in soil chemistry and aggregate stability induced by fertilizer application, burning and trash retention on a long-term sugarcane experiment in South Africa. *Eur. J. Soil Sci.* 53: 589–598.

Graham, M.H., Haynes, R.J., Zelles, L., and Meyer, J.H. 2001. Long-term effects of green cane harvesting versus burning on the size and diversity of the soil microbial community. *Proc. S. Afr. Sugar Technol. Assoc.* 75: 228–233.

Granatstein, D.M., Bezdicek, D.F., Cochran, V.L., Elliott, L.E., and Hammel, J. 1987. Long-term tillage and rotation effects on soil microbial biomass, carbon and nitrogen. *Biol. Fert. Soils* 5: 265–270.

Gregorich, E.G., Liang, B.C., Ellert, B.H., and Drury, C.F. 1996. Fertilization effects on soil organic matter turnover and corn residue C storage. *Soil Sci. Soc. Am. J.* 60: 472–476.

Guggenberger, G., Frey, S.D., Six, J., Paustian, K., and Elliott, E.T. 1999. Bacterial and fungal cell-wall residues in conventional and no-tillage agroecosystems. *Soil Sci. Soc. Am. J.* 63: 1188–1198.

Hartemink, A.E., and Wood, A.W. 1998. Sustainable land management in the tropics: The case of sugarcane plantations. Proceedings 16th International Congress of Soil Science., Montpellier, France.

Haynes, R.J., Dominy, C.S., and Graham, M.H. 2003. Effects of agricultural land use on soil organic matter status and the composition of earthworm communities in KwaZulu-Natal, South Africa. *Agric. Ecosyst. Environ.* 95: 453–464.

Haynes, R.J., and Hamilton, C.S. 1999. Effects of sugarcane production on soil quality: A synthesis of world literature. *Proc. S. Afr. Sugar Technol. Assoc.* 73: 45–51.

Insam, H., and Domsch, K.H. 1988. Relationship between soil organic carbon and microbial biomass on chronosequences of reclamation sites. *Microb. Ecol.* 15: 177–188.

IUSS Working Group WRB 2014. *World Reference Base for Soil Resources. International Soil Classification System for Naming Soils and Creating Legends for Soil Maps. World Soil Resources.* Report No. 106. Food and Agriculture Organization of the United Nations (FAO), Rome, Italy.

Jenkinson, D.S., and Rayner, J.H. 1977. The turnover of soil organic matter in some of the Rothamsted classical experiments. *Soil Sci.* 123: 298–305.

Jones, A., Brenning-Madsen, H., Brossard, M., Dampha, A., Deckers, J., Dewitte, O., Gallali, T., Hallett, S., Jones, R., Kilasara, M., Le Roux, P., MIcheli, E., Montanarella, L., Spaarzaren, O., Thiombiano, L., Van Ranst, E., Yemefack, M., and Zougmoré, R. (eds) 2013. *Soil Atlas of Africa.* European Commission Publication Office of the European Union, Luxembourg.

Kögel-Knaber, I., and Amelung, W. 2014. Dynamics, chemistry, and preservation of organic matter in soils. In: *Treatise on Geochemistry*, eds. H.D. Holland and K.K. Turekian, pp. 157–215. Elsevier, Oxford, UK.

Kösters, R. 2018. *Regeneration of degraded cropland by secondary pasture management: A humus-chemical case study in the South Africa Highveld.* Bonner Bodenkundliche Abhandlungen. Band 71, University of Bonn, Bonn, Germany.

Kösters, R., Du Preez, C.C., and Amelung, W. 2018. Lignin dynamics in secondary pasture soils of the South African Highveld. *Geoderma* 319: 113–121.

Kösters, R., Preger, A.C., Du Preez, C.C., and Amelung, W. 2013. Re-aggregation of degraded cropland soils with prolonged secondary pasture management in the South African Highveld. *Geoderma* 192: 173–181.

Kotzé, E., and Du Preez, C.C. 2007. Influence of long-term wheat residue management on organic matter in an Avalon soil. *S. Afr. J. Plant Soil* 24: 114–119.

Kotzé, E., Loke, P.F., Akhosi-Setaka, M.C., and Du Preez, C.C. 2016. Land use change affecting soil humic substances in three semi-arid agro-ecosystems in South Africa. *Agric. Ecosyst. Environ.* 216: 194–202.

Land Type Survey Staff. 2004. *Land Types of South Africa.* Agricultural Research Council – Institute for Soil, Climate and Water, Pretoria, South Africa.

Lauer, F., Kösters, R., Du Preez, C.C., and Amelung, W. 2011. Microbial residues as indicators of soil restoration in South African secondary pastures. *Soil Biol. Biochem.* 43: 787–794.

Lobe, I., Amelung, W., and Du Preez, C.C. 2001. Losses of carbon and nitrogen with prolonged arable cropping from sandy soils of the South African Highveld. *Eur. J. Soil Sci.* 52: 93–101.

Lobe, I., Bol, R., Ludwig, B., Du Preez, C.C., and Amelung, W. 2005. Savanna-derived organic matter remaining in arable soils of the South African Highveld after long-term mixed cropping - Evidence from ^{13}C and ^{15}N natural abundance. *Soil Biol. Biochem.* 37: 1898–1909.

Lobe, I., Du, Preez C.C., and Amelung, W. 2002. Influence of prolonged arable cropping on lignin compounds in sandy soils of the South African Highveld. *Eur. J. Soil Sci.* 53: 553–562.

Lobe, I., Sandhage-Hofmann, A., Brodowski, S., Du Preez, C.C., and Amelung, W. 2011. Aggregate dynamics and associated soil organic matter contents as influenced by prolonged arable cropping in the South African Highveld. *Geoderma* 162: 251–259.

Loke, P.F., Kotzé, E., and Du Preez, C.C. 2012. Changes in soil organic matter indices following 32 years of different wheat production management practices in semi-arid South Africa. *Nutr. Cycl. Agroecosyst.* 94: 97–109.

López-Garrido, R., Madejón, E., Murillo, J.M., and Moreno, F. 2011. Soil quality alteration by mouldboard ploughing in a commercial farm devoted to no-tillage under Mediterranean conditions. *Agric. Ecosyst. Environ.* 140: 182–190.

Low, A.B., and Rebelo, A.G. (eds) 1996. *Vegetation of South Africa, Lesotho and Swaziland.* Department Environmental Affairs and Tourism, Pretoria, South Africa.

Macvicar, C.N., De Villiers, J.M., Loxton, R.F., Verster, E., Lambrechts, J.J.N., Merryweather, F.R., Le Roux, J., Van Rooyen, T.H., von, M., and Harmse, H.J. 1977. *Soil Classification: A Binomial System for South Africa.* Department of Agriculture Technical Services, Pretoria, South Africa.

Martens, R. 1995. Current methods for measuring microbial biomass C in soil: Potentials and limitations. *Biol. Fert. Soils* 19: 87–99.

Milne, R.K. 2002. An investigation of factors contributing to soil degradation under dairy farming in the Tsitsikamma. M.Sc. Agric. Dissertation. University of KwaZulu-Natal, Pietermaritzburg, South Africa.

Moore, J.M., Klose, S., and Tabatabai, M.A. 2000. Soil microbial biomass carbon and nitrogen as affected by cropping systems. *Biol. Fert. Soils* 31: 200–210.

Mthimkhulu, S., Podwojewski, P., Hughes, J., Titshall, L., and Van Antwerpen, R. 2016. The effect of 72 years of sugarcane residues and fertilizer management on soil physico-chemical properties. *Agric. Ecosyst. Environ.* 225: 54–61.

Paul, E.A. 2016. The nature and dynamics of soil organic matter: Plant inputs, microbial transformations, and organic matter stabilization. *Soil Biol. Biochem.* 98: 109–126.

Phohlo, M.P. 2016. Soil quality of kikuyu, ryegrass and clover pasture mixtures in the Tsitsikamma. M.Sc. Agric. Dissertation. University of the Free State, Bloemfontein, South Africa.

Preger, A.C., Kösters, R., Du Preez, C.C., Brodowski, S., and Amelung, W. 2010. Carbon sequestration in secondary pasture soils: A chronosequence study in the South African Highveld. *Eur. J. Soil Sci.* 61: 551–562.

Preger, A.C., Rillig, M.C., Johns, A.R., Du Preez, C.C., Lobe, I., and Amelung, W. 2007. Losses of glomalin-related soil protein under prolonged arable cropping: A chronosequence study in sandy soils of the South African Highveld. *Soil Biol. Biochem.* 39: 445–453.

Qonggo, L.L., and Van Antwerpen, R. 2000. Effect of long-term sugarcane production on physical and chemical properties of soils in KwaZulu-Natal. *Proc. S. Afr. Sugar Technol. Assoc.* 74: 114–121.

Rantoa, N.R. 2009. Estimating organic carbon stocks in South African soils. M.Sc. Agric. Dissertation. University of the Free State, Bloemfontein, South Africa.

Rantoa, N.R., Van Huyssteen, C.W., and Du Preez, C.C. 2015. Organic carbon content in the soil master horizons of South Africa. *Vadose Zone J.* 14: 1–12.

Schulze, R.E., Maharaj, M., Lynch, S.D., Howe, B.J., and Melvil-Thompson, B. 2001. *South African Atlas of Agrohydrology and Climatology, Beta 1.002.* University of KwaZulu-Natal, Pietermaritzburg, South Africa.

Schumann, R.A., Meyer, J.H., and Van Antwerpen, R. 2000. A review of green manuring practices in sugarcane production. *Proc. S. Afr. Sugar Technol. Assoc.* 74: 93–100.

Skjemstad, J.O., Lefeuvre, R.P., and Prebble, R.E. 1990. Turnover of soil organic matter under pasture as determined by ^{13}C natural abundance. *Aust. J. Soil Res.* 28: 267–276.

Skjemstad, J.O., Spouncer, L.R., Cowie, B., and Swift, R.S. 2004. Calibration of the Rothamsted organic carbon turnover model (RothC ver.26.3), using measurable soil organic carbon pools. *Aust. J. Soil Res.* 42: 79–88.

Smit, M.L. 2004. The effect of cropping systems and tillage practices on some fertility indicators of a Swartland soil near Malmesbury. M.Sc. Agric. Dissertation. University of the Free State, Bloemfontein, South Africa.

Smith, J.D.V. 2014. The effect of long-term no-till crop rotation practices on the soil organic matter functional pools. M.Sc. Agric. Dissertation. University of Stellenbosch, Stellenbosch, South Africa.

Soil Classification Working Group. 1991. *Soil Classification. A Taxonomic System for South Africa.* Memoirs on the Agricultural Natural Resources of South Africa No. 15. Department of Agricultural Development, Pretoria, South Africa.

Soil Survey Staff. 2014. *Keys to Soil Taxonomy,* 12th edition. USDA Natural Resources Conservation Service, Washington, D.C.

Solomon, D., Lehmann, J., Kinyangi, J., Amelung, W., Lobe, I., Pell, A., Riha, S., Ngoze, S., Verchot, L., Mbugua, D., Skjemstad, J., and Schäfer, T. 2007. Long-term impacts of anthropogenic perturbations on dynamics and speciation of organic carbon in tropical forest and subtropical grassland ecosystems. *Glob. Change Biol.* 13: 511–530.

Solomon, D., Lehmann, J., Lobe, I., Martinez, C.E., Tveitnes, S., Du Preez, C.C., and Amelung, W. 2005. Sulphur speciation and biogeochemical cycling in long-term arable cropping of subtropical soils: Evidence from wet-chemical reduction and S K-edge XANES spectroscopy. *Eur. J. Soil Sci.* 56: 621–634.

Stevenson, F.J., and Cole, M.H. 1999. *Cycles of Soils: Carbon, Nitrogen, Phosphorus, Sulfur, Micronutrients.* John Wiley & Sons, New York.

Swanepoel, C.M., van der Laan, M., Weepener, H.L., Du Preez, C.C., and Annandale, J.G. 2016. Review and meta-analysis of organic matter in cultivated soils in southern Africa. *Nutr. Cycl. Agroecosyst.* 104: 107–123.

Swanepoel, P.A. 2014. Soil quality of kikuyu-ryegrass pastures in the southern Cape. Ph.D. Thesis. University of the Free State, Bloemfontein, South Africa.

Swanepoel, P.A., Botha, P.R., Du Preez, C.C., and Snyman, H.A. 2013. Physical quality of a podzolic soil following 19 years of irrigated minimum-till kikuyu-ryegrass pasture. *Soil Till. Res.* 133: 10–15.

Swanepoel, P.A., Botha, P.R., Snyman, H.A., and Du Preez, C.C. 2014a. Impact of cultivation method on productivity and botanical composition of a kikuyu-ryegrass pasture. *Afr. J. Range For. Sci.* 31: 215–220.

Swanepoel, P.A., Du Preez, C.C., Botha, P.R., Snyman, H.A., and Habig, J. 2014b. Soil quality characteristics of kikuyu-ryegrass pastures in South Africa. *Geoderma* 232–234: 589–599.

Swanepoel, P.A., Habig, J., Du Preez, C.C., Botha, P.R., and Snyman, H.A. 2014c. Biological quality of a podzolic soil after 19 years of irrigated minimum-till kikuyu-ryegrass pasture. *Soil Res.* 52: 64–75.

Swanepoel, P.A., Botha, P.R., Du Preez, C.C., Snyman, H.A., and Labuschagne, J. 2015a. Managing cultivated pastures for improving soil quality in South Africa: Challenges and opportunities. *Afr. J. Range For. Sci.* 32: 91–96.

Swanepoel, P.A., Du Preez, C.C., Botha, P.R., and Snyman, H.A. 2015b. A critical view on the soil fertility status of minimum-till kikuyu-ryegrass pastures in South Africa. *Afr. J. Range For. Sci.* 32: 113–124.

Swanepoel, P.A., Du Preez, C.C., Botha, P.R., Snyman, H.A., and Habig, J. 2015c. Assessment of tillage effects on soil quality of pastures in South Africa with indexing methods. *Soil Res.* 53: 274–285.

Swanepoel, P.A., Habig, J., Du Preez, C.C., Snyman, H.A., and Botha, P.R. 2017. Tillage effects, soil quality and production potential of kikuyu-ryegrass pastures in South Africa. *Grass For. Sci.* 72: 308–321.

Swift, M.J., Seward, P.D., Frost, P.G.H., Quereshi, J.N., and Muchena, F.N. 1994. Long-term experiments in Africa: Developing a database for sustainable land use under global change. In: *Long-Term Experiments in Agricultural and Ecological Sciences*, eds. R.A. Leigh and A.E. Johnston, pp. 229–251. CAB International, Wallingford, UK.

Twomlow, S.J., Steyn, J.T., and Du Preez, C.C. 2006. Dryland farming in southern Africa. In: *Dryland Agriculture*, 2nd edition., eds. G.A. Peterson, P.W. Unger and W.A. Payne, pp. 769–836. American Society of Agronomy, Madison, WI.

Van Antwerpen, R., and Meyer, J.H. 1996. Soil degradation under sugarcane cultivation in northern KwaZulu-Natal. *Proc. S. Afr. Sugar Technol. Assoc.* 70: 29–33.

Van Antwerpen, R., and Meyer, J.H. 1998. Soil degradation. Effect of trash and inorganic fertilizer application on soil strength. *Proc. S. Afr. Sugar Technol. Assoc.* 72: 152–158.

Van Antwerpen, R., Meyer, J.H., and Turner, P. 2001. The effects of cane trash on yield and nutrition from the long-term field trial at Mount Edgecombe. *Proc. S. Afr. Sugar Technol. Assoc.* 75: 235–241.

Von Sperber, C., Stallforth, R., Du Preez, C.C., and Amelung, W. 2017. Changes in soil phosphorus pools during prolonged arable cropping in semi-arid grasslands. *Eur. J. Soil Sci.* 68: 462–471.

Weil, R.R., Islam, K.R., Stine, M.A., Gruver, J.B., and Samson-Liebig, S.E. 2003. Estimating active carbon for soil quality assessment: A simplified method for laboratory and field use. *Am. J. Altern. Agric.* 18: 3–17.

Weil, R.R., and Magdoff, F. 2004. Significance of soil organic matter to soil quality and health. In: *Soil Organic Matter in Sustainable Agriculture*, eds. F. Magdoff and R.R. Weil, pp. 1–43. CRC Press, New York.

Wiltshire, G.H., and Du Preez, C.C. 1993. Long-term effects of conservation practices on the nitrogen fertility of a soil cropped annual to wheat. *S. Afr. J. Plant Soil* 10: 70–76.

8 Rangeland Management and Soil Quality in South Africa

E. Kotzé, H. A. Snyman, and C. C. Du Preez

CONTENTS

8.1 INTRODUCTION

Rangelands cover half the world's land area, and in the arid to semi-arid environments of South Africa, they comprise even more than 75% of the agricultural land surface that can be utilized only by livestock and game farming (Van der Westhuizen and Snyman 2014). Rangelands form the basis of the broader livestock industries in South Africa, and also happen to be the cheapest feeding source stock farmers can use. Concerns about the effects of rangeland management practices on ecosystem functioning have initiated interest in soil quality/health. Soil quality is defined by scientists in various ways, but it can be generally seen as the capacity of a soil to perform a specific function within natural or managed ecosystem boundaries. Soil properties that change due to management or climate leads to changes in the soil's capacity to achieve a specific function. Due to the high variability that exists in rangelands and the multiple use demands, they represent a unique challenge for soil quality assessment (Snyman and Du Preez 2005). In order to utilize rangelands as a natural resource more sustainably, soil responses to overgrazing, drought and fire on degradation need to be quantified in developing suitable rangeland management practices. Overgrazing is worldwide considered the most important cause of rangeland degradation, which then leads to soil degradation such as erosion and compaction (Snyman 2009a). When the production potential of rangelands is overestimated the subsequent overgrazing will lead to a decrease of palatable perennial plants in favor of less palatable, undesirable vegetation (Van der Westhuizen et al. 1999; Snyman 2004). Such changes in rangeland condition usually lead to increased soil compaction (Chanasyk and Naeth 1995; Snyman and Du Preez 2005), reduced soil aggregate stability (Lal and Elliot 1994; Russell et al. 2001), soil fertility (Dormaar and Willms 1998; Ingram 2002), soil organic matter (SOM) content (Du Preez and Snyman 1993; Du Preez and Snyman 2003; Snyman 1999; Snyman and Du Preez 2005) and soil-water content (Snyman 2009b).

Rangeland condition and soil quality are dependent on each other. The functioning of both the soil and the plant communities is crucial for rangeland condition. Rangeland condition can be seen as the level to which the integrity of the soil, vegetation, water and air, as well as ecological processes of rangeland ecosystem functioning, are balanced and sustained (USDA 2016). In Figure 8.1 a simplified model proposed by Greenwood and McKenzie (2001) shows the interaction between soil, vegetation and livestock in a grazing system, integrating both direct and indirect effects of livestock on soil properties.

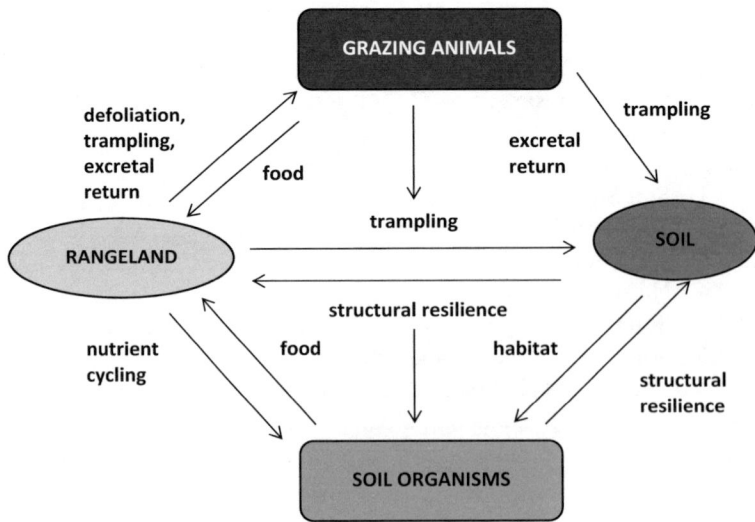

FIGURE 8.1 Interaction between soil, vegetation and livestock in a rangeland grazing system (modified from Greenwood and McKenzie 2001).

Soil quality is the capacity of a soil to function within natural or managed ecosystem boundaries, sustain plant and animal productivity, maintain or enhance the quality of water and air, and support human health and habitation (Karlen et al. 1997). Changes in the capacity of soil to function are reflected in soil properties that change in response to management or climate. Soil quality affects environmental processes, such as the capturing, storing and redistribution of water; plant growth; and plant nutrient cycling. For example, when physical soil crusting increases, the infiltration capacity of the soil will decrease, and consequently the amount of plant available water (PAW) will be affected. This then has an influence on plant productivity, where some plant species may even disappear, and less desirable species may become abundant. These shifts in vegetation composition may precede or follow changes in soil properties and processes. Substantial changes in vegetation are normally connected with changes in soil properties and processes. In some cases, this shift may be irreversible, such as a change in the soil profile due to accelerated erosion, while recovery might be possible in other cases.

Against this background, the chapter will first deal with topography, climate and soil as physical determinants of rangeland productivity, followed by the contribution and ecological impact of different biomes. The rangeland management systems are then discussed to review the historical background, current status and classification. Lastly, the effects of rangeland management on soil quality are addressed.

8.2 PHYSIOGRAPHY, CLIMATE AND SOILS

8.2.1 PHYSIOGRAPHY

Several periods of uplifting, erosion and deposition in geological time have formed complex landforms determined by the underlying geology. Physical barriers provided by the Great Escarpment and the Drakensberg mountains determine the climate and vegetation of most of the rangeland regions of South Africa. South Africa has five main physiographic regions, with the first region being the Fold mountains in the southwest, which determine the climate and vegetation patterns of the southern Cape Province (see Figure 8.2 for the provinces of South Africa). The second region is the coastal plain, extending from the Namibian border, all along the coast to southern Mozambique. Most intensive livestock production occurs in this narrow plain between the ocean and the Great Escarpment, with its fertile soils and moderate to high rainfall. The Great Escarpment forms a major barrier to moisture reaching the interior areas. Most of the high elevation rangelands occur in this region together with the central Highveld, which occurs at 1,600–1,700 m (Palmer and Ainslee 2006). Major urban, mining and agricultural activities take place in the central Highveld. The great Karoo basin, a semi-arid region situated at 1,400–1,600 m, contains steppe-type vegetation, with fertile aridic soils. The Kalahari region borders on Namibia and Botswana and represents a very important, extensive livestock producing area. This southern part of the continental scale basin is covered by sands of varying depth (sometimes >200 m), with an arid savanna vegetation and low grazing capacity.

8.2.2 CLIMATE

South Africa has a predominant arid to semi-arid climate (59%) with an average annual rainfall of about 464 mm, compared to the world annual average of about 860 mm (GCIS 2015). Wide regional variation in annual rainfall occurs, with less than 50 mm in the Richtersveld on the border with Namibia and more than 3,000 mm in the mountains of the south Western Cape Province; however, 72% of South Africa receives less than 600 mm per annum (Table 8.1). This relatively low rainfall is exacerbated by the extremely high annual potential evaporation, ranging from less than 1,800 mm in the east to more than 3,000 mm in the northwestern regions of South Africa. In most parts of South Africa, rainfall occurs mainly in the summer months, in the form of brief afternoon thunderstorms – with the exception of the Western Cape Province, which has a typical Mediterranean

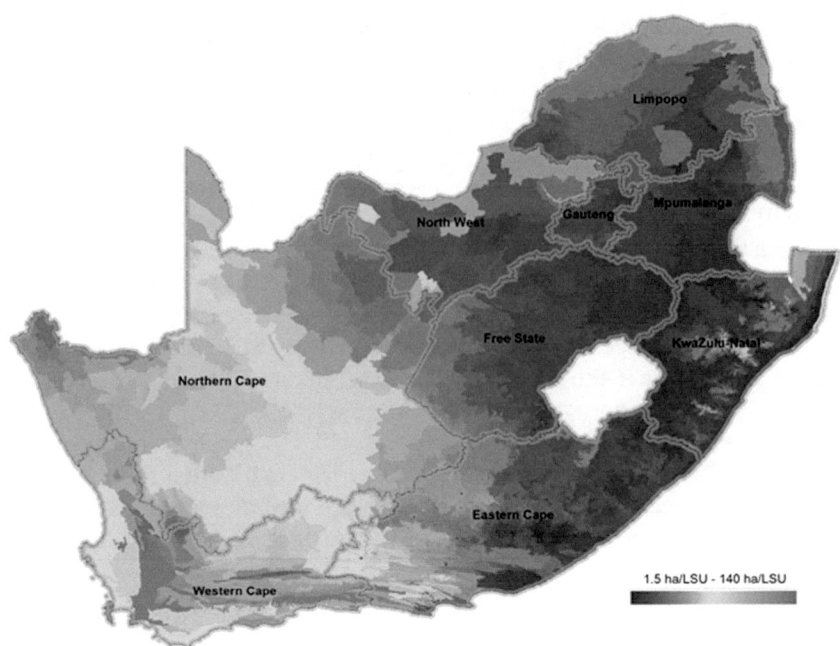

FIGURE 8.2   **(See color insert.)** Long-term grazing capacity map for South Africa – 2016 (DAFF 2016a).

TABLE 8.1
Climatic Classification and Regional
Annual Rainfall Distribution in South
Africa

Classification	Rainfall (mm)	Land surface (%)
Desert	<200	22.8
Arid	201–400	24.6
Semi-arid	401–600	24.6
Sub-humid	601–800	18.5
Humid	801–1,000	6.7
Super-humid	>1,000	2.8

Source: Adapted from Palmer and Ainslee 2006.

climate, with rainfall concentrated in the winter months. Snow is uncommon in South Africa, but can occur in the winter months on the mountain peaks of the south Western Cape Province and Drakensberg. South Africa has cold winters (June–August), with minimum temperatures dropping below freezing, primarily due to the altitude, and warm summers (November–February). Elevation and distance from the sea strongly determine temperatures in South Africa. Inland regions with higher elevations (1,500–1,700 m) normally experience warm summer mean daily maximum temperatures (26–28°C) and cool winter mean daily minimum temperatures (0–2°C), with frost during the coolest months (Schulze 1997). Along the east coastal zone between East London and Mozambique, the warm Agulhas current plays a strong role in temperature, with strong subtropical climates being experienced in the northern parts of the coastal zone. The vast interior, embodied by the Kalahari basin and the Nama-Karoo, has a more extreme climate, with low winter mean daily minimum temperatures (0–2°C) and high mean daily summer maximum temperatures (32–34°C).

The southern and southwestern coastal zone experiences moderate winter mean daily minimum temperatures (6–8°C) (Palmer and Ainslee 2006). The temperatures on the west coast, from Cape Town to Port Nolloth, are influenced by the cold Benguela current, which favors the development of fog during the winter months, bringing cold, moist air onto the coastal plain. This arid region experiences mean daily minimum temperatures of 6–8°C in July, but little or no frost (GCIS 2015).

Three major rainfall regions within South Africa exist: the winter rainfall region of the western, southwestern and southern Cape Province; the bimodal rainfall region of the Eastern Cape Province; and the summer rainfall region of the central Highveld and KwaZulu-Natal Province (Palmer and Ainslee 2006). The summer rainfall region is strongly influenced by the frontal systems developing in the Southern Ocean, which bring cool, moist air during the winter season. Generally, the natural vegetation of these regions is less beneficial for livestock production. Because of the varying rainfall seasonality, growing periods differ throughout the country. In the north, east and along the coastal belt, summer seasonality encourages C_4 grasses and the main focus is cattle and sheep farming. In the semi-arid central and western regions C_3 grasses and shrubs predominate, and this favors sheep and goat farming. Recently game farming is also becoming more popular in South Africa.

8.2.3 SOILS

South African soils have been classified using a hierarchical system due to the unique soils found in the country and the ease of use of the system (Soil Classification Working Group 1991). The South African system currently classifies soils into 74 soil forms at the highest level and numerous soil families at the lower level. Soil forms are identified based on a unique sequence of diagnostic horizons, which are in turn based on the identification of master horizons. Fey (2010) has grouped the soil forms into 14 soil groups, which are determined mainly by the geology underneath them (these have been summarized in Table 8.2). A more detailed description of the soils of South Africa can be found in Chapter 7 (Changes in SOM Content and Quality in South African Arable Land).

The distribution of soils in South Africa is mostly determined by parent material and climate, while topography, organisms and time play secondary roles (Fanning and Fanning 1989). Unique soils in distinctive environments are formed in this way, making them suitable for specific agricultural activities. South African geology is fairly young and active, and has produced relatively high

TABLE 8.2
Main Soil Groups Found in South Africa, with Associated Land Use (Fey 2010)

Soil group	Inherent properties	Associated land use
Organic	>10% organic C in topsoil	Wetlands under natural vegetation
Humic	Organic C-enriched, highly weathered, acidic topsoil	Crop and vegetable production
Vertic	Clayey, swelling, well-structured topsoil	Extensive rangelands
Melanic	Dark, fertile well-structured topsoil	Rangeland with irrigation
Silicic	Silica-cemented subsoil, shallow depth, high pH, sandy soil	Arid rangelands
Calcic	Soft or cemented carbonate subsoil, shallow depth, high pH	Extensive rangelands
Duplex	Clayey, well-structured subsoil, erodible	Permanent pastures
Podzolic	Fe and Al and organic C-enriched subsoil, sandy, nutrient deficient	Annual and perennial cultivated pastures
Plinthic	Fe-enriched (soft or cemented) subsoil, poor drainage	Marginal crop production
Oxidic	Bright (red or yellow), uniformly colored subsoil	All land uses
Gleyic	Gleyed subsoil	Wetlands under conservation
Cumulic	Young, unconsolidated deep soil	Cultivation
Lithic	Rocky subsoil	Rangelands, game and livestock farming
Anthropic	Anthropogenic material in subsoil	Crops and vegetables

nutrient status soils. The central regions of the country predominantly contain mudstones and sand-stones of the Karoo Supergroup, which give rise to shallow silicic or calcic soils, with a calcareous hardpan layer usually present in the soil profile. These sedimentary rocks were intruded by dolerites during the Jurassic period, and can be seen across the landscape as typical dykes and sills. These plagioclase-containing dolerites are responsible for high clay content soils, and contain many grass species and associated woody shrubs. They also provide summer grazing, whereas the nutrient-rich calcareous plains provide abundant, high quality winter forage. High nutrient status soils of basalt and dolerite origin are also common in the rangelands of the Highveld. The Mpumalanga Lowveld savannas are found on the gabbros and granites of the Bushveld igneous complex. The granites are responsible for sandy soils of moderate nutrient status and the gabbros for nutrient-rich rocky soils. The biggest surfaces that intrude above the African planes are the siliceous rocks of the Cape Fold mountains giving rise to immature podzol soils, and the basaltic Lesotho Highlands giving rise to melanic soils (Palmer and Ainslee 2006).

8.3 CONTRIBUTION AND ECOLOGICAL IMPACT OF VEGETATION

With an especially rich plant life, South Africa is ranked sixth in the world, having more than 20,000 different plant species, which add up to about 10% of all known plants on Earth. In fact, Mucina and Rutherford (2006) recognized 435 vegetation types. The classification of vegetation indicates the differentiation of land, usually according to a general vegetation structure or a specific plant species composition. The composition of the vegetation is in turn determined by the local climate, soil and terrain. The further classification of vegetation into biomes provides valuable information regarding the type of animals that will be suitable or adapted for the type of vegetation. Several biomes occur in South Africa and the latest vegetation classification by Mucina and Rutherford (2006), recognizes nine biomes (Figure 8.3).

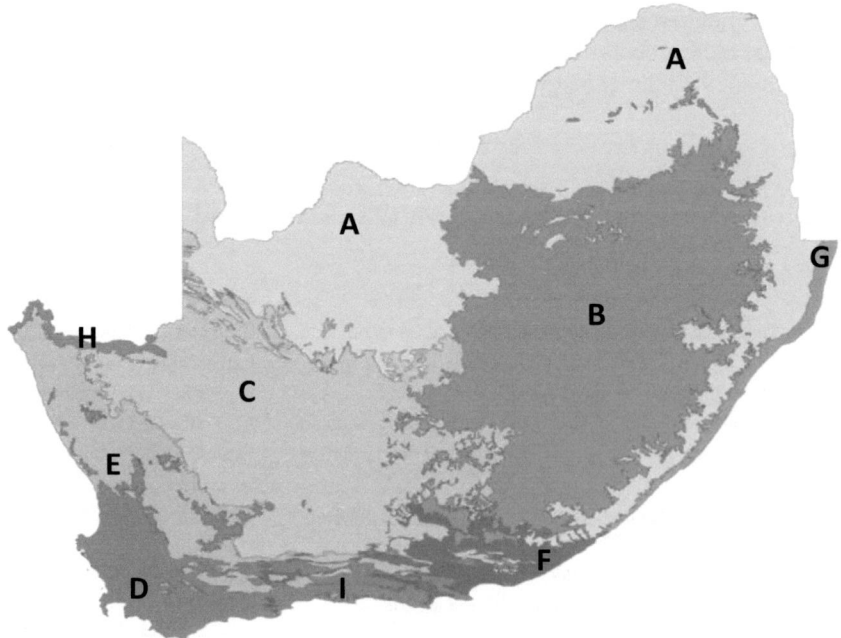

FIGURE 8.3 Most current classification by Mucina and Rutherford (2006) of the nine biomes of South Africa (A – Savanna, B – Grassland, C – Nama-Karoo, D – Fynbos, E – Succulent Karoo, F – Albany Thicket, G – Indian Ocean Coastal Belt, H – Desert, I – Forest) (adapted from Van Oudtshoorn 2015).

A. The Savanna Biome is the largest biome covering about 33% of the land area in South Africa, situated in the Northern Cape, North West, Mpumalanga and Limpopo Provinces. This biome occurs from about sea level up to about 1,500 m, covering an area of roughly 959,000 km². The climate in this biome is largely subtropical, with rainfall almost entirely in the summer. The annual rainfall varies from 220 mm in the west to 960 mm in the east. The term bushveld is used to describe this biome, which can be referred to as grass in combination with denser trees and shrubs. The biome is mostly used for livestock farming and game ranching and well-known game parks such as the Kruger National Park are located in this biome. Maize, groundnuts and sunflower are also grown in this biome.

B. The Grassland Biome is the second largest biome, comprising 28% and spanning over six provinces in the higher central plateaus of South Africa. The Grassland Biome covers an area of about 343,000 km², with an altitude that varies from about 1,000 to 3,400 m above sea level. The climate is temperate, with summer rainfall (400 to 2,500 mm per annum) and cold, dry winters with frost. Fires ignited by lightning are a common occurrence in this biome. The vegetation consists of short herbaceous plants, dominated by grasses, with woody plants being rare and confined to areas protected from frost and fire. Non-grassy herbaceous plants make a significant contribution to the botanical diversity. The biome is extensively used for cropping maize, cherries and sunflower, forestry and livestock farming. Game ranching is also becoming important in this biome.

C. The Nama-Karoo Biome is the third largest biome in South Africa, comprising 19.6% and covering about 451,000 km². It is situated mainly in the western part of the high plateau of South Africa, with a mostly flat topography and scattered hills. Hot, dry summers with cold winters are common in this summer rainfall area. Rainfall is erratic and ranges from 70 to 500 mm per annum. The vegetation in this biome consists of small shrubs intermixed with grasses and bulbs, and the land is mainly used for small livestock farming such as sheep and angora. Ostrich as well as game ranching also are practiced in this biome.

D. The Fynbos Biome is the fourth largest biome in South Africa and is internationally known for its high plant diversity. The Fynbos Biome comprises about 6.6% of the land surface area, covering roughly 70,000 km² of South Africa, with an altitude that varies from sea level up to 2,325 m. The climate is typically Mediterranean, with hot, dry summers and cool, wet winters. Winter rainfall becomes less prominent toward the east, with a range of 175 to 1,350 mm per annum. During summer, mist is common in the mountainous regions. The Fynbos vegetation is determined largely by the seasonality rather than by the amount of rainfall. The term "fynbos" is used to describe the dominant vegetation in this biome with small, generally fine leaves. The vegetation can be described as a combination of open to closed grasses, shrubs and smaller trees. Fire plays an important role in this biome, because many of the plants are dependent on fire to complete their life cycle. The grazing capacity is very low and the land is used mainly for cultivated pastures. Many areas in this biome are now considered endangered due to a transformation of undisturbed land to the cultivation of wheat. The production of rooibos tea is common, as well as certain fruits and vegetables.

E. The Succulent Karoo Biome is roughly the size of the Fynbos Biome, comprising about 6.5% of the land surface area. It is situated along the western part of the Northern and Western Cape Provinces, with an altitude ranging from sea level to about 1,500 m above sea level. This semi-desert region has a strong coastal influence on the mild climate, with a long-term average rainfall ranging from 60 to 315 mm per annum, falling mostly in winter. Typical vegetation in this biome consists of a variety of succulents, with a huge diversity of arid-zone plants. Very low grazing capacity vegetation is grazed by sheep and goats, and agricultural practices are very limited.

F. The Albany Thicket Biome is relatively small, only comprising 2.2% of the land surface area of South Africa, and limited to the Eastern Cape Province plus a smaller part of the

Western Cape Province. The biome has a semi-arid to subtropical climate, with a rainfall ranging from 200 to 650 mm per annum. Droughts of several months each are quite common in this biome due to unseasonal and unreliable rainfall. The vegetation is described as about 2 to 3 m tall, dense, woody, semi-succulent and thorny. Livestock farming and game ranching is mainly practiced, with some cropping in regions of higher rainfall.

G. The Indian Ocean Coastal Belt Biome is relatively small comprising about 1.1% of the land surface area, occurring as a coastal belt of about 800 km along the eastern coast of KwaZulu-Natal and Eastern Cape Provinces. The climate is near tropical, with an average rainfall of 1,000 mm per annum and relatively high humidity. Vegetation generally has a tropical character, including various types of coastal forests and coastal, sandy rangelands. This biome is known for pineapple, banana and mango production, as well as sugarcane farming.

H. The Desert Biome is the second smallest biome in South Africa, comprising only 0.5% of the land surface area. It consists of a narrow, 20 km wide belt along the Orange River valley beside the southern border of Namibia. The climate is one of extreme aridity, with an average long-term rainfall of below 70 mm per annum. It is characterized by erratic rainfall patterns and high temperatures. The vegetation consists of less than 10% cover and mostly consists of perennial species that are well adapted to hyper-arid climates. Communal farming with sheep and goats is very common, with especially low grazing capacity in this biome.

I. The Forest Biome is the smallest biome in South Africa, comprising a total of about 0.3% of the land surface area. These biomes are evergreen and range from 3 to 30 m in height, scattered in patches in mountain ranges and coastal lowlands along the eastern and southern parts of South Africa. The climate is associated with cold harsh winters and rainfall throughout the year (about 525 mm in winter and 725 mm in summer). Forestry is the main land use in this biome, and the rural communities use the wood for firewood and medicine.

These nine biomes do not have clear boundaries and the transition from one to another is very gradual.

Furthermore, natural variation exists within each biome and they are thus subdivided broadly into two different vegetation types, namely sweetveld and sourveld, with an in-between class known as mixed veld. Sweetveld occurs mainly in low-lying, frost-free areas, with low and erratic rainfall. Most grasses are palatable and offer grazing throughout the year, however they are sensitive to overgrazing with relatively sparse cover. Sourveld occurs mainly at higher altitudes where frost can be severe. Rainfall and grazing capacity is higher than in sweetveld, with palatable grazing only during the growing season. Sourveld can tolerate mild overgrazing, but is slow in recovery after overgrazing. A combination of the geology and soil type, rainfall, temperatures and terrain determines whether an area will be sweet- or sourveld. Mixed veld is an intermediate form that occurs between the sweetveld and sourveld extremes, with characteristics of both. In mixed veld, the sweetveld grasses occur in protected areas with fertile soil, while the sourveld grasses occur in open areas.

This unusually high botanical diversity can be attributed to the variety of environments caused by the different terrain and climatic conditions in South Africa. For a better understanding of the natural resources (soil, vegetation and climate) knowledge of the type of vegetation present on a rangeland is of extreme importance, as different plants have different forage values, which can lead to higher potential animal production. The evaluation of ecological processes on rangelands is made through the assessment and monitoring of soil and vegetation indicators, which provide information about the functional status of soil and rangelands. Classifying and managing the vegetation is therefore important if rangeland management is being considered. Table 8.3 illustrates some soil quality properties that are related to rangeland health indicators and can be used to monitor and assess rangelands.

TABLE 8.3

Soil Quality Indicators Related to Rangeland Health Indicators for Assessment and Monitoring Purposes

Rangeland health indicators	Soil quality indicators	Associated soil degradation processes
Rills and gullies	Slaking	Water erosion
Water flow patterns	Infiltration	Water erosion
Pedestals and/or terracettes	Slaking	Water erosion, wind erosion
	Soil crusts	
Bare gound	Bulk density	Water erosion, wind erosion
	Soil crusts	
Wind-scoured areas	Soil structure	Wind erosion
Litter movement	Total organic C	Water erosion, wind erosion
Soil surface resistance to erosion	Aggregate stability	Physical and biological soil crusts
	Bulk density	
Soil surface loss or degradation	Soil crusts	Water erosion, wind erosion
Plant community composition and distribution relative to infiltration and runoff	Available water capacity	Water erosion
	Infiltration	
Compaction layer	Bulk density	Compaction
Functional/structural groups	Soil biology	Loss of soil life
Plant mortality/decadence	Soil biology	Loss of soil cover
Litter amount	Soil organic matter	Loss of soil cover
Invasive plants	Soil cover	
Reproductive capability or Perennial plants	Soil cover	

Source: Adapted from USDA 2016.

8.3.1 RANGELAND MANAGEMENT SYSTEMS

According to the *Abstract of Agricultural Statistics*, there are roughly 13.7 million cattle, 1.1 million sheep and 1.9 million goats in South Africa, with a smaller number of pigs, poultry and farmed ostriches, although these numbers can fluctuate depending on whether rainfall is high or low (DAFF 2016b). More cattle are in the communal compared to the commercial sector, however the communal sector contributes minimally to the beef industry (Table 8.4). In South Africa, beef production is the most significant livestock-related activity, followed by small stock (sheep and goat) production. The entire livestock sector contributes about 75% of the total agricultural output (DAFF 2016b). A constant increase in the production of chicken meat has taken place since 1992, with a general decline in beef and veal production.

TABLE 8.4

Livestock Numbers (million) in Commercial and Communal Sectors of South Africa According to the 1999 Census (DAFF 2016b)

Agricultural sectors	Cattle	Sheep	Goats
Commercial	6.3	19.3	2.1
Communal	6.9	9.3	4.2
Total	13.1	28.6	6.3

Almost 10% of South Africa is designated as National Parks and formal conservation areas, due to a rich and diverse wildlife resource, although a large segment of the wildlife exists outside formally proclaimed conservation areas. Numerous livestock farmers currently derive some or all of their income from game ranching.

8.3.2 Historical Background

From 1652 to 1870, farming in South Africa was aimed at self-sufficiency. The transport network and markets were inadequate and poor and many farmers were nomadic and exclusively stock farmers who were focused on providing in their own needs. Land was abundantly available, unspoiled and only a small percentage of it was cultivated. Generally, the agricultural resources were well-preserved, though the natural processes of geologic or erosion did take place (Snyman 2014).

The overutilization of South Africa's natural agricultural resources only started on a large scale around 1875, with the discovery of gold and diamonds. This resulted in disorderly development, with many detrimental effects on agricultural resources, some of which may still be evident and experienced today. For example, immigrants flocked into the country, transport networks were built and many new inland markets were established. Of the many people who chose farming as a vocation at that stage, few had the necessary knowledge. The agricultural resources were consequently subjected to unrealistic demands accompanied by improper grazing and cultivation practices, which initiated the beginning of the large-scale deterioration of agricultural resources in South Africa. In time, land became more scarce and more expensive, in turn leading to the indiscriminate subdivision of agricultural land (Snyman 2014).

As the markets grew and the transport network improved, this destruction of resources continued and gathered momentum. Until 1925, the farmers had to manage without extension services and mainly learned from their own mistakes. Due to their lack of finances and knowledge, farmers could not always comprehend the underlying causes of soil erosion. The interaction between the different ecosystem components was not well understood in those days. Soil and rangeland were often viewed as an inexhaustible resource available to farmers. In many cases, the soil-plant-water equilibrium was consciously or unconsciously radically disturbed, leading to major losses in terms of soil, soil fertility and overall productivity of the rangeland.

This initial stage of wasteful exploitation of resources is also common in other areas, such as when the United States, Australia and South America were colonized by Europeans in the 17th and 18th centuries (Dregne 2002). After this stage of exploitation and disturbance of the grazing ecosystems, the dangers thereof became apparent and with the growth in knowledge, attempts were made to control and manage the degradation of rangelands. Today three main types of rangeland management systems can be found in South Africa, namely commercial livestock farming systems, communal livestock farming systems and game ranching (Smet and Ward 2005). These systems differ in management structure (multiple vs. single farmers), animal diversity, management of grazing resources (grazing system and grazing pressure, vegetation diversity/uniformity) and products (quantity, quality, diversity, market) (Smet and Ward 2005).

8.3.3 Communal Grazing Systems

About 17% of the total farming area of South Africa is occupied by communal grazing systems (Table 8.5), mainly confined to the eastern and northern parts of the country, holding roughly 52% of cattle, 72% of goats and 17% of sheep total populations (Table 8.4) (DAFF 2016b). Communal grazing contrasts noticeably with commercial grazing in the production systems, objectives and property rights, with grazing areas normally being shared by community members. The communal sector is also more densely human-populated compared to the commercial sector, with state involvement in decision-making being very low, and maintenance of infrastructure (e.g., access roads, fences, water provision, power supply, dipping facilities) neglected. These communal areas

TABLE 8.5

Land Utilized (million ha) in South Africa (DAFF 2016b)

	Total area	Farm land	Potential arable land	Arable land used	Grazing land	Nature conservation	Forestry	Other[1]
Commercial sector	105.2	86.2	14.2	12.9	71.9	11.0	1.2	6.8
Communal sector	17.1	14.5	2.5	N/A	11.9	0.8	0.2	1.6

Note: [1]e.g., urban, mining and industry

are usually based on pastoralism and are labor intensive, subsistence-based, have an inadequate use of technology and external inputs, and tend to have a negative effect on overall rangeland soil quality, due to the management strategies applied. The outputs and objectives of livestock ownership in communal systems are distinctly different from those in the commercial systems, and include aspects such as draft power, milk, dung, meat, cash income and capital storage as well as socio-cultural factors. These objectives, in combination with a strategy of herd maximization rather than turnover, lead to even the larger herd owners only selling livestock to meet cash needs. Between and within regions herd sizes differ greatly, with a small number of people owning large herds and the majority owning just a few animals or none at all (Vetter 2013).

Mixed livestock ownership is more common in communal than commercial areas, with cattle normally being preferred. Goats, followed by sheep are also widely distributed in the communal areas, while pigs and poultry are mainly kept for commercial purposes (Vetter 2013). Communal systems tend to avoid the use of fire, which has encouraged bush encroachment in some areas of South Africa and ultimately led to the impairment of the grazing potential of the rangeland.

8.3.4 COMMERCIAL GRAZING SYSTEMS

Commercial production systems in South Africa are accountable for about 75% of the national agricultural output and derive from 52% of the grazing land (Table 8.5) (DAFF 2016b). The rangelands that have a higher grazing capacity, especially in the eastern parts of the country, are grazed predominantly by cattle. The main contributor to commercial farming income is beef cattle. Sheep are bred primarily for wool and meat production and are mostly concentrated in the drier western and southeastern parts of South Africa, while goats are more widely distributed across South Africa. Over the last decade, a decrease in small stock farming has taken place due to big predator and theft problems occurring in South Africa. Extensive ranching conditions that rely on natural pasture, with occasional provided supplements, are used for grazing livestock. Ostrich farming takes place in the southern parts of South Africa, where natural vegetation is supplemented by fodders and concentrates.

Commercial grazing systems are generally practiced on fenced farms that are subdivided into a number of paddocks, where the normal practice is some form of rotational grazing with conservative stocking rates. The objective of these management strategies typically aims to improve the overall rangeland soil quality. Fire is used especially in the higher elevated rangelands to remove low quality material that remained after winter, in order to encourage the growth of short green grass during the early growing spring season (Tainton 1999). Game ranching and ecotourism has increased significantly in the commercial grazing systems, because of the complications and consequences of farming with monospecific (grazer) domestic livestock.

8.3.5 GAME FARMING/RANCHING

Both continuous and rotational grazing systems are applied in game ranching, with rotational systems being the more common choice. However, moving the game from one camp to another can be challenging, and Todd et al. (2009) suggested the following strategies: occasional closing of water points to force animals to move to another area; moving supplemental feeding around camps; selective burning of certain areas, as game tend to prefer palatable plants arising from recently burned areas; using electrical fences to enclose certain areas, where resting periods can then be applied. Game farming/ranching has become an increasingly important component of several farming enterprises in South Africa. However, the use of inappropriate game species, overstocking and poor management can pose the same threats to ecosystem biodiversity as livestock farming.

8.4   OBJECTIVES AND PRINCIPLES OF RANGELAND MANAGEMENT

Although the objectives and principles of rangeland management are agreed upon, researchers, extension officers and farmers differ on the relative merits of different rangeland management systems, regarding the fulfilling of these principles. The first principle is the keeping of an appropriate number of animals on the area. The second principle involves effective resting and utilization of rangeland. Stocking rate management in combination with grazing capacity is therefore the most important principle for sustainable rangeland utilization. A rangeland management system is consequently a combination of resting- and grazing periods, in which the principles and objectives of rangeland (vegetation and soil), livestock and game management are incorporated.

8.4.1   Stocking Rate, Grazing Capacity and Rangeland Condition

The first principle in rangeland management is establishing how many animals can be supported by the different vegetation types without degrading any natural resource. Stocking rate (number of animals per unit area maintained on the farm) must be considered the single most important determinant of the ecological sustainability of the farming enterprise (Tainton 1999). Stocking rates primarily determine the amount of plant production that will be removed by livestock, as well as the impact that livestock will have on the vegetation and soil on a farm. Over a long period, overstocking in combination with drought is a major contributor to rangeland degradation (Snyman 2009a).

The grazing capacity (the production potential of a specific area of rangeland) depends on rangeland condition (species composition and soil health) (Table 8.6), with considerable variation existing between farms and even between paddocks. In South Africa, there is a high priority to conduct more specific grazing capacity assessments at farm level to achieve sustainable rangeland utilization (Van der Westhuizen et al. 1999). The Department of Agriculture in South Africa (DAFF 2016a) has thus developed grazing capacity norms with accompanying maps for rangelands in good condition (Figure 8.2).

Rangeland condition, grazing capacity and stocking rate are certainly the most important factors affecting the sustainability and profitability of a farm over the long-term (Table 8.6). Rangeland condition is an indicator of both vegetation and soil health, which determines animal production and the ability of rangeland to handle risks associated with grazing and droughts (Snyman 2004). Rangeland condition is broadly determined by the ratio of palatable vs. less palatable plants, basal cover and productivity. The composition of these species is linked to soil quality over time (Snyman 2004). Without thorough rangeland and soil evaluation it is impossible to determine the state of health of the rangeland. The present condition of a rangeland is an indication of how that rangeland was managed in the past, and helps the farmer to make future management decisions. Table 8.6 demonstrates the implications of rangeland condition on vegetation dynamics and some soil properties in a semi-arid rangeland in South Africa (Snyman 2014). Adaptive rangeland management depends on effective monitoring to detect possible vegetation and soil degradation in time (Tainton 1999).

Rangeland condition determines the grazing capacity that contributes to the sustainable utilization of the ecosystem (Table 8.6). Without monitoring the state of the rangeland, it is impossible to determine whether the stocking rate (the farmer's estimate of the number of animals to be kept on the farm) is in accordance with the grazing capacity (actual production potential of a farm) (Van der

TABLE 8.6

Vegetation Dynamics and Soil Properties of Three Rangeland Condition Classes (Good, Moderate and Poor) in a Semi-Arid Rangeland of South Africa, Measured over 40 Seasons

Vegetation and soil properties	Rangeland condition		
	Good	Moderate	Poor
Rangeland condition (%)	88	59	31
Basal cover (%)	8.3	6.4	2.9
Dry matter production (kg/ha)	1,238	768	368
Root mass (kg/ha)	2,341	1,088	615
Water-use efficiency (kg/ha/mm)	2.5	1.58	0.78
Grazing capacity (ha/LSU)[1]	5.22	8.34	19.38
Evapotranspiration (mm/day)	1.73	1.61	1.55
Highest soil temperature in 0–50 mm (°C)	44	47	52
Highest soil temperature in 50–100 mm (°C)	36	39	43
Surface runoff (% of annual rainfall)	3.5	5.6	8.7
Sediment loss (t/ha/a)	0.41	1.20	2.55
Soil compaction (kg/cm)	4.40	6.12	11.00
Loss in organic C (kg/ha)	0	2,659	5,225
Loss in total N (kg/ha)	0	180	331
Income (beef cattle production) ($/ha)[2]	39.2	24.6	10.6
($/ha/mm rainfall)	0.07	0.04	0.02

Source: adapted from Snyman 1998, and Snyman 2014.
Notes: [1] LSU – Large Stock Unit;
[2] $205 gross marge/LSU – based on 2019 values.

Westhuizen and Snyman 2014). The continuous evaluation of grazing (both vegetation and soil) is essential to quantify the trend of rangeland condition.

It is also important to remember that the best-adapted vegetation for any particular ecosystem will naturally come back if the line of "no return" has not been crossed. Additionally, landowners should work together with nature by following a scientific-based grazing management system on the farm (nutritional needs of livestock, forage quality) that will increase profitability without natural resource degradation (soil and vegetation).

8.4.2 Effective Utilization and Resting of Rangeland

Effective utilization (intensity, frequency and season) is an important challenge for successful rangeland management. This must also be accompanied by long-term rangeland improvement or maintenance, in order to be sustainable. Although the correct stocking rate is the most important component, certain decisions, such as when animals will be moved from one paddock to another, are also essential for the effective utilization of the grazing ecosystem. In addition, it is also important to consider aspects such as production of the maximum quantity of digestible nutrients from the rangeland, optimal utilization of the produced forage on the rangeland, and efficient conversion of the produced and utilized forage into animal products like meat, milk and fiber.

In South Africa, the effects of grazing on the vigor of particularly highly desired plants vary drastically over the growing season. Repeated heavy grazing during spring, just after the first rains, and during early summer is one of the most detrimental types of treatment that can occur on rangelands. This causes a rapid decline in the vigor of palatable plants, while less palatable plants and pioneers are usually stronger when competing for water. Subsequently, with severe defoliation, growing reserves are continuously withdrawn from the roots and the vigor of these grasses is reduced.

The more frequently grazing occurs on rangeland during a growing season, the more detrimental the effect is on above- and below-ground vigor. Contrastingly, no grazing for several years can also be harmful, because ungrazed tufts tend to collect a lot of dry material that harms vigor and seed production. These tufts will eventually die off and be replaced by annual herbs or less desirable plants. The ideal number of times a paddock should be utilized during a growing season, ranges between two and three defoliations. Currently scientists and farmers are emphasizing the incorporation of long rest periods (about every third or fourth year).

Grazing intensity refers to the amount of material to be removed during grazing. It is largely determined by the amount of older, dry fodder, with easier short grazing of paddocks during the spring than winter months. From a quality point of view, it is good to utilize paddocks until they are fairly clean, only every third year. When this practice occurs annually, rangeland conditions will decline over the long-term. It is important to acknowledge that clean rangeland grazing requires longer resting periods, to make up for both above- and below-ground recovery.

The period that livestock can normally graze a paddock without harming the rangeland depends on factors such as the season of use and the type of livestock, as well as the paddock and herd size. If the grazing of paddocks during the growing season is longer than three months, selective grazing will occur where palatable plants are overgrazed due to repeated utilization. These long periods of grazing should preferably be avoided, and only applied if the number of paddocks is limited, in which case it is better to extend grazing periods in order to effectively lengthen rest periods as well. It is also beneficial to vary grazing intensity over several years.

Rangelands must be rested with the intention of improving the productivity of grazing plants as much as possible, so that animal production can be maintained at peak levels over the long-term. Additional aims for rangeland resting include: the provision of seed production and seedling establishment; the accumulation of above-ground growth; the changing of botanical composition; and ensuring fast-growing and vigorous plants to sustain soil health/quality.

8.5   GRAZING SYSTEMS ON RANGELAND

Rangeland is and remains the cheapest source of feed for animals and should therefore be protected, conserved and utilized sustainably, at all costs. Until the 1960s, grazing systems in South Africa were largely based on a conservation approach. With the introduction of the Acocks/Howell system, utilization began forming part of the conservation approach (Zietsman 2014). Various techniques emerged to quantify rangeland condition and grazing capacity.

Overstocking, continuous grazing, grazing periods that are too long, repeated grazing at the same time of the year, stock breeds or game species not adapted to the rangeland type, as well as long-term and indiscriminate provision of supplementary feeding, can be seen as the most important detrimental grazing practices (Van der Westhuizen and Snyman 2014).

The main objective of a grazing system is to minimize selective grazing and control frequency and intensity of defoliation of grasses, shrubs and trees. A grazing system refers to the pattern with which paddocks are alternatively grazed and rested, and in South Africa, these grazing systems can roughly be divided into continuous grazing and rotational grazing.

8.5.1   CONTINUOUS GRAZING

Continuous grazing is the simplest grazing system that can be applied and is mainly applied in the communal areas and on certain game farms. It is also known as fixed stocking where one or more groups of animals remain in one undivided grazed area for more than one year. Animals have free access to the whole area and are not limited in terms of their movement and choice of grazing, with stocking rate being the only aspect that is adjusted. Continuous grazing can only be sustainable if stocking rate is below the grazing capacity of the rangeland (Smith 2006).

8.5.2 Rotational Grazing

Rotational grazing suggests a system where a group of animals on a farm is relocated to apply rest to a rangeland area that has been grazed. Rotation among paddocks normally takes place in a cycle in order to apply the essential principle of rangeland rest, similar to migrating grazers in an undisturbed ecosystem. Any form of rotational grazing is known to have a better long-term effect on a rangeland compared to continuous grazing, and aids in the establishment of a sustainable grazing system.

Over the years, different approaches to rotational grazing were developed by grassland scientists, botanists and farmers in order to improve rangelands. These changed from the so-called conventional systems, with two to three paddocks per herd (1938–1950), to the multi-camp systems (1960–1990), with usually more than five paddocks per herd. The multi-camp systems can be seen as a development or intensification of the conventional system. These systems were designed to fulfil the physiological requirements of the grazing plant with the emphasis on systematic resting periods. In rotational grazing, the season, frequency and intensity of grazing (defoliation) are manipulated by the number of paddocks, the period of grazing in and the period of absence from a paddock (Snyman 2014). There are two main approaches regarding the degree of utilization and its effect on the plant community, namely controlled selective grazing (CSG – high production utilization, more applicable in sweetveld) (1968–1990) (Drewes 1991) and non-selective grazing (NSG – high utilization grazing, more applicable in sourveld) (Savory 1988; Zietsman 2014).

In the past, less emphasis was put on root production (indirectly soil quality), when selecting a grazing system (Snyman 2009a). In South Africa, the full growing season's rest system is most popular at the moment, where a third or quarter of the farm will, on a rotational basis, receive a resting period for a full growing season each year (Van Pletzen et al. 1995; Kirkman 1995). The purpose of this long resting period is to fully recover the plants' vigor (above- and below-ground), which ensures sufficient root growth. A detailed discussion of above-mentioned grazing systems applied in South Africa can be found in Snyman (2014).

8.6 DROUGHT

Drought in especially arid and semi-arid rangeland areas can be responsible for substantial damage to rangeland condition, as plants are put under pressure, available forage decreases and soil cover is significantly reduced. Drought management is therefore a vital part of rangeland management, and can include strategies such as vegetation recovery before restocking, establishment of fodder crops, supplement feed, prevention of overgrazing and good rangeland management. It is recommended that livestock numbers be reduced during extended drought periods. In most cases it is also important to continue with lower stocking rates for a certain period after a drought, to ensure the rangeland recovers as quickly as possible. Contrastingly, rangeland degradation (vegetation and soil) due to animal impact is creating its own droughts, so called man-made droughts (Table 8.6). Parts of grass tufts mostly die off during a severe drought, producing a lower plant cover and grazing capacity. Furthermore, high grazing pressures during drought periods have detrimental effects on rangeland recovery after drought (Van der Westhuizen and Snyman 2014). However, a positive effect of tuft tillers that die off is that the dead roots of the tillers can improve soil quality. Figure 8.4 illustrates the impact of a severe drought on root die-off for red grass (*Themeda triandra*) after defoliation, in a drought-stricken semi-arid rangeland.

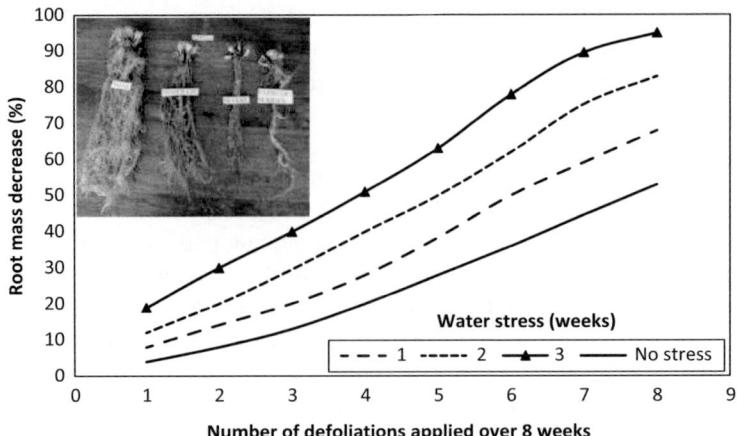

FIGURE 8.4 Root mass decrease (%) for *T. triandra* applying different defoliation frequencies (weekly, every second, fourth and eighth weeks) at 25 mm height (average sheep grazing height), then subjected to different water stress levels (1, 2 and 3 weeks) (Snyman 2016).

8.7 WILDFIRES

Fire management is a much debated rangeland management practice, and if used incorrectly, the fires can be devastating to people and properties. The effects of burning depend on a variety of factors such as wind, temperature and amount of plant material. Two aspects are important with fire management: the control and prevention of wildfires by using firebreaks; and the use of prescribed burning as an important rangeland management tool, for improvement of the grazing value, and to control undesirable plants such as woody or invasive species (Tainton 1999).

Burning can only form part of an existing rangeland management system, and is not a direct method of utilization of vegetation in the way that grazing is. As in the case of grazing, the season and frequency of fire can be controlled, except for natural wildfires caused by lightning. Under normal circumstances, perennial grasses do not burn to death during the dormant season, because the growing points of grasses are located lower down. Burning must never be applied in an attempt to rectify the results of poor rangeland management (Trollope 2009). The main reason for rangeland burning should be to remove unwanted grass material and to prevent the encroachment of undesirable invader plants.

In most of the arid and semi-arid rangelands of South Africa, there is no need for burning, because the rangelands can be utilized efficiently without fire. However, fire is usually part of the management system in the sourveld due to problems with rangeland quality. Intensive research done in the semi-arid rangelands of South Africa show that at least two full growing seasons of rest is needed for rangelands (soil and vegetation) recovering after a fire due to tuft die back (Snyman 2015a). The first season following the fire, above-ground plant production can be half than that of unburned rangeland, with almost two full growing seasons needed to catch up from unburned rangeland. The sooner the rangeland is grazed after the fire, the longer the recovery period and the higher the drought risk. The relevance of fire and soil quality in terms of lower basal cover and tuft die-off for one and two seasons following the fire can be seen in Table 8.7, with higher soil temperatures and soil compaction, and lower water content and soil nutrients, which are all detrimental aspects for the sustainability of the rangeland ecosystem in drier rangeland areas (Snyman 2015b).

8.8 EFFECT OF RANGELAND MANAGEMENT ON VEGETATION AND SOIL

Rangeland management can affect vegetation and soil in various ways (Table 8.6). Direct effects are related to animal trampling and excretion, while indirect effects are involved with changes in vegetation structure and function, like defoliation, burning and drought (Figure 8.1) (Taboada et al. 2011).

TABLE 8.7

Influence of Burning on Vegetation Cover and Soil Properties in a Semi-Arid Rangeland, One and Two Seasons Following the Fire

Soil properties	Unburned	Burned First season	Burned Second season
Basal cover (%)	8.14	4.56	6.91
Plant density (plants/m²)	87	78	80
Soil compaction (Mpa)	6.2	9.1	9.1
Soil temperature(°C) – up to 100 mm depth	32.5	36.8	32.8
Soil-water content (mm) – up to 300 mm depth	60.2	49.2	55.3
Soil nutrient content – up to 50 mm depth			
Organic C (g/kg)	10.2	6.9	8.0
Total N (g/kg)	1.0	0.7	0.7
pH	5.9	5.9	5.6
Ca (mg/kg)	564	852	714
Mg (mg/kg)	140	202	158
K (mg/kg)	138	274	220
Na (mg/kg)	28	30	30
P (mg/kg)	2.22	1.14	1.93

Source: Adapted from Snyman 2015a.

Both beneficial and detrimental effects of rangeland management on soil properties have been found in several studies (Sandhage-Hofmann et al. 2015; Kotzé et al. 2013; Kotzé et al. 2017). Livestock grazing are known to significantly change almost every aspect of soil structure and function, including soil porosity, chemistry, microbiology, nutrient cycles, productivity and erosion rates (Petersen et al. 2004; Snyman and Du Preez 2005; Du Toit et al. 2009). It has been reported that high grazing intensity increases soil compaction and erosion, while reducing soil aggregate stability and decreasing soil nutrient and organic matter levels (Table 8.6). Drought and fire in arid and semi-arid rangelands generally reduce SOM content and the soil seed bank, which can lead to reduced porosity and increased bulk density, causing further soil compaction (Snyman 2015b). However, the decrease in root mass due to drought can also add organic matter to the soil (Figure 8.4). These effects, especially when combined, lead to rangeland degradation in the form of compaction or erosion (Du Preez and Snyman 1993; Roberson 1996; Snyman 1999).

8.8.1 VEGETATION

8.8.1.1 Defoliation and Decreased Plant Cover

Plant roots fulfill a crucial role in rangeland ecosystem functioning by being the link between above-ground (vegetation) and below-ground parameters (soil). Their role is to anchor the plant to the soil and take up water and nutrients, and, in turn, to provide structure and nutrients to the soil organisms. Livestock grazing applies pressure on certain palatable plant species, and may decrease depending on the grazing system and period of exclusion, leading to a possible reduction in overall grass biodiversity (Taboada et al. 2011). Decreases in plant cover, caused by either drought, fire or grazing, can trigger soil erosion (Table 8.6) as well as rapid evaporation from the soil (Teague et al. 2011). The frequency of defoliation in rangelands has a direct influence on root growth, which, in turn, influences properties such as organic matter and organisms in the soil. The impact of different defoliation frequencies (weekly, every second, fourth and sixth weeks) on the root growth of

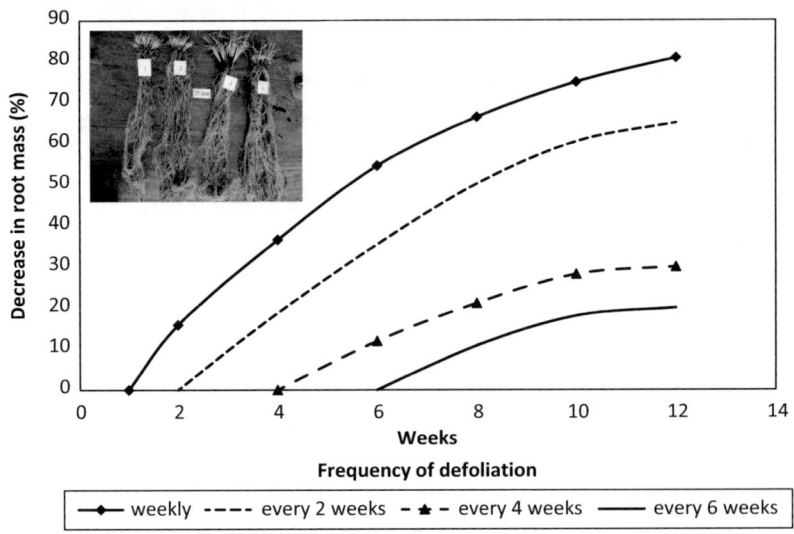

FIGURE 8.5   Root mass decrease (%) for *T. triandra* by applying different defoliation frequencies, at 25 mm height (average sheep grazing height), applied over twelve weeks (Snyman 2016).

T. triandra (the most significant grass species contributor to beef production in South Africa) is illustrated in Figure 8.5 (Snyman 2014). Defoliation can also be responsible for more bare soil, which leads to increases in the fluctuation of daily temperatures. This is illustrated in Table 8.8 where the results from Lavado and Taboada (1987) and Snyman and Du Preez (2005) display how soil temperatures and water evaporation are increased by the effects of defoliation caused by grazing in semi-arid rangelands. The areas that were ungrazed caused an increase of litter and plant biomass, which reduced topsoil temperatures during the day and led to lower evaporation from the soil. Table 8.8 further demonstrates the effect of plant litter and biomass in conserving soil water, where evaporation rates were 4.6 to 11.2 times higher in the grazed compared to ungrazed areas. Additionally, when salinized deep soil horizons are present, higher evaporation rates due to lower plant cover can stimulate the accumulation of salts in the topsoil, which is due to the upward movement of water and soluble salts, leading to the sporadic salinization of the topsoil. This is demonstrated in the Northern Cape Province of South Africa in studies done by Peterson et al. (2004) and Haarmeyer et al. (2010) on grazing affecting soil properties and vegetation types. They show that salinization leads to a significant reduction in rangeland forage production, but can be controlled by rangeland management strategies such as increased ground cover through changes in grazing regimes, increased resting periods or shifting from continuous to rotational grazing.

TABLE 8.8

Variations in Topsoil Temperatures and Evaporation Rates as Affected by Rangeland Grazing

	Soil temperature variations (°C)		Evaporation rates (mm d⁻¹)	
Months	**Ungrazed**	**Grazed**	**Ungrazed**	**Grazed**
January	22.8–25.5	26.4–32.7	0.19	2.13
February	17.7–23.9	21.5–30.4	0.39	1.79

Source: Adapted from Lavado and Taboada 1987; Snyman and Du Preez 2005.

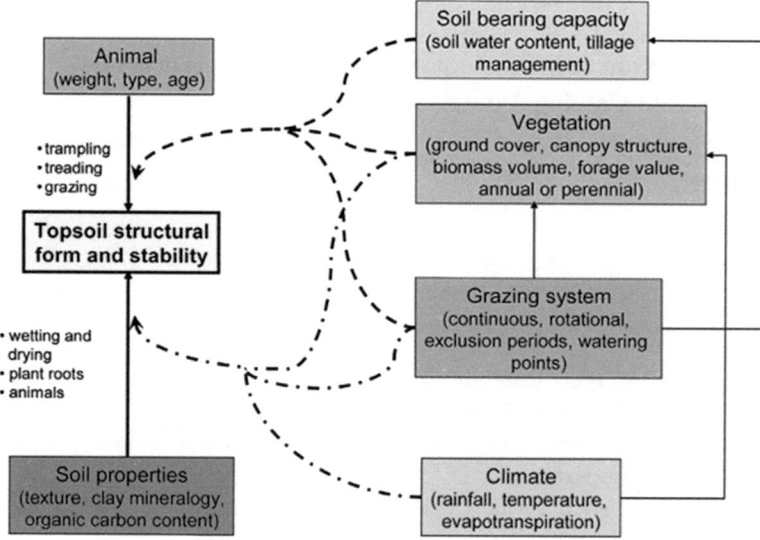

FIGURE 8.6 Diagram from Taboada et al. (2011), indicating factors affecting soil structure and stability in rangeland systems. Solid lines indicate direct influences, dotted lines indicate indirect influences.

Bare ground areas caused by overgrazing, drought or fires can trigger desertification, and this is linked to litter removal and soil organic carbon (C) decreases (Haarmeyer et al. 2010). Defoliation and excretal returns also impact C and nutrient cycling in soil, where soil organisms are particularly involved. These soil organisms live in the soil pores, and are responsible for biotic mechanisms of soil structural resilience (Young and Ritz 2005).

8.8.1.2 Animal Trampling

Trampling by the hooves of grazing animals can severely disturb topsoil structure due to the mechanical effect on the soil surface. This changes the form and stability of soil aggregates, which then changes other properties such as soil bulk density, pore size distribution, soil strength and water and oxygen movement. Various factors are responsible for this change in aggregate stability. Taboada et al. (2011) proposed a conceptual model integrating both direct and indirect effects of grazing animals on soil physical properties (Figure 8.6). By contrast, Beukes and Cowling (2003) investigated non-selective grazing impacts on soil properties in the arid Nama Karoo of South Africa, and found that while soils are loosened and aerated by animal trampling, they were not necessarily compacted. It is known that grazing animals can maintain a rangeland by breaking soil crusts that interfere with seed germination and rainfall absorption, by mulching the soil surface and protecting it against erosion and evaporation, and by composting the plant material and fertilizing the soil (Livingwell 2010).

8.9 ANIMAL TYPE AND STOCKING RATE

The trampling of grazing animals breaks up soil aggregates, and the impact of this depends on the type of animal, the soil bearing capacity, the vegetation and the type of grazing system. The destruction of soil structure can be counteracted by factors such as soil type and recovery times on climatic, vegetation and grazing systems. The pressure that grazing animals apply on soil is determined by the animal's mass and hoof size, as well as movement. A comparison between different grazing animals is demonstrated in Table 8.9, with averaged static pressures exerted specifically by cattle (138 kPa), sheep (66 kPa) and goats (67 kPa). Despite sheep and goats applying similar lower

TABLE 8.9

Comparison of Mass, Foot Area and Pressure Exerted by Different Grazing Animals while Standing

Grazing animal	Animal mass (kg)	Total foot area (cm²)	Pressure (kPa)
Cattle	306–612	264–460	98–192
Sheep	40–54	55–84	48–83
Goats	40	55	60–73

Source: Adapted from Taboada et al. 2011.

static pressures than cattle, their trampling may also lead to rangeland degradation in semi-arid areas of South Africa (Du Toit et al. 2008; 2009, 2011).

Stocking rates normally increase bulk density and decrease soil structure stability, leading to decreased water infiltration and hydraulic conductivity. Rangeland management systems where continuous grazing is applied often decrease vegetation quantity and quality, especially around gateways and feeding troughs (Du Toit et al. 2011).

8.9.1 GRAZING SYSTEMS

Rotational grazing systems aim to avoid selective and overgrazing, where livestock is allowed to graze during short time periods at moderate to high stocking rates. Rotational grazing was found to be more successful in increasing vegetation cover, and had a positive effect on overall soil quality in a Grassland and Savanna Biome in South Africa (Kotzé et al. 2013; Sandhage-Hofmann et al. 2015; Kotzé et al. 2017). Additionally, Snyman (2014) found that rangeland condition as affected by grazing, decrease basal cover and dry matter production on a rangeland (Table 8.6).

8.9.2 SOIL

8.9.2.1 Soil Structure

Soil structure is the soil property most immediately affected by grazing due to compaction, which reduces water and air infiltration into the soil and restricts plant root growth. Damage to soil structure by animal trampling is influenced by soil-water content, with larger soil pores being more vulnerable when soil is moist, which ultimately leads to an increase in bulk density by compaction (Greenwood and McKenzie 2001). Water infiltration rate is a very sensitive parameter for identifying soil physical damage by grazing. Petersen et al. (2004) found that grazing decreased water infiltration by nearly 75% when compared to ungrazed areas in South Africa. This then leads to increased runoff and erosion – associated with organic matter losses and lower PAW (Table 8.6). The decrease in water infiltration taking place in grazed pastures is due to the loss of macropores, as well as by vegetation removal. Table 8.6 also demonstrates the effect of rangeland condition on surface runoff, sediment loss and soil compaction associated with grazing. Burning causes a reduction in organic matter, which can change the structure of the soil, leading to reduced porosity and increased soil compaction (Snyman 2015b).

8.9.2.2 Soil Fertility and Nutrient Dynamics

Grazing animals greatly affect both soil fertility and chemistry indirectly, through the effects on plant species composition and soil structure. This happens through herbivory digestion and excretion, which dramatically increase the decomposition rate and thus directly alter the distribution of nutrients stored in the soil and the availability of those nutrients to plants. Grazing also affects soil pH, where grazed areas generally have significantly lower pH than ungrazed areas (Roberson 1996).

Drought and fire have an indirect influence on soil fertility due to the negative effect it has on the potential source for soil nutrients originating from vegetation. Snyman (2015b) also reported that fire can increase pH as well as concentrations of exchangeable potassium, calcium, magnesium, and sodium in soil, while substantial losses of organic nitrogen (N) to the atmosphere can take place in semi-arid rangelands.

8.9.2.3 Total Soil Nutrients

Grazing livestock strongly affect the dynamics of soil nutrients in a rangeland. Nutrients are removed by livestock consuming plants and convert them into livestock biomass, which is transported off site. Around 60 to 95% of the nutrients contained in the consumed forage are returned back to the soil in a different chemical form, thus significantly affecting the rates of nutrient cycling (Roberson 1996; Taboada et al. 2011). Grazers return nutrients to the ecosystem as dung and urine, forming localized excretal patches, leading to changes in the spatial distribution of soil nutrients, especially the horizontal distribution. Excretal patches cover only a small proportion of the pasture area, which increases the spatial heterogeneity of nutrient inputs to soil and availability to plants. For example, grazing animals spend much of the hot daylight hours under shade and around water, where substantially increased deposition of feces and urine takes place, leading to accumulation of N, phosphorus (P), and potassium, especially in the pastures closest to shade, water and feeding troughs (Roberson 1996). These excretal patches also have a significant effect on soil biological properties, mostly by being a food source for soil microorganisms (Kotzé et al. 2017).

Grazing also has an effect on nutrient losses, which are concentrated in these excretal patches and comprise both gaseous and leaching outputs. These nutrients are lost through increased erosion of nutrient-rich surface soil, accelerated decomposition of litter and organic matter, and leaching. Generalizing about nutrients removed by grazers is quite challenging, as this is mainly determined by the total amount of consumed forage, which in turn is controlled by factors such as the availability of forage, the farming management system and the rate of forage utilization. As a consequence, the total amount of nutrients consumed by grazing animals, and hence the role of grazing on nutrient budgets, may be difficult to predict accurately (Taboada et al. 2011). Direct comparisons of grazed and ungrazed soils generally show that grazing reduces total soil nutrient levels. In long-term studies, researchers found significantly more total soil N and C up to 1 m depth in ungrazed compared to grazed areas (Snyman and Du Preez 2005).

8.9.2.4 Nutrient Availability

While nutrient availability is more essential to ecosystem function than total soil nutrient levels, the effects of livestock grazing on the availability is not that well-known. Just because a nutrient is present in the soil, does not necessarily make it available to plants. Huge amounts of nutrients like N and P are stored in relatively unavailable forms in vegetation and SOM. It is only when vegetation and SOM are decomposed by soil organisms that nutrients are released into the soil solution and become available to plants and other organisms (Steinfeld and Wassenaar 2007).

Livestock grazing is responsible for the rate at which vegetation and SOM are decomposed (Laurenroth et al. 1994). Forage is rapidly decomposed during digestion by grazed livestock and nutrients are returned to the soil in readily available forms as feces and urine, which generally leads to increased short-term soil nutrient availability.

8.9.2.5 Soil Carbon Dynamics

As much as two-thirds of C stored in rangelands are found in SOM pools below ground (McSherry and Ritchie 2013). The C stocks in rangelands are largely influenced by climate, with total C increasing with precipitation due to vegetation growth, and decreases when temperatures rise (Delgado-Baquerizo et al. 2017). Grazing has a major influence on rangeland C cycling, affecting transfers between vegetation and soils, and therefore contributes to the regulation of the global C cycle. Table 8.6 confirms the effect of rangeland condition on C and N dynamics, with poor condition contributing to a loss of about 50% C and N, compared to moderate condition. Any rangeland management system (grazing, fire, drought) that influences SOM, consequently influences total soil C.

8.9.2.6 Soil Organic Matter and Litter

SOM and litter are responsible for improved soil structure, water holding capacity and infiltration, and are important sources for soil nutrients. SOM and litter also increase water availability to plants throughout the growing season by their water absorption capabilities, and help soils to resist erosion, compaction and deformation (Roberson 1996). Large amounts of plant biomass are normally removed by grazing livestock, and these areas tend to have lower SOM levels compared to ungrazed areas (Table 8.6) (Snyman and Du Preez 2005). SOM is also greatly affected by burning and drought, because SOM is normally concentrated at or near the soil surface, where it is directly exposed to heating by radiation produced during the combustion of above-ground fuels and elements of nature. Organic matter and soil organisms also change irreversibly at relatively low temperatures (DARD 2014).

8.10 CURRENT STATUS OF RANGELANDS IN SOUTH AFRICA

Palmer and Ainslee (2006) reported that soil degradation (erosion and compaction) due to rangeland management in South Africa is substantially higher in communal than in commercial rangelands, due to the variations in their approach to management. The most affected areas are steep, sloping land in the eastern parts of South Africa primarily used for grazing, and land in KwaZulu-Natal and Northern and Eastern Cape Provinces also being badly affected by soil degradation.

Palmer and Ainslee (2006) further reported that rangeland degradation (loss of plant cover and change in species composition – from loss of biodiversity, bush encroachment, alien plant invasions and deforestation) were also higher in communal than in commercial rangelands. Bush encroachment and alien plant invasions are generally worse in commercial compared to communal areas. Rural poverty and communal land-use systems, are closely correlated with rangeland degradation, because poor rural areas are reliant on natural resources for their energy and food requirements, and the responsibility for sustainable management practices on the part of local land users is diminished. Rangeland degradation seems to be the highest in the KwaZulu-Natal and Limpopo Provinces of South Africa. The eastern Karoo is no longer seen to be severely degraded by agricultural experts, as it appears to have benefited considerably from rangeland management suggestions from people such as John Acocks in the middle of the 20th century. The rate of rangeland degradation is decreasing in commercial areas, largely due to state intervention, strategies and schemes, while it is perceived to be increasing in communal districts.

Rangeland management, specifically rangeland condition, significantly affects the structure, composition, fertility, chemistry and functioning of soils, often in ways that compromise both short- and long-term productivity. It is therefore important that rangeland management includes strong, effective and specific measures to prevent degradative effects and to repair soil damage where it has already occurred. Soils and vegetation (rangeland condition) must be thoroughly evaluated for rangeland damage in order to develop rangeland management programs that will promote soil quality/health.

Changes in rangeland management, especially applying long enough resting periods, can be very effective in controlling soil compaction, infiltration and erosion. The most effective approach to protect a rangeland ecosystem is therefore to link rangeland management directly to soil quality/ health through monitoring and adaptive management. This can include the following:

- Rotational grazing aims to give grazeable plants the opportunity to produce viable seeds, by matching resting periods with growing periods.
- Long-term improvement of rangeland condition, which will lead to an increase in not only grazing capacity, but also soil quality.

- Stocking rate should be as close as possible to the grazing capacity.
- Livestock production systems must be adapted to the environment.
- Resting periods between grazing periods must coincide with the grazing pressure applied.
- Supplementary feeding must increase energy and mineral shortages, instead of subsidizing food requirements.
- Adapting grazing and resting periods to unforeseen climatic occurrences such as drought and fires caused by climate change.

It is also important that a more interdisciplinary research approach should be followed to achieve successful sustainable rangeland management, especially in communal areas, where rangeland degradation is the highest. Natural resources have been put under more pressure in the last decade, due to the popularity of game ranching/farming. Due to the variations that exist in grazing preferences, habitats and relocation problems of the different game species, more intensive skills are needed in future to ensure sustainable utilization of the rangeland ecosystem.

8.11 CONCLUDING REMARKS

There is no doubt that rangeland management affects soil quality, and that these effects vary across different climate conditions, soil types, grazing systems and kinds of livestock. In general, grazing changes soil physical properties by treading, which in turn decreases ground cover, leading to increased topsoil temperatures and hence evaporation rates, and can also cause salinization in warmer, drier areas. The treading of animals often leads to surface compaction when moist soil is trampled and puddling when wet soil is trampled. The restoration of damage to soil structure by grazing animals can be done by natural regenerative practices. This may however take many years in warmer, drier climates, whereas faster responses are possible in cooler, wetter climates, due to the action of frequent wetting and drying cycles, freezing and thawing cycles, and vigorous pasture root growth.

Although the balance between soil C inputs and outputs varies widely among grazed systems, the size of C pools and fluxes in rangelands can be strongly controlled by grazing. This takes place through the decomposition of litter as well as plant community turnover, which, to a large extent, mediates the soil microbial community. Grazing furthermore modifies N and P cycling by the return of animal dung and urine to the soil, as well as by increasing N losses through volatilization and leaching.

Protecting the soil by vegetative cover is especially important in semi-arid rangelands. This can be done by introducing rest periods by excluding livestock from grazing. Rotational grazing can also effectively increase forage utilization efficiency and reduce the fraction of dung and urine deposited on the soil, thus reducing field heterogeneity. Due to a lack of information about rangeland management effects on soil quality in South Africa, further research is warranted.

REFERENCES

Beukes, P. C., and Cowling, R. M. 2003. Non-selective grazing impacts on soil properties of the Nama-Karoo. *Journal of Range Management* 56:547–552.

Chanasyk, D. S., and Naeth, M. A. 1995. Grazing impacts of bulk density and soil strength in the foothill fescue grasslands of Alberta Canada. *Canadian Journal of Soil Science* 75:551–557.

DAFF. 2016a. *Long-Term Grazing Capacity Map for South Africa 2016*. Pretoria, South Africa: Department of Agriculture, Forestry and Fisheries.

DAFF. 2016b. *Abstract of Agricultural Statistics*. Pretoria, South Africa: Department of Agriculture, Forestry and Fisheries. Available at: www.daff.gov.za/Daffweb3/Portals/0/Statistics%20and%20Economic%20 Analysis/Statistical%20Information/Abstract%202016%20.pdf.

DARD. 2014. *The Effect of Fire on Soil and Vegetation*. Potchefstroom, South Africa: Pasture Division, Department of Agriculture and Rural Development.

Delgado-Baquerizo, M., Eldridge, D. J., Maestre, F. T., Karunaratne, S. B., Trivedi, P., Reich, P. B., and Singh, B. K. 2017. Climate legacies drive global soil carbon stocks in terrestrial ecosystems. *Science Advances* 3(4):e1602008.

Dormaar, J. F., and Willms, W. D. 1998. Effect of forty-four years of grazing on fescue grassland soils. *Journal of Range Management* 51:122–126.

Dregne, H. E. 2002. Land degradation in the drylands. *Arid Land Research and Management* 16:99–132.

Drewes, R. H. 1991. Potch system: An approach to the management of semi-arid grasslands in South Africa. *Journal of the Grassland Society of South Africa* 9:24–29.

Du Preez, C. C., and Snyman, H. A. 1993. Organic matter content of a soil in a semi-arid climate with three long-standing veld conditions. *African Journal of Range and Forage Science* 19:108–110.

Du Preez, C. C., and Snyman, H. A. 2003. Soil organic matter changes following rangeland degradation in a semi-arid South Africa. *Proceedings of the VII International Rangeland Congress*, Durban, South Africa, pp. 476–478.

Du Toit, G. V. N., Snyman, H. A., and Malan, P. J. 2008. Physical impact of grazing by sheep in the Nama Karoo subshrub/grass rangeland of South Africa on litter and dung distribution. *South African Journal of Animal Science* 38(4):320–330.

Du Toit, GvN., Snyman, H. A., and Malan, P. J. 2009. Physical impact of grazing by sheep on soil parameters in the Nama Karoo subshrub/grass rangeland of South Africa. *Journal of Arid Environments* 73:804–810.

Du Toit, G. V. N., Snyman, H. A., and Malan, P. J. 2011. Physical impact of sheep grazing on arid Karoo subshrub/grass rangeland, South Africa. *South African Journal of Animal Science* 41:281–287.

Fanning, D. S., and Fanning, M. C. B. 1989. *Soil Morphology, Genesis and Classification*. New York: John Wiley & Sons.

Fey, M. V. 2010. *Soils of South Africa: Their Distribution, Properties, Classification, Genesis, Use and Environmental Significance*. New York: Cambridge University Press.

GCIS. 2015. Government. In: *South Africa Yearbook 2015/2016*, 23rd edition, Government Communication and Information System. Available at: www.gcis.gov.za/content/resourcecentre/sa-info/yearbook 2015-16.

Greenwood, K. L., and McKenzie, B. M. 2001. Grazing effects on soil physical properties and the consequences for pastures: A review. *Australian Journal of Experimental Agriculture* 41:1231–1250.

Haarmeyer, D. H., Schmiedel, U., Dengler, J., and Bösing, B. M. 2010. How does grazing intensity affect different vegetation types in South Africa? Implications for conservation management. *Biological Conservation* 143:588–596.

Ingram, L. J. 2002. Growth, nutrient cycling and grazing of three perennial tussock grasses of the Pilbara region of NW Australia. PhD Thesis, Department of Botany, University of Western Australia, Perth, Australia.

Karlen, D. L., Mausbach, M. J., Doran, J. W., Cline, R. G., Harris, R. F., and Schuman, G. E. 1997. Soil quality: A concept, definition, and framework for evaluation. *Soil Science Society of America Journal* 61:4–10.

Kirkman, K. P. 1995. Effects of grazing and resting on veld productivity. *Bulletin of the Grassland Society of Southern Africa* 6:9–11.

Kotzé, E., Sandhage-Hofmann, A., Amelung, W., Oomen, R. J., and du Preez, C. C. 2017. Soil microbial communities in different rangeland management systems of a sandy savanna and clayey grassland ecosystem, South Africa. *Nutrient Cycling in Agroecosystems* 107:227–245.

Kotzé, E., Sandhage-Hofmann, A., Meinel, J. -A., Du Preez, C. C., and Amelung, W. 2013. Rangeland management impacts on the properties of clayey soils along grazing gradients in the semi-arid grassland biome of South Africa. *Journal of Arid Environments* 97:220–229.

Lal, R., and Elliot, W. 1994. Erodibility and erosivity. In: *Soil Erosion Research Methods*, 2nd edition, R. Lal (ed.), pp. 181–208. Ankeny, IA: Soil and Water Conservation Society.

Lauenroth, W. K., Milchunas, D. F., Dodd, J. L., Hart, R. H., Heitschmidt, R. K., and Rittenhouse, L. R. 1994. Effects of grazing on ecosystems of the Great Plains. In: *Ecological Implications of Livestock Herbivory in the West*, M. Vavra, W. A. Laycock, and R. D. Pieper (eds.), pp. 69–100. Denver, CO: Society for Range Management.

Lavado, R. S., and Taboada, M. A. 1987. Soil salinization as an effect of grazing in a native grassland soil in the flooding Pampa of Argentina. *Soil Use and Management* 3:143–148.

Livingwell, J. 2010. *Animal Impact: How Trampling and Disturbance Benefit Grassland Ecosystems*. Available at: https://managingwholes.com/animal-impact.htm/

McSherry, M. E., and Ritchie, M. E. 2013. Effects of grazing on grassland soil carbon: A global review. *Global Change Biology* 19(5):1347–1357.

Mucina, L., and Rutherford, M. C. 2006. *The Vegetation of South Africa, Lesotho and Swaziland*. Pretoria, South Africa: South African National Biodiversity Institute.

Palmer, T., and Ainslee, A. 2006. Country Pasture/Forage Resource Profiles: South Africa. Prepared by Palmer and Ainslee in May 2002; livestock data updated by S.G. Reynolds in August 2006.

Petersen, A. E., Young, E. M., Hoffman, M. T., Musil, C. F., and van Staden, J. 2004. The impact of livestock grazing on landscape biophysical attributes in privately and communally managed rangelands in Namaqualand. *South African Journal of Botany* 70(5):777–783.

Roberson, E. 1996. *Impacts of Livestock Grazing on Soils and Recommendations for Management*. California Native Plant Society Land Management. Available at: https://184.168.207.85/cnps/archive/letters/soils.pdf.

Russell, J. R., Betteridge, K., Costall, D. A., and Mackay, A. D. 2001. Cattle treading effects on sediment loss and water infiltration. *Journal of Range Management* 54:184–190.

Sandhage-Hofmann, A., Kotzé, E., van Delden, L., Dominiak, M., Fouché, H. J., Van der Westhuizen, H. C., Oomen, R. J., Du Preez, C. C., and Amelung, W. 2015. Rangeland management effects on soil properties in the savanna biome, South Africa: A case study along grazing gradients in communal and commercial farms. *Journal of Arid Environments* 120:14–25.

Savory, C. A. R. 1988. *Holistic Resource Management*. Covelo, CA: Island Press.

Schulze, R. E. 1997. South African atlas of agro-hydrology and -climatology. Report TT82/96. Water Research Commission, Pretoria, South Africa.

Smet, M., and Ward, D. 2005. A comparison of the effects of different rangeland management systems on plant species composition, diversity and vegetation structure in a semi-arid savanna. *African Journal of Range and Forage Science* 22:59–71.

Smith, B. 2006. *The Farming Handbook*. Pietermaritzburg, South Africa: Interpak Books.

Snyman, H. A. 1998. Dynamics and sustainable utilization of rangeland ecosystems in arid and semi-arid climates of southern Africa. *Journal of Arid Environments* 39:645–666.

Snyman, H. A. 1999. Soil erosion and conservation. In: *Veld Management in South Africa*, N. M. Tainton (ed.). Scottsville, South Africa: University of Natal Press.

Snyman, H. A. 2004. Soil seed bank evaluation and seedling establishment along a degradation gradient in a semi-arid rangeland. *African Journal of Range and Forage Science* 21:37–47.

Snyman, H. A. 2009a. Root studies on grass species in a semi-arid South Africa along a soil-water gradient. *Agriculture, Ecosystems and Environment* 131:247–254.

Snyman, H. A. 2009b. Root studies on grass species in a semi-arid South Africa along a degradation gradient. *Agriculture, Ecosystems and Environment* 130:100–108.

Snyman, H. A. 2014. *Gids tot Volhoubare Produksie van Weiding*, 2nd edition. Cape Town, South Africa: Landbouweekblad and Landbou.com, Media 24, 539pp.

Snyman, H. A. 2015a. Short-term responses of Southern African semi-arid rangelands to fire: A review on impact on plants. *Arid Land Research and Management* 29:237–254.

Snyman, H. A. 2015b. Short-term responses of Southern African semi-arid rangelands to fire: A review on impact on soils. *Arid Land Research and Management* 29:222–236.

Snyman, H. A. 2016. Beweiding en wortelgroei by weiplante. *Agriforum* 29(7):110–116.

Snyman, H. A., and Du Preez, C. C. 2005. Rangeland degradation in a semi-arid South Africa II: Influence on soil quality. *Journal of Arid Environments* 60:483–507.

Soil Classification Working Group. 1991. In: *Soil Classification—A Taxonomic System for South Africa*. Memoirs on the Agricultural Natural Resources of South Africa No. 15. Pretoria, South Africa: Department of Agricultural Development.

Steinfeld, H., and Wassenaar, T. 2007. The role of livestock production in carbon and nitrogen cycles. *Annual Review of Environment and Resources* 32:271–294.

Taboada, M., Rubio, A. G., and Chaneton, E. J. 2011. Grazing impacts on soil physical, chemical, and ecological properties in forage production systems. In: *Soil Management: Building a Stable Base for Agriculture*, J. L. Hatfield and T. J. Sauer (eds.), pp. 301–320. Madison, WI: American Society of Agronomy and Soil Science Society of America.

Tainton, N. M. 1999. *Veld Management in South Africa*. Pietermaritzburg, South Africa: University of Natal Press.

Teague, W. R., Dowhower, S. L., Baker, S. A., Haile, N., Delaune, P. B., and Conover, D. M. 2011. Grazing management impacts on vegetation, soil biota and soil chemical, physical and hydrological properties in tall grass prairie. *Agriculture, Ecosystems and Environment* 141:310–322.

Todd, S., Milton, S., Dean, R., Carrick, P., and Meyer, A. 2009. *Ecological Best-Practice Livestock Production Guidelines for the Namakwa District*. Project Proposal Botanical Society of South Africa as Part of the Namakwa District Project Funded by the Critical Ecosystem Partnership Fund, Northern Province, South Africa.

Trollope, W. S. W. 2009. Fire management of African Grassland and Savanna ecosystems. *Proceedings of the Association Fire Ecology, Fourth International Congress: Fire as Global Process*. Savannah, GA.

USDA. 2016. *Grazing Management and Soil Health. Keys to Better Soil, Plant, Animal and Financial Health. National Resources Conservation Services*. Washington, DC: United States Department of Agriculture.

Van der Westhuizen, H. C., and Snyman, H. A. 2014. Evaluering van onderskeie veldbestuursbenaderings. In: *Gids tot Die Volhoubare Produksie van Weiding*, 2nd edition, H. A. Snyman (ed.), pp. 500–529. Cape Town, South Africa: Landbouweekblad en Landbou.com, Media 24.

Van der Westhuizen, H. C., Van Rensburg, W. L. J., and Snyman, H. A. 1999. The quantification of rangeland condition in a semi-arid grassland of South Africa. *African Journal of Range and Forage Science* 16:49–61.

Van Oudtshoorn, F. 2015. *Veld Management: Principles and Practices*, 1st edition. Pretoria, South Africa: Briza Publications.

Van Pletzen, H. W., Becker, J. D., De Villiers, M., and Kemp, J. H. 1995. Nuwe veldbestuurstelsel vir Wildebeesfontein Proefplaas. *Bulletin of the Grassland Society of Southern Africa* 6:22–25.

Vetter, S. 2013. Development and sustainable management of rangeland commons – Aligning policy with the realities of South Africa's rural landscape. *African Journal of Range and Forage Science* 30:1–9.

Young, I. M., and Ritz, K. 2005. The habitat of soil microbes. In: *Biological Diversity and Function in Soils*, R. D. Bardge, M. B. Usher, and D. W. Hopkins (eds.), pp. 31–41. Cambridge, UK: University Press.

Zietsman, J. 2014. *Man, Cattle and Veld*, 1st edition. Wyoming: Beefpower LLC.

9 The Fertilizer Dilemma

Amit Roy

CONTENTS

9.1   INTRODUCTION

Fertilizers provide plants with the nutrients they need for their growth and development. Plants live, grow, and reproduce by taking up water and nutrients, carbon dioxide from the air, and energy from the sun. Apart from carbon, hydrogen, and oxygen, which collectively make up 90–95% of the dry matter of all plants, other nutrients needed by plants come from the media in which they grow – essentially from the soil. The other nutrients are subdivided into primary nutrients (nitrogen, N; phosphorus, P; and potassium, K) and secondary nutrients (calcium, magnesium, and sulfur). In addition, plants also need other nutrients in much smaller amounts, and they are referred to as micronutrients (boron, chlorine, copper, iron, manganese, molybdenum, and zinc).

To maintain soil fertility and productivity and prevent land degradation, nutrients taken up by crops must be replenished through the application of fertilizers, which are both manufactured and naturally available, including manure. The use of fertilizer results in many benefits to producers, consumers, and the environment, starting with increased agricultural outputs (mainly food and fiber) to contributing to soil organic matter maintenance, water-holding capacity, biological nitrogen fixation, soil erosion control, other physical and chemical properties, and less extensive land use. These benefits contribute to increased agricultural growth and agribusiness activities, which are catalysts for broad-based economic growth and development in most developed and developing economies; agriculture's links to the non-farm economy generate considerable employment, income, and growth in the rest of the economy.

9.2 HISTORY OF FERTILIZER USE

The beginning of our dependence on inorganic fertilizer can be traced back to the nineteenth century when Justus von Liebig articulated the theoretical foundations of crop production and when John Bennett Lawes began producing fertilizers containing phosphorus (Smil 1997). However, only since the 1960s, when global starvation became a real possibility, have fertilizers assumed a predominant role in increasing agricultural productivity. Fertilizer was an integral part of technological trinity – seed, water, and fertilizer – responsible for bringing about the "Green Revolution" that helped many densely populated countries, including India, China, and Indonesia, achieve food self-sufficiency in a short span of 20–25 years.

Since the 1970s, global cereal production has nearly doubled, increasing from 1,400 million metric tons (mt) in 1970 to over 2,600 million mt in 2013, with developing countries accounting for nearly 70% of the increase (Figure 9.1) (FAO 2002). Most of the increased cereal production in South Asia was through higher yields, but increases in sub-Saharan Africa were mostly through area expansion (Figure 9.2) (FAO 2002). While recognizing the importance of advances in breeding and crop protection products and the expansion of irrigation, there is evidence that 40–60% of the production gains are directly attributable to fertilizer use (Figure 9.3) (Stewart et al. 2005). Consequently, cereal production closely parallels fertilizer use, increasing from less than 50 million mt in 1970 to 190 million mt in 2013 (Figure 9.4) (FAO 2002). The only exception is sub-Saharan Africa where per capita food production has decreased since the 1970s. This decline can be attributed to several factors, including low soil fertility, agroclimatic conditions, and low fertilizer use. Compared with a world average of 132 kg of fertilizer nutrients per hectare, sub-Saharan Africa uses only 18 kg of fertilizer nutrients per hectare, resulting in a significant mining of inherent nutrients from soils that are already low in nutrient status (Figures 9.5 and 9.6).

In the coming decades, growth in demand for food and feed will be influenced by changes in several forces, but mainly by growth in population, income levels, and economic development, and changes in the food preferences of consumers. While the global population is projected to grow by about 30% by 2050 (United Nations 2013), food demand is predicted to increase by nearly 70% as emerging regions urbanize, develop economically and consume a nutritious diet. However, the absolute increase in food required by 2050 will be as large as the increase since the Green Revolution was launched in the 1960s – as available arable land and water become scarcer. Hence the need for more plant nutrients.

Since the 1960s, the population has more than doubled and is estimated to reach 9.8 billion by 2050 (Figure 9.7). While the population increase in developed regions is expected to remain

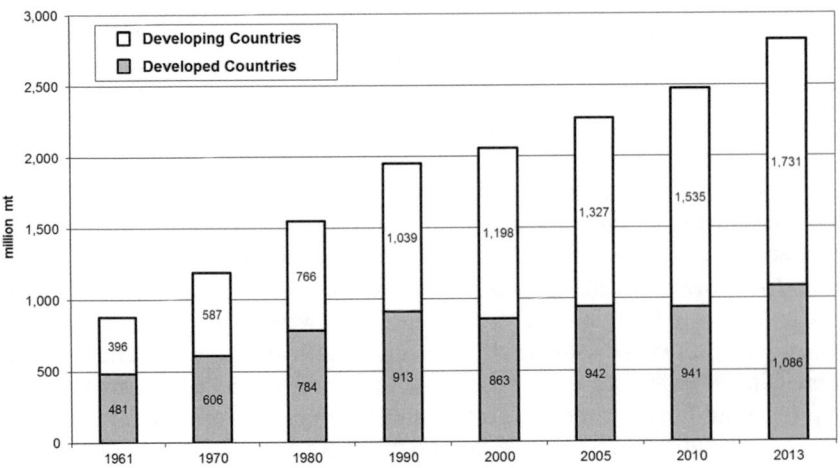

FIGURE 9.1   Cereal production in developed and developing countries, 1961–2013.

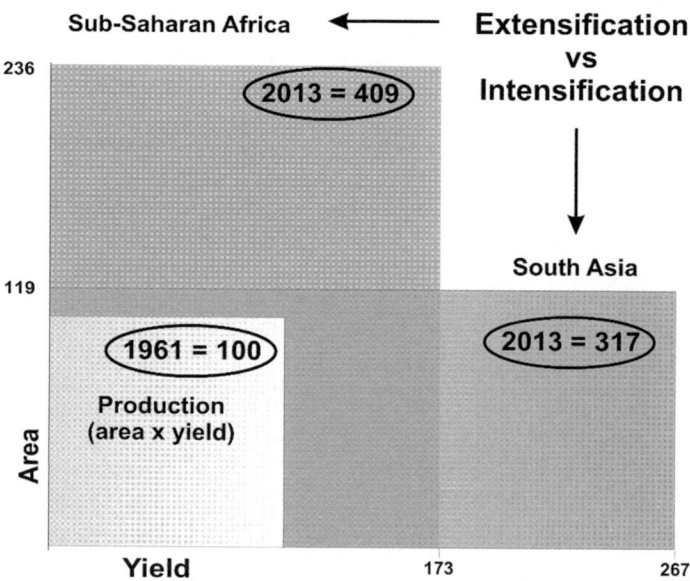

FIGURE 9.2 Growth of cereal yield and area in cereal production in South Asia and sub-Saharan Africa between 1961 and 2013 (1961 = 100 for yield and area).

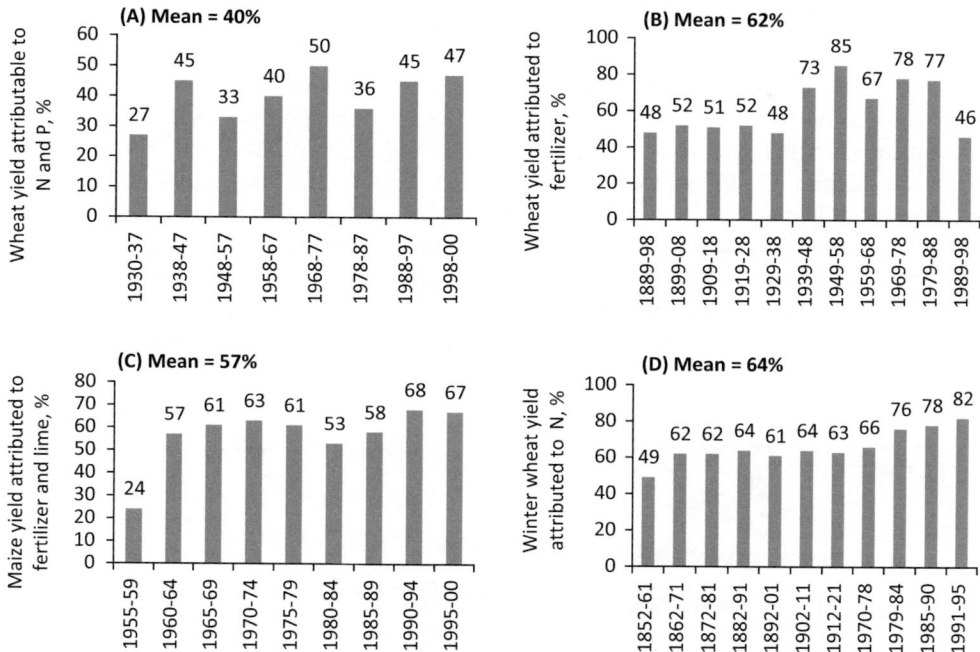

FIGURE 9.3 Yield attributed to fertilizer: (A) N and P from 1930 to 2000, in the Oklahoma State University Magruder plots; (B) N, P, and K from 1989 to 1998 in the University of Missouri Sanborn Field plots; (C) N, P, K, and lime from 1955 to 2000 in the University of Illinois Morrow plots; and (D) N with adequate P, K vs. P, and K alone from 1852 to 1995 (years between 1921 and 1969 excluded because part of the experiment was fallowed each year for weed control) in the Broadbalk Experiment at Rothhamsted, United Kingdom (Stewart et al. 2005).

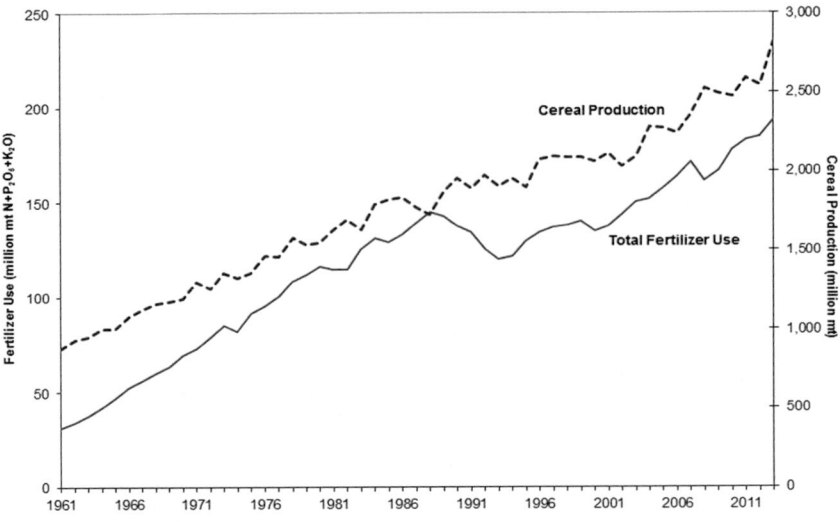

FIGURE 9.4 World total cereal production and total fertilizer use, 1961–2013.

virtually unchanged at 1.3 billion, developing regions will see an increase from 5.7 billion in 2010 to 8.3 billion by 2050.

The dramatic increase in global food production is partly attributable to fertilizers. However, overuse, misuse and underuse of fertilizers leads to air and water pollution. For example, the biggest issues facing the use of chemical fertilizers are air pollution and surface and groundwater contamination. Urea, which is the most commonly used nitrogen fertilizer, breaks down in the soil and results in the release of reactive nitrogen in the air and nitrates that travel easily through the soil (Figure 9.8). Because it is water-soluble and can remain in groundwater for decades, the addition of more nitrogen over the years has an accumulative effect. Groundwater contamination has been linked to some forms of cancer and birth defects. Similarly, excess phosphates in rivers and tributaries cause algae growth, depleting water of oxygen and resulting in fish kill.

Hence the dilemma faced by fertilizer researchers is that too little fertilizer leads to lower crop productivity, poor human nutrition, and soil degradation, but too much fertilizer leads to environmental pollution and its related consequences. There is no question that the scientific community should aim to produce more food while reducing or at least holding constant fertilizer application rate. This chapter reviews the recent advances in fertilizers, particularly new products and their application and management. And finally lays out a path to develop more efficient, less polluting, and affordable fertilizers.

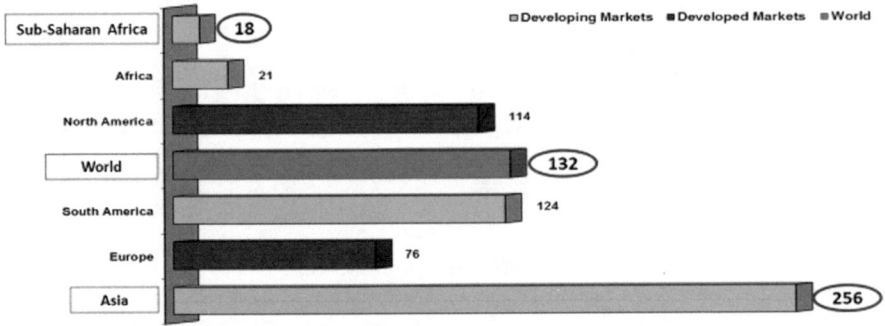

FIGURE 9.5 **(See color insert.)** Per hectare fertilizer use by regions and markets, 2013 (kg/ha).

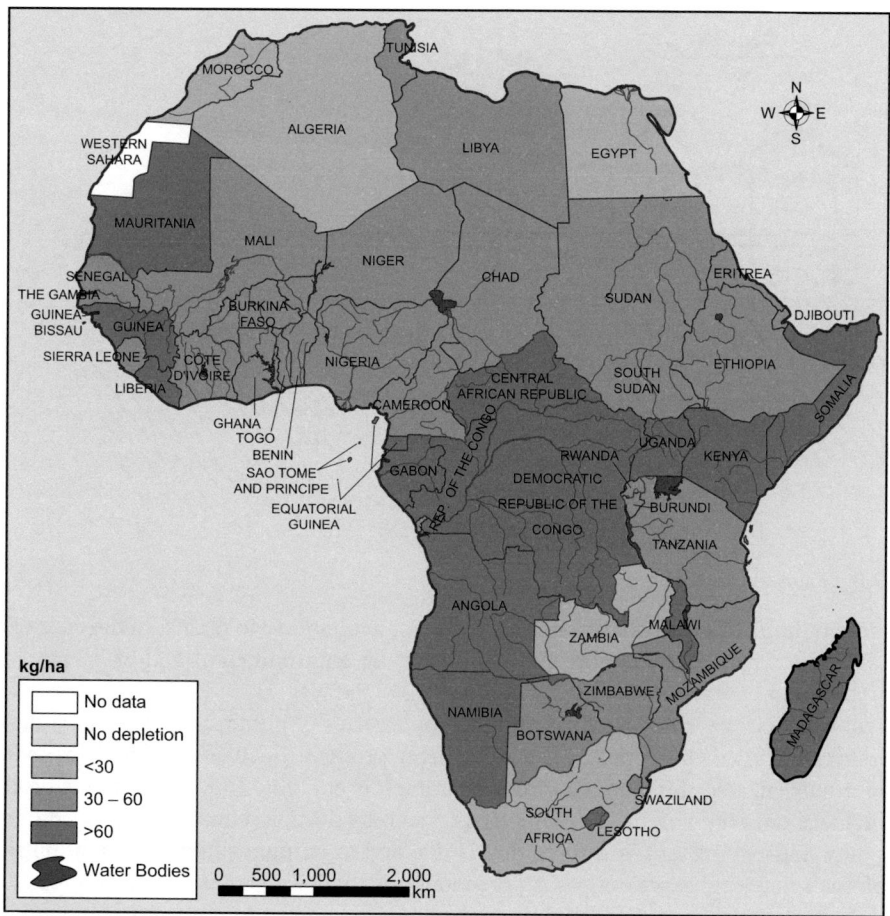

FIGURE 9.6 Average nutrient depletion (NPK) in Africa for 2010.

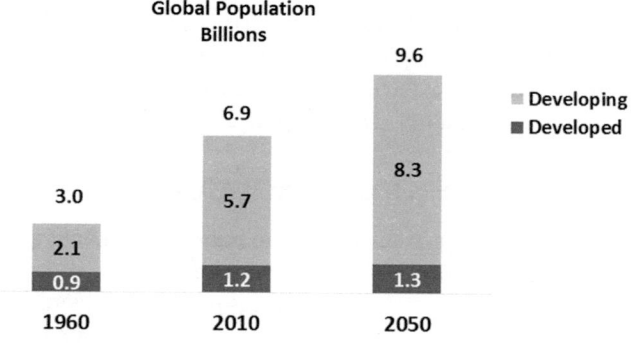

FIGURE 9.7 World population projection.

9.3 WHAT IS A FERTILIZER?

A commercial fertilizer is a material that contains at least one of the plant nutrients in chemical form that, when applied to the soil, is soluble in the soil solution phase and assimilable by or "available" to plant roots. Most often, this implies chemical forms that are water soluble. In the case of phosphorus, solubility in special reagent solutions (citric acid, neutral ammonium citrate, or

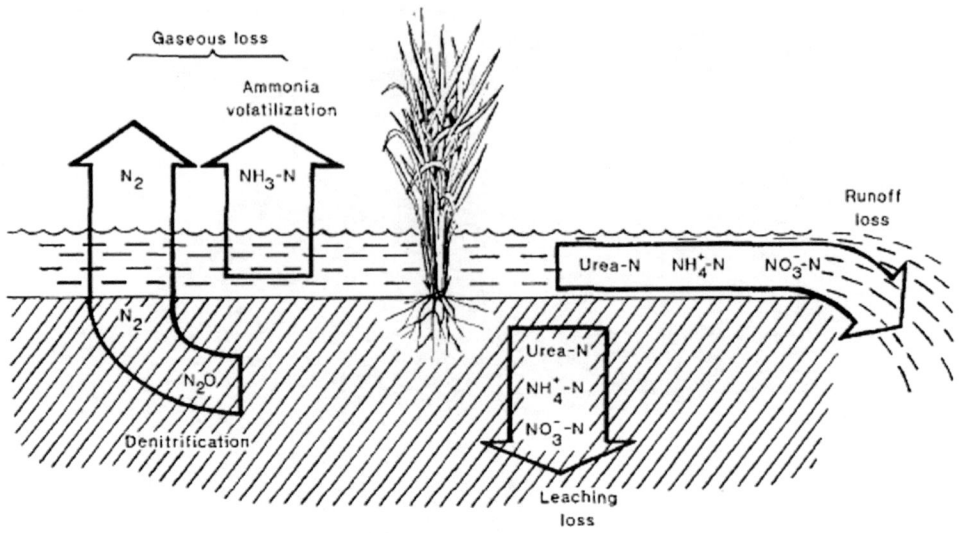

FIGURE 9.8 Pathways to N losses from urea fertilizer.

alkaline ammonium citrate) is often used as a guide for availability to plants. In the case of nitrogen, slow solubility in water may be more desirable from an environmental and efficiency standpoint than easy solubility.

Fertilizer products are customarily designated by a series of numbers separated by dashes. This set of numbers is called the "grade" of the fertilizer product. Each of the numbers indicates the amount of a nutrient that the manufacturers guarantee is contained in the fertilizer product. This number includes only the amount of the nutrient found by accepted analytical procedures, thereby excluding any nutrient present in a form that is deemed to be unavailable for plant nutrition. The content of each nutrient, expressed as a percentage of total weight, is the guaranteed minimum rather than actual, which is usually slightly higher.

Usually, three numbers are used when giving the grade of the fertilizer product, and these three numbers always refer, in order, to the content of the primary nutrients: nitrogen, phosphorus, and potassium. If other nutrients are present, their content can also be indicated in the grade of the fertilizer product; each extra number is followed by the chemical symbol of the nutrient it represents. Many countries indicate the content of phosphorus and potassium not in the elemental form but in the oxide form, P_2O_5 and K_2O. Thus, a fertilizer product with a grade of 12–6–22–2MgO is guaranteed by the manufacturer to contain: 12% N, 6% P_2O_5, 22% K_2O, and 2% MgO.

9.4 RAW MATERIALS FOR FERTILIZER PRODUCTION

The primary raw materials for nitrogen fertilizers are natural gas, naphtha, fuel oil, and coal. The manufacturing of phosphate fertilizers most often requires phosphate rock. Naturally occurring potassium salts form the basis of the production of most potash fertilizers.

Natural gas, naphtha, fuel oil, and sulfur (for production of phosphatic fertilizers via the sulfuric acid route) have definable specifications. In contrast, phosphate rock and coal are products that can vary significantly in composition and other characteristics. These variations can affect the processes used to upgrade the "as-mined" ores or the processes for manufacture of fertilizers from beneficiated products. Potash ores also vary greatly in composition depending on origin; however, the end products of mining, beneficiation, and processing generally have relatively constant compositions.

The adequacy of the requisite raw materials is the most obvious concern when facing a substantial increase in future demand. Two separate yet intertwined issues in the case of fertilizers are:

the sufficiency of raw materials and the availability of energy to convert them into final products. Potassium is of least concern among the three primary nutrients. Not only is this element abundantly present in the Earth's crust, but it can also be found in conveniently concentrated deposits in both deeply buried and near-surface sediments. Potassium deposits in descending order of known reserves are in North America (Canada and the United States), Germany, Russia, Belarus, Brazil, Israel, Ethiopia, and Jordan. Even the most conservative reserve base estimates indicate more than 400 years of reserves at the projected rate of use.

Phosphate rock is a finite nonrenewable resource. There is no substitute for phosphorus. Phosphate deposits are known from every continent of the world except for Antarctica. Recently published total world phosphate rock reserves range from 60 billion tons of producible concentrate to 65 billion tons of undifferentiated phosphate rock (Van Kauwenbergh 2010). Total world phosphate rock resources may range from 290 billion tons to 460 billion tons of phosphate rock at varying P_2O_5 grades (Herring and Fantel 1993). Phosphate deposits are not evenly distributed around the globe. The most abundant known reserves, in descending order of abundance, are in Morocco, China, the United States, Jordan, and Russia. At the current rates of extraction, these reserves are available for 300–400 years. Of course, in the future, based mainly on requirements due to population growth, the increased demand for phosphorus-based products will require increased rates of phosphate rock production, and the rate of depletion of known reserves may increase. This time horizon can be extended by exploiting known, incompletely explored, and as yet unknown phosphate deposits at higher costs. The conservation and efficient use of this valuable natural resource and recycling of P is receiving increased attention through sustainable P management initiatives (Scholz et al. 2014).

Nitrogen fertilizers via ammonia synthesis account for more than 90% of the world's nitrogen fertilizers. The nitrogen supply for ammonia synthesis is truly inexhaustible since the atmosphere contains 3.8 quadrillion tons of the element. Various feedstocks can be used to obtain hydrogen, and during the last several decades the focus has been to improve the energy efficiency of ammonia synthesis. Natural gas is the preferred feedstock and the best natural gas-based plants now use less than 30 GJ/t N. The global mean, which is affected by more energy-intensive reforming of heavier hydrocarbons (naphtha and fuel oil) and coal, is now between 49 and 55 GJ/t N, roughly half of the level prevailing during the early 1950s.

Even if all the energy needed to fix the fertilizer nitrogen were to come from natural gas, it would still be less than 7% of the recent annual global consumption of the fuel and less than 2% of all energy derived from fossil fuels (British Petroleum-Amoco 2015). Clearly, there is little reason to be concerned about either the current needs or the future supplies of energy for producing nitrogenous fertilizers via the ammonia synthesis route. Moreover, there is no doubt that higher absolute energy needs for nitrogen fertilizers will be partially offset by improved efficiency of ammonia synthesis and by higher efficiencies of fertilizer use. Because today's low-income countries will experience a much faster growth of energy needs in sectors other than the fertilizer industry, the share of global fossil fuel consumption for ammonia synthesis by the middle of the twenty-first century may be only marginally higher than it is today.

Global natural gas resources are considered abundant. Generally, proven reserves of natural gas are considered to be those quantities that geological and engineering data indicate can be recovered with reasonable certainty in the future from known reservoirs under presently existing operating and economic conditions. Total world proved reserves were 187.10 trillion cubic meters at the end of 2014 (British Petroleum 2015). The reserve to production ratio (R/P ratio) at the end of 2014 was 54.1. That is, the amount of proven reserves divided by the production that year indicates the number of years of reserves remaining at that rate. It should be noted that gas reserves have increased over the last 20 years from 122.4 trillion cubic meters in 1989 to the present figure of 187.10 trillion cubic meters. Like many other natural resources, natural gas reserves are a dynamic quantity, determined by exploration activity, technology, and economics. In the absence of natural gas, naphtha, or fuel oil, ammonia synthesis could be accomplished (albeit a more costly process) by tapping the world's coal resources or by using a variety of biomass feedstocks.

9.4.1 SHORTCOMINGS OF N AND P FERTILIZERS

N, P, and K fertilizers are the most widely used fertilizers, primarily because virtually all crops require significant amounts of these nutrients to maximize production. legume crops are only exceptions because they have symbiotic bacterial colonies associated with their roots, and therefore can meet plant N needs through biological N fixation. However, legumes do require large amounts of P and K. Of the primary nutrients, N and P fertilizers are subjected to more chemically and energy-intensive production processes, are available in several formulations and forms (solid blends and liquids), and often "carry" much smaller (often minute) amounts of secondary and micronutrients (SMNs) that are essential for maximum economic yields and improved nutritional value. The production processes for the manufacture of fertilizers are in the *Fertilizer Manual* (UNIDO 1998). These fertilizers, however, are largely unchanged from the formulations and forms that were manufactured in the 1970s and 1980s and, in the absence of best management practices, result in significant loss of nutrients to the ecosystem. These losses produce negative economic and environmental impacts.

9.4.2 ECONOMIC AND ENVIRONMENTAL SUSTAINABILITY OF FERTILIZERS IN USE

The availability of fertilizer-N to plants is largely controlled by soil microbial processes. The N cycle in soils is complex, and under certain conditions, large amounts of plant available N can be lost from the soil to the atmosphere or in surface and sub-surface water bodies. The N lost to the atmosphere is in various forms of nitrous oxide gases (collectively referred to as NOx gases), whereas the N entering water bodies via runoff or leaching is in the form of nitrates. Most nitrates found in subsurface and surface waters result from crop production. The other primary nutrients (P and K) are not readily lost from soils, although runoffs containing P nutrients from crop production and animal waste can be significant pollutants in some bodies of water. Importantly, the amount of applied nutrient that is taken up by plants (referred to as "nutrient use efficiency" or NUE) is a function of the type of fertilizer(s) and how well a farmer can match the placement, timing, and quantity of applied nutrient to the plant's needs throughout its growing cycle. In 2010, the NUE of N (NUE-N) ranged between 25% and about 70% depending on the region and country (Figure 9.9). For example, in Canada and the US, the NUE-N is about 65%, while in China and India it is 25%

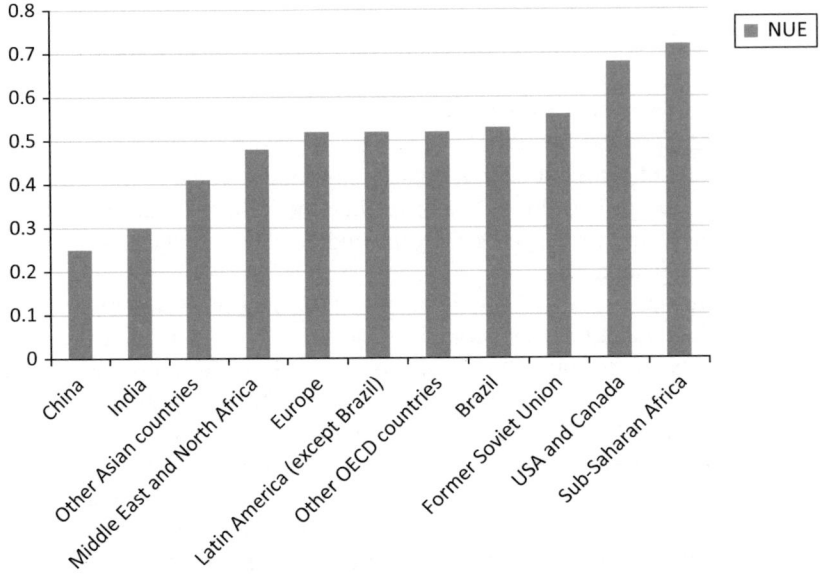

FIGURE 9.9 Nutrient use efficiency of N (NUE-N) in selected countries and regions (derived from data in Zhang et al. 2015).

and 30%, respectively. In contrast the NUE-N in sub-Saharan Africa (SSA) is about 70% There, despite a fertilizer application rate that is nearly an order of magnitude lower than the world average of 130 kg nutrients per hectare of arable land, the NUE-N remains high due to inherently low soil nutrient content and nutrient need of the plant. Even though SSA's NUE-N is comparably high, SSA contributes to greenhouse gas (GHG) emissions through slash and burn practices.

The economic impact of low NUE-N is significant. For example, assuming an annual fertilizer-N consumption of 120 mmt and an NUE-N of 50%, up to $35 billion of the total price paid by farmers is lost annually. In addition to the economic impact of low NUE-N, air and water pollution from unused N is a major factor. Manure also contributes to air and water pollution. In 2014 manure-N on cropland and pastures equaled manufactured fertilizer-N application, around 100 million tons of N per year (FAO 2018). The other major fertilizers (P and K) typically have higher NUE efficiencies, because they are considerably less mobile, easily recycled (K) or have residual nutrient benefits for subsequent crops. For P fertilizers, the NUE can reach 90% under best management practices, in which applied P is made slowly available for several crops over several years (Syers et al. 2008). But the NUE of P and K will be much lower, particularly if soil erosion is a problem or if sandy soils are being cropped. The nutrient runoff from agricultural land causes eutrophication and dead zones in water bodies and excessive algae growth.

In addition to the three primary nutrients, viz. N, P, and K, there are 14 other elements identified by Justus von Liebig, a nineteenth-century German scientist, as essential for plant growth. In soil studies in India, there is evidence of the gradual reduction of essential secondary and micronutrients required by plants (Figure 9.10), suggesting that these essential elements may now be limiting yields and reducing the efficiency of primary nutrients.

There are several reasons for these unintended nutrient deficiencies, but two critical factors appear to be (Smil 1997) fertilizers not containing enough secondary and micronutrients and (FAO 2002) imbalanced fertilizer applications. The latter, in some instances, relates to policies that disproportionately subsidize one nutrient more than the other. For example, in India, N, P, and K have been highly subsidized. To rein in the subsidy, the government implemented the nutrient-based subsidy scheme to progressively deregulate the prices of N, P, and K fertilizers. Unfortunately, the scheme was only implemented for P and K fertilizers, saving some subsidy but creating further imbalanced use of N, P, and K (Gulati 2014). A recent decision by the government of India to introduce specialty urea for agricultural application is designed to increase NUE-N, thereby increasing crop yield per unit of applied nitrogen.

9.5 N FERTILIZER PRODUCTION: TOTAL DEPENDENCY ON HABER–BOSCH PROCESS

The contribution of the Haber–Bosch process in dramatically increasing world food production is well documented. An estimated 35–40% of the world's population would not have any food

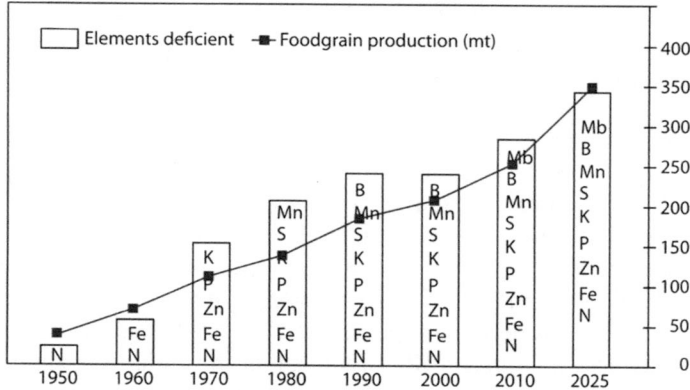

FIGURE 9.10 Emerging deficiencies of plant nutrients in relation to increased food-grain production.

FIGURE 9.11   Burrup Fertilizers' ammonia plant in Australia.

without the Haber–Bosch process of ammonia synthesis, which "fixes" nitrogen from the air to make it usable as a nutrient for crops. Unfortunately, the process has two attributes that are increasingly problematic, considering the marked shift in future food demand toward developing regions and the continuing concerns about energy sufficiency and environmental sustainability (Figure 9.11):

- The process uses a fossil fuel-based source for energy and hydrogen, which is subsequently reacted with nitrogen from air to produce ammonia. Currently natural gas is the most economical source of energy and hydrogen. Thus, N fertilizer production is dependent on a feedstock that is location-centric (natural gas sources) and subject to global pricing and supply-driven by its alternative use (as an energy source).
- The process is operationally complex and requires robust production facilities that can withstand high process temperatures and pressures. As a result, individual production sites are capital-intensive, which also makes them location-centric and in need of large-scale markets to justify their cost.

9.6   P FERTILIZER PRODUCTION: A CHALLENGE TO HANDLE BYPRODUCT

Unlike N, P fertilizers are produced from a finite resource – phosphate rock – which is converted into fertilizer, starting with beneficiation to remove unwanted materials and followed by chemical reaction with primarily sulfuric acid to produce phosphoric acid, which is generally reacted with ammonia or additional phosphate rock to produce phosphatic fertilizer. The whole chain of mining the rock to producing fertilizer to use and crop production is quite inefficient. Less than 10% of mined phosphate is in the food and feed (Scholz et al. 2014). In addition, the reaction of phosphate rock with sulfuric acid generates a considerable amount of phosphogypsum. For every ton of phosphate rock converted to phosphoric acid 1.6–1.7 tons of phosphogypsum is produced. In the US, the phosphogypsum is stacked in piles that are more than 200 feet high, covering a large area. It is unsuitable for use, because it is mildly radioactive, enough to exceed a level that the US Environmental Protection Agency (US EPA) has deemed safe for humans. However, the International Atomic Energy Agency (IAEA 2013) has determined, based on scientific studies, that phosphogypsum should be considered a co-product, with no radiological impediments for its use, making it more socially and environmentally sustainable than indefinite storage. As a result, phosphogypsum is used in several countries, including China, India, Brazil, and Tunisia, for soil remediation of calcareous soils, wallboard and roadbuilding materials.

Extremely alarming
Alarming
Serious
Moderate
Low
Insufficient data, significant concern*
2018 Global Hunger Index by Severity
GLOBAL HUNGER INDEX
About
2018 GHI Results
Issues in Focus
Country Case Studies
Policy Recommendations
Download
* See Box 2.1 for details
**See About section for details
EMBED

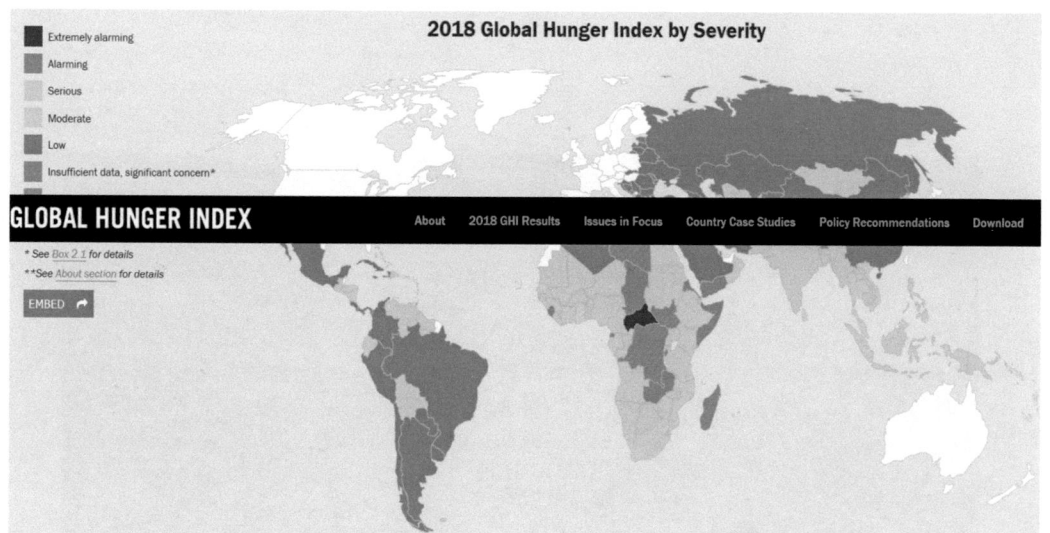

FIGURE 3.1 World hunger hot spots defined as those regions in the world where at least 20% of children under the age of five are stunted and underweight (http://www.globalhungerindex.org/results/ accessed on October 19, 2018).

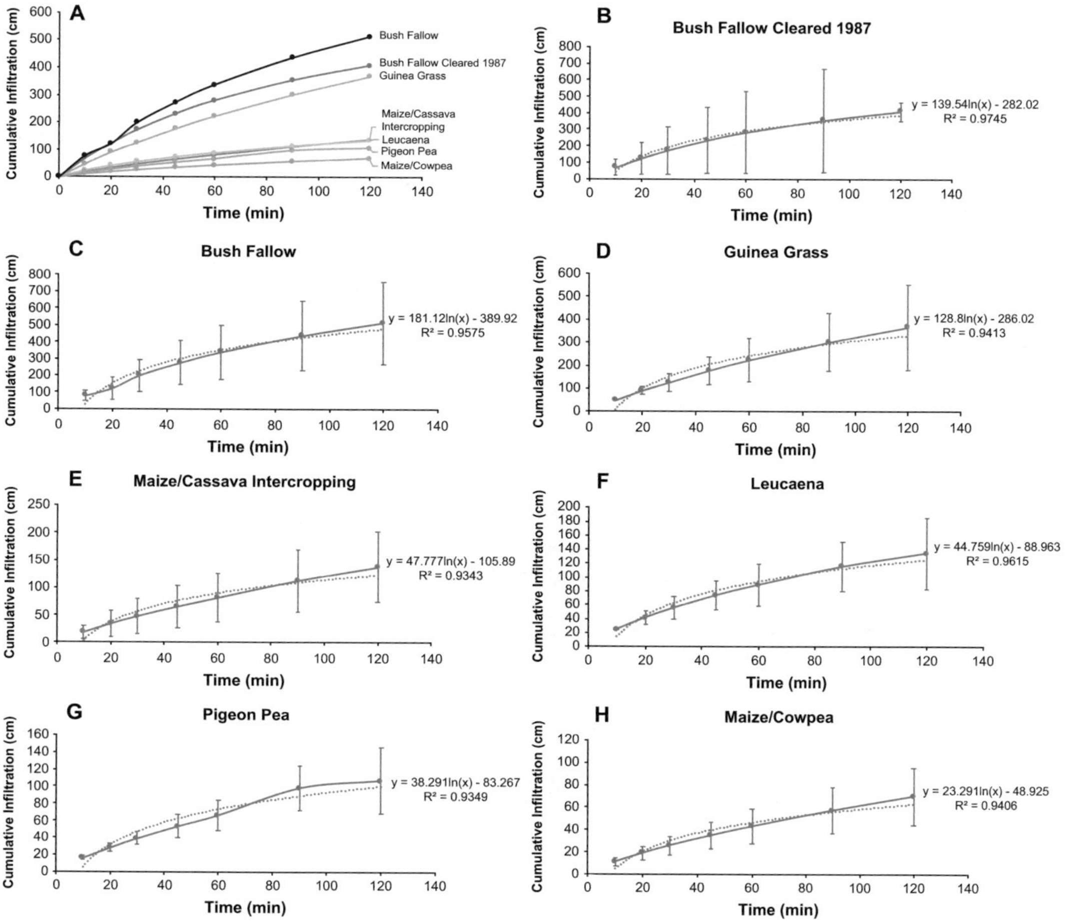

A
Bush Fallow
Bush Fallow Cleared 1987
Guinea Grass
Maize/Cassava Intercropping
Leucaena
Pigeon Pea
Maize/Cowpea
B
Bush Fallow Cleared 1987
y = 139.54ln(x) - 282.02
R² = 0.9745
C
Bush Fallow
y = 181.12ln(x) - 389.92
R² = 0.9575
D
Guinea Grass
y = 128.8ln(x) - 286.02
R² = 0.9413
E
Maize/Cassava Intercropping
y = 47.777ln(x) - 105.89
R² = 0.9343
F
Leucaena
y = 44.759ln(x) - 88.963
R² = 0.9615
G
Pigeon Pea
y = 38.291ln(x) - 83.267
R² = 0.9349
H
Maize/Cowpea
y = 23.291ln(x) - 48.925
R² = 0.9406
Cumulative Infiltration (cm)
Time (min)

FIGURE 4.1 Land-use effects on cumulative infiltration into Alfisols in western Nigeria. Different land-use and cropping system treatments are marked A–H.

FIGURE 6.1 Examples of typical salt-affected patches in the Vaalharts irrigation scheme.

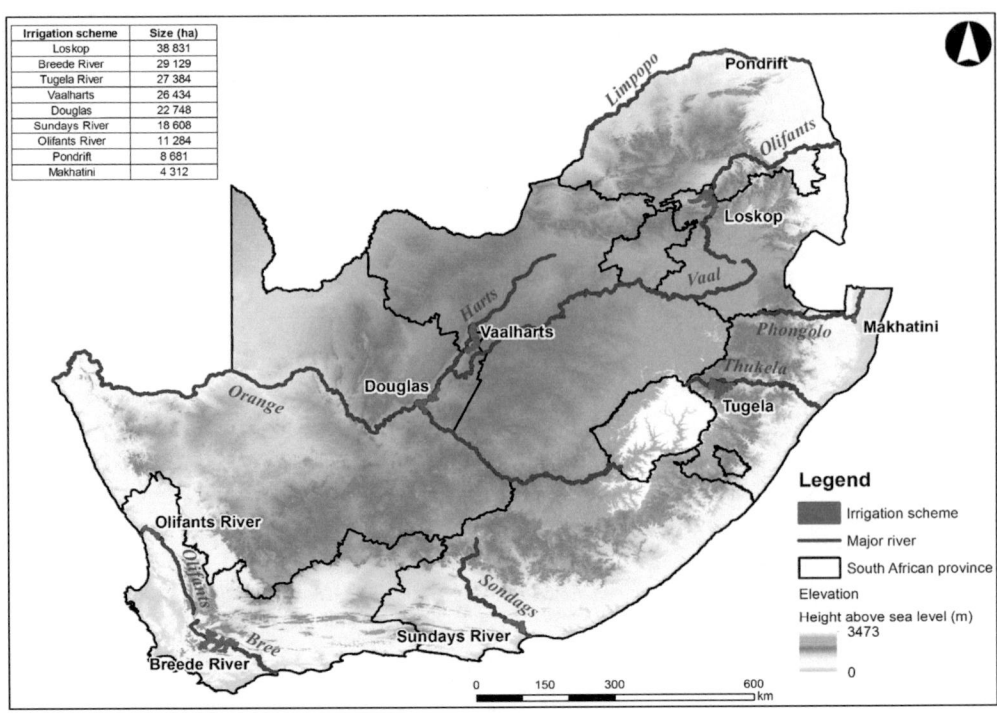

Irrigation scheme	Size (ha)
Loskop	38 831
Breede River	29 129
Tugela River	27 384
Vaalharts	26 434
Douglas	22 748
Sundays River	18 608
Olifants River	11 284
Pondrift	8 681
Makhatini	4 312

Legend

■ Irrigation scheme
— Major river
□ South African province

Elevation
Height above sea level (m)
3473

0

FIGURE 6.2 Geographical distribution of the nine irrigation schemes across South Africa, with elevation data as backdrop.

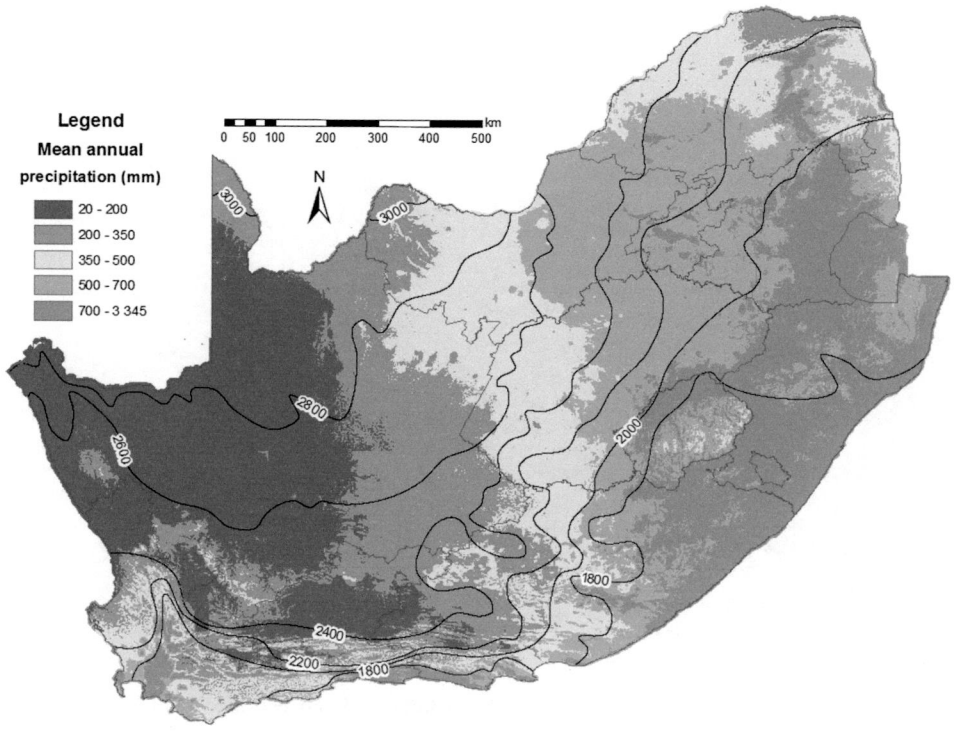

FIGURE 7.2 Average annual precipitation and average annual evaporation isohyets (mm) as A-pan equivalent in South Africa (interpolated from Schulze et al. 2001).

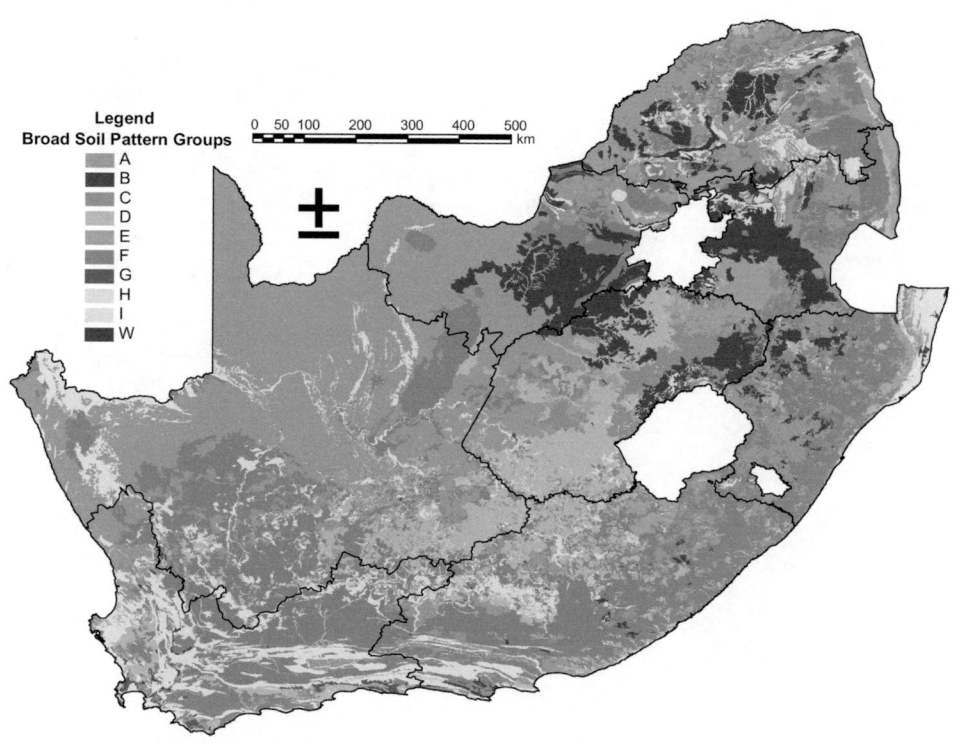

FIGURE 7.3 Generalized broad soil pattern groups map (adapted from Land Type Survey Staff 2004).

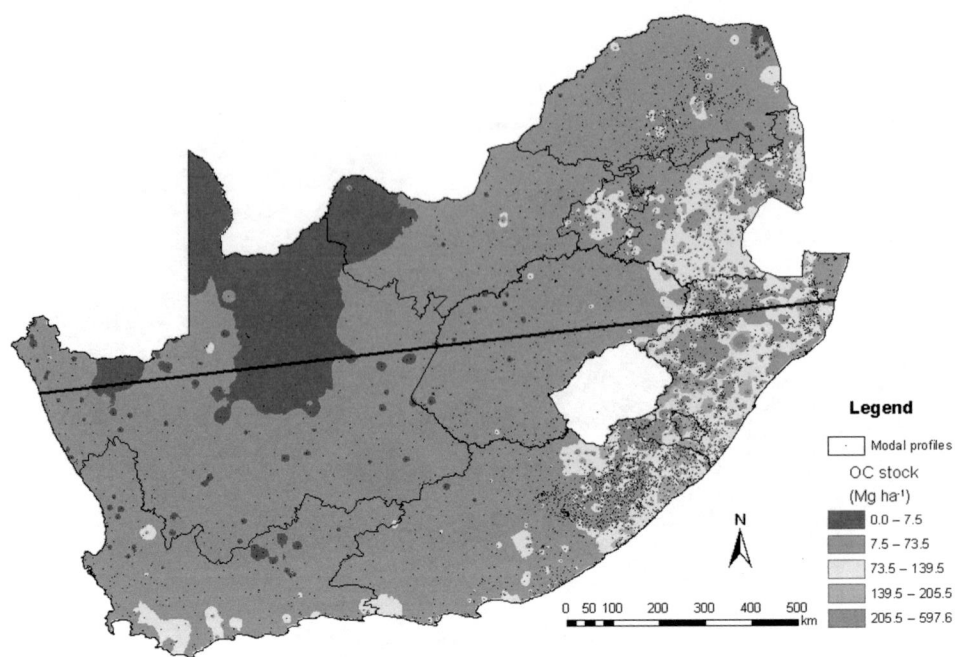

FIGURE 7.4 Map showing soil organic carbon stocks to 300 mm depth, estimated with a bulk density of 1.5 Mg m^{-3} (adapted from Rantoa 2009). The cross section is given in Figure 7.5.

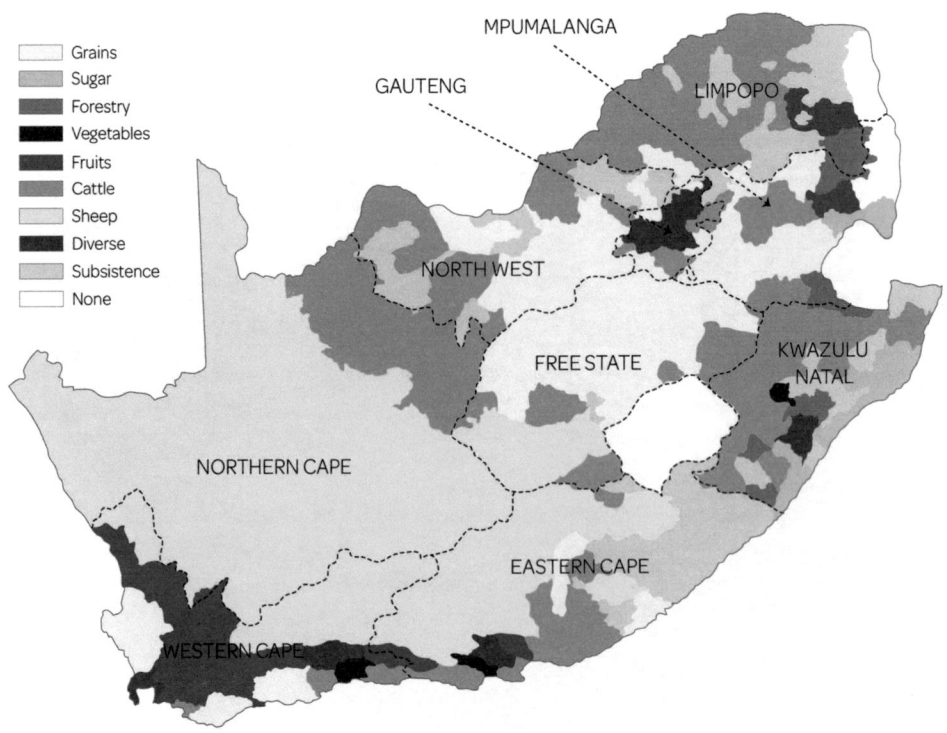

FIGURE 7.6 Agricultural production areas in South Africa (adapted from FAO 2005).

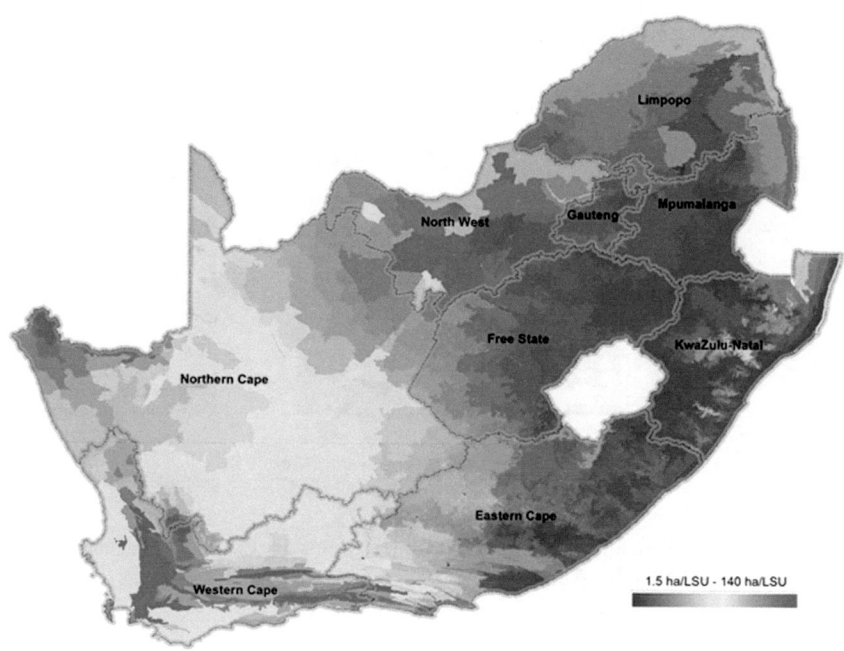

FIGURE 8.2 Long-term grazing capacity map for South Africa – 2016 (DAFF 2016a).

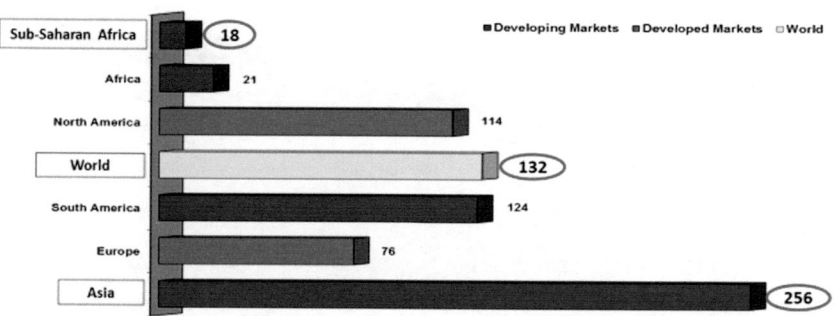

FIGURE 9.5 Per hectare fertilizer use by regions and markets, 2013 (kg/ha).

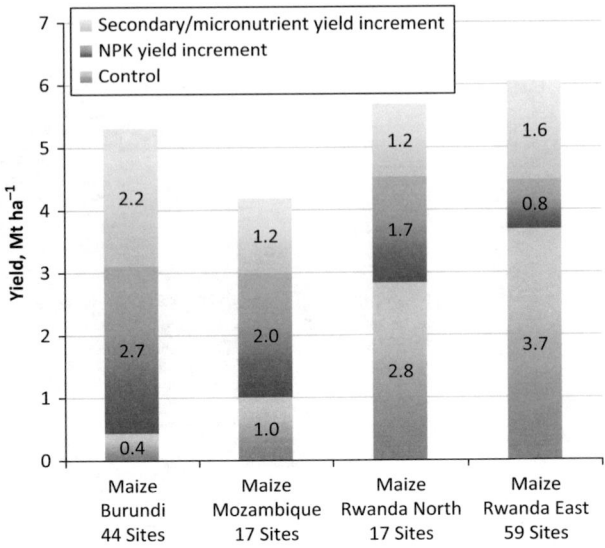

FIGURE 9.14 Effect of balanced fertilization on maize yields in Burundi, Mozambique, and Rwanda (Personal communication 2017, John Wendt).

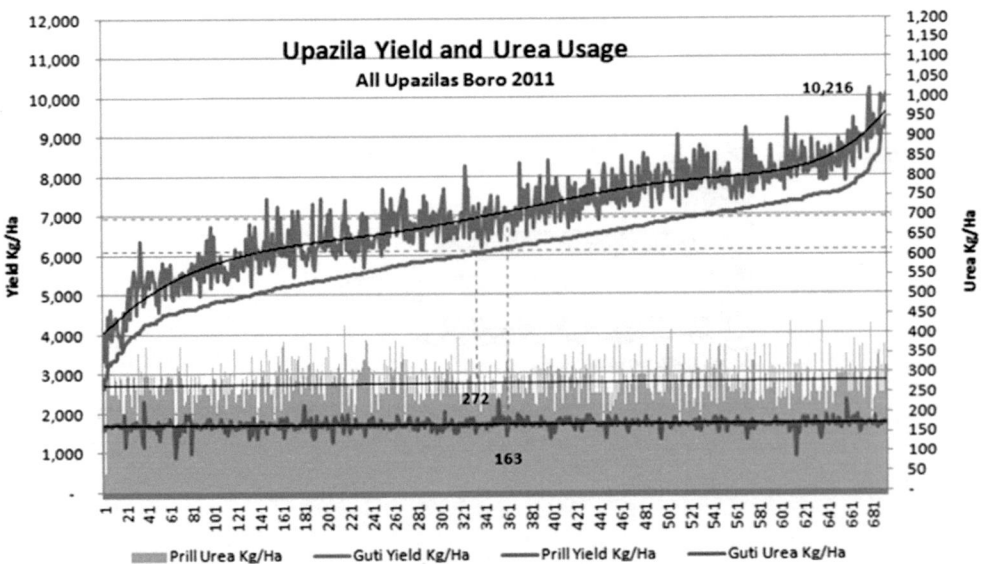

FIGURE 9.16 The deep placement of urea reduces N losses to the environment.

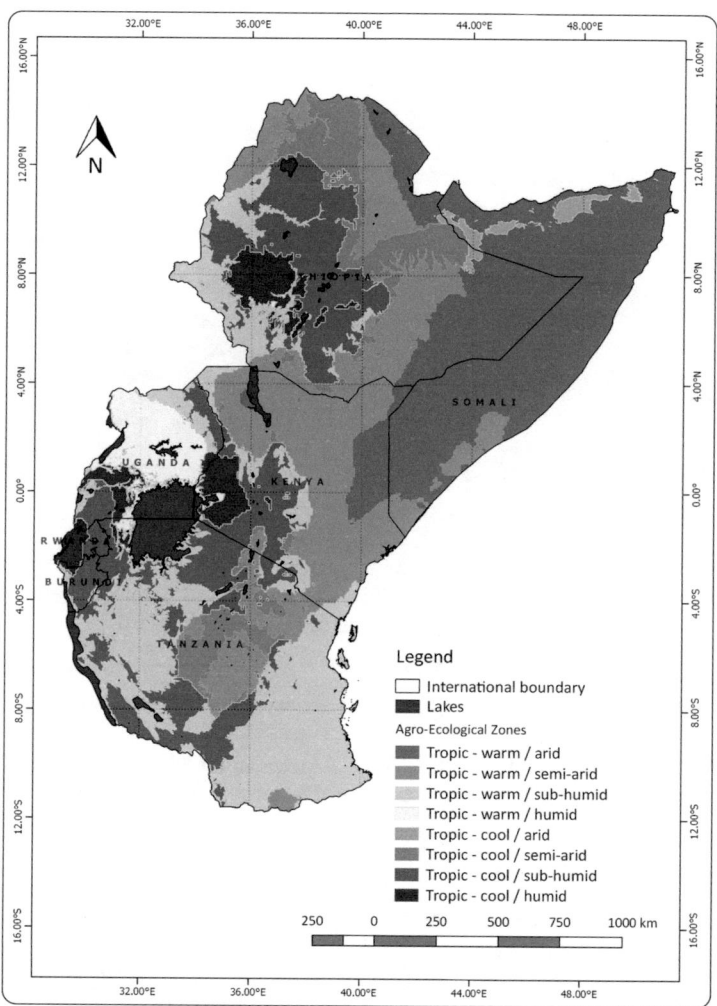

FIGURE 12.2 The ecoregions of East Africa. (Data Source: HarvestChoice, 2015).

FIGURE 12.5 Illustrations of water and wind erosion in Kenya. (Photo Credit: authors).

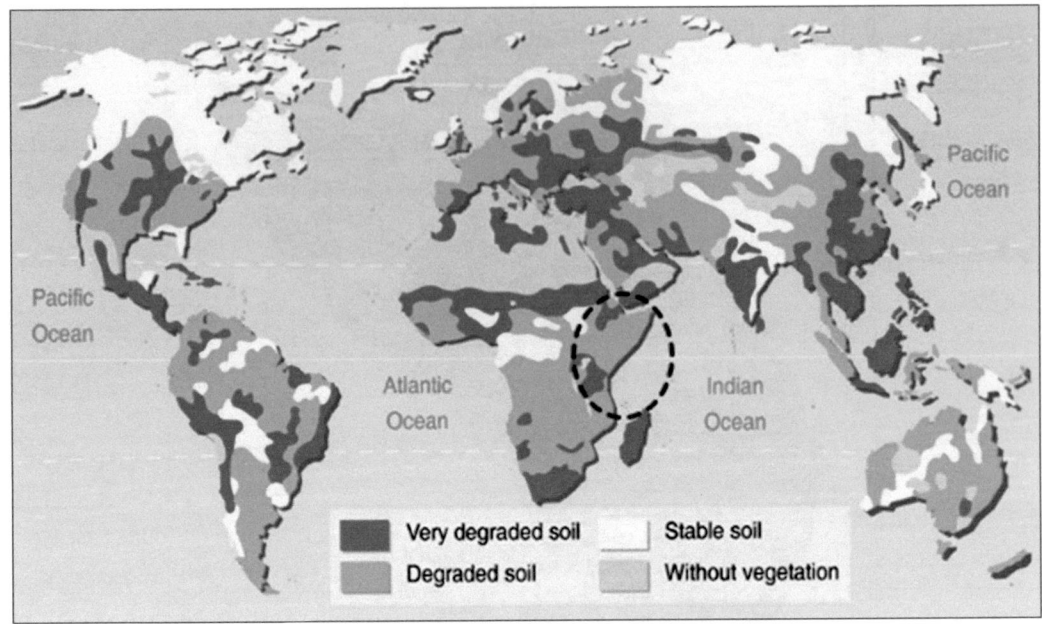

FIGURE 12.6 Human-induced soil degradation in East Africa and the world. (Source: UNEP, International Soil Reference and Information Centre (ISRIC), and *World Atlas of Desertification*, 1997).

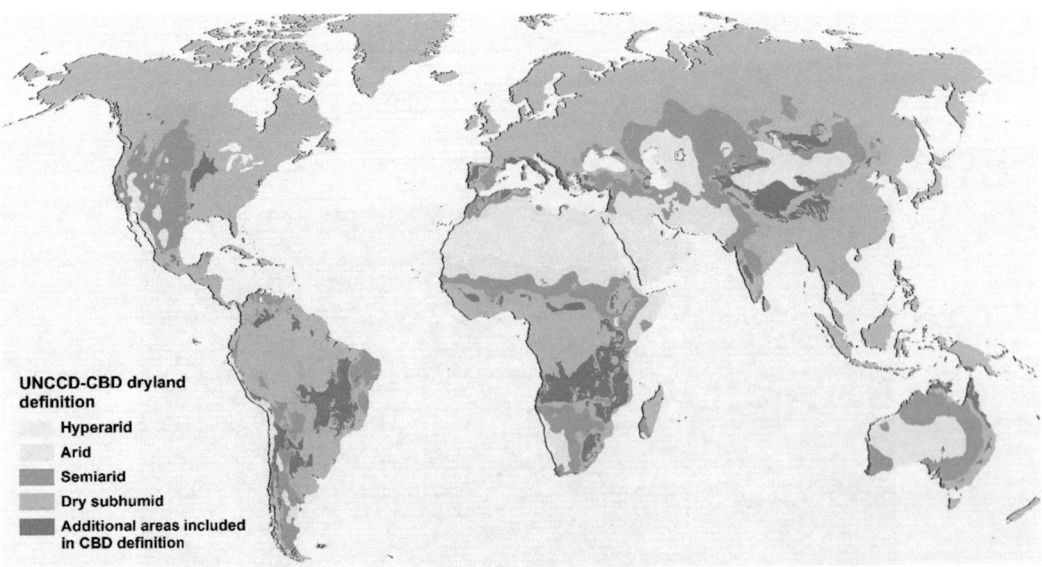

FIGURE 13.1 World distribution of drylands areas. (Source: UNEP-WCMC (2007); drawn using GIS data available at www.unep-wcmc.org.)

9.7   FERTILIZER INDUSTRY PERSPECTIVES

In the context of agriculture, fertilizer is seen as a commodity that is produced, transported and marketed by global private companies. In that regard, the fertilizer industry has addressed the shortcomings discussed above with both educational support and improvements in production processes and application methods to improve farmer's productivity and profit. Factors shaping the industry response include:

- Significant economic and environmental issues attributed to current fertilizer use became much more pressing from the 1990s, when several factors took on greater prominence concurrently. These included: global environmental and land and water scarcity concerns; persistent and large cost increases of commodities; and the economic development surge in major developing countries (such as China, India, and Brazil).
- With large up-front capital costs for mining and production, and the need to meet sharply rising fertilizer demand in the developing regions, the industry's focus has been to meet demand, and subsequently follow up with in-plant and in-field yield optimization and problem mitigation.
- Technically, today's fertilizers are essentially unchanged from those introduced at the advent of the Green Revolution in the 1960s and 1970s. The evolution from those fertilizers focused on the development of compound fertilizers that simultaneously deliver several nutrients in one product (e.g., diammonium phosphate [DAP]) or by physical addition and/or mixing (micronutrients sprayed on major nutrient granules, or blended fertilizers with multiple nutrients of equal-sized granules bagged together). Most of this evolution had occurred by the 1980s. As a result, current research and development spending by the fertilizer industry, for the next generation of fertilizers, is negligible (less than 1.0% of total sales).
- In contrast, the seed industry – with fundamentally different economics and environmental considerations – has followed a path of technical innovation: continued development of higher-yielding varieties; the addition of new features, such as resistance to pests and adverse climatic conditions; and, most recently, the use of genetic optimization to increase yield and "hardiness." Companies in the seed industry spend about 9% of sales on research and development.

Consequently, the magnitude of the demand for fertilizers in the future now requires a more fundamental response – technology needs to be applied again to the industry in order to develop a new generation of intelligent fertilizers that minimize the economic and environmental drawbacks of current fertilizers and can ensure maximum crop production to meet future food demand.

9.8   ADDRESSING EFFICIENCY THROUGH MANAGEMENT

In most countries, fertilizer application is based on optimum economic return based on crop response data averaged over large areas. But studies have shown that farmers' fields show large variations in terms of nutrient-supplying capacity and crop response to nutrients. Thus, blanket fertilizer application recommendations may lead farmers to overfertilize in some areas and underfertilize in others, or to apply an improper balance of nutrients for their soil or crop. Site- and crop-specific production techniques and practices is one approach to maximize economic, social, and environmental benefits. This includes soil and crop management to fit the different conditions encountered in cropping areas. Sometimes this is also referred to as site-specific nutrient management (SSNM), which, among other things, includes a balanced use of all essential crop nutrients. Long-term experiments in the Great Plains in the US show twice as much N was taken up by the corn crop with the application of fertilizers containing P (Figure 9.12). The SSNM approach has proven to be an efficient way to reduce NOx emissions to the atmosphere and leaching losses to surface and subsurface waters (Richards et al. 2015).

In Europe a combination of balanced fertilizer application and management more than doubled cereal yield per unit fertilizer-N. For example, at an application rate of 200 kg fertilizer-N/ha yield increased from 4t/ha to more than 8t/ha (Bindraban et al. 2008). In SSA, soil fertility is very low, and it is, beside NPK, also highly deficient in secondary and micronutrients and organic matter. Addressing all soil fertility constraints, including macronutrients (NPK), as well as secondary and

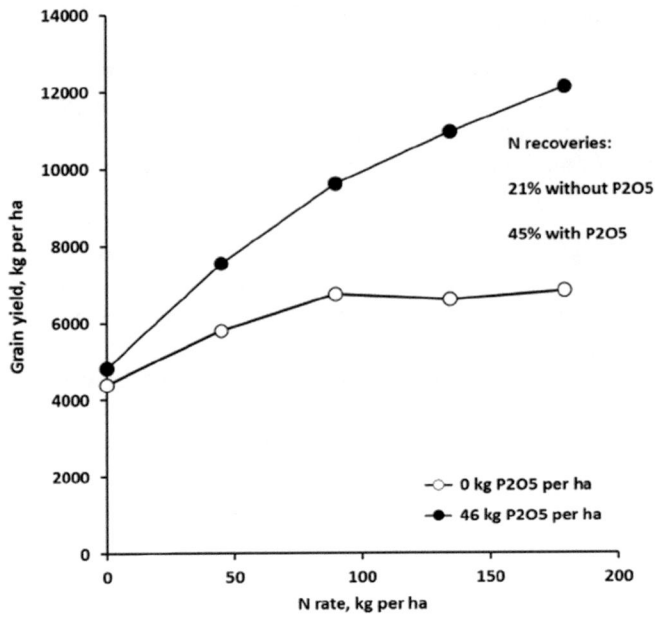

FIGURE 9.12   Improved N recoveries with addition of fertilizers containing P. (From Schlegel & Havlin 1995.)

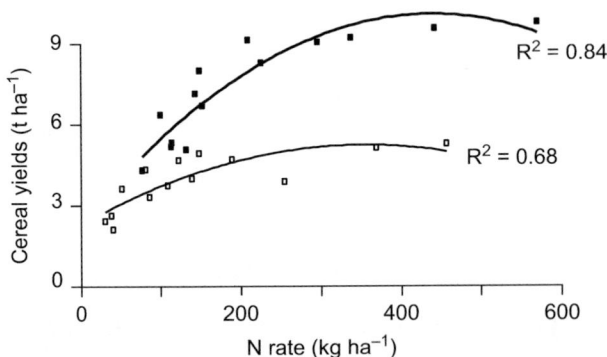

FIGURE 9.13   Impact of balanced fertilizer application and improved management on cereal yield in Europe. (From Bindraban, Loffler & Rabbinge.)

micronutrients and pH correction, is vital to achieving optimal fertilizer-use efficiency and economic returns. Integrated soil fertility management (ISFM) that includes the use of organic matter, soil amendments, and balanced fertilizers, is an integral part of revitalizing the SSA soils and building up soil organic carbon that in turn improves NUE (Figure 9.13).

In Burundi, Mozambique, and Rwanda, average field trial results clearly show that balanced fertilizers (NPK + secondary + micronutrients), based on soil analyses, performed better than control plots and better than those with the application of fertilizers containing only NPK (Figure 9.14). In all cases, adding secondary and micronutrients significantly increased maize yield by more efficient uptake of nutrients, resulting in less loss to the environment. In Mozambique, the SMNs were Mg, S, Zn, and B. In Rwanda, the SMNs were S, Zn, and B. In Burundi, the SMNs were Ca, Mg, S, Zn, and B (personal communication 2017). Certainly, the application of the right type of fertilizer, in the right amount, at the right time, and in the right place ensures increased crop production, better

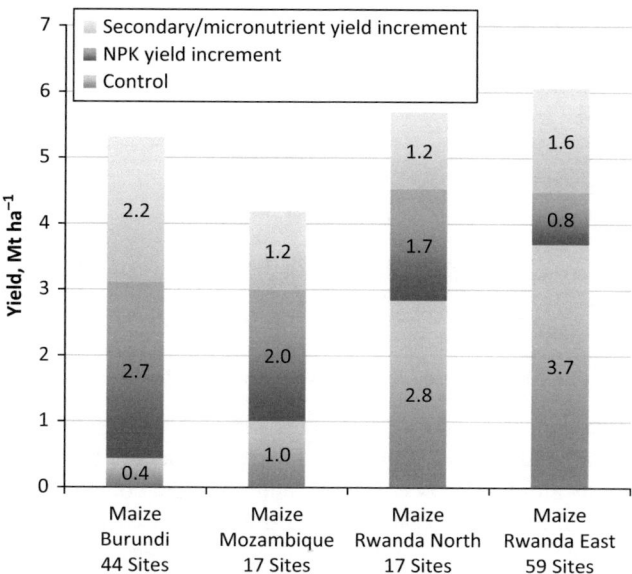

FIGURE 9.14 (See color insert.) Effect of balanced fertilization on maize yields in Burundi, Mozambique, and Rwanda (Personal communication 2017, John Wendt.)

utilization of nutrients and reduced losses of nitrogen to the environment. This strategy, referred to as "4Rs" is implemented by (Bruulselma et al. 2012):

Right product: providing essential nutrients starting with the application of organic nutrient sources and supplemented by mineral fertilizers tailored for site- and crop-specific conditions;

Right rate: applying recommended fertilizer rates based on crop needs and considering the current supply of nutrients in the soil;

Right time: ensuring nutrients are available when crops need them by assessing crop nutrient dynamics;

Right place: ensuring that the fertilizer sources are close to the plant roots to optimize uptake.

The 4R principles have been adopted in several countries with excellent results. An advancement of the 4R principles is "precision farming" (also known as precision agriculture), particularly popular in North America. Precision farming employs detailed, site-specific information to precisely manage production inputs. The idea is to know the soil and crop characteristics unique to each part of the field, and to only apply inputs (seed, fertilizer, chemicals, etc.) within small portions of the field. Precision farming includes a set of technologies that combines sensors, information systems, advanced machinery, and information management in order to optimize production and allow better use of resources to maintain the quality of the environment while improving the sustainability of the food supply.

An adaptation of 4R principles is the subsurface application of urea, which is the prime nitrogen fertilizer for the cultivation of rice and accounts for about 85% of rice fertilizer N (Gregory et al. 2010). Urea is prone to high losses (as much as 70%) under the present application practice of broadcasting, particularly in developing regions. However, research over several decades has shown that if urea, instead of being broadcast, is point-placed below the soil surface (subsurface) near the root zone of the rice plant, the N losses are lowered, with a simultaneous increase in yield. This application method involves producing larger urea granules (1–3 g), which are inserted below the soil surface (7–10 cm), approximately equidistant from four rice plants, two weeks after transplanting of rice seedlings. This is done once during the growing season. The weight of the urea granules depends on the rate of application. This technology, known as urea deep placement (UDP), makes N available to the rice plant throughout its growth cycle, thereby reducing losses and increasing

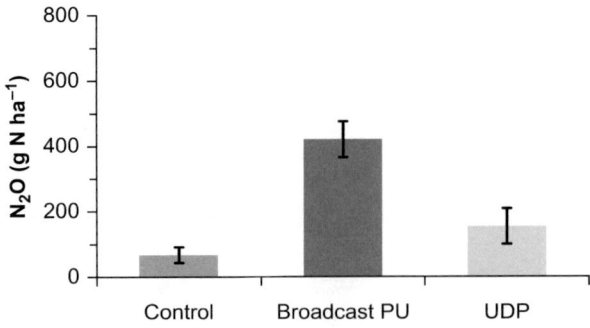

FIGURE 9.15   The deep placement of urea reduces N losses to the environment.

yields. Results in the greenhouse and farmers' fields show a consistent increase in yield compared with broadcasting (Savant and Stangel 1990). The deep placement of urea also significantly reduces N concentration in floodwater and NOx emissions under certain conditions (Figure 9.15).

A survey among 681 farmers in Bangladesh showed an average rice yield increase of 15%, while applying 35% less urea (Roy 1990) (Figure 9.16). This results in increased income of about $150 per hectare for Bangladeshi rice farmers; about 2 million of them have adopted this technology, and the numbers are growing. In Bangladesh, larger urea granules are produced by compacting commercially available prilled urea in the shape of a briquet in a briquetting machine (Figure 9.17). These machines range in capacity from 250 kg/hour to 450 kg/hour, and the process of manufacturing the briquettes is relatively simple.

The briquette application is backbreaking work, as it involves applying the briquets by hand; however, briquets are applied once during the planting season compared with two to three applications using conventional fertilizers. Nevertheless, a mechanical applicator is essential for widespread acceptance of UDP technology, particularly by smallholder farmers. Over the years, several hand application prototypes have been developed and tested, but for wider adoption of this technology a powered applicator for larger planting areas is essential.

While UDP technology continues to be adopted by farmers in Bangladesh, several countries in Asia and sub-Saharan Africa are promoting this technology based on field trial results. Further,

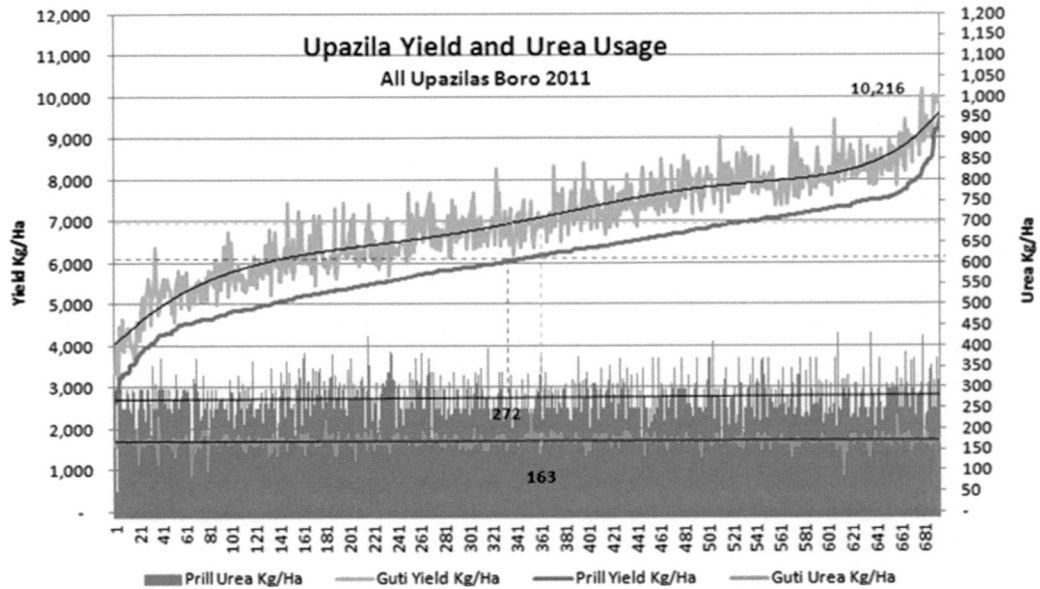

FIGURE 9.16   **(See color insert.)** Urea deep placement: 15% yield increase with 35% less nutrient applied.

Prilled Urea   Urea Briquettes

FIGURE 9.17   Commercial urea (prilled) and urea briquets produced by compacting prilled urea.

briquets containing N, P, and K are being used in Bangladesh by vegetable producers, particularly women, because of an additional income of more than $100 per hectare.

9.9   ADDRESSING EFFICIENCY THROUGH TECHNOLOGY

Improving NUE in agriculture has been a concern for decades (Dobermann 2005), and numerous new technologies have been developed to achieve this. Increasing the efficiency of mineral nitrogen (N) fertilizer use is complicated by the fact that plants take up N normally as nitrate or ammonium ions through their roots from the soil solution. However, ammonium-N, unlike nitrate-N, can be retained in the soil where soil and plants compete for available ammonium-N (Trenkel 1997). This competition for nitrogen is the main problem when it is added as mineral fertilizer to feed plants. Only a certain proportion of the N is taken up by the growing plants. The unused portion is lost to the environment mainly through ammonium volatilization, denitrification, and leaching. So, researchers in academia and the fertilizer industry have been challenged to develop special types of fertilizers that will overcome such losses. Two important special types of fertilizers are:

1. slow- and controlled-release fertilizers with the release of nutrients over several months
2. stabilized fertilizers (fertilizers associated with nitrification or urease inhibitors) delaying either the nitrification of ammonia or the ammonification of urea

These special types of fertilizers release, either by design or naturally, their nutrient content over an extended period to match the nutrient requirements of the crop (Figure 9.18). Because of economic and environmental considerations, the slow, controlled-release, and stabilized N fertilizers are much more important than phosphate (or potash), particularly under certain soil and climatic conditions. In most cases unutilized phosphate and potash remain available for subsequent crops. Nevertheless, there are a few controlled-release water-soluble phosphate and potash fertilizers in the market for specialty application.

9.9.1   Slow-Release Nitrogen Fertilizers

Slow-release nitrogen fertilizers (SRNFs) release their nutrients at a slower rate than urea and ammonium nitrate, but the rate, pattern, and duration of release are controlled by soil and climatic conditions and cannot be predicted with any accuracy. Example of SRNFs are organic-N low-solubility compounds such as urea formaldehyde (UF) and isobutylidene-diurea (IBDU).

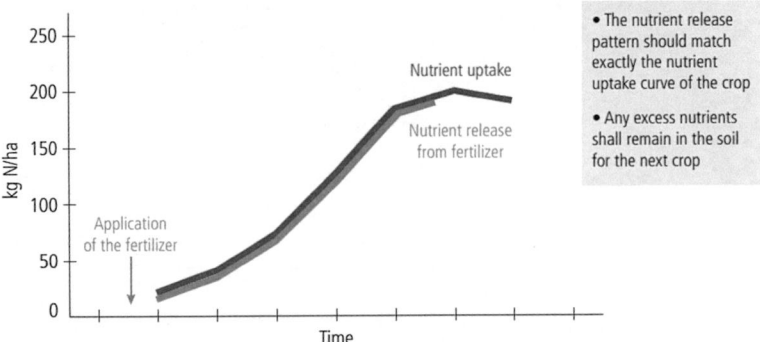

FIGURE 9.18 The ideal slow- and controlled-release fertilizers match the crop's nutrient requirements (Trenkel 1997).

9.9.2 CONTROLLED-RELEASE NITROGEN FERTILIZERS

Controlled-release nitrogen fertilizers (CRNFs) are those whose rate, release pattern, and duration of release are predictable and controllable through a physical barrier applied onto granules. These coatings can be hydrophobic polymers or as matrices in which the soluble active material restricts the dissolution of the fertilizer. The coated fertilizers can be further divided into fertilizers with organic polymer coatings – that are either thermoplastic or resins – and fertilizers coated with inorganic materials such as sulfur (S) or mineral-based coatings (Figure 9.19). Additional information regarding various coating materials is discussed by Shaviv (2005).

The sulfur-coated urea (SCU), one of the early CRNFs, was developed by the Tennessee Valley Authority (TVA) in the early 1960s, and involved spraying molten sulfur onto urea granules in a rotating drum. The applied sulfur oxidized over time and was an important secondary nutrient, but the coating had many imperfections that hindered the prediction of the precise release rate, and additional paraffin coating was applied to cover the imperfections. The relatively high price of SCU compared with conventional fertilizers makes it uneconomical for use in crop production, but has seen acceptance in non-agricultural markets such as golf courses and high-value specialty crops.

The SCU provided the clue for the industry to develop improved coatings using cross-linked thermosetting and thermoplastic polymers, resulting in a polymer-coated commercial product known as Osmocote. Subsequently, recent technology enhancements have resulted in new slow-release products that are cost-effective for corn (*Zea mays*), wheat (*Triticum aestivum*), potatoes (*Solanum tuberosum*), rice (*Oryza staiva*) and others. For example, Nutrien (a company formed through the merger of Agrium Inc. and Potash Corp.) manufactures a product, sold mainly in North America under the trade name ESN®, which is a polymer-coated urea N product whose release mechanism is

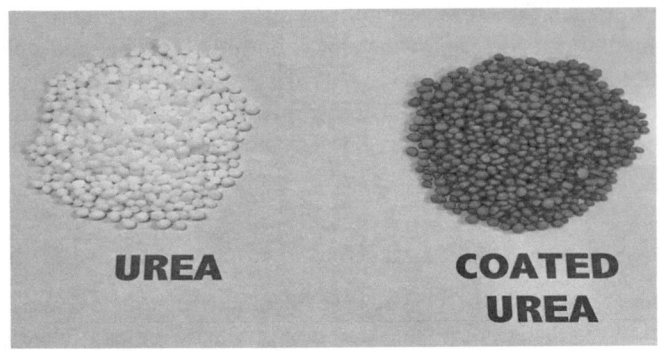

FIGURE 9.19 Uncoated commercial urea and polymer-coated urea.

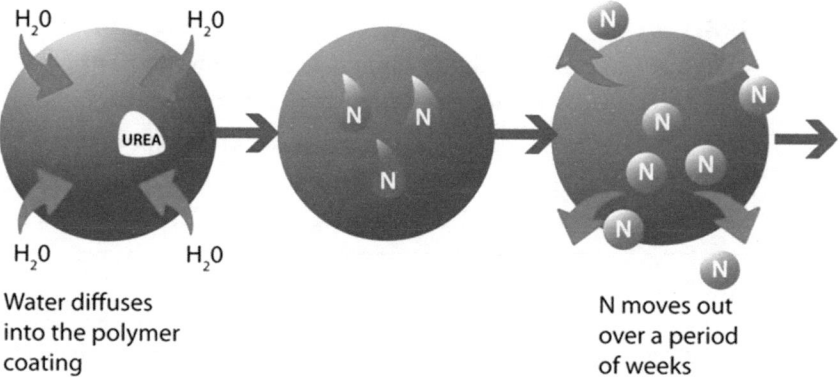

FIGURE 9.20   Release of N from polymer-coated urea.

temperature- and moisture-controlled (Figure 9.20) (Blaylock 2005). Typically, ESN® shows a yield increase of about 10% compared with conventional fertilizers (Killorn et al. 2003) (Figure 9.21).

In Japan, with its intensive rice cultivation, N fertilizer application rate is high. In order to maximize yield and reduce N release to air and water, the farmers use a polymer-coated CRNF, sold under the trade name Meister®, and realize a NUE-N of nearly 80% compared with only 30% under broadcast application (Shoji 2005) (Table 9.1). Although not currently practical for use in developing economies, the concept of polymer coating as used in ESN® and Meister® has spurred on other companies to research alternative chemicals to produce an affordable slow-release product for use in developing economies.

9.9.3   Stabilized Nitrogen Fertilizers, Urease Inhibitors

Stabilized nitrogen fertilizers (urease inhibitors) are compounds that prevent or suppress the transformation of amide-N in urea to ammonium hydroxide and ammonium through the action of the enzyme urease. By slowing down the rate at which urea is hydrolyzed in the soil, volatilization losses of ammonia to the air (as well as further leaching losses of nitrate) are either reduced or

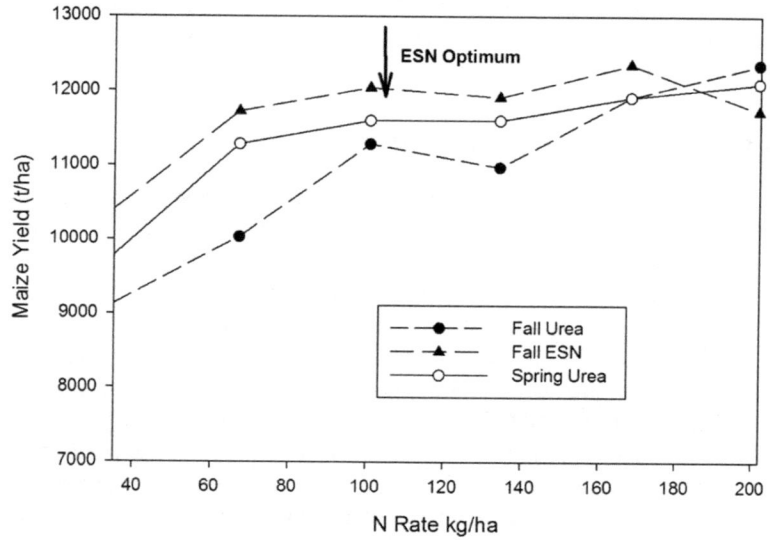

FIGURE 9.21   Maize response to urea vs. ESN® (polymer-coated urea).

TABLE 9.1

Effect of Controlled-Release Urea (Meister®) on Nutrient Use Efficiency-N for Rice Production in Japan

Application	N rate, kg/ha	N uptake by rice, kg N/ha	NUE-N %
Urea	100	30	30
Meister®	40	32	80

avoided. Thus, the efficiency of urea and of N fertilizers containing urea is increased, and any adverse environmental impact from their use is decreased. There has been considerable research to develop an effective urease inhibitor. Among many, N-(n-butyl) thiophosphoric triamide (NBPT) is the only urease inhibitor, marketed under the trade name Agrotain®, that has gained practical and commercial importance in agriculture.

9.9.4 STABILIZED NITROGEN FERTILIZERS, NITRIFICATION INHIBITORS

These are compounds that delay the oxidation of the ammonium ion (NH_4^+) by the *Nitrosomonas* bacteria in the soil. These bacteria transform ammonium ions into nitrite (NO_2^-), which is further transformed into NO_3^- by *Nitrobacter* and *Nitrosolobus* bacteria. Thus, the nitrification inhibitors control the loss of NO_3^- by leaching, or the production of nitrous oxide (N_2O) by denitrification, resulting in increasing NUE-N. Nitrification inhibitors, including dicyandiamide (DCD), 3,4-dimethylpyrazole phosphate (DMPP), and 2-chloro-6-(trichloromethyl)pyridine (nitrapyrin), are commercially produced and used in certain markets, mainly in developed economies.

The neem (*Azadipachta indica*) tree is indigenous to many tropical and semi-arid countries. Meliacins, a component of neem oil, has the characteristic of retarding the nitrification process. Neem cake has been used to coat urea to improve NUE-N. The efficacy of neem-coated urea (NCU) has been confirmed through several years of field trials in India. Compared with commercial urea, NCU results in 2–10% higher rice yields. In some cases, the rice yields with 80% NCU are comparable to that by application of 100% commercial urea. In general, the farmers apply the same amount of NCU as uncoated urea but realize 5–10% yield increases and subsequently lower N losses. The total global market for SRNF, CRNF, and stabilized N fertilizer is about 10% of the global nutrient consumption.

9.10 ADDRESSING EFFICIENCY THROUGH POLICY

New technologies and management practices are essential for improving productivity and NUE, increasing farmers' income, and reducing environmental degradation. But without conducive policies, neither technologies nor management practices can be effective.

In Europe, NUE-N has progressively increased since 1990s. In 2016, 30% more crop was produced with half the fertilizer-N application rate resulting in an increase of NUE-N by 20 percentage points. The European Union (EU) Common Agricultural Policy partly triggered this increase in NUE-N by reducing crop subsidies and producing the EU Nitrates Directive, which limited manure application rates on agricultural land (Zhang et al. 2015) (Figure 9.22).

China and India collectively account for more than 30% of global nitrogen consumption and have a relatively low NUE-N (ranging between 25% and 30%). In China, an input subsidy reached $17 billion in 2014, and farmers' use of fertilizer-N exceeded the recommended amount by 20–60% in cereals and even more in vegetables, whose NUE-N is around 15%. In China, farmers made a conscious decision to shift to fruit and vegetable production, partly because the crop-to-fertilizer price

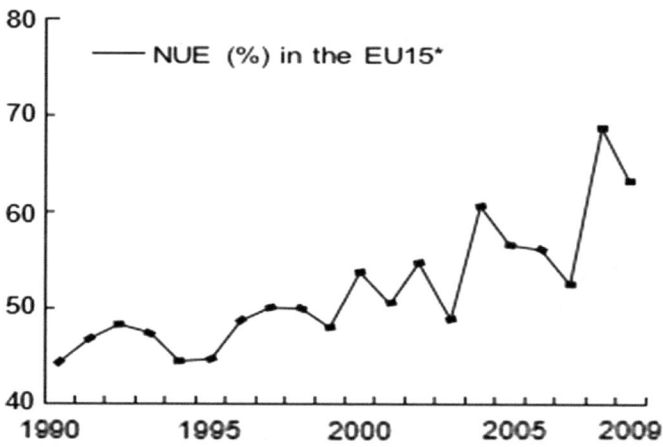

FIGURE 9.22 Progressive increase in NUE-N in Europe (EU-15) driven by agricultural policies.

ratio is more attractive than cereal production (particularly soybean, which is imported mainly from the US and has much higher NUE-N, 70–80%). The Chinese government, driven by low NUE-N and environmental challenges, has decided to cap fertilizer consumption at the 2020 level, which will see an increased use of more efficient products and better management practices (Reuters 2015).

In India, fertilizers are highly subsidized, and nitrogen more so than other nutrients, which results in the overuse of nitrogen-based fertilizers, distorting nutrient (nitrogen-phosphorus-potassium) ratios, all of which result in a low NUE-N. Recent field trials have shown that a balanced fertilization that includes all essential plant nutrients can increase NUE-N by an average of 20% (Role of Fertilizers 2016). To correct nutrient ratio distortion, the Indian government introduced a nutrient-based subsidy in 2010 (personal communication 2016). The subsidy had only a minimal impact on improving NUE-N since urea was excluded from the scheme (Figure 9.23). To combat the continued overuse of nitrogen, the government attempted to make fertilizer-N more efficient by requiring all agricultural urea be coated with an extract from neem cake, a nitrification inhibitor. Although early, the policy appears to be yielding the desired results of increasing NUE-N by 5% to 7%.

9.11 SUSTAINING FOOD PRODUCTION – NEXT-GENERATION FERTILIZERS

As mentioned earlier, the fertilizers in use were developed several decades ago for temperate agriculture with nutrient contents (mainly NPK) as high as possible (high analyses), with the intent of

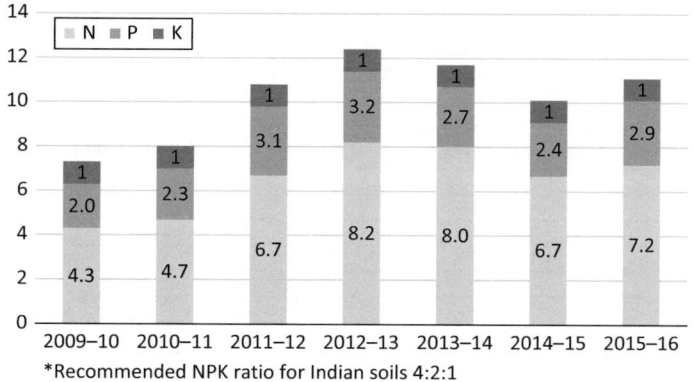

FIGURE 9.23 Imbalance in NPK use ratio in India (Shoji 2005).

delivering them at the lowest cost to farmers. The application and management of these fertilizers, whose nutrients were readily available to plants, dramatically increased the agricultural productivity of countries in temperate zones, leading to economic development that increased gross domestic product (GDP) per capita. From the early 1940s until the mid-1960s fertilizer consumption was mainly in developed regions. With the advent of the Green Revolution in the 1960s and early 1970s, developing regions began to use the same fertilizers that were developed for temperate zones to increase their agricultural production. Farmers broadcasted fertilizers according to blanket fertilizer application recommendations. Production increased but it was not proportional to the increased fertilizer application rate and resulted in a progressive decrease in NUE-N. Hence there was a need to understand the behavior of these highly soluble fertilizers in the tropics and subtropics, where climate and soils are very different from those in the temperate regions. This resulted in the creation of the International Fertilizer Development Center (IFDC) in 1974, with the specific mandate of developing fertilizers for the tropics and subtropics. Initial focus was on: (1) identifying pathways for the loss of nitrogen from applied urea on flooded rice in Asia; and (2) assessing the effectiveness of use of "as mined" phosphate rock (direct application) instead of expensive, commercially available phosphate fertilizers on acid soils in Latin America. Nitrogen research led to the development and evaluation of several SRNF, CRNF, and stabilized fertilizers and the subsurface application of urea. The effectiveness of direct application phosphate rock depended on its solubility in neutral ammonium citrate (NAC): the higher the solubility, the more reactive the rock and the more effective as a fertilizer (Figure 9.24). Grinding the rock to micron particle sizes (nanoparticles) increases its solubility in NAC and thus makes it more effective as a phosphate fertilizer (Shaviv 2005).

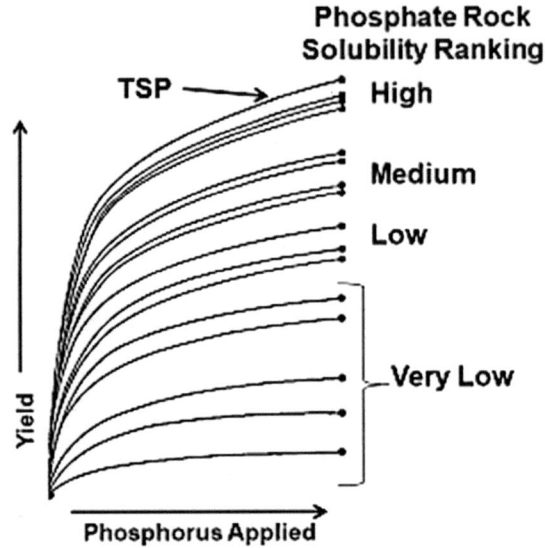

Soluble P$_2$O$_5$ in Neutral Ammonium Citrate	RAE	Solubility Ranking
(% P$_2$O$_5$)	(%)	
>5.9	>90	High
3.4–5.9	90–70	Medium
1.1–3.4	70–30	Low
<1.1	<30	Very low

$$RAE\ \% = \frac{(yield\ of\ ground\ PR) - (yield\ of\ check)}{(yeld\ of\ TSP) - (yield\ of\ check)} \times 100$$

FIGURE 9.24 Effectiveness of direct application rock as a function of its reactivity.

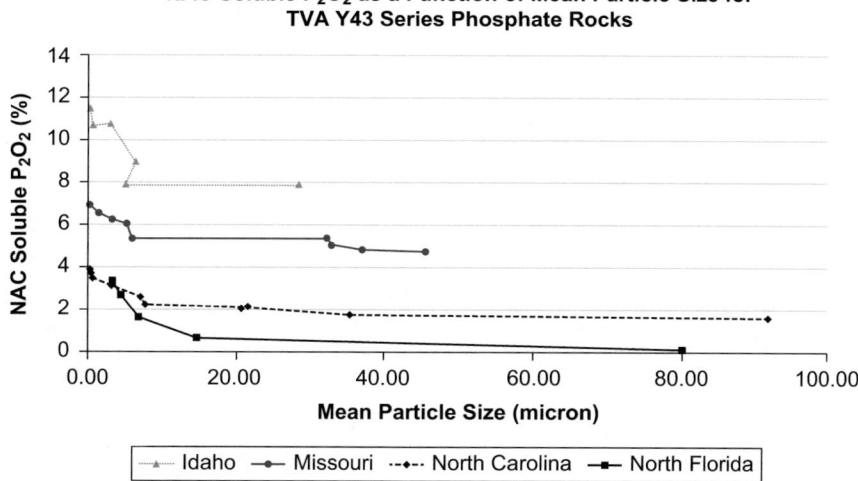

FIGURE 9.25 Effect of phosphate rock fineness (grind) on solubility in NAC.

Reflecting on: (1) the magnitude of the future food supply challenge in the face of increasing land and water scarcity and uncertain climatic conditions; (2) the increasing role that smallholder farmers will have to continue to play in the food supply chains in developing countries; and (3) the inherent flaws in current fertilizers – advanced technologies need to be applied to introduce new fertilizers and improved production. But fertilizer alone is not the answer. It must be a part of a package that is anchored in nutrient management practices, conducive policies, extension, and information and knowledge management. With the vast amount of data that are generated through research and satellite systems, information must be converted into knowledge and packaged for easy access by smallholder farmers through new and emerging communication tools. Nevertheless, manufactured fertilizers will be essential to increasing food production (Figure 9.25).

Fertilizer consumption in developing regions is more than 60% of global consumption and this percentage will continue to increase to meet expanding food demand. Consequently, developing regions will need a new generation of intelligent N and P fertilizers, which will be safer and more adaptive, eco-sensitive, and economical. In the development of new intelligent products, improvements or alternatives to current production methods are needed, starting with the production of ammonia and conversion of phosphate rock into phosphate fertilizer.

The Haber–Bosch is a high-temperature (~500°C) and high-pressure process (~200 atm) to produce ammonia that is subsequently reacted with CO_2 to produce urea. Natural gas is the energy and feedstock of choice. Even though the new plant's energy consumption is close to theoretical, the energy requirement (in simple terms) to produce one ton of urea is that contained in nearly four barrels of crude oil. So, the recent research focus should be to produce ammonia as close to the ambient temperature and atmospheric pressure as possible. If successful, lowering the operating temperature and pressure would reduce the price of fertilizer produced using ammonia.

Phosphate rock is a finite resource and the production of current fertilizers involves several steps where losses occur (Scholz et al. 2014). The current phosphate fertilizer production plants, located close to the phosphate deposits, are expensive (more than $1.5 billion for a basic plant) and this investment is made based on the potential/projected market for phosphate fertilizers. At present there is an adequate supply of phosphate fertilizers to meet global demand, but these products may not be optimum for the soils of the tropics. So, there is a need for research to develop a phosphate fertilizer with a soluble phosphate component, a slow-release component, and a soil-treatment component that inhibits fixation. At a more fundamental level there is a need to evaluate the effect of phosphate-solubilizing bacteria in making the nutrient available more predictably.

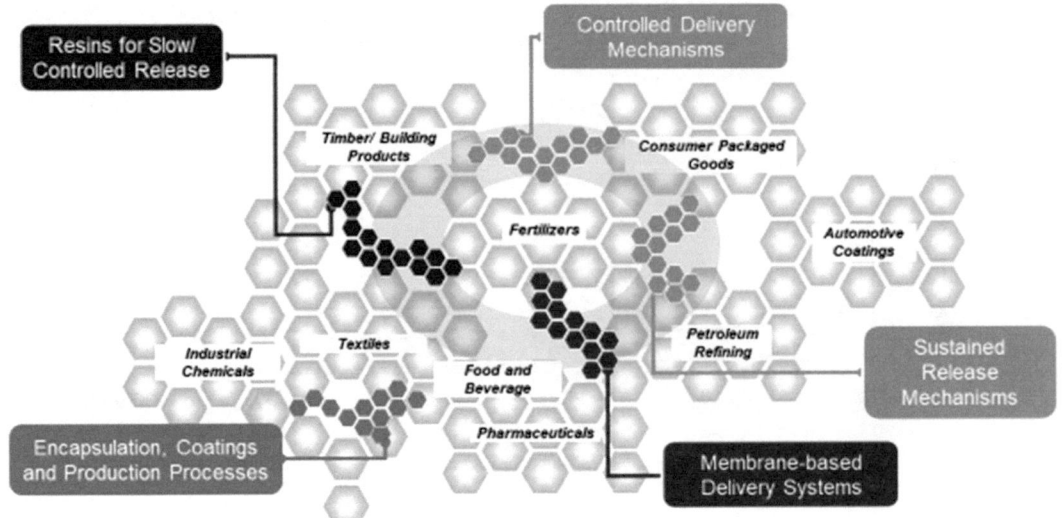

FIGURE 9.26   Making connection with other sectors for technologies to improve NUE-N.

To develop more efficient products with the desired attributes it is essential to use multidisciplinary sciences. The three foundational sciences for crop production (plant, soil, nutrient) need to be combined with other advanced technologies (e.g., molecular biology, nanotechnology, coatings sciences, genetic modification) to deliver the targeted outcomes. An initial view of the areas likely to benefit most from the application of multidisciplinary know-how include the following:

- Manipulating the fundamental mechanisms that govern N and P NUE behavior and reducing the costs of controllers and inhibitors to improve yield, economic impact, and environmental impact .
- Initial developments for N fertilizers might focus on off-patent coatings. Proprietary versions might be considered for later development for higher NUE performance, or to widen the performance spectrum vs. earlier coatings (Figure 9.26). Longer-term developments for both N and P fertilizers would benefit from the application of plant sciences to influence nutrient release and availability by plant-generated signals.
- Tapping unconventional, locally available nutrient sources to enhance nutrient self-reliance, improve economic considerations, and enable easier access for smallholder farmers. Two specific opportunities include nutrient recovery from waste streams and the utilization of locally available phosphate rock (present in several developing regions but often too low-grade [e.g., deposit size, reactivity of the phosphate rock] and/or remote for use in prevalent production processes).

9.12   SUMMARY

Fertilizers have played a pivotal role in increasing global food production and will continue to do so as the population is projected to reach 9.8 billion by 2050. The raw materials needed to produce fertilizers containing primary nutrients – nitrogen, phosphorus, and potassium – are adequate for the foreseeable future. However, phosphorus and potassium are finite resources and need to be sustainably managed to ensure availability for the future generations. The current suite of fertilizers developed several decades ago for temperate regions has characteristics that make them prone to losses to air and water unless managed. More than 90% of the projected population increase will be in the developing regions, where more food will be needed, and hence requiring a greater use of fertilizers with a resultant adverse impact on the environment. To reduce this impact, adoption of technologies and management practices at the farm-scale is needed. Some common principles include the 4Rs approach of applying the right source, at the right rate, at the right time, in the right place. However, the technologies and management practices needed to achieve the 4Rs vary, depending on the cropping systems, soil types, climate, and socioeconomic situation. Improvements in plant breeding, irrigation, and application of available 4R technologies have already made large gains in improving NUE; however, new technological developments may be needed to achieve further gains, such as more affordable slow- and controlled- release fertilizers, nitrification, and urease inhibitors. Precision agriculture, which uses a combination of information technology, remote sensing, and ground measurements, is very promising, but for developing regions to use this technology effectively, information should be readily available, accessible, affordable, and site-specific. However, the progress made so far is insufficient to achieve the projected 2050 goals of both food security and environmental stewardship. Hence the development of the next generation of fertilizers is vital to achieving the twin objectives of producing adequate nutritious food while protecting the environment by reducing losses of nutrients to it. There are several pathways to developing new fertilizers, but it will require a strong public–private partnership, where the public sector should invest in developing fundamental knowledge of nutrient movements in the soils and their uptakes by plants, and the private sector should develop products that are affordable, eco-sensitive, and result in improving farmers' incomes. This approach also requires the fertilizer industry to invest more on research to develop new product.

REFERENCES

Bindraban, P.S., Loffler, H., and Rabbinge, R. 2008. *International Journal Techniques & Global* 4(3), 276–295.

Blaylock, A. 2010. *Enhanced Efficiency Fertilizers.* Colorado State University Soil Fertility Lecture, October 20, 2010. Agrium Advanced Technologies, Loveland, CO. Available at: http://soilcrop.colostate.edu/westfall/documents/Dr%20Blaylock_CSUsoil-fertility-lecture_10-20-10.pdf.

British Petroleum 2015. *BP Statistical Review of World Energy.* Available at: http://www.bp.com/statistical review.

British Petroleum–Amoco 2015. *Statistical Review of World Energy.*

Bruulselma, T.W., Fixen, P.E., and Sulewski, G.D. (Eds). 2012. *4R Plant Nutrition Manual: A Manual for Improving the Management of Plant Nutrition.* International Plant Nutrition Institute (IPNI), Norcross, GA.

Dobermann, A. 2005. Nitrogen use efficiency – State of the art. IFA International Workshop on Enhanced Efficiency Fertilizers, Frankfurt/Main. International Fertilizer Association, Paris, France.

Food and Agriculture Organization of the United Nations (FAO). 2002. *FAOSTAT Database.* Rome, Italy. Available at: http://faostat.fao.org.

Food and Agriculture Organization of the United Nations (FAO) 2018. *Nitrogen Inputs to Agricultural Soils from Livestock Manure. New Statistics.* Rome, Italy. Available at: www.fao.org/3/I8153EN/i8153en.pdf.

Gregory, D.I., Buresh, R.J., Haefele, S.M., and Singh, U. 2010. Fertilizer use, markets and management. In *Rice in the Global Economy: Strategic Research and Policy Issues for Food Security*, S. Pandey, D. Byerlee, D. Dawe, A. Dobermann, S. Mohanty, S. Rozelle and B. Hardy, (Eds.). International Rice Research Institute, Manila, Philippines, pp. 231–263.

Gulati, A. 2014. Fertiliser pricing and subsidy policy in India: Exploring options for change. Paper presented at the FAI Annual Seminar "Unshackling the Fertiliser Sector," December 10–12, Aerocity, New Delhi, India.

Herring, J.R., and Fantel, R.J. 1993. Phosphate rock demand into the next century: Impact on world food supply. *Nonrenewable Resources* 2(3), 226–246.

International Atomic Energy Agency (IAEA) 2013. *Safety Reports Series No. 78.* Available at: www-pub.iaea.org/MTCD/publications/PDF/Pub1582_web.pdf.

Killorn, R., Gonzalez, M., and Rueber, D. 2003. *Effect of a Slow-Release N Fertilizer on Corn Yield.* Available at: http:/lib.dr.iastate.edu/farms reports/1258.

Mangat, G.S., and Narang, J.K. 2004. Agronomical trial for efficacy of NFL neem-coated urea. *Fertiliser Marketing News* 35(11), 1–3, 5, 7.

Mr. John Wendt, Senior Soil Scientist, International Fertilizer Development Center (IFDC), Muscle Shoals, Alabama. 2017. Personal Communications.

Mr. Satish Chander, Director General. 2016. Fertilizer Association of India, New Delhi, India. Personal communications.

Reuters News Service 2015. *China Targets Zero Growth in Chemical Fertilizer Use in 2020.* Available at: www.producer.com/daily/china-targets-zero-growth-in-chemical-fertilizer-use-in-2020.

Richards, M.B., Butterbach-Bahl, K., Jat, M.L., Lipinski, B., Ortiz-Monasterio, I., and Sapkota, T. 2015. *Site-Specific Nutrient Management: Implementation Guidance for Policymakers and Investors.* Global Alliance for Climate-Smart Agriculture. Food and Agriculture Organizations of the United Nations, Rome, Italy.

Roy, A.H. 2015. Global fertilizer industry: Transitioning from volume to value. The 29th Francis New memorial Lecture. Proceedings 769. International Fertilizer Society, London, UK.

Savant, N.K., and Stangel, P.J. 1990. Deep placement of urea supergranules in transplanted rice: Principles and practices. *Fertilizer Research* 25(1), 1–83.

Schlegel, A. J., and Havlin, J.L., 1995. Corn response to long term nitrogen and phosphorus fertilization. *Journal of Production Agriculture* 8, 181–185.

Scholz, R.W., Roy, A.H., Band, F.S., Hellums, D.T., and Ulrich, A.E. (Eds). 2014. *Sustainable Phosphorus Management – A Global Transdisciplinary Roadmap.* Springer.

Shaviv, A. 2005. Controlled release fertilizers. IFA International Workshop on Enhanced-Efficiency Fertilizers, Frankfurt/Main. International Fertilizer Association, Paris, France.

Shoji, S. 2005. Innovative use of controlled availability fertilizers with high performance for intensive agriculture and environmental conservation. *Science in China, Series C: Life Sciences* 48 supplement 2, 912–920.

Smil, V. 1997. *Cycles of Life.* Scientific American Library, New York.

Stewart, W.M., Dibb, D.W., Johnston, A.E., and Smyth, T.J. 2005. The contribution of commercial fertilizer nutrients to food production. *Agronomy Journal* 97(1).

Syers, J.K., Johnston, A.E., and Curtin, D. 2008. *Efficiency of Soil and Fertilizer Phosphorus Use.* FAO fertilizer and plant nutrition Bulletin 18. Food and Agriculture Organization of the United Nations, Rome, Italy, p. 108.

The Role of Fertilizers in Climate-Smart Agriculture. 2016. Contribution of the International Fertilizer Association to the UN Climate Change Conference in Marrakesh, Morocco – COP 22.

Trenkel, M.E. 1997. *Improving Fertilizer Use Efficiency: Controlled-Release and Stabilized Fertilizers In Agriculture.* International Fertilizer Association, Paris, France.

United Nations Department of Economic and Social Affairs, Population Division. 2013. *World Population Prospects: 2012 Revision, Key Findings and Advance Tables.*

United Nations Industrial Development Organization (UNIDO) and International Fertilizer Development Center (IFDC). 1998. *Fertilizer Manual.* Kluwer Academic Publishers, Dordrecht, The Netherlands.

Van Kauwenbergh, S.J. 2010. *World Phosphate Rock Reserves and Resources.* IDFC, Technical Bulletin T-75. International Fertilizer Development Center (IDFC), Muscle Shoal, AL.

Zhang, X., Davidson, E.A., Mauzerall, D.L., Searchinger, T.D., Dumas, P., and Shen, Y. 2015. Managing nitrogen for sustainable development. *Nature* 528(7580), 51–59.

10 Conservation Agriculture in Tanzania

Peter W. Mtakwa, Ndelilio N. Urio,
Faith Milkah Wakonyo Muniale, Alpha P. Mtakwa,
Rattan Lal, and Bal Ram Singh

CONTENTS

10.1   INTRODUCTION

Conservation agriculture, commonly abbreviated as CA, is promoted to increase crop production and environmental sustainability. It is proposed to be a solution for the problem of poor agricultural yields and soil degradation in most sub-Saharan African countries. Agriculture is the mainstay of Tanzania's economy. The agriculture sector faces various challenges and has been the government's top priority for development in order to reduce poverty. Farming efficiently has been a challenge for many farmers. Lack of finance and inadequate education in farming practices have limited many farmers to subsistence farming.

Tanzania has a total area of 94.5 million hectares (Mha) with inland lakes covering 5.9 Mha (6%). The country is home to the highest point in Africa, Mount Kilimanjaro, and the lowest point on the continent, Lake Tanganyika. Tanzanian soil types vary drastically throughout the country and each region gets varied amounts of rainfall.

There are six main soil types in the country, namely (Deckers et al. 2002):

- Andisols – volcanic soils, predominantly in the northern highland regions
- Umbrisols – sandy soils, predominantly in the coastal regions, mainly used for grazing
- Alfisols – granite/gneiss soils, predominantly in the northern regions of Mwanza and Tabora
- Ferralsols – red soils, predominantly in the central plateau
- Ultisols with plinthite: predominantly in the western regions such as Kagera and *Kigoma*
- Vertisols – called *mbuga* black soils/black cotton soils, spread across most of the country

Agriculture employs about 75% of the Tanzanian labor force. It contributes 24.7% of gross domestic product (GDP), about 20% of traditional export farming, and provides 95% of food requirements in the country (URT 2001 and 2003). In spite of this, the majority (91%) of the cultivators of agricultural land in Tanzania are smallholder farmers who rely on traditional methods of cultivation. Yields are generally low, for example, averaging below 1 Mg/ha for maize (*Zea mays*), being held back by such factors as low and generally declining soil fertility, soil and water loss through erosion, erratic and unreliable rainfall. Conventional farming practices such as burning or removing crop residue and intensive tillage often worsen these problems (Msolla et al. 1997; Kakeya et al. 1998). Soil erosion has been an issue of growing concern in Tanzania since the 1920s. It has been identified as one of the major causes of the decline of agricultural productivity, and soil conservation measures have been strongly encouraged. Soil erosion is still prevalent, however, particularly in the Dodoma, Shinyanga, Mtwara, Singida, Iringa, Tanga, Mwanza, Mara, Morogoro, Tabora, Kilimanjaro and Arusha regions. Soil erosion by water is the major problem in the mountainous areas of Tanzania, leading to environmental hazards, low land productivity, low income and increased poverty (Msita et al. 2012).

In many arable lands, nutrient mining is severe, with cropping activities estimated to be depleting nutrients at rates six to seven times greater than the rate at which they are being replenished. Increasingly, farmers are pointing to soil degradation as a key issue among the factors constraining crop production (Taruvinga 1995; BACAS 1996; Temu 1996). Poor and declining farm outputs, and especially the instability in yields that even minor climatic changes bring on, virtually immediately affect food security and farm incomes adversely.

Tanzania recognizes that managing its natural resources sustainably needs to be an integral part of its agenda for agricultural productivity (URT 2001 and 2003). It is, therefore, promoting CA especially in the arid and semi-arid areas and in hilly slopes, as a combination of crop and crop–livestock production practices that make the land more productive even as it improves the resilience of natural resources. Conservation agriculture is gaining recognition as a farming approach that boosts agricultural performance.

The term CA encompasses many techniques for capturing and storing water as well as improving soil quality that has degraded due to poor agronomic practices and artificial chemical fertilization (Mkonda and He 2017). As explained by both Duru et al. (2015a) and Duru et al. (2015b), CA can therefore optimize yields and profits by improving soil structure, conserving water, reducing agricultural inputs and enhancing environmental conservation.

In general, CA can go hand-in-hand with soil and water conservation measures that may include but are not limited to:

Terracing – sections of a hill are levelled or grassed to reduce runoff and conserve water and nutrients.

Conservation tillage – crop rotation and minimal tilling of the land that helps to maintain the quality of the soil cover.

Chololo pits – large pits are dug in a row. The pits are 22 cm in diameter, 30 cm deep, and spaced 60 cm apart. Organic matter is deposited and covered with soil for planting two or three single plants. The pits retain water and provide vital nutrients for the crop.

Trenches – furrows are dug along slopes and crop residues are deposited in them to increase the fertility and water-holding capacity. The trenches are covered with soil and crops are planted along them.

Cover cropping – crops such as lablab (*Lablab purpureus* or *Dolichos lablab*) are planted in between the main crop to help reduce evaporation during dry spells.

Ridges – earth is banked up in rows and crops are planted in between to help conserve water.

Miraba – an indigenous soil conservation method made of widely spaced, rectangular, grass strip bounds across the slope that do not necessarily follow contours.

What really do we mean by the term CA?

The practice of CA was developed in response to continuously declining land productivity under "conventional" soil tillage systems. CA farming practices revolve around three principles (FAO 2010):

(i)   minimizing soil disturbance;
(ii) maintaining a permanent organic soil cover; and
(iii) a system of crop rotations.

Simultaneous application of these principles allow farmers to better manage available soil, water and biological resources as well as farm inputs and labor, and make more effective use of natural ecological processes. CA contributes both to environmental conservation and to enhanced, sustained agricultural production. CA offers solutions for smallholder and larger-scale farmers alike. It significantly increases and stabilizes crop yields while reducing production costs. Full benefits and sustainability from CA are derived when all three principles are applied simultaneously.

CA can, therefore, be described as a set of soil management practices that minimize the disruption of the soil's structure, composition and natural biodiversity. CA is, in fact, a process that starts with reduced tillage. Conservation farming before CA is attained when there is no more than 20% tilling of soil in order to preserve soil structure and aggregate stability, and there is at least 30% organic cover throughout the year (so as to minimize heat fluxes in the soil, protect the soil from splash erosion that is normally severe at the onset of rains, and smother weeds before cover crops establish). Finally, in addition to the above two prerequisites, CA as a process becomes complete when there is an established system of crop rotations that permit nutrient injection and cycling in the soil, aggregate stability enhancement and pest management. To sum up, conservation agriculture entails minimal soil disturbance (reduced/minimum till to no-till, NT) and permanent soil cover (mulch, or with cover crop) combined with rotations, as a more sustainable cultivation system for the future. CA aims to conserve, improve and make more efficient the use of natural resources through integrated management of available soil, water and biological resources combined with external inputs. It contributes to environmental conservation as well as to enhanced and sustained agricultural production. It can also be referred to as resource efficient or resource effective agriculture.

Due to the scope of CA in terms of the principles and approaches involved, CA has been defined differently by different authors.

The Food and Agriculture Organization of the United Nations (FAO) has provided a generic definition, which we can call a working definition:

CA is a concept for resource-saving agricultural crop production that strives to achieve acceptable profits together with high and sustained production levels while concurrently conserving the environment. CA is based on enhancing natural biological processes above and below the ground. Interventions

TABLE 10.1
CA and Non-CA Practices

Practices within the concept of CA	Practices outside the concept of CA
Subsoiling/chisel ploughing	
Ripping	Ploughing (disc/moldboard plough)/harrowing
Tied ridges	Ploughing/harrowing
Vibro-flex cultivator	Roller tillers
Pitting	Ploughing/harrowing
Direct seeding	Planting after plough
Crop rotation	Monocropping
Contours	Cultivation on sloping land
Cover crops	Incorporating green manure
Intercropping	Monocropping
Incorporating residues in mulch	Removing residues
Partial removal of crop residues	Crop residues burned or used as fodder
Uprooting and leaving weeds on the soil surface	Uprooting weeds and removing them from the field
Agroforestry	Monocropping
Zero-grazing	Post-harvest grazing
Improved pasture	Crop residues as fodder

Source: Adapted from Triomphe 2005 and revised by Löfstrand 2005.

such as mechanical soil tillage are reduced to an absolute minimum, and the use of external inputs such as agrochemicals and nutrients of mineral or organic origin are applied at an optimum level and in a way and quantity that does not interfere with, or disrupt, the biological processes.

From this definition, we can infer that CA is not an actual technology; rather, it refers to a wide array of specific technologies that are based on applying one or more of the three main CA principles (IIRR and ACT 2005): reduce the intensity of soil tillage or suppress it altogether; cover the soil surface adequately, if possible completely and continuously throughout the year; diversify crop rotations.

Ideally, what we call "conservation agriculture systems" comprise a specific set of components or individual practices that, combined in a coherent, locally adapted sequence, allow these three principles to be applied simultaneously (Erenstein 2003; Lal 2015). When such a situation is achieved consistently, we speak of "full conservation agriculture," as illustrated by the practices of many farmers in southern Brazil (do Prado Wildner 2004; Bolliger et al. 2006) and other Latin American countries (Scopel et al. 2004).

The Table 10.1 helps to differentiate between conservation agriculture and non-conservation agriculture practices.

10.2 INDIGENOUS SOIL CONSERVATION SYSTEM IN TANZANIA

The term CA may be new to Tanzanians, but it is not new to most farmers and rural communities. It is a concept that demands that the land must not be allowed to degenerate while it is producing crops. Farming operations are based on this understanding. Fallowing and using organic matter are practices that farmers traditionally use to maintain or restore soil fertility.

Since the 1920s, numerous reports have warned against the disastrous effects of increasing erosion, land degradation, desertification and the mismanagement of natural resources due to increasing demographic pressure, and, as a result, soil conservation emerged at the end of the 1930s as a central concern in East Africa (Anderson 1984). In many African countries considerable efforts have been made during and since colonial times to conserve soil and water resources. African cultivators apply a wide range of techniques, such as crop rotation, crop mixtures, the application of manure, the protection of nitrogen-fixing trees, terrace-building, pitting systems, drainage ditches, small dams in valley floors and so on, to conserve soil and water and to prevent soil degradation.

From the 1950s, Tanganyikan, and now Tanzanian, government agricultural extension programs have promoted soil and water conservation practices to control surface water runoff, such as by the use of stone and earth bunds, ridging, pitting, infiltration or cut-off drains, bench terraces and contours (Shetto and Lyimo 2001). Vegetable growers as well as coffee (*Coffea arabica*) and banana (*Musa acuminata*) growers in Arusha, Kagera, Kilimanjaro and Mbeya regions commonly mulched their fields. The system in Kilimanjaro region is known as *the Chagga home garden*. In the Uluguru mountains, ladder terraces were used. Permanent farming systems with soil conserving practices, for example, terracing, manuring and stabling, were also applied to grow a variety of food crops by various local communities in Tanzania including: Matengo, Makonde, Kinga, Sandawe, Iraqw, Fipa, Nyaturu, Gogo, Mbugu, Shambala, Pare, Meru, Taita and Wakara (Shetto and Lyimo 2001).

Most traditional soil and water conservation practices, however, have turned out to be ineffective or simply impossible to apply. Land pressures in most cases have rendered fallowing simply impossible.

10.3 EVOLUTION OF CONSERVATION AGRICULTURE IN TANZANIA

CA has great potential in Africa, particularly in Tanzania, because it can control erosion, produce stable yields and reduce labor needs. The story of CA in Africa is not new. Across wide areas of Africa, CA principles used to be normal practice, before ploughs were introduced. Farmers would cultivate by hand, often with hoes, rotate crops and fallow fields for several years. Rising populations and ploughs changed all that. European settlers and colonial regimes introduced ploughs, and they quickly came to dominate farming because they enabled farmers to open up more land quickly and cheaply. But, just as in the United States, the plough has gradually eroded Africa's soils. Fertility and yields have fallen, and many countries now face critical food shortages. But not all Africa's farmland was put to the plough, or to the deep-till hoe, and pockets of conservation-friendly farming still remain.

CA re-emerged in several different places at about the same time in Africa. Initiatives by government research and extension agencies, donors and the private sector promoted CA for smallholder farmers in Cameroon, Ghana, Kenya, Madagascar, Malawi, Namibia, Tanzania, Uganda, Zambia, Zimbabwe, to mention just some of them. In addition, today, various institutions conduct research on or promote conservation agriculture.

In Tanzania, increased livestock and human activity led to the collapse of the conventional soil conservation system and increased land degradation evidenced by soil compaction, depletion of nutrients and organic matter, lowering of the soil's water-holding capacity and microbial activity, among other problems. In the late 1980s the government of Tanzania initiated programs to address the situation, mostly through mechanical and biological measures, such as reforestation activities, agroforestry, protection of water catchments, improved land husbandry and environmental conservation in general (Shetto and Lyimo 2001). CA is now promoted in Tanzania as an instrument for a sustainable and productive agriculture.

10.4 PRINCIPLES AND TYPES OF CONSERVATION AGRICULTURE IN TANZANIA

CA is carried countrywide in different forms and types depending on the topography, soil and available indigenous knowledge (IK). In some regions, CA practices are sponsored by projects from developmental partners such as US Agency for International Development (USAID), Department for

International Development (DFID), Swedish International Development Agency (Sida). Principally, CA operates on minimal soil disturbance, permanent soil organic matter to cover the soil and then diversified crop rotations as alluded to earlier in this chapter. According to the literature, agroforestry is mainly done in Kilimanjaro, Shinyanga, Arusha and Mbeya regions while crop rotation, intercropping, addition of manure (straw and animal manure) and mulching are done in several areas in the country (Löfstrand 2005; Batjes 2011).

These agronomic practices have improved soil fertility and increased crop yields in about all places where they have been applied. In areas with indigenous knowledge, modifications have been introduced in order to integrate with scientific knowledge and better results have been achieved. The Conservation Agriculture for Sustainable Agriculture and Rural Development (CA-SARD) project conducted in the northern part of Tanzania, aimed at improving food security and rural livelihoods for small- and medium-scale farmers through the use of proper CA practices (CA-SARD Project 2009). The project was participatory in nature as it involved small farmers in Conservation Agriculture practices aiming to improve their socioeconomic livelihoods (FAO 2012). The project achieved good results as more people adopted CA and greater crop yields were obtained.

10.5 ADOPTION OF CONSERVATION AGRICULTURE IN TANZANIA

CA may be difficult for people to accept because it goes against many of their acquired and cherished beliefs. The biggest obstacle is to get people to understand that ploughing destroys soils and the environment. Once they accept this, they may be willing to try something new. It is not just farmers who have to change their thinking: universities, agricultural training institutions, extension providers, researchers and the farming community itself must also change.

In Tanzania, CA has been in practice for many years, in different forms, albeit in some areas it is still in the adoption stage (Kimaro et al. 2015). Agronomic practices such as mulching, crop rotation, terraces, no-tillage and agroforestry, to mention a few, are the most applicable soil organic management practices in Tanzania representing CA. Maize, millet (*Pennisetum glaucum*) and sorghum (*Sorghum bicolor*) are good examples of crops involved in conservation agriculture because their straws are recommendable in organic matter formation.

The adoption of soil conservation is influenced by a knowledge of conservation measures and a perception of the effectiveness of the measures to control soil erosion and, especially, to improve productivity. The rate of adopting CA in Tanzania has a temporal and spatial variation. Areas with stressed environments are more likely to adopt than those with less environmental stress. Central and northern parts of Tanzania seem to have implemented more CA than other parts, probably because they are semi-arid (Branca et al. 2013). Dodoma, Manyara and Arusha are good examples of the region where CA is well adopted and has brought a significant contribution to crop yields. A number of projects have been implemented in those areas to stimulate and instill a sense of CA adoption.

According to the report by the FAO (2006), about 85% of farmers had adopted CA practices to meet the aims of Sustainable Agriculture and Rural Development (SARD). The majority of these farmers had about 0.2 hectares under CA (Shetto and Owenya 2007). Some farmers adopted reduced tillage in small areas (20 m × 20 m) while others adopted live cover crops, such as lablab (*Dolichos lablab*) or velvet bean (*Mucuna pruriens*), in slightly larger land areas compared to the former. In the southern highlands, planting basins, as a CA practice, has been adopted by farmers, facilitated by projects like Southern Agricultural Growth Corridor (SAGCOT). In Arusha about 60% of farmers have adopted planting basins as a well-established CA practice. This helps to retain water around the plant root for longer and reduces the magnitude of crop drying and wilting during dry spells (Lobell and Burke 2008). Subsequently, about 70% of farmers have adopted organic fertilization (crop straw and animal – farm yard and compost – manure) in the area. Crop residues are left in the farm after harvest to allow their decomposition and soil fertilization (Kimaro et al. 2015). Similarly, burning and grazing on the harvested farm is discouraged.

Some of the existing CA processes are indigenous, based on the indigenous knowledge of the local communities. For instance: organic farming in Ukara Island (Mwanza region), rotational grazing in Mbulu (Arusha region), *Matengo* pits or *Ngoro* system in (Ruvuma region), *Ngitiri* pasture conservation (Shinyanga and Mwanza regions), terracing and contouring (Arusha, Kilimanjaro, Tanga and Morogoro regions), mounds and ridges in Rukwa region, stone barriers on slopes in Korogwe district (Tanga region), and intercropping with trees in Arusha and Kilimanjaro regions. This CA management has made a tremendous contribution to the conservation of soil fertility (Liaudanskiene et al. 2013). However, these practices are insufficient to curb food insecurity in Tanzania (FAO 2012)and the overall adoption is still low. Thus, the further adoption of CA practices should be encouraged.

CA in Tanzania has been promoted in many ways including;

Training – Training is necessary for various groups, including farmers, implement makers, input suppliers and extension personnel.

Extension – The government extension service advises farmers on farming technologies. It provides advice and training, and manages demonstrations and field days. It can be an important promoter of CA.

Demonstrations – Demonstrations enable farmers to see CA in practice before they try it out themselves.

Farmer field schools – This is a participatory extension approach in which farmers get together to study farming in their own fields.

Farmers do not all adapt to the different aspects of CA at the same time. Therefore, the adoption process should also be adaptable to the farmers in different regions. Some of the ideas in the process of adapting to CA by farmers in Tanzania include regional assessments with the sets of questions outlined in Table 10.2.

Despite the efforts, the rate of adopting CA in Africa (including Tanzania) is insufficient compared to other parts of the world (Mkonda and He 2017). Based on the benefits attached to CA, there is need to emphasize its importance and speed up its adoption and utilization in Africa.

10.5.1 Case studies of conservation agriculture in Tanzania

This section presents studies that have been published (and a few yet to be published) and have been carried out in various parts of Tanzania. It synthesizes experiences and results obtained by applying

TABLE 10.2
Farmers' Considerations before Adapting to CA Practices

Approach	Rationale	Considerations
Crops and combinations	Different regions in the country have different crop varieties that grow best in their conditions. The main crop and cover crop are planted at different times in different regions.	Is there enough soil moisture to plant earlier? Is it better to plant closer or further apart? Which crops, and cover crops grow best together? How much compost or fertilizer is needed in this season?
Soil cover	Different regions have different soil types and topography that determine the type of soil cover required.	What is the best way to maintain soil cover? Are there dual-purpose cover crops that provide both food and fodder? Can traditional local cover crops be used instead of exotic types?
Equipment	CA equipment can be hard to find, too expensive, or not suited to local conditions.	Can local artisans adapt or fabricate existing designs? Are the materials locally available?

and using CA principles and technologies in a specific region in past or ongoing efforts and projects. The cases covered are from Kondoa, Karatu, Mkoji, Arumeru, Babati, Chamwino, Lushoto and Njombe districts.

The case studies analyzed are as follows.

1. **CA in Tanzania, a case study produced by the Oakland Institute and co-published by the Oakland Institute and the Alliance for Food Sovereignty in Africa (AFSA) at www.oaklandinstitute.org and www.afsafrica.org. (Omwenya and Semlowe 2006)**

The case focuses on a group of farmers in Likamba area, Arusha region, which launched a self-help organization in 2001, the Eotulelo Farmer Field School (FFS). Eotulelo translates as "come and join us" in the local Maasai language. The objectives of the Eotulelo FFS were to involve members in collective activities including soil erosion control, environmental protection (reduction of gullies), income-generating activities and improving traditional agriculture to increase yield.

2. **Soil management and agrodiversity: a case study from Arumeru, Arusha, Tanzania, documented by F.B.S. Kaihura, M. Stocking and E. Kahembe for the symposium on Managing Biodiversity in Agricultural Systems, Montreal, 8–12 November 2001. (Kaihura et al. 2001)**

The paper focuses on land-transforming operations that influence the behavior of physical and chemical aspects of the soil, surface and near-surface physical and biological processes, hydrology, microclimate, etc., and its impact on agrodiversity. It examines the different methods of managing the land, water and biota for crop production and the maintenance of soil fertility and structure. It also considers the diverse soil management methods used by farmers in Arumeru to address diverse constraints all aimed at soil productivity improvement, agrodiversity enhancement and conservation.

3. **CA in Babati district, Tanzania. Impacts of conservation agriculture for small-scale farmers and methods for increasing soil fertility. A master's thesis by Fredrik Löfstrand and supervised by Professor Erasmus Otabbong, presented at Swedish University of Agricultural Science, Department of Soil Sciences. (Löfstrand 2005).**

This study – carried out in the Babati district in Northern Tanzania – is one of several case studies conducted in Africa on CA during 2005. This was planned and facilitated by the FAO, the French Agricultural Research Centre for International Development (CIRAD), the Regional Land Management Unit of the World Agroforestry Centre (RELMA) in International Centre for Research in Agroforestry (ICRAF) and the African Conservation Tillage network (ACT) in preparation for the Third World Congress on Conservation Agriculture in Nairobi, October 2005. The major objectives of these case studies were to improve the understanding and documentation of past and current CA experiences.

4. **Social perception of soil conservation benefits in Kondoa eroded area of Tanzania. A case study by Rajendra P. Shrestha and Paul J. Ligonja focusing on a soil conservation project that was implemented in Tanzania for over 30 years. The study applied a socioeconomic approach to examine and analyze the benefits of soil conservation in the Kondoa eroded area of Tanzania by conducting a household survey of 240 households. (Shrestha & Ligonja 2015).**

Following past government efforts on soil conservation activities initiated in the early 1970s through decentralization, institutional collaboration, socioeconomic support to farmers and continuous local community participation in restoring the degraded ecosystem of Kondoa, this case study took place 30 years later to examine how that contributed to ensure environmental and socioeconomic sustainability in the area. The main land-use practices in the area are conservation tillage, integrated farming and use of organic fertilizers, controlled stall feeding, agroforestry, construction of cut-off drains, contour bunds and contour ridges cultivation.

5. **Soil fertility and crop yield variability under major soil and water conservation technologies in the Usambara mountains, Tanzania by S.B. Mwango, B.M. Msanya, P.W. Mtakwa, D.N. Kimaro, J. Deckers, J. Poesen, J.L. Meliyo and S. Dondeyne (Mwango et al. 2015)**

This study evaluated the variability of chemical soil fertility and crop yields under bench terraces, micro-ridges and miraba in order to explore and compare their strengths and limitations in smallholder farming conditions in Majulai village, west Usambara mountains, Lushoto, Tanzania.

6. **Extrapolating effects of conservation tillage on yield, soil moisture and dry spell mitigation using simulation modelling by Z.J. Mkoga, S.D. Tumbo, N. Kihupi, J. Semoka (Mkoga et al. 2010)**

This case study used a calibrated and validated Agricultural Production Systems Simulator (APSIM) crop simulation model to simulate long-term scenarios based on two treatments: conventional tillage practice (using ox-plough) with slash and burn treatment of crop residues commonly practiced by farmers, and the ripping tillage practice with crop residues left on the soil surface. Simulations were done for the soil and weather conditions of Igurusi (representing middle Mkoji subcatchment). Simulation outputs of daily soil moisture content and annual grain yield were plotted to give trends of soil moisture and yield for the two different tillage practices over a period of 23 years (1985–2008).

7. **Case study of CA in Karatu District as documented in CA in Africa series by Dominick E. Ringo, Catherine Maguzu, Wilfred Mariki, Marietha Owenya, and Njumbo, Frank Swai and published by African Conservation Tillage (ACT) Network (Ringo et al. 2002)**

This case study presents the status of CA in Tanzania and is one in a series of eight case studies about CA in Africa, which were developed within the framework of a collaboration between CIRAD (French Agricultural Research Centre for International Development), FAO (Food and Agriculture Organization of the United Nations), RELMA in ICRAF and ACT (African Conservation Tillage Network).

8. **Case study of SUA's crop and livestock integrated production model, culminating in conservation agriculture in Njombe and Wanging'ombe districts, Njombe region, by P.W. Mtakwa, N.A. Urio, A.P. Mtakwa, L. Eik, L. Kurwijila et al. 2017 (in preparation)**

Over the last 14 years, a team of Sokoine University of Agriculture (SUA) scientists, in collaboration with their Norwegian counterparts, have been engaged in action research to integrate livestock in a crop farming system that has experienced a decline in productivity from the 1980s when fertilizer subsidies were withdrawn due to the economic difficulties of the prestructural adjustment era (1970s). This brief presentation describes the approaches that have been used to engage farmers as co-researchers in addressing the challenges of declining soil fertility and seasonality of feed availability for stall-fed dairy cattle production systems in Njombe district. The results of the interventions indicate how integrating livestock in crop farming can catalyze socioeconomic, liveli-hood and environmental outcomes for smallholder production systems that are comparable to high input-high output commercial systems. The inherent characteristics of Njombe region are: inher-ently low fertile soil in humid and sub-humid zones; low organic carbon content with low cation exchange capacity (CEC) thus high leaching on plant nutrient; and low pH that fixes phosphate fer-tilizer. In order to produce enough food to feed the family, farmers have to expand their cultivatable land. With unreliable rainfall, several years face crop failure in arid areas. The conversion of dry-lands to cropland is a major cause of high carbon losses through water and wind erosion, especially when grazing crop residues instead of leaving the crop residues on the soil surface to minimize splash erosion and sequester carbon. Dairy farming was introduced in Njombe in the 1970s through the Swiss Government Assisted Smallholder Dairy Development Programme (SHDDP) and the heifer gift program of the Heifer Project International (HPI). The low productivity of dairy cattle became apparent a few years after the SHDDP project ended in the mid-1990s. SUA and Norwegian project intervention started in early 2000 to address initially the dry season feed scarcity challenge by introducing multipurpose fodder trees and high yielding grass species such as Guatemala grass and Napier/Bana grass.

While better feeding improved milk production, sustaining fodder crops required more soil nutrients. Compositing of manure by mixing fresh cow dung and feed remains provided a good solution. Production of biogas was later introduced. Farmers found the biogas digester effluent to be an even better organic fertilizer, especially for vegetable farming. Farmers kept on experimenting with the help of scientists from SUA, the Norwegian University of Life Sciences (NMBU) and the Tanzania Domestic Biogas Program (TDBP).

And in the last three years, the biogas digester effluent had become not only a good organic fertilizer but also a crop pest repellent as well. High population, declining soil fertility leads to deforestation and consequential soil erosion and degradation. Biogas production solved the energy problem for cooking and lighting, and saved the forest trees. Further experimentation showed that a combination of select industrial fertilizers formulated to address specific soil nutri-ent deficiencies gives much higher yields than either fertilizer applied alone. In one particular case, yields were raised from 1 Mg per hectare following conventional agriculture with only top-dressing with urea, to 5 Mg/ha with a full dose of basal fertilization, two top dressings and one foliar application of industrial fertilizer, to 8 Mg/ha when compost manure, bio-slurry and half the recommended industrial fertilizer rate was used (Mtakwa et al. 2017, in preparation). This meant that the area under maize cultivation could be reduced to a conservative half of the area if CA and the combination of organic and half the recommended industrial fertilizer was adopted. This would then free labor from agriculture and make it available to other economic ventures. One aspect that the team found striking was the shift of labor for weeding, something that is primarily carried out by women, became a man's job. Whereas one woman can weed 0.4 ha in 7 days, the man now can use herbicides and spray the same area in one hour to one hour twenty minutes. Over the last five years, this has led to significant improvements in households, where the woman now has enough time to cater for the children as well as the cattle that provide the family with milk, biogas and fertilizer in form of compost and farm yard manure (FYM) as well as bio-slurry. There are testimonies from husbands who say that now they can prepare their own tea, instead of waiting for the wives to prepare the same for them. The reason? Biogas has made kitchen work attractive to men!

In a nutshell, minimum tillage (strip tillage/double digging), use of compost manure and bio-slurry emanating from the bio-digester for maize cropping has improved the general household economy in Njombe region. Further, biogas saves labor and forests. Compost and FYM sustain soil fertility through improving soil organic carbon (SOC), giving rise to higher fodder, maize and vegetable yields, while milk improves both the nutritional as well as the economic status of the individuals keeping dairy cattle. The SUA Crop-Livestock-Integrated-Production-System (CLIPS)

Model is based on maize crop, dairy cattle and CA practices in Njombe region and it facilitates out-scaling and up-scaling of innovations that have been tested in Njombe.

Through the Enhancing Pro-Poor Innovations in Natural Resources and Agricultural Value-Chains (EPINAV) program (sponsoring the conservation project in Njombe and Wanging'ombe districts, Njombe region), SUA has set up a learning center at Lunyanywi village. The center is used to train farmers from different villages in Njombe region and beyond. Apart from the generous funding from the Norwegian government through the Norwegian Agency for Development Cooperation (NORAD), SUA has cooperated with the Norwegian University of Life Sciences (NMBU), the Tanzania Domestic Biogas Program (TDBP), Yara (Tanzania), Syngenta (Tanzania) and PANNAR Seed Company (Tanzania). This public–private farmer partnership has made co-learning both enjoyable and profitable.

10.6 REASONS FOR CHOOSING CONSERVATION AGRICULTURE

Declining yields and environmental problems are associated with many agricultural systems. According to the *FAO Year Book 1984* (FAO 1985), crop production levels in Tanzania are generally below potential, averaging about 905 kg ha^{-1} for maize and 458 kg ha^{-1} for beans. Many farmers associate these low yield levels with poor inherent soil fertility and continuous cultivation with few, if any, inputs. The work of Stoorvogel and Smalling (1990) indicated that Tanzania's arable soils lost nutrients at an average net rate of 27, 9 and 21 kg N, P$_2$O$_5$ and K$_2$O, respectively, per ha per annum in 1983. The rate of loss was projected to increase to 32, 12 and 25 kg N, P$_2$O$_5$ and K$_2$O, respectively, per ha per annum by the year 2000 if the production trends of 1983 were not reversed.

Most of the losses were associated with the harvesting of crops, the removal of residues and soil erosion. The amount of nitrogen (N) and phosphorus (P) removed from the soil every year by the main crops was estimated to be 251,448 Mg N, and 115,112 Mg P$_2$O$_5$ by the year 2000. Only 21% and 14% of the N and P removed, respectively, was projected to be replaced through fertilization. This implies a continued removal of fertility with a concomitant decline in soil productivity.

According to Benites and Ashburner (2003), one of the main causes of land degradation is the use of plough-based and hoe-based agricultural practices. These practices make soil denser and more compact, leading to decreases in organic matter content, while water runoff and soil erosion increase. Intensive tillage can destroy soil aggregates or clods and make the soil more vulnerable to erosion (Celander et al. 2003). Years of ploughing has led to the formation of a plough pan in many fields and this plough pan hinders both water and roots from penetrating deeper soil layers. In 1994/1995 a survey was carried out by Land Management Programme (LAMP) and revealed that compacted soils were a major problem in Babati District in Arusha, Tanzania, and that this was limiting for both water infiltration and crop growth. Another problem was sheet erosion that occurred between contour bunds due to the low infiltration capacity. This erosion caused low yields in the upper part of the field and higher yields in the lower part where sediments that had been washed on and carried from the upper part were deposited.

During the 1980s, poor harvests and total crop failures occurred frequently in Babati district due to regular crop water stress. The maize yields, for example, dropped by more than half. The water stress was caused by inadequate soil moisture even though the district has rather fertile soils and an average rainfall of 794 mm per annum (Figures 10.1 through 10.6).

In the late 1990s, northern Tanzania's Arusha region was encountering serious threats to farming and food security. As a result of crop failures and erratic rainfall, communities had increasing difficulties meeting their food needs through the year and were unable to pay for social services such as school fees and medicines. The fragile nature of the soil and land formation made the area highly vulnerable to erosion, which was eating into the fields. Herds of cattle roamed the area, eating

FIGURE 10.1   A Njombe farmer being trained by a Norwegian student to cut Rhodes grass using a scythe during the rainy season. Farmers can now harvest and dry the grass three times instead of the usual once per year. (Photograph by NA Urio.)

FIGURE 10.2   Composting using bio-slurry and feed remains. (After Mtakwa et al. 2017.)

FIGURE 10.3   Biogas and bio-slurry for a better life. (Source: Mtakwa et al. 2017.)

FIGURE 10.4   Compare maize production under CA on the right vs. normal practice left, in Njombe. (After Mtakwa et al. 2017.)

FIGURE 10.5   Yield transformation due CA adoption: on the left, maize before field was converted to CA (one year later) at Ibumila village in Njombe district. (Source: Mtakwa et al. 2017.)

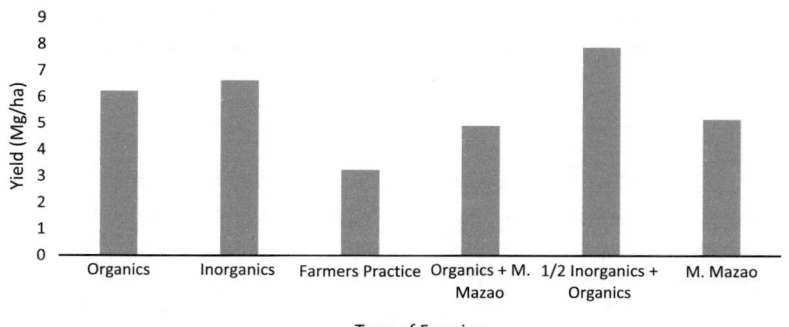

FIGURE 10.6   Average maize production over six villages where CA was introduced in Njombe region, indicating that using compost manure, bio-slurry and half the recommended rate of industrial fertilizer out-performed the other fertilizer regimes. (Source: Mtakwa et al. 2017.)

FIGURE 10.7 Loss of vegetation cover is a key trigger for soil erosion, which can be caused by natural factors (periods of drought) and human factors (land management decisions). Complex interactions between these two issues can lead to catastrophic soil loss. (Photograph: Carey Marks/Plymouth University.)

FIGURE 10.8 Conservation agriculture may be part of a solution to soil erosion, but for many communities this will require a steep change in land management practice and may question traditions that go back generations. Soil erosion is a complex, global challenge. (Photograph: Carey Marks/Plymouth University.)

whatever they could find. There were few trees left: the rest had been cut down for firewood. Poor farming practices, which failed to yield rewards for farmers, made soil erosion worse while the area experienced heavy soil structure destruction, soil fauna depletion, and soil moisture loss. The Jali Ardhi Project (Jali Ardhi is a Swahili word that means "take care of [or respect] the land"), funded by the UK government's Global Challenges Research Fund through the Natural Environment Research Council, explored the complex impact of soil erosion in East Africa. The photographs in Figures 10.7 through 10.8 10.9 10.10 demonstrate the soil issues in Northern Tanzania. Similar severe erosion features are also depicted in Figure 10.11, a picture taken by P.W. Mtakwa in 2017 in Kongwa district, Dodoma region (central Tanzania).

FIGURE 10.9 Conservation agriculture is not without its own issues. Here, the collapse of historic terraces from the colonial period has contributed to bad land formation. (Photograph: Carey Marks/Plymouth University.)

FIGURE 10.10 The global threats of climate change and population growth are key catalysts of the soil erosion challenge that is pushing these East African landscapes towards a tipping point. (Photograph: Carey Marks/Plymouth University.)

FIGURE 10.11 Gully erosion at Mlali, in Kongwa district, Dodoma. (Photograph: P.W. Mtakwa.)

10.7   APPROACHES FOR RESPONDING TO SOIL DEGRADATION

Various communities and projects in the different regions in Tanzania use different approaches to respond to their specific soil degradation challenges. Since this is a relatively new intervention, the promoters would first mobilize the community members. Often the promoters are government extension teams or belong to projects working in collaboration with government extension teams. Community members that voluntarily join the initiative are then involved in collective activities, including soil erosion control, environmental protection (reduction of gullies), income-generating activities and improving traditional agriculture to increase yield. Specific techniques may need intensive on-farm training, for instance in erosion-control technologies, such as tree planting (including creating tree nurseries to raise and care for tree seedlings), construction of soil and water conservation structures, such as collecting ponds, and building of contour bunds.

The initiatives also promote the adoption of sustainability measures such as rainwater harvesting techniques, crop rotation, and intercropping – legume intercropping, especially of sweet beans (*Lathyrus odoratus*) and lablab (*Dolichos lablab* – was new for the farmers and was highly popular, as it provided a very profitable and easy to sell crop). In some regions and projects, the approaches are more holistic and include offering training on, for example, how to construct improved cattle and goat sheds, or fodder-growing techniques including planting Guatemala grass on contours to serve as livestock feed and to stabilize the contours. To support livelihoods, the communities are helped to develop new income sources, such as beekeeping, vegetable production, and raising chickens. Beekeeping is also adopted as a coping mechanism to prevent livestock keepers from grazing in selected forest lots where the beehives are kept, and the bees scare the livestock as well as their herders away.

A common approach is the use of a central site where farmers can learn together called a field school. After learning collectively at a field school, farmers gradually adopt the information, and start by implementing at least one of the three principles of CA on their own land. The most popular practices are minimum soil disturbance (ripping, using no-till direct planters or jab planters) and keeping the soil covered (not burning crop residues, not allowing animals to graze freely, but planting lablab).

Some of the conservation agriculture practices that have been commonly adopted in Tanzania are as follows.

10.7.1   AGROFORESTRY

Agroforestry is an approach that has been commonly used in Tanzania to reduce land degradation. Tree planting in and around fields and around homesteads should give the households a source of fuel wood, fodder, shade and timber. Trees along the field can also mark the boundary, reducing the risk of others claiming the right to the land. The trees can increase the amount of organic matter in the soil and the roots bind the soil and can thereby decrease soil erosion. Trees can also act as a nutrient pump, bringing nutrients from greater depths up to the surface. In Northern Tanzania, farmers plant a variety of trees species, including silky oak (*Grevillea robusta)*, Egyptian river-hemp (*Sesbania sesban)*, apple-ring acacia, also known as white acacia *(Faidherbia albida)*, and Australian pine tree (*Casuarina equisetifolia)*, along contours and on field borders. Sometimes they also plant different kinds of fruit trees. In most villages in Babati district there are by-laws demanding that villagers must plant a certain number of trees per year on their land.

10.7.2   CONTOUR BUNDS

In hilly areas where fields are situated on slopes, erosion remains a great problem. Contour bunds have been an important start in improving the agriculture and conserving the soils. The first contour bunds occurred in the 1950s at the colonial farms and were also introduced by missionaries. In some areas villagers have tried to stop the erosion by constructing trenches and trying to build contours bunds, but these efforts have often not been enough. In the early 1990s, the Land Management

Program began to form village teams consisting of extension officers and farmers who were taught how to measure and build good contours, *Fanya juu* terraces where a trench is dug across the slope, throwing the soil upslope. The teams went around and measured contours and after that it was up to the farmer to build the contour bund. It was recommended to plant trees or grasses on the contour bunds to stabilize them. It was also important to keep cattle out of the field as they could easily destroy the contour bunds.

10.7.3 Subsoiling

Subsoiling is done to break the plough pan and by doing so increase infiltration of water into the soil and allow plant roots to penetrate the deeper soil layers. Subsoiling is often the entry-point to CA in most regions as the plough pan has to be broken before reduced tillage methods can be introduced. If a farmer does not use minimum tillage, the recommendation from the extension service is to subsoil the land every third year. Subsoiling is done with both oxen- and tractor-drawn implements. There are two tractor-drawn implements that farmers can access by hiring and the cost of subsoiling one hectare with a tractor is approximately 17,000 TSH (USD 8.5). Subsoiling can also be done with an implement that fits a standard plough or ripper beam (Bwalya 2003). The cost for using the animal drawn subsoiler is about half that of tractor-drawn implements. The subsoiling should be done when the soil is dry, or when only slightly moist, and, in most regions in Tanzania, subsoiling is usually done after harvest in September up till planting in December and January.

Formation of plough pan is related to the soil type and subsoiling is not recommended on all fields. Black cracking clays or sandy soils are two soil types where subsoiling is not recommended. Instead, the structure of the cracking clay soils should be improved by increased organic matter content and, if subsoiling is done on sandy soils, a water deficiency can occur as the water level is lowered when macro pores are formed.

10.7.4 Ripping

To reduce the soil disturbance, and keep and improve the soil structure, ripping has been introduced to replace conventional ploughing. The benefit with the ripper is that it only opens a narrow furrow every 75 cm and does not turn the soil, nor does it incorporate crop residues into the soil. In conventional farming the land is ploughed one to three times to get rid of weeds and prepare the seedbed. To use the plough, the farmer has to wait for the first rains to be able to start the activity. By using the ripper, the farmer can begin to prepare the land earlier, when the soil is still dry. A negative effect is that the reduced tillage leads to greater problems with weeds, and, to suppress the weeds, herbicides or soil cover are needed. Planting is done by dropping seeds in the furrow that the ripper creates. A planter attachment to the Magoye ripper (Figure 10.12) has now been introduced in Tanzania and has made direct planting possible. The ripper is also used for weeding between the rows and is then equipped with extended wings that cover the weeds with soil. As the ripper does not turn the soil, the draught power needed is also lower and only two oxen are needed instead of up to six as needed when ploughing with moldboard ploughs.

10.7.5 Soil Cover

There are many ways to create a soil cover and protect the soil from heavy rains and the heat and to suppress the weeds. A soil cover can be established with crop residues, cover crops, mulch and use of intercropping.

10.7.5.1 Intercropping

The most common cropping system in most of the case studies is intercropping, Figure 10.13, which is practiced by about 60% of the farmers in Tanzania. Pigeon pea (*Cajanus cajan*) is often

FIGURE 10.12   The Magoye ripper.

FIGURE 10.13   A healthy maize stand intercropped with sunn hemp under CA.

intercropped with maize and gives cover over a long time period; it is also a nitrogen-fixing crop and can therefore improve the soil fertility. Maize and pigeon pea are planted at the same time, either in separate rows or in the same row. The pigeon pea is also used as a weapon to prevent post-harvest grazing, as it is a crop that stays in the field for a long time, and cattle are barred from entering a field that has a crop. The introduction of pigeon pea intercropping came spontaneously as there was a good market for it. The common bean (*Phaseolus vulgaris*) is also used in intercropping, both with maize and with sunflower. When sunflower and beans are intercropped, sunflowers are planted with eight rows of beans in between. Some farmers in Babati district plant sunn hemp (*Crotalaria juncea*) around the fields with claims of benefits such as reduced attacks from maize stalk borer as the insect attacks sunn hemp instead of the maize.

FIGURE 10.14 CA weed-free maize at two weeks with crop residues of the previous maize crop used as mulch.

10.7.5.2 Crop Residues

After harvest, CA farmers now leave the crop residues in the field, Figure 10.14, to establish a soil cover to protect the soil from the sun and heavy rains. The soil cover is also said to preserve moisture and recycle nutrients. Usually the crop residues would be used as fodder for livestock and not left in the fields. However, the practice of leaving crop residues in the field is becoming more common, though there is still a demand for the residues as fodder.

The incorporation of the crop residues can also lead to an immobilization of nitrogen when the stover is decomposing. Farmers usually have to clear rows or parts of the field before tillage with the ripper as the crop residues otherwise clog the ripper. Other farmers slash the residues with a *panga* (machete) into smaller pieces to overcome the problem. A knife roller has been tried in some of the cases but the result was not satisfying as it only pressed down the residues and did not chop them.

10.7.5.3 Cover Cropping

The use of cover crops has many advantages. For example, a cover crop can give a soil cover, green manure and/or give the farmer a more secure yield, as two crops are cultivated. After the introduction of reduced tillage, mechanical weeding is reduced, and the farmer has to suppress weeds using other means: cover crops are one such way. Some cover crops are used traditionally, for example, pumpkin, calabash and watermelon. Today lablab (*Lablab purpureus*, previously called *Dolichos lablab*) and velvet bean (*Mucuna pruriens*) are promoted in most regions in Tanzania. For most small-scale farmers cover crops and green manure will be the least expensive way to increase the organic matter content of the soil significantly. Important characteristics of a good cover crop include vigorous growth, with a payload of up to 50 Mg biomass per hectare per year; high ability to control weeds, ability to grow well even under poor soil quality, ability to fix nitrogen and, where possible, serve as human food or animal feed. Additionally, a cover crop should neither constitute an added cost, nor should it bring major changes to the farmer's existing farming system. Apart from lablab and velvet bean, other popular cover crops in Tanzania include jack bean (*Canavalia ensiformis*), tephrosia (*T. vogelii*) and cowpea (*Vigna unguiculata*).

FIGURE 10.15 A vegetable farm covered with straw mulch.

10.7.5.4 Mulching

Mulching, in the meaning of bringing plant material into the field, is used mostly for small portions of land, for instance in seed nursery beds. It is also common in banana fields. Straws from maize, rice, millet and sorghum are suitable for soil cover as shown in Figure 10.15. Today the use of rice husks as a mulch material is gaining momentum in Tanzania. Mulch helps conserve soil moisture and, in the long run, soil fertility is improved. Mulching is done to curb the problems of drought and reduced soil fertility, which are the major problems for crop production in Tanzania, especially in the semi-arid areas of the country.

10.7.6 RIDGING

Ridging is mainly used for some root crops, for example, sweet potatoes and cassava. The ridges are done with hand hoes as demonstrated by Figure 10.16.

10.7.7 JAB PLANTING

The jab planter was introduced into Tanzania in 2003, through the Land management program (LAMP) funded by the Embassy of Sweden in Tanzania. However, in most of the case studies, the farmers used tractors and draught animals for seed drilling. In Babati district, the farmers consider the use of jab planters a step backwards because they already use planters driven by either oxen or tractors. Therefore, there is low adoption of the jab planter.

10.8 THE OUTCOME

The soil conservation efforts made by farmers in the various case studies demonstrate three key points:

(a) With good soil management, it is possible to improve soil fertility on steep slopes and grow crops sustainably.
(b) Although commercial farming is believed to reduce agro-biodiversity, conservation farming can still support commercial agriculture and conserve agro-biodiversity as is the case for Irish potato (*Solanum tuberosum*) (round) production in Olgilai/Ngiresi site (Kaihura et al. 2001).

FIGURE 10.16   A photo of F.M. Muniale making hand hoe ridges in a field trial site at Sokoine University for an experiment comparing ecosystem services between conservation agriculture and conventional tillage. (Photograph: P.W. Mtakwa.)

(c) Conservation farming structures cost little labor and have uses that include social welfare that is beyond the commonly advocated purposes of introduced erosion control measures. They create a good environment for the multiplication of soil biota and fauna, soil fertility and crop/cropping systems diversity. Although some farmers claim that such well-developed structures harbor mice and rats, their advantages far exceed the disadvantages.

10.8.1   Impacts of Conservation Agriculture on Yields

The majority of farmers in Tanzania have farming as their only source of income, therefore they are highly dependent on the yields that they produce. Changes in yields could therefore have a great impact on the farmer's life, especially for the small-scale farmers. When farmers changed from conventional agriculture to CA they often started by subsoiling their fields. Once the plough pan was broken the infiltration rate of water improved and plant roots could penetrate the deeper soil layers. After subsoiling the yields increased according to interviewed farmers and Jonsson et al. (2000). When the plough pan was removed the plant could assimilate nutrients from the deeper soil layers that had not been available for the crops earlier. After some seasons the yields tended to decrease as these nutrients were used up, something that underlines the importance of adding fertilizers to the fields (Jonsson et al. 2000).

The case studies show that CA has led to an increase in food security. In Northern Tanzania; Arumeru, Karatu and Bukoba districts, about 900 families were reached directly and 300 indirectly with CA initiatives. Food availability data analyzed for 144 households impacted by the CA-SARD project between 2004 and 2006 showed that the number of households suffering severe food short-age declined from 75% in 2004 to just over 50% in 2006. The overall food deficit period declined from four months to three months. Participating farmers found that their CA fields produced 50% more than their conventionally ploughed fields.

Crop production now provides a more viable business. Livestock produce manure (an average household has one cow, five goats and/or sheep, five chickens, and one donkey) that can be sold for quick cash needs. A maize-plus-lablab intercrop provides food for the household as well as income. In an average season, a household harvests 1,500 kg to 2,000 kg of maize grain from one hectare. A family with 1.2 ha harvests at least 4,500 kg of maize grain and sells about 40% of this harvest. The same family would harvest up to 600 kg of lablab beans. Lablab is ranked first among legumes because it is highly drought tolerant and grows well as an intercrop. In addition, the young leaves can be eaten as a green vegetable, whereas lablab beans can be made into *loshoro* (mixed with pounded maize plus milk) or *makande* (mixed with pounded maize only). Lablab is also used as a medicine to cure various illnesses, and by pregnant and lactating women (to produce more milk). Finally, lablab is a readily available fodder that women can harvest daily in fields to feed animals, saving time for other activities.

Crops grown with CA were less vulnerable to drought than those grown in the conventional fields. Rainwater seeped into the soil through the ripped lines, retaining soil moisture for longer. During the 2004 drought in Arumeru, even though adequate cover was not attained, farmers who had ripped their land and planted lablab with maize were able to harvest at least 500 kg to 2,000 kg of maize per hectare, while conventional farmers harvested nothing or less than 250 kg per hectare.

Efficient use of water was important for the farmers especially in the dry areas of northern Tanzania as they often saw scarcity of water as one of the main problems in their farming. When the ripper was used, farmers said that water infiltration increased and that they got higher and more secure yields. Contour bunds also had an impact on the yields, as water and nutrients were kept in the field, as the runoff was lower. Soil covered with crop residues or cover crops reduced the evaporation from the soil and kept it humid for a longer period. One farmer mentioned that when he began to intercrop maize with pigeon pea and started to use the ox-plough, maize yields increased from 750 kg to 2,000 kg/ha. He had now realized that this could be due to the N-fixation and that the pigeon pea taproots could improve the soil structure. Other farmers had seen that the yields of pigeon pea decreased due to disease and therefore used crop rotation and cultivated beans or other crops for some years and then cultivated pigeon pea again.

In the sociological survey in Kondoa district, 56% of 240 households surveyed perceived that crop yield had increased due to improved soil fertility in their farms. Maize yield per ha was 1.5 Mg and per capita maize production was 120 kg, which is above the Tanzanian national per capita average of 112 kg. The number of people with a relative surplus maize production is high because only 28% of respondents obtained 0.5–1.5 Mg/ha, while 49% of respondents obtained 2.0–3.0 Mg, 22% obtained 3.5–4.9 Mg/ha. The majority of households (68%) have sufficient food balance throughout the year. It was observed that there is a significant association between household food balance and the rate of reduction of soil erosion. Household food balance and soil fertility increase were also found to have a significant association. An increase in food production and income are also attributed to there being a growing number of livestock keepers over time. For example, in 1989 only 13% of people owned livestock in Kondoa's eroded areas (Mung'ong'o 1995) compared to 68% in 2010, however increased livestock numbers could have negative consequences for environmental degradation if the livestock population is not managed.

Yield prediction based on the common rainfall scenarios: below normal, normal and above normal in Mkoji, Mbeya, indicates that even in below normal seasons, an average yield as high as 4.0 Mg/ha and 4.5 Mg/ha with conventional tillage (CT) and conservation tillage (ox-ripping, with surface crop residues [ripping and residue, RR] treatments), respectively, can be obtained. However, the yield with CT treatment is much less stable, as indicated by a higher standard deviation (0.81 Mg/ha) as compared to (0.55 Mg/ha) in RR treatment, suggesting the suitability of RR treatment during low rainfall seasons. In normal rainfall scenarios it is expected to get yield of 5.0 Mg/ha regardless of a tillage treatment at about 0.44 Mg/ha standard deviation. This applies also for above normal seasons in which same level of yields (about 4.5 Mg/ha) are expected for both CT and RR treatments. However, the standard deviations for yields in RR treatment in such seasons vary more widely with a standard deviation of 0.52 Mg/ha as compared to 0.31 Mg/ha for CT treatment. Similarly, the question of the sensitivity of maize crop to waterlogging conditions in RR treatments may explain the variability.

Improved agricultural practices were recorded as one reason, among others, that farmers mentioned for improved yields. For instance, all the following had a positive impact on the yield: the use of certified and improved seeds; using crop varieties that were appropriate in the respective agro-ecological zones where the farmer was cultivating; planting with proper spacing instead of zigzag as before; and planting three seeds at each spot for a more secure germination. Earlier the use of farm yield manure had been limited by both knowledge and the fact that the livestock had been grazing over a wide area. Today zero-grazing stables makes the collection of farm yield manure much easier, and its use has improved the yields. Farmers who kept their livestock in zero-grazing stables also said that the milk production had increased and that the animals were healthier. The production increased as the animals were given better fodder, and the farmers also had better control than before over how much fodder each animal got.

10.8.2   Impacts of Conservation Agriculture on Labor

CA has resulted in less work. Ripping requires less labor than ploughing; it uses two people instead of three; and it can be done a lot faster, in only 2.5 to 5 hours per hectare versus two days for the same surface for ploughing. Without ploughing, the farmers can quickly sow their seed after the first heavy rain, which means an earlier harvest, thus avoiding the risk of drought at the end of the growing season. Those who practice conventional farming have to wait until the soil allows them to plough, resulting in a three -to seven-day delay. Before CA, the farmers in the reference case studies did most of the work on the farm themselves together with their family; during weeding and harvesting some farmers hired labor; and many farmers hired tractors or draught animals during land preparation. Laborers were either paid with money or given local brew at the pub in the village. When the crop was planted in straight lines, weeding and thinning had to be done by more skillful workers who were paid with money. Some of the newly introduced practices were more time-consuming than the old ones; for instance, planting in straight lines and with proper spacing made the planting more time-consuming than before.

In conventional farming, weeding was usually done twice in the case study areas. Now the first weeding was more laborious, as thinning had to be done. The second weeding was however done more easily as the crop stood in straight lines. With proper soil cover the impact of weeds was low and weeding could be done easily just by picking the weeds by hand according to a farmer. For fields infested with some of the more difficult weeds, for example, couch grass (*Digitaria abyssinica*), soil cover was not enough, and these fields had to be ploughed or sprayed. The Magoye ripper was used for weeding between the lines, and this made weeding less burdensome. It also gave the women less work, as the husband and oxen took over the weeding. Some farmers also got an extra income from weeding other people's fields.

For land where soil cover has been achieved, for instance in some farms in Karatu, weeding has been reduced from three times to once, and the task is eased as simply uprooting the weeds is sufficient in some fields. Before soil cover is established, managing weeds remains challenging. To avoid tilling farmers will scrap weeds or use herbicides. Table 10.3 shows the difference between conventional and conservation agriculture and their labor requirement.

TABLE 10.3

Labor Requirement by Different Activities in Conventional and Conservation Agriculture

Activity	Conventional agriculture		Conservation agriculture	
	Method	**Labor required**	**Method**	**Labor required**
Land preparation	Slash, heap burn, plough	3 persons × 6 days	Slash, spread	3 persons × 2–3 days
Seeding	Hand hoe	4 persons × 3 days	Jab planter	2 persons × 2 days
Weeding	Hand hoe	Not mentioned but easier	Scraping by *panga*	Not mentioned but takes more time

10.8.3 ENVIRONMENTAL IMPACTS OF CONSERVATION AGRICULTURE

Deforestation and erosion had been the main environmental problems in the case study areas. When agroforestry was introduced, the pressure on the natural forests decreased and the forests began to recover in most of the district. Using more economical stoves also reduced the demand for fuelwood. Farmers had seen less erosion after the establishment of contour bunds and after subsoiling, as the surface runoff decreased and soils were hindered from eroding from the fields.

Sedimentation in Lake Babati, which had become larger but shallower during the past decades, is expected to be under control following decreased erosion. Gully erosion had also damaged roads and fields in the area. Recently many villages introduced by-laws about tree planting, encouraging every household to establish on-farm tree lots. Soil conservation initiatives have contributed significantly in improving soil fertility and reducing soil erosion and, in turn, have improved crop productivity and livelihoods. Farmers in Kondoa Dodoma (68%) confirmed decreased rates of soil erosion, and another 70% of farmers there perceived that soil fertility has improved. Main indicators of increased soil fertility described by farmers included high yields, healthy vegetation, increased soil organic matter contents and soil depth.

The study in Lushoto district shows that soil fertility levels differ between various soil and water conservation technologies applied by farmers on degraded soils. Higher pH was observed under bench terraces and micro-ridges than under *miraba* and sections without any conservation measures. The SOC was higher under miraba than under bench terraces, micro-ridges and sections without any conservation measures. Total N content was higher under micro-ridges than under sections without any conservation measures.

The higher organic carbon content under miraba can be explained by the presence of grass strips that form miraba, which, on decomposition, contribute substantially to the organic carbon content in the soil. P content was higher under bench terraces and micro-ridges than under *miraba* and sections without any conservation measures, while Ca^{2+} content was higher under bench terraces than under miraba and sections without any conservation measures, and was higher under micro-ridges than under the sections without any conservation measures. Mg^{2+} content was higher under bench terraces and miraba than under micro-ridges and sections without any conservation measures. Generally, the soil fertility status in the studied soil and water conservation technologies followed the trend: bench terraces > micro-ridges > miraba > sections without any conservation measures. The studied soil and water conservation technologies in each slope position were on the same soil type – *Chromic Acrisol Profondic, Cutanic*). Thus the observed differences in soil properties have developed as a result of the conservation technologies' intervention. The higher content of most of nutrients under bench terraces is probably due to the fact that bench terraces (Figure 10.17)

FIGURE 10.17 Major soil and water conservation technologies in the Usambara mountains, Lushoto, Tanzania. (i) A & B = miraba; (ii) C & D = stone bunds; (iii) E = micro-ridges and bench terraces; (iv) F = bench terraces. (Courtesy: S.B. Mwango 2016.)

are nearly level surfaces supported by grass barriers that prevent soil nutrients from being washed out by runoff.

A similar observation was reported by Kyaruzi (2013) where bench terraces influenced soil chemical properties such as pH, total N, SOC, CEC, Ca^{2+} and Mg^{2+}. Micro-ridges (Figure 10.17) are spaced closely together, and are too small to resist heavy runoff in areas with very steep slopes, for example, the Usambara mountains in Tanzania. However, the furrows associated with micro-ridges act as reservoirs that prevent soil nutrients from being washed out by runoff. The higher soil fertility status under bench terraces and micro-ridges, when compared with miraba and sections with no conservation measures, can partly be explained by land-use and management practices, where bench terraces and micro-ridges are usually used for the cultivation of vegetables in which fertilizers such as urea and DAP are frequently applied. The low soil fertility status under miraba can be explained by the fact that the surfaces under miraba are not levelled, while the wide spacing of grass strips (Figure 10.17) provides a running track that accelerates runoff velocity, thereby washing away soil nutrients. This is strongly supported by Kaswamila (2013), who hypothesized that grass strip spacing is an important aspect in soil conservation planning; in other words, the closer the strips, the more effective they become and vice versa (Figures 10.18 and 10.19).

10.8.4 SOCIOECONOMIC IMPACTS OF CONSERVATION AGRICULTURE

The majority of households in the area where the studies were conducted were poor or middle level (Ringo et al. 2002). Many of the surveyed households were given support of different kinds from the

FIGURE 10.18   Reduced tillage in northern Tanzania using oxen plough. This reduces the total disturbance of the soil. It leaves the soil more compact with reduced erosion and increased soil organic matter. It also increases the N-mineralization and soil aggregation. The nearby plant leaves are kept between the furrows for further decomposition and formation of organic matter. Therefore, reduced tillage improves and conserves soil fertility. (Adapted from FAO 2006.)

FIGURE 10.19   Researchers from Norway, Tanzania and Zambia at a CA site in Zambia. Dr. Mtakwa is in the foreground explaining the project. (Photograph: NA Urio.)

promoting projects to adopt CA. The probability of poor households adopting CA seems to be low as the adoption of CA demands knowledge, time and money. In order to change to CA, the farmer has to invest in subsoiling, building of contour bunds, cover crops and equipment. To raise funds or take loans to make the investments poses a big challenge for most farmers. The only opportunity for a farmer to get a loan was to be a member of a registered group of some kind, and, through this group the farmer could then take a loan. This was possible only for farmers in certain villages, where savings and credit cooperative (SACCO) groups had been established.

CA also resulted in savings, as farmers reduced their use of inputs, and saved money by not having to buy expensive fertilizers and herbicides. Several farmers in Karatu district reported reduced costs of production by using CA technologies. No tilling and low weed infestations have lowered the cost of production. However, at the time of study, the socioeconomic effects of CA in the community were not significant. This is because adoption was still at an initial stage, and, in particular, the number of individuals who had adopted the technologies was still low and the adoption was partial in nature.

10.8.5 Conservation Agriculture and Improved Livelihood

An improved livelihood for small-scale farmers depended highly on the yields and the market for the agricultural products. A higher yield gave the farmer an opportunity to sell the surplus and so their income increased. The increased income was, for example, used for secondary education for the children, building new houses or buying oxen. Higher production also led to fewer worries about getting food, and the social climate improved as people became friendlier towards each other. The way that the new agricultural techniques were taught, in groups, made the relations between farmers better and now they were able to discuss agriculture with each other to a greater extent. Improved agricultural production as an outcome of extension officers' advice increased farmers' trust for the extension service and, as a result, farmers were now more frequently going to the extension officers for advice.

With regard to energy, the study in Kondoa revealed that 98% of households obtained fuel-wood from their own farms and homesteads, following efforts to establish agroforestry for soil conservation. In sub-Saharan Africa, women walk very long distances to obtain firewood, and the minimizing of that allows time to engage in other development activities. There is also a significant relationship between tree planting and a source of construction poles, indicating that increased tree planting also influences the increased availability of construction poles, and, therefore, improved housing. Due to intercropping, most households use fodder from their own farms while a few get fodder from their farms as well as off-farm sources. Only a small proportion of households completely collect fodder for livestock from off-farm or from the wilderness.

10.9 GENDER AND CONSERVATION AGRICULTURE IN TANZANIA

The success of CA depends not only on whether it can produce high yields or good profits. It also depends on local customs and culture, the way people think of farming and the unwritten rules they follow. CA is, therefore, affected by the different roles and special needs of women and men in Tanzanian communities.

In many villages, and in most families, men and women tend to be responsible for different things. They do different things, know different things and have different interests, priorities and needs. They have different backgrounds and experiences, and their families, neighbors and society as a whole expect men to fulfil certain roles and women to perform other roles. With regard to labor, in some places, for example, men may plough the land and sell the produce, leaving the women to do the planting, weeding and harvesting. In other places, women may do the land preparation. Men and women may each tend different fields and grow different crops. Men are often responsible for the crops that can be sold for cash, while women deal with the lower status crops the family eats every day. It may be

TABLE 10.4
Labor Division by Gender in Tanzania

Activity	Women (%)	Men (%)	Both (%)
Site cleaning	45.6	43.5	10.9
Land preparation by hoe	38.3	42.5	19.2
Land preparation by oxen	32.5	50.0	14.5
Sowing and planting	40.4	34.1	25.5
Weeding	69.3	12.3	18.4

culturally unacceptable (or at least unusual) for a woman to handle draught animals or drive a tractor. In general, women do more farm manual work than men, though this is often not recognized.

Decision-making at the household level continues to be male-dominated in all farming-related activities, even in those where women contribute the majority of the labor. However, joint decision-making is commonplace.

According to an FAO factsheet on agriculture in Tanzania, women are generally more involved in farming activities. Table 10.4 shows labor distribution by gender based on the Mogabiri extension project in Tanzania.

Once CA has been established on a farm, crop cover will slowly but surely eliminate weeding. In the establishment period, some work removing the weeds is required where herbicides are not used. Therefore, there is a significant labor reduction for women during weeding. Land preparation by hoe or oxen is mainly carried out by men. CA changes the labor input because there is only a little preparation needed that might use oxen or ridgers.

10.10   CHALLENGES OF CONSERVATION AGRICULTURE IN TANZANIA

Attaining food security and development goals at the household, national, regional and global levels requires a shift from conventional to more efficient and sustainable food production practices. The common farming practices by the majority of farmers in sub-Saharan Africa are characterized by extensive soil disturbance through ploughing, low use of inputs and exploitation of fragile lands. These land management practices lead to soil mining, severe soil degradation and declining levels of soil organic matter (World Bank 2008; ACT 2008; Benites 2008). As land continues to degrade, livelihood options for at least 485 million Africans also dwindle with it (TerrAfrica 2009). While soil and vegetation on the Earth's land surface store three times the carbon present in the Earth's atmosphere, land clearing and degradation in the quest for virgin land or conventional tillage turns this valuable carbon sink into a major source of greenhouse gas emissions. In Tanzania, there are a number of challenges facing the land and soil, including, and not limited to: increasing land pressure; poor or low adoption of various soil technologies over the years due to weak linkages between research, extension and farmers; limited soil data update and information; ineffective partnerships by both public and private actors, and fragmented efforts; lack of appropriate dissemination approaches; and inadequate market access to farmers' produce.

Major challenges are also seen in the lack of sustainability of the conservation activities that are introduced in the communities because of a recent policy decision to withdraw conservation investment by the central government of Tanzania. Since local capacity was not built to take over management responsibility, it has resulted in overgrazing, overcultivation and uncontrolled commercial logging. This has undermined the past successes and is now negatively impacting the sustainability of both environment and livelihood. Despite that, it can be concluded that soil conservation activities, including agriculture transformation in favor of agroforestry, and crop diversification have contributed to an improvement in land cover and ensuring environmental and socioeconomic sustainability in the area. The breakdown of centralized conservation activities can be addressed by enhancing local community-based participatory approaches, and this requires effort to enhance the socioeconomic factors that determine increased participation in soil conservation, such as extension services, education and awareness.

The FAO identified the following needs and priorities for sustainable soil management in Tanzania.

(1) Land use/soil and water management.
(2) Validation of appropriate technologies including soil and water conservation, soil fertility management, agroforestry, water harvesting, conservation agriculture and promotion of indigenous knowledge for enhancing productivity and production.
(3) Strengthening soil mapping and land resources inventory.
(4) Update of soil data and information.
(5) Environmental management and climate change resilience, which entails development of new tools for risk-based, sustainable soil management planning for different agroecological zones or farming systems, and research to address the loss of soil biodiversity due to commercial overexploitation and other factors.
(6) Strengthening research infrastructure (improved soil laboratories), funding and human resources.
(7) Enhancing technology transfer and partnership with: promotion and creation awareness of stakeholders on sustainable soil management practices and formation of stakeholder platforms at national level, for example, Tanzania Soil Health Consortium (TASHCO) to enhance linkage among stakeholders and networking within and outside the country.
(8) Regular monitoring, evaluation of impact assessment of land degradation and interventions, as well as adoption of sustainable soil management practices.

10.11 CONCLUSION

Although CA is being slowly adapted and or adopted in Tanzania, it seems to be the only way forward if the country is to realize sustainable agricultural production.

It is high time we did more to safeguard and strengthen our capacity for securing the health of our soils in order to attain sustainable and quality production systems through conducive policies, strong partnerships, increasing funding as well as wise use of other resources. Understanding of the soil and addressing its problems could help us deal with other major global challenges including climate change and its impacts and the accelerating loss of soil biodiversity. Tanzania will continue establishing innovative platforms at different levels in order to raise awareness of communities and individual users of soils about the importance of healthy soil and to advocate sustainable soil management practices. More specifically, conservation farming has great potential in Tanzania for soil management and food production. Integrating crops and livestock production systems seems to be a plausible way to realize sustainable conservation agriculture: a key to combatting land degradation in Tanzania. Through that integration, Tanzania will ensure affordable clean energy, reduce inequality, improve livelihoods and foster economic growth, thus contributing to the sustainable development goals as suggested, in part, by the FAO (2018).

REFERENCES

ACT. 2008. *Linking production, livelihoods and conservation.* Proceedings of the Third World Congress on Conservation Agriculture, African Conservation Tillage Network, Nairobi, Kenya, 3–7 October, 2005. *www.act-africa.org/lib.php?com=5&com2=20&res_id=62*

Anderson, D. 1984. Depression, dust bowl, demography, and drought: The colonial state and soil conservation in East Africa during the 1930's. *African Affairs* 83(332):321–344.

BACAS (Bureau for Agricultural Consultancy and Advisory Services). 1996. Land conservation programme for Southern Highlands zone. In: *Consultancy Report Prepared for IFAD (International Fund for Agricultural Development) Southern Highlands Extension and Rural Financial Services Project 2007.* Morogoro, Tanzania: Sokoine University of Agriculture.

Benites, J.R.2008. Effect of No-Till on conservation of the soil and soil fertility, pp. 59–72. In: *No-Till Farming Systems.* Special publication, no. 3, Goddard, T., Zoebisch, M.A., Gan, T.T., Ellis, W., Watson, A., Sombatpanit, S. (eds.). Bangkok, Thailand: World Association of Soil and Water Conservation.

Benites, J.R., and Ashburner, J.E. 2003. FAO's role in promoting conservation agriculture. In: *Conservation Agriculture–Environment, Farmers Experiences, Innovations, Socio-Economy, Policy,* García-Torres, L., Benites, J., Martínez-Vilela, A. and Holgado-Cabrera, A., (eds.). Dordrecht: Netherlands: Kluwer Academic Publishers, pp. 139–154.

Batjes, N.H., 2011. Soil organic carbon stocks under native vegetation–Revised estimates for use with the simple assessment option of the Carbon Benefits Project system. *Agriculture, Ecosystems & Environment, 142*(3-4), pp.365–373.

Bolliger, A., Magid, J., Carneiro, A.T.J., Skorra, N.F., dos Santos Ribeiro, M.F., Calegari, A., Ralisch, R., and de Neergaard, A. 2006. Taking stock of the Brazilian 'zero-till revolution': A review of landmark research and farmer's practice. *Advances in Agronomy* 91:48–110.

Branca, G.L., Lipper, L., McCarthy, N., and Jolejole, M.C. 2013. Food security, climate change, and sustainable land management. A review. *Agronomy for Sustainable Development* 33(4):635–650.

Bwalya, M. 2003. *Conservation Tillage Implements for Smallholder Farming Systems.* ACT Information Series No. 6. Harare, Zimbabwe: African Conservation Tillage Network. www.act.org.zw/infoseries. html.

CA SARD Project (2009). Six monthly report July- December 2009; Conservation Agriculture for Sustainable Agriculture and Rural Development. Selian Agricultural Research Institute (SARI), Arusha, Tanzania.

Celander, T., Sibuga, K., and Lunogelo, P. 2003. *Completion of a Success Story or an Opportunity Lost? An Evaluation of the Soil and Water Conservation Programme in Arusha Region (SCAPA) Sida Evaluation 03/13.* Stockholm, Sweden: Department of Natural Resources and Environment. Report available at www.sida.se

Deckers, J.A., Driessen, P.M., Nachtergaele, F.O., and Spaargaren, O.C. 2002. *World Reference Base for Soil Resources.* Encyclopedia of Soil Science. New York: Marcel Dekker - ISBN 0824705181 - p. 1446–1451.

do Prado, W.L., Hercilio de, F.V., and McGuire, M. 2004. Use of green manures/cover crops and conservation tillage in Santa Catarina, Brazil. In: *Green Manure/Cover Crop Systems of Smallholder Farmers. Experiences from Tropical and Subtropical Regions,* Eilitta, M., Mureithi, J., Derpsch, R. (eds.). Dordrecht, Netherlands: Kluwer Academic Publishers, pp. 1–36.

Duru, M., Therond, O., and Fares, M. 2015a. Designing agroecological transitions. *Agronomy for Sustainable Development* 35(4):1237–1257.

Duru, M., Therond, O., Martin, G., Martin-Clouaire, R., Magne, M., Justes, E., Journet, E., Aubertot, J., Savary, S., Bergez, J., and Sarthou, J.P. 2015b. How to implement biodiversity-based agriculture to enhance ecosystem services. *Agronomy for Sustainable Development* 35(4):1259–1281.

Erenstein, O. 2003. Smallholder conservation farming in the tropics and sub-tropics: A guide to the development and dissemination of mulching with crop residue and cover crops. *Agriculture, Ecosystems and Environment* 100(1):17–37.

FAO. 1985. *FAO Yearbook 1984—The State of Food and Agriculture.* FAO Agriculture Series 18. Available at: www.fao.org/3/a-ap664e.pdf.

FAO. 2006. *Conservation Agriculture for SARD and Food Security in Southern and Eastern Africa (Kenya and Tanzania).* Rome, Italy: Food and Agriculture Organization of the United Nations (FAO). Available at:http://www.ipcinfo.org/fileadmin/user_upload/oed/docs/GCPRAF390GER_2005_ER.pdf

FAO. 2010. *What Is Conservation Agriculture.* Rome, Italy: Food and Agriculture Organization of the United Nations (FAO). Available at: www.fao.org/conservation-agriculture/overview/what-is-conservation-agriculture/en/.

FAO. 2012. *Conservation agriculture and sustainable crop intensification.* Integrated Crop Management. Vol. 15 . Rome: Food and Agriculture Organization of the United Nations. http://www.fao.org/3/a-i2643e.pdf

FAO. 2018. *Transforming Food and Agriculture to Achieve the SDGs: 20 Interconnected Actions to Guide Decision Makers.* Rome: Food and Agriculture Organization of the United Nations, 76 pp.

IIRR and ACT (International Institute of Rural Reconstruction and Africa Conservation Tillage Network). 2005. *Conservation Agriculture: A Manual for Farmers and Extension Workers in Africa.* Nairobi, Kenya: IIRR.

Jonsson, L.O., Singish, M.A., and Mbise, S.M.E. 2000. Dry land farming in Tanzania. Experiences from the Land Management Program. In: *Conservation Tillage for Dryland Farming. Technological Options and Experiences in Eastern and Southern Africa*, Biamah, E.K., Rockstrom, J., Okwach, G.E. (eds.), RELMA, Nairobi. Workshop Report number 3, pp. 96–113.

Kaihura, F.B.S., Stocking, M., and Kahembe, E. 2001. Soil management and agrodiversity: A case study from Arumeru, Arusha, Tanzania. In Proceedings of the Symposium on Managing Biodiversity in Agricultural Systems, Montreal, QC, 8–12 November. International Institute for Sustainable Development (IISD, Montreal, Canada

Kakeya, M., Takamura, Y., Mattee, A.Z. 1998. *Integrated Agro-Ecological Research of the Miombo Woodlands in Tanzania: Final Report*. Japanese International Cooperation Agency (JICA), 413 pp. Tokyo, Japan.

Kaswamila, A.L. 2013. Assessment of the effectiveness of soil erosion control measures using soil surface micro topographic features in the West Usambara Mountains, Tanzania. *International Journal of Marine, Atmospheric & Earth Sciences* 1(2):68–80.

Kimaro, A., Mpanda, M., Rioux, J., Aynekulu, E., Shaba, S., Thiong'o, M., Mutuo, P., Abwanda, S., Shepherd, K., Neufeldt, H., and Rosenstock, T.S. 2015. *Is Conservation Agriculture 'Climate-Smart' for Maize Farmers in the Highlands of Tanzania?* Berlin/Heidelberg: Springer.

Kyaruzi, L.A. 2013. Relationship between soil and land form derived land qualities and conservation agriculture practices in West Usambara Mountains, Tanzania. MSc dissertation, Sokoine University of Agriculture, Morogoro. 140 pp.

Lal, R. 2015. A system approach to conservation agriculture. *Journal of Soil and Water Conservation* 70(4):82A–88A.

Liaudanskienė, I., Šlepetienė, A., Šlepetys, J., and Stukonis, V. 2013. Evaluation of soil organic carbon stability in grasslands of protected areas and arable lands applying chemo-destructive fractionation. *Zemdirbyste* 100(4):339–348.

Lobell, D.B., Burke, M.B., Tebaldi, C., Mastrandrea, M.D., Falcon, W.P., and Naylor, R.L. 2008. Prioritizing climate change adaptation needs for food security in 2030. *Science* 319(5863):607–610.

Löfstrand, F. 2005. Impacts of conservation agriculture for small-scale farmers and methods for increasing soil fertility. MSc Dissertation, Swedish University of Agricultural Science, Uppsala 84 pp.

Mkoga, Z.J., Tumbo, S.D., kihupi, N., and Semoka, J. 2010. Extrapolating effects of conservation tillage on yield, soil moisture and dry spell mitigation using simulation modelling. *Physics and Chemistry of the Earth, Parts A/B/C* 35(13–14):686–698.

Mkonda, M., and He, X. 2017. *Conservation Agriculture in Tanzania*. Springer International Publishing. Switzerland 2017 309 E. Lichtfouse (ed.), *Sustainable Agriculture Reviews, Sustainable Agriculture Reviews* 22, DOI 10.1007/978-3-319-48006-0_10 pp.309-324

Msita, H.B., Kimaro, D.N., Mtakwa, P.W., Msanya, B.M., Dondyene, S., Poesen, J., and Deckers, J. 2012. Effectiveness of Miraba an Indigenous Soil and Water Conservation Measures on Reducing Runoff and Soil Loss in Arable Land of Western Usambara Mountains. *Geophysical Research Abstracts* Vol. 14, EGU2012-14122-2, 2012 EGU General Assembly of the European Geosciences Union (EGU) 2012, Vienna, Austria.

Msolla, M., Hansen, J., and Bjoemshave-Hansen, A. 1997 Sustainable agriculture as a component of watershed management. In: *Proceedings of Danida's 2nd International Workshop on Watershed Development*, 26 May–5 June. Power, S., Makungu, P., Qaraeen, A. (eds.), Iringa, Tanzania. pp465-482

Mtakwa, P.W., Urio, N.A., Mtakwa, A.P., Eik, L., and Kurwijila, L. 2017. *Crop Production and Livestock Keeping Can Work together: Case of the Crop and Livestock Integrated Production Model in Conservation Agriculture in Njombe and Wanging'ombe Districts*, Njombe Region, Tanzania, in preparation.

Mung'ong'o, C.G. 1995. *Social Processes and Ecology in the Kondoa Irangi Hills Central Tanzania*. PhD dissertation. Stockholm, Sweden: Department of Human Geography, Stockholm University.

Mwango, S. B., Msanya, B.M., Mtakwa, P.W., Deckers, J., Poesen, J., Meliyo, J.L., and Dondeyne, S. 2015. Soil fertility and crop yield variability under major soil and water conservation technologies in the Usambara Mountains, Tanzania. *Journal of Scientific Research and Reports* 5(1):32–46.

Owenya, M., and Semlowe, M. 2006. *The Eotulelo Farmer Field School: Learning and Promoting Conservation Agriculture*. Rome, Italy: Food and Agriculture Organization of the United Nations (FAO). Available at:.http://afsafrica.org/wp-content/uploads/2015/11/Conservation_Agriculture_Ta nzania.pdf (accessed October 13, 2014).

Ringo, D.E., Maguzu, W.C., Mariki, W., Owenya, M., and Njumbo, F.S. 2002. Conservation agriculture as practised in Tanzania – Karatu District case study. In: Shetto, R., Owenya, M. (eds.). *Conservation Agriculture as Practised in Tanzania: Three Case Studies*. Nairobi, Kenya: African Conservation Tillage Network; Centre de Coopération Internationale de Recherche Agronomique pour le Développement; Food and Agriculture Organization of the United Nations.

Scopel, E., Triomphe, B., Séguy, L., dos Santos Ribeiro, M.F., Denardin, J.E., and Kochhann, R.A. 2004. Direct seeding mulch-based cropping systems (DMC) in Latin America. Communication Presented at the 4th International Crop Science Congress, Brisbane, Australia, 26 September–1 October, https://www.food grainsbank.ca/uploads/Direct%20seeding%20mulch-based%20cropping%20systems%20(DMC) %20in%20Latin%20America,%20Eric%20Scopel%20et.%20al.,%202001.pdf.

Shetto, M.R., and Lyimo, M.G. 2001. Conservation agriculture in Tanzania. LAMP Planning Workshop on Conservation Tillage, Babati, Tanzania, pp. 19–21. African Conservation Tillage Network (ACT), Nairobi.

Shetto, R., and Owenya, M. (eds.). 2007. *Conservation Agriculture as Practised in Tanzania: Three Case Studies*. Nairobi, Kenya: African Conservation Tillage Network, Centre de Coopération Internationale de Recherche Agronomique pour le Développement, Food and Agriculture Organization of the United Nations.

Shrestha, R.P., and Ligonja, P.J. 2015. Social perception of soil conservation benefits in Kondoa eroded area of Tanzania. *International Soil and Water Conservation Research* 3(3):183–195.

Stoorvogel, J.J., and Smaling, E.M.A. 1990. *Assessment of Soil Nutrient Depletion in Sub-Saharan Africa: 1983–2000*. Wageningen, The Netherlands: The Winand Staring Centre for Integrated Land, Oil and Water Research. Report 28, 4 Volumes; Volume 1, pp. 96–97.

Taruvinga, R.P. 1995. Report on an environmental profile of the Southern Highlands, Tanzania. In: *Consultancy Report Prepared for International Fund for Agricultural Development (IFAD)*, Southern Highlands Extension and Rural Financial Services Project, Mbeya, Tanzania, pp. 42–245.

Temu, A.E. 1996. Baseline survey for the Njombe District Natural Resources Conservation and Land Use Management Project. Mbeya, Tanzania: Agriculture Research Institute Uyole.

Terr Africa. 2009. Sustainable land management in Africa: Opportunities for increasing agricultural productivity and greenhouse gas mitigation. TerrAfrica Climate Brief No2. PP. 1–2.

Triomphe, B. 2005. CA in theory verses in practice, as a set of technologies verses as an innovation process: Lessons, gaps and challenges from selected experiences around the world. Published in; ACT, 2008, Linking Production, Livelihoods and Conservation; Proceedings of the Third World Congress on Conservation Agriculture, 3–7 October, 2005, Nairobi. African Conservation Tillage Network, Nairobi.

URT (United Republic of Tanzania). 2001. *Agricultural Sector Development Strategy*. Dar es Salaam, Tanzania: Tanzania Government.

URT (United Republic of Tanzania). 2003. *Agricultural Sector Development Programme: Framework and Process Document. Final Draft*. Dar es Salaam, Tanzania: Tanzania Government.

World Bank. 2008. *World Development Report, 2008. Agriculture for Development*. Washington, D.C.: Oxford University Press for the World Bank.

11 The Storage of Organic Carbon in Dryland Soils of Africa
Constraints and Opportunities

Brahim Soudi, Rachid Bouabid, and Mohamed Badraoui

CONTENTS

11.1 INTRODUCTION

By definition, drylands include arid, semi-arid and dry sub-humid zones. In the continent of Africa, drylands represent about 43% of the total surface area and cover more than 70% of agricultural land. About 50% of the African population lives in these areas and is significantly fragile in terms of food insecurity. Climate change, which is expected to increase the frequency and severity of extreme weather events, will exacerbate the vulnerability of these lands if effective adaptation and resilience action is not undertaken. The livelihoods of most dryland populations depend on natural resource-based activities, such as agriculture and livestock. Forced to meet urgent short-term needs, households resort to unsustainable practices, resulting in strong pressure on the natural resources, loss of biodiversity and severe soil degradation. One of the crucial factors considered as a cause and a consequence of agroecosystem vulnerability in dry areas is soil degradation (Biancalani et al. 2015). Various interacting processes of such degradation include water and wind erosion, salinization and loss of soil fertility, mainly the decrease of its organic matter (OM) content. These processes lead to a decline of the soil health and productivity, as well as its capacity to contribute to the reduction of carbon (C) emissions into the atmosphere.

Studies on the understanding of the dynamics of soil organic matter (SOM) in different soil and climate contexts of the world have been extensively addressed, and their scientific bases remain current (Hénin and Depuis 1945; Hénin et al. 1959; Jenny 1941; Laudelout 1993; Bremer et al. 1995; Janzen et al. 1997). Hénin and Dupuis (1945) developed, for the first time, balance equations for SOM decomposition. Hénin and Monnier (1959) addressed the physical and biochemical determinants of the dynamics of SOM, and Laudelout et al. (1960) established quantitative relations between the content of SOM and climate. These and many other research studies were motivated by the need to understand soil genesis, to improve soil properties and productivity, and to elucidate the interrelationships among plant nutrition and C and nitrogen (N) biogeochemical cycles (Campbell et al. 1984). Studies included laboratory and field experiments involving long-term trials and modeling approaches. More recently, Campbell and Keith (2015) conducted a fairly comprehensive literature review on developments in SOM modeling. The current concerns related to climate change drive the interest to mitigate greenhouse gases (GHG) and adaptation to climate change in soils, more precisely through soil organic carbon (SOC) sequestration (Lal 2004).

This chapter discusses the status of SOC in the dryland agricultural soils of Africa and looks at some of the underlying issues, which remain relevant despite the abundant literature on this topic.

11.2 SOIL ORGANIC CARBON STATUS

Organic matter plays various important roles in soil properties and behavior. It is a major soil fertility and quality indicator and has long been closely related to the capacity of soils to sustain crop growth (FAO 2005). OM is the main source of energy for sustaining the soil living organisms and is an important source of nutrients for crops. It improves soil chemical, physical and biological properties and acts as a protective factor against sealing and erosion. OM is an important soil moisture and temperature conditioner, which make it a valuable soil component under dryland conditions (Plaza-Bonilla et al. 2015). In the context of global climate changes, SOM represents a major reservoir of C and is advocated as an important sink for atmospheric C sequestration (Lal 2004).

The increase and maintenance of "adequate" amounts of SOM, that is, SOC, is of major concern for farmers and scientists, especially in the context of the recent factual climatic changes. In arid and semi-arid conditions, low SOM contents are related to low biomass productivity and low returns as a result of limiting climate conditions, low water and nutrient use and efficiencies, and non-appropriate farming practices (Jarecki and Lal 2003; Lal 2004; Lal 2005). It is estimated that SOC content in drylands is lower than 1%, and in many areas, it does not exceed 0.5% (Lal 2002; 2004). The stock of SOC pools is mainly present in the top layer of the soil and declines rapidly with depth. Such conditions unavoidably contribute to soil fertility decline and alter the various soil functions. In dryland African soils, the level of SOC is, in general, low to very low (Figure 11.1A) and exhibits a close relationship with the aridity of the climate (Figure 11.1B).

Enhancing OM levels in agricultural lands is a major issue, in general, and it is even more serious in dryland soils. OM decomposition depends on two natural processes, humification and mineralization. The importance of each of these processes depends greatly on climate and soil conditions, the nature of OM, and the agricultural practices (Bot and Bernites 2005). Humification leads to the

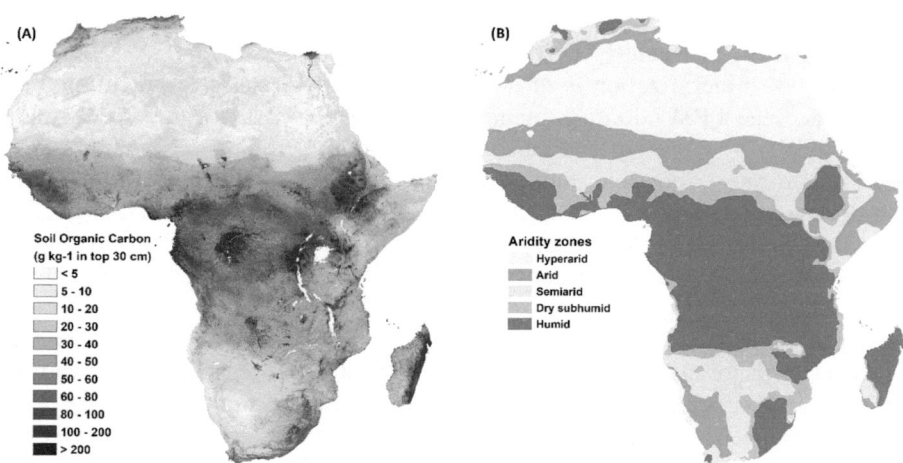

FIGURE 11.1 (A) Soil organic carbon in topsoil in Africa, (B) aridity zones in Africa. (Drawn using data from Soil Grid/ISRIC, 2018; www.soilgrids.org; Adapted using data from Millennium Ecosystem Assessment, 2018; www.millenniumassessment.org.)

formation of stable humus, while mineralization leads to the rapid decomposition of the labile fraction and part of the stable humus, with the release of various mineral constituents (Guggenberger 2005). The processes of humification and mineralization coexist, and the importance of one over the other depends on several factors, mainly soil moisture and aeration, temperature conditions, and C/N ratio.

The crucial issue for drylands lies in the predominance of mineralization over humification. The question is how to reverse the trend of loss of SOM occurring at variable rates and magnitudes? Scientifically, how to act on these antagonistic mineralization-humification processes in order to favor more stable humus, that is, C sequestration, while taking into account the temperature rise (favorable to mineralization) due to climate change, extreme events (mainly droughts), low nutrient input (N and P), low biomass production, low residue return, etc.? It is theoretically accepted and scientifically proven that the storage of C in the soil contributes to reducing GHG emissions (Lal 2004). Although global carbon mass balances are well established, in order to move from theory to practice and to make this concept a reality, it would be necessary to make drastic changes in terms of farming systems as well as in terms of farmer attitudes and practices. In parallel, a series of questions still under discussion at international level, need to be objectively addressed: (i) What is the maximum level of SOC storage that is biochemically feasible in dry agroecological zones of Africa? (ii) What is the time required for C sequestration taking into consideration the amount of fresh OM incorporated in the soil and the simultaneous half-life of humification and mineralization processes? (iii) What are the most adapted agroecological innovations in favor of C sequestration in the conditions of African agriculture (physical and socioeconomic)? and (iv) What are the perception and the degree of acceptance of small farmers with respect to these innovations?

Humification vs. mineralization under dryland farming is a "dilemma" that needs particular attention. When the process of humification is dominant, SOM decomposition results in appreciable amounts of stable humus that play a major role in the edification of the soil fertility; humus build-up will contribute to C accumulation and therefore to C sequestration, depending on the rate of return to an equilibrium state, which depends in turn on various local climatic and biophysical factors. When the process of mineralization is dominant, SOM decomposition results in high amounts of mineral constituents, which represent an important source of nutrients for crops, even though, on the other hand, a continuous depletion of OM takes place. This is particularly true when the organic residues returned to the soil are rich in N (i.e., of low C/N). The same process applies for organic amendments, such as N-rich farm manures, as well as for residues from legume crops. In drylands, where crop seasons are characterized by high temperatures and appreciable precipitation, annual SOM depletion is very important (FAO 2004).

Humification vs. mineralization is of a major debate with regard to potential C storage in soils. For instance, the 4 PM initiative (mentioned above) advocates for the objective to increase SOC stocks globally by 0.4% per year, as a way to mitigate climate change and improve food security. If such a rate is possible under humid and temperate regions, it is very difficult to achieve in dryland agricultural soils of Africa, mainly due to the low recycling of OM in most areas, as well as to the dominance of the mineralization process. Carbon storage is potentially possible under various African agro-ecosystems, but its rate might be very variable (Corbeels et al. 2018).

Therefore, SOM management as a key component of soil fertility and as a potential sink of C sequestration faces major challenges in the dryland agriculture of Africa from the various dimensions of climate and pedoclimate conditions, cropping systems and farmers' practices. The interactions among these dimensions vary from one region to another and create a multitude of situations that need to be considered at the local scale. Managing SOM in arid, subsistence smallholdings of Africa would be completely different from that of arid, large farms of North America. As the agroecosystems in the African continent are very diverse, the rates of C increase in African soil expected from the innovations advocated by various initiatives need to be adapted to the local contexts.

11.3   KINETICS OF SOIL ORGANIC CARBON EVOLUTION

The evolution of SOC is not a linear process, but operates according to first-order kinetics. The basic equation established in early studies (Hénin and Depuis 1945; Hénin et al. 1959; Jenny 1941) and still in use in recent models is expressed as follows:

$$dC / dt = -k.C \qquad (11.1)$$

where C is the carbon content, t is time and k is the evolution rate (accumulation or depletion). The integration of this equation from t_0 to t, respectively corresponding to C_0 and C, gives the following equation:

$$C = C_0 . e^{(-kt)} \qquad (11.2)$$

The half-life $t_{1/2}$ corresponds to the time required for the accumulation or the depletion of 50% of the initial C_0. $t_{1/2}$ can be calculated through the log transformation of equation (2). For a situation of depletion, at $t_{1/2}C = C_0/2$, therefore,

$$\ln(C / C_0) = -k. t_{1/2} \text{ and } t_{1/2} = \ln(2)/k = 0,69/k \qquad (11.3)$$

This equation shows that the theoretical humification or mineralization half-life of the initial C content (C_0) occurs in terms of years depending on soil and climate conditions. Various studies found that $t_{1/2}$ can vary from a few years to several decades. Table 11.1 shows that values of $t_{1/2}$ ranged from about 2 years for a k rate of 0.33 yr^{-1}, in the conditions of Zaire for a rice (*Oryza* spp.) maize (*Zea mays*) rotation (Laudelout and Meyer 1951), to about 13 years for a k rate of 0.0608 yr^{-1}, in the conditions of Ohio for a continuous maize (Jenny 1941). Estimated values of half-lives from other studies of steady-state SOC decomposition are in the range of 5 to 20 years (Hans and Evans 1957; Mann 1986; Davidson and Ackerman 1993; Bremer et al. 1995;, West and Post 2015). In drylands, the SOC evolution half-life $t_{1/2}$ would vary from 5 to 10 years depending on climatic and biophysical conditions.

TABLE 11.1

SOM Decomposition Rates (k) and Corresponding Half-Lives for Various Cropping Systems

k yr^{-1}	Half-life (years)	Cropping system/region
0.33	2.1	Maize rice/Zaire
0.0608	11	Continuous maize/Ohio, US
0.052	13	Quinquennial rotation/Ohio, US
0.07	10	Continuous and alternate cereals with fallow/Kansas, US
0.22	3.1	Fallow/Puerto Rico

Source: Laudelout, 1993.

11.4 SCENARIOS OF SOIL ORGANIC CARBON EVOLUTION

Three scenarios are possible for the evolution of SOC as illustrated by Figure 11.2.

SCENARIO 1

This scenario corresponds to the situation where an exponential decrease in SOC takes place as a result of factors causing low OM return and/or intense mineralization. This is the situation that applies most for the drylands of Africa. The depletion of SOC starts when negative disturbance factors take place (e.g., cultivation after deforestation, intensive cultivation) and tends toward a minimum steady state level ($C_{(b)}$) if practices are kept to "business as usual." This evolution follows an exponential trend, and the rate of decrease following disturbance depends on the type and intensity of the depleting factors. In this case, mineralization becomes the dominant process, especially with favorable soil moisture and temperature conditions allowing high microbial activity. Such a trend is described and corroborated by several studies of SOM mineralization and its seasonal variations in arid irrigated rainfed areas (Berdai et al. 2002; Soudi et al. 1990). This scenario is the most likely to occur in dryland farming of Africa, and is expected to intensify even more with global warming if appropriate measures are not undertaken to reverse it.

SCENARIO 2

This scenario corresponds to a situation of a "natural" equilibrium where the SOC balance remains relatively stable and the C stock stays in the range of initial C_0 before disturbance. This equilibrium

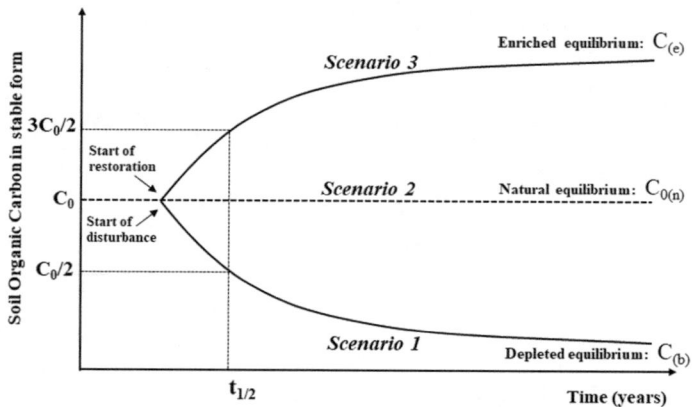

FIGURE 11.2 Scenarios of soil organic carbon evolution in soils. $C_{(b)}$ = f(climate, soil properties, management practices business as usual, disturbance factors); $C_{0(n)}$ = f(climate, soil properties, basic management practices adopted, no disturbance factors); $C_{(e)}$ = f(climate, soil properties, appropriate management practices adopted).

is driven by climate conditions, soil properties and basic agricultural practices that are not inducive of any significant loss. This situation corresponds to a relatively stable soil quality and to a zero growth of SOC. The amplitudes of change with time are small and the average balance $C_{0(n)}$ stays, over time, around the initial C_0. This scenario can be encountered in agroecosystems that have not been subject to much pressure or in cases where anthropic activities are conducted in harmony with available resources.

SCENARIO 3

In the contexts that favor the formation of stable humus, related to land-use changes, innovative and adapted agricultural practices (residues input and management, rotations, etc.), and/or soil properties (high charge clays and oxyhydroxides, calcium carbonates, etc.) (Lal 2016), an increase of SOC is expected. This scenario also corresponds to an exponential increase of stable SOC as a result of the dominance of humification, and a new enriched steady-state (equilibrium) $C_{(e)}$ level is reached depending on the factors involved. The $C_{(e)}$ would also depend on the climatic conditions, but much more on the quantities and the nature of biomass or crop residues returned to the soil and the activities that favor their accumulation. This situation is more likely to occur in humid temperate regions with high SOM return, but is unlikely to occur in dryland cultivated soils (such as those of Africa), where the restitution of OM is generally very low (a few tons per hectare), crop residue management is often inappropriate or absent and the rates of mineralization are high.

11.5   WHAT ARE POSSIBLE MARGINS FOR CARBON STORAGE IN SOILS OF DRYLANDS?

The majority of dryland soils in Africa are low in SOC and are even experiencing an increasing loss of OM (Lal 2004; FAO 2004; 2005). This situation corresponds to "trend a" (Figure 11.3), where the residual SOC is often stabilizing at a critical minimum ($C_{(b)}$). Assuming that a recovery of SOC is triggered from a given point on the curve of depletion, the increase of SOC (rate and maximum storage capacity) would depend on the various factors given earlier. This recovery is a major challenge that would require sustained efforts in terms of overcoming factors of disturbance, adopting appropriate management practices, operating significant changes on farmers' attitudes and behaviors, as

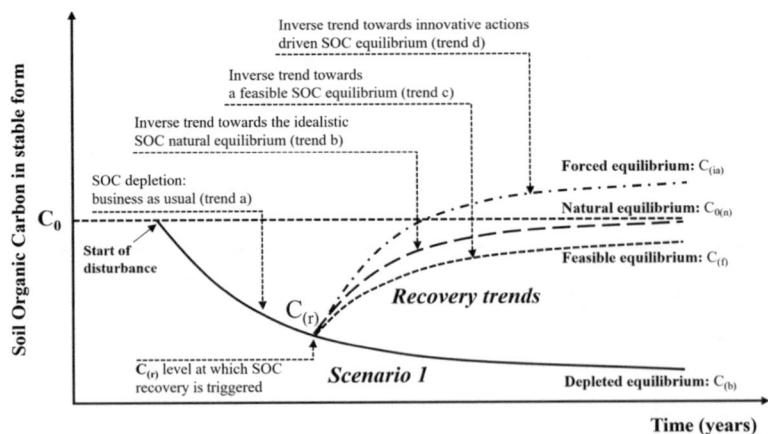

FIGURE 11.3   Idealistic and feasible options for soil organic carbon recovery and storage under dryland conditions. $C_{(r)}$: carbon level from which recovery is triggered; $C_{0(n)}$ = f(climate, soil properties, basic management practices adopted, no disturbance factors); $C_{(b)}$ = f(climate, soil properties, management practices business as usual, disturbance factors); $C_{(f)}$ = f(climate, soil properties, feasible management practices adopted); $C_{(ia)}$ = f(climate, soil properties, innovative management actions and practices adopted).

well as on socioeconomic factors and policies. The *recovery trend* would follow first-order kinetics and would require several years. The reconstitution half-life of the lost C would be slow, in general – from 5 to10 years depending on the local context.

Ultimately, if all conditions are favorable, the recovery of SOC would rise back toward the natural initial storage capacity $C_{(n)}$ (trend b). However, this trend seems challenging, knowing that, in most cases, it would be difficult to guarantee all the necessary conditions for a positive and increasing annual balance of SOC. For instance, a soil that has undergone significant degradation by water erosion as a result of land cover degradation and subsequent organic matter loss would be difficult to restore to its initial situation despite rigorous efforts.

The alternative trend for SOC sequestration when adopting a set of "feasible" agricultural practices that favor SOC storage would be a new state of equilibrium $C_{0(f)}$ that depends on the local context (trend c), and which would be less than $C_{0(n)}$. Therefore, it is emphasized here that reversing the trends of C storage in dryland soils would depend greatly on the local conditions (climate, soil properties and management practices). Farmers can only move forward to a set of feasible management practices according to their local conditions (farm size, cropping systems, level of input, productivity, income, advisory, government support, degree of adoption of innovations, etc.). The SOC would eventually be enhanced from the "business as usual" level ($C_{(b)}$) to the "maximum feasible level" ($C_{(f)}$), but would be unlikely to go back to the initial natural level ($C_{0(n)}$).

Under special conditions relying on forced action measures, it is possible to increase SOC to a level ($C_{(ia)}$) beyond the idealistic initial natural level. This situation is optimistically achievable, but would require the introduction of innovative management practices and assumes no limiting factors to their implementation (adaptation, sufficient long-term funding, acceptance by farmers, capacity-building, advisory, policy, etc.). Afforestation and reforestation under special conditions could be a good example to such recovery.

11.6 FACTORS DETERMINING CARBON STORAGE CAPACITY IN DRYLAND SOILS

11.6.1 QUANTITY OF CROP RESIDUES

Crop residues are the main source of SOC and improve the chemical, physical and biological properties of the soil. According to Power and Legg (1978), degraded soils of North Africa are a good example of the effect of off-field residues exports on SOM. Ayanlaja et al. (1991) showed that the SOC content is proportional to the amount of residues recycled. Soudi et al. (2000) reported that the amount of residues returned to the soil in arid zones of Morocco are low (Figure 11.4), because

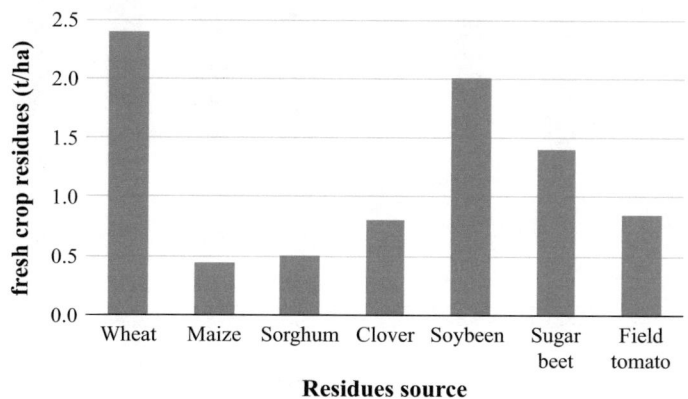

FIGURE 11.4 Quantities of residues returned by different crops in arid irrigated areas in Morocco. (Adapted from Soudi et al. 2000.)

off-field exports are used for livestock feed during the dry season. This is a typical situation in small-holders' mixed cropping in Africa (FAO 2005). Naman and Soudi (1999) and Traoré et al. (2007) underlined that improving management of crop residues is the best way to regulate the SOC levels.

11.6.2 NATURE OF CROP RESIDUES

According to Mustin (1987), the resistance to biodegradation of SOM is correlated to the bio-chemical composition of the plant tissues. Similar findings were reported by studies carried out in irrigated drylands in Morocco (Naman et al. 2015; Naman et al. 2018; Bouajila et al. 2014). The biochemical fractionation of the OM returned to the soils from seven crop residues revealed that their tissues contained more cellulose, hemicellulose and soluble fractions than lignin (Figure 11.5). Wheat residues are characterized by the highest concentration of cellulose (40.6%) compared to the other residues. Sorghum (*Sorghum*) and maize (*Zea mays*) contained moderate amounts of lignin. The hemicellulose concentration varies between 10.0% for tomato residues and 23.5% for wheat residues. The highest lignin concentration was found in sorghum residues (9.6%). In terms of soluble fractions, tomato residues had the highest fraction (65.3%) and those of wheat had the lowest fraction (29.8%). Crop residues with high concentrations of fiber generate more stable C in the soil. Lignin is degraded almost entirely to phenolic compounds with polymerization properties leading to stable humic substances (Soltner 2003). According to Bahri et al. (2006), organic residues decompose differently depending on their biochemical composition. Residues rich in lignin are difficult to decompose due to the recalcitrance of these macromolecules. Cellulose, the main constituent of the cell wall structure, is easily biodegradable, but becomes less biodegradable if coated with hemicellulose, which has a higher degree of polymerization. When hemicellulose becomes encrusted in lignin, it also becomes protected from rapid decomposition.

11.6.3 C/N RATIO: A STILL PRACTICAL INDICATOR FOR ORGANIC MATTER DECOMPOSITION

The C/N ratio is an indicator of the humification potential of returned OM (plant residues, manure, etc.) to the soil. It is unanimously accepted that the higher the C/N ratio of organic residues incorporated into the soil, the slower is the decomposition and the more stable is the humus produced (Waksman 1924; Jensen 1929; Allison 1955; Fog 1988). The slow decomposition is essentially attributed to the highly polymerized humus macromolecules that are difficult to degrade (Swift et al. 1979; Monties 1980).

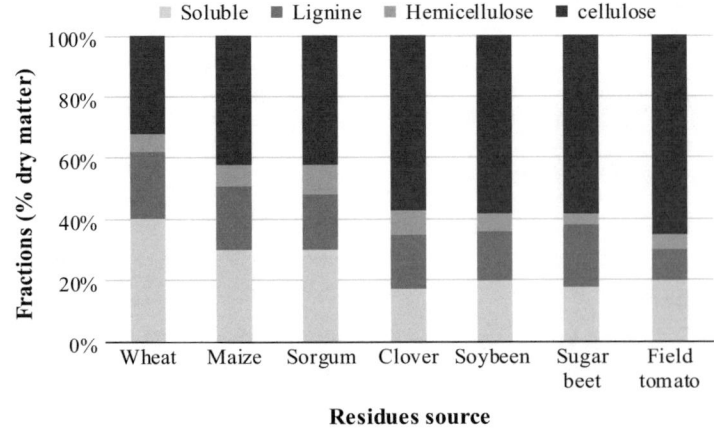

FIGURE 11.5 Biochemical fractions of crop residues in irrigated dryland soils in Morocco. (Adapted from Soudi et al. 2000.)

To better understand the processes of humification and mineralization of the SOM and the consequent supply of mineral N potentially absorbed by the crops, it is essential to consider the interactions between the C and N cycles. In fact, a high C/N ratio favors the production of stable C in humus, whereas a low C/N ratio favors the mineralization of OM and the production of CO_2, mineral-N and other mineral constituents. The breakpoint value that drives the dominance of one process over the other is related to the assimilation needs of the microorganisms present in the soil (to satisfy the C/N ratio of their tissues), and varies with their population type (Janssen 1996). C/N breakpoints varying from 20 to 25 were reported by Tate (1995) and Bengtston et al. (2003). Vigil and Kissel (1991) reported a higher value of 40. Chen et al. (2014) ranked data for C/N in relation to mineralization/immobilization for residues from various crops, and showed that mineralization is dominant for C/N values less than 22.7, that concurrent mineralization-immobilization occurs for C/N values exceeding 30, while immobilization becomes dominant for high C/N values (>78).

11.6.4 MINERALIZATION VS. HUMIFICATION IN DRYLANDS

In the drylands of Africa, the low OM maintenance in the soils, as a result of low restitution of crop residues, is often aggravated by rapid and intense mineralization because of the high temperatures favoring microbial activity. The OM mineralization rate is amplified in irrigated soils in semi-arid and arid areas because of the combined effect of temperature and moisture content. Estimated rates of OM mineralization under Mediterranean conditions in Morocco range from 1.9 to 3.3% yr^{-1} (Soudi et al. 2000).

Carbon fluxes resulting from the decomposition of fresh OM added to the soil concurrently follow the process of humification with a rate of k_h and that of mineralization with a rate of k_m. The latter involves primary mineralization (k_{m1}) and secondary mineralization (k_{m2}) (Figure 11.6).

In drylands, although secondary mineralization is slower compared to primary mineralization, given the molecular complexity of the humic compounds, both processes greatly exceed the humification rate in terms of intensity. Furthermore, in these areas, the rate of mineralization is amplified when the soil moisture content is greater than 50% of the soil water-holding capacity, particularly during the warm rainy season or in the case of irrigation (Soudi et al. 1990; Li 1990; Ju and Li 1998).

Using the values reported by Soudi (2000) and Naman et al. (2015) for Mediterranean conditions namely: (i) an average quantity of residues returned of 1.5 Mg ha^{-1}, (ii) an average humification rate (k_h) of 15% for most crop residues, (iii) an average annual OM mineralization rate (k_{m2}) of 2.5%, and (iv) an average stable SOC content of 0.7%, and applying equation (11.4) proposed by Henin and Dupuis (1945), the apparent annual deficit of SOC from OM degradation is estimated to about 0.3 Mg ha^{-1}.

$$dC/dt = \left(k_h * fOC\right) - k_{m2} * Cs \qquad (11.4)$$

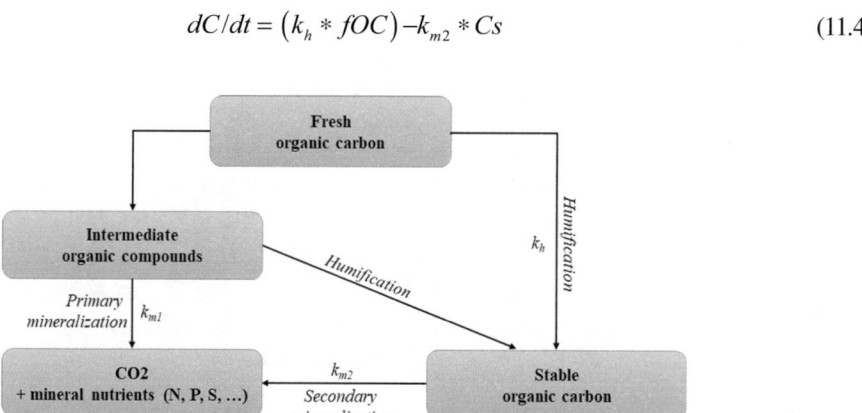

FIGURE 11.6 Soil organic carbon evolution in the soil (k_h: rate of humification; k_{m1}: rate of primary mineralization; k_{m2}: rate of secondary mineralization).

where C: organic carbon; k_h: humification rate; fOC: fresh organic carbon in the crop residues; k_{m2}: secondary mineralization; Cs: stable SOC.

This clearly shows that current farming systems in drylands are part of a trend of SOC losses that are not expected to be counterbalanced by stable C production. A study by Naman et al. (2015) showed that the rate of compensation of C depleted by mineralization in the case of arid irrigated areas in Morocco varies from 4% to 32% depending on soil type and the nature of residue returned (Figure 11.7). Wheat (with the highest C/N ratio) showed the highest compensation rates among the five crops, and the calcium-rich mollisol showed the highest values among the three soil types.

11.6.5 Effect of Soil Texture

As noted above, after going through a partial or total humification process, SOM develops close bonds with clays. Ladd et al. (1996) and Hassink et al. (1994) showed that the decomposition of fresh crop residues and the mineralization of native SOM are very fast in sandy soils compared to clay soils. This is partly attributed to a strong physical or physicochemical protection of SOM compounds by adsorption on the clay minerals' surface or by their encapsulation in the very small pores of aggregates that are inaccessible to microorganisms (Elliot and Coleman 1988). In addition, the protecting effect depends on the kind of clay minerals. High charge 2:1 clays have more bounding forces than 1:1 clays. Martin and Haider (1986) reported that smectites are very effective protectors of organic compounds, whereas kaolinites are rather weak. This protective effect is generally related to the importance of the charge, the location of the charge and to the swelling-shrinking properties. Theng et al. (1986) stated that SOM structures can be incorporated in the interlayers of clay minerals. Nguyen (1982) showed that the extracellular enzymes are adsorbed by the clays and therefore are not able to reach their C substrates in the micropores. A characterization study of C content in different particle-size fractions conducted in irrigated semi-arid soils in Morocco (Naman et al. 2002) showed that the three particle-size fractions, sands, silts and clays, comprised 15 to 37%, 19 to 40% and 24 to 66% of the total carbon, respectively. Similar results were reported by Quattara et al. (2006).

Zech et al. (1997) studied correlations among the C stock and various soil parameters in different regions of Africa. They reported that in semi-arid soils of Senegal, soil C stock was significantly correlated to fine earth fraction (correlation coefficient, $r = 0.51^*$), clay content ($r = 0.79^{***}$), cation exchange capacity (CEC) ($r = 0.8^{***}$), N reserves ($r = 0.93^{***}$) and P reserves ($r = 0.63^{***}$). Eliminating

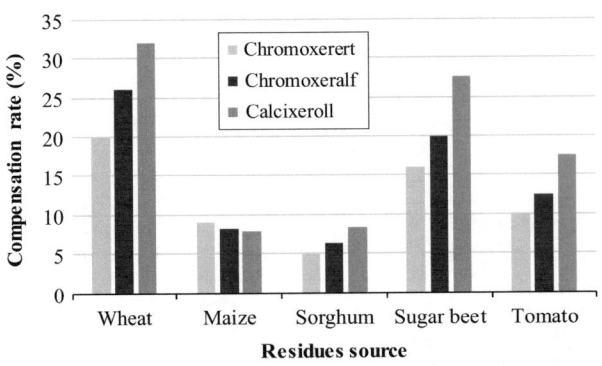

FIGURE 11.7 Compensation rate of the mineralized organic carbon by humification for selected crop residues in three different soil types. (Adapted from Naman et al., 2015.)

variables that were autocorrelated, the authors found that the regression can be simplified to contain only clay content as follows:

$$C\ \text{stock} = 15.9 + 0.017 * Clay; (R = 0.62^{***})$$

with C and clay contents in 10^6 g ha^{-1} in 1 m soil depth.

11.6.6 Carbon Status and Intensive Cultivation in Dry Irrigated Areas

The history of farming systems and the degree of intensive cultivation have marked effects on SOC dynamics. Unsound agricultural intensification can cause negative effects to the soil, mainly SOM loss, as well as to the environment (Naman et al. 2001; Condron et al. 2014; Liao 2016). However, if intensification is accompanied by sustainable management practices (sustainable intensification), positive effects can be obtained, including carbon sequestration. Lal (2002) reported that, in China, agricultural intensification and the adoption of a set of recommended appropriate management practices on cropland, forest land and grazing land have a potential to sequester 59–106 Tg C yr^{-1}, with 25–37 Tg C yr^{-1} in the croplands.

Comparing two vertic calcixerolls under arid Mediterranean conditions in Morocco that have undergone similar pedogenesis, Soudi et al. (2000) observed that the soil under irrigated intensive agriculture (Tadla) is poorer in organic C, labile organic N forms and clay-fixed (non-exchangeable) ammonium than that under rainfed agriculture (Chaouia) (Table 11.2). These differences were attributed mainly to the intensive cultivation practices that are not accompanied by adequate management of crop residues. Indeed, in most irrigated arid zones, temperature and irrigation ensure optimum thermal and moisture conditions for the mineralization process. This phenomenon is amplified by frequent tillage that increases the accessibility of OM to biodegradation. The low levels of chemically hydrolyzable-N and aminoacids-N in the irrigated soil showed a tendency to depletion of the readily biodegradable organic-N forms. In fact, the lack of appropriate management of crop residues does not allow replenishment of these pools of SOM. The low non-exchangeable ammonium content in the irrigated soil, compared to that in rainfed soil, is also a worthy indicator of the effect of intensive cultivation. The intense nitrification process under irrigation and the high crop mobilization of mineral N shift the equilibrium toward the release of the clay fixed-ammonium. These results confirm that soil type alone cannot explain the trends in SOM evolution and that the degree of agricultural intensification and soil use have a significant impact. Similar conclusions were drawn by Quattara et al. (2006) in soils subject to intensification in Burkina Faso.

Other studies have reported opposite trends with agricultural intensification when residues are appropriately managed. Liao et al. (2016) reported that intensification in northern China engendered a significant increase of SOC. Their results indicate that from 1982 to 2011, SOC content and C

TABLE 11.2

Comparison of Some Dynamic Parameters of Organic Matter in Topsoil Layer (0–10 cm) in Two Contrasting Arid Zones of Morocco: Chaouia (rainfed) and Tadla (intensively irrigated)

Parameters	Chaouia (rainfed)	Tadla (intensively irrigated)
Organic C (g/kg)	23.3	13.0
Organic N (g/kg)	2.2	1.4
Hydrolyzable-N (mg/kg)	1192	915.6
Amino-acid-N (mg/kg)	603.4	428.8
Clay fixed-ammonium (mg/kg)	120.8	71.3

Source: Soudi et al. 2000.

stock in the surface (0–20 cm) layer of the cropland increased from 7.8 to 11.0 g kg^{-1} (41%) and from 21 to 33.0 Mg ha^{-1} (54%), respectively. The estimated SOC stock (0–20 cm) of the farmland for the entire country increased from 0.75 to 1.2 Tg (59%). Correlation analysis revealed that under intensification conditions SOC was increased significantly with crop residues, while it was decreased with increased mean annual temperature. This shows that unless proper management practices are adopted, intensification will cause a decline of SOC.

11.7 TREES OF CONSTRAINTS AND SOLUTIONS FOR CARBON SEQUESTRATION IN DRY AREAS

11.7.1 Constraints Hierarchy

Figure 11.8 shows the relationship between the multiple causes related to climate, soil and management practices, their effects on low SOC storage, low productivity and consequent food insecurity. Such effects are contributing to high C emissions and to aggravating climatic changes, which, in turn, will impact on exacerbating the causes. Considering the aspects discussed above, it can be deduced that the storage of C in dryland soils in general, and in those of Africa in particular, is difficult and challenging, given the farming systems characterized by low input, low biomass production and low return of crop residues. Intense mineralization attributed to temperature and soil moisture factors during the rainy season, or to irrigation, is a driving force for organic C decline. This is aggravated when no or low exogenous organic amendments (manure, compost, etc.) are used. In addition, and as shown in Figure 11.8, the cropping systems do not always include crops

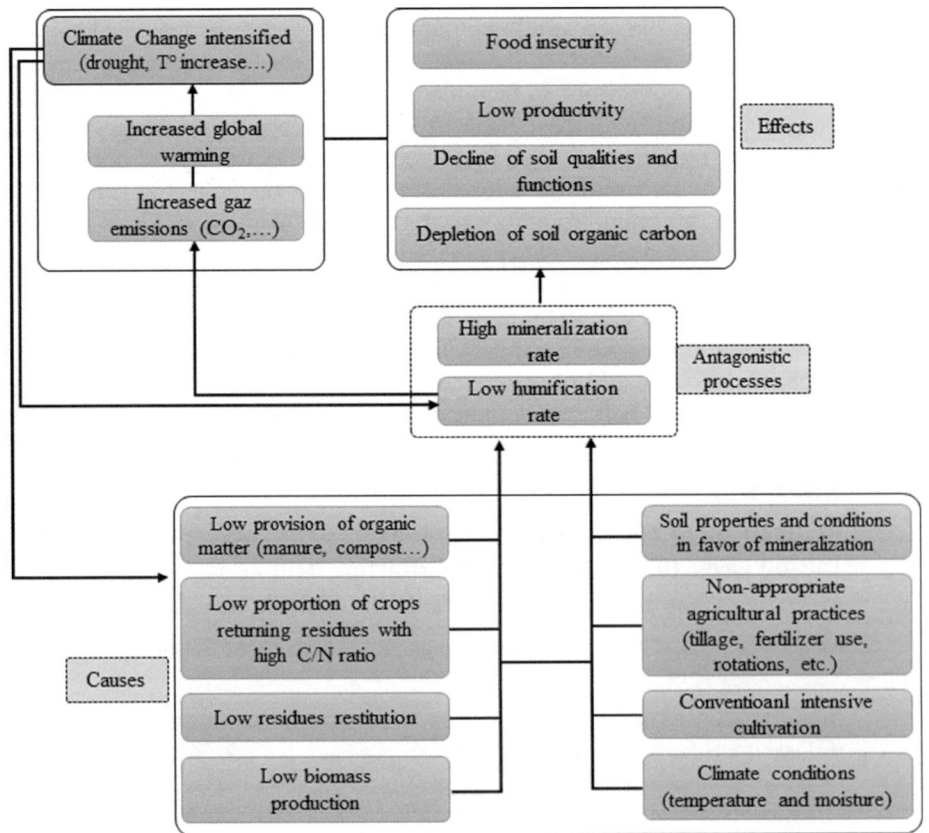

FIGURE 11.8 Hierarchy tree of constraints to stable carbon production (causes are not in order of importance).

generating residues with a high C/N ratio and rich in carbon compounds with high potential of carbon stabilization.

It is important to note that while C storage is possible in dryland farming in Africa, it can be increased only up to a level where SOC equilibrium is established with regard to climate and appropriate farming practices. In such conditions, it would be too ambitious to strive for an increase of SOC beyond the level that is considered a feasible balance (equilibrium: C_0) as discussed earlier. The more realistic option for C storage in these areas would be to close the gap between the current low levels and the $C_{0(n)}$ level (Figure 11.3), which is equivalent to the maximum C storage capacity of the soil in its local context.

On the other hand, and as stated earlier with regard to the influence of soil type, the physical and physicochemical links between OM and soil mineral particles mean that soils do not have the same ability to store stable SOC. For instance, it is well established that the OM content of sandy soils is always lower than that of clay soils. Calcium-saturated soils (calcareous soils) have often more stable SOC than acid soils. Moreover, it is prudent to avoid comparing, in absolute terms, the values of the C or OM contents between such contrasting soils. In fact, a sandy soil with an OM content of 2% could be considered richer than a clay soil with the same content. This argues further for: (i) the effect of fine clay particles on the level of protection of C against mineralization; and (ii) the difficulty of trying to exceed the maximum C storage capacity of the soil ($C_{0(n)}$). It is somewhat comforting that these observations are based on "stable carbon" rather than "total C," including that of fresh OM.

11.7.2 SOLUTIONS HIERARCHY

It is crucial to underline that, even if it is difficult to close the gap between higher levels of SOC in African soils, their potential for C storage is quantitatively important. Africa has a vast area of arable (205 million ha) and potentially arable (870 million ha) land. The amount of C sequestration potential in African drylands, estimated by current soil carbon stocks for continental Africa, are about 80.1 Gt C for the 0–30 cm depth and 74.5 Gt C for the 30–100 cm depth, a total of 154.6 Gt C for 0–100 cm depth (Jones et al. 2013). By adopting feasible and viable agricultural practices, a solutions tree (Figure 11.9) can be deduced by reversing the constraints in Figure 11.8.

Implementing these solutions in the real world, through agroecological innovations that promote carbon storage, requires drastic changes in African farming systems, particularly in rainfed arid and semi-arid areas. Solutions also need to target changes of attitudes and behavior, not only at the farmer level, but also at all levels of decision-making (policy, research/development, advisory, etc.). The main agricultural practices likely to maximize the socioeconomic benefits of farmers and, at the same time, promote the C stock in soils are briefly described below (not in order of importance).

11.7.3 MANAGEMENT PRACTICES AFFECTING SOC STORAGE IN DRYLANDS

Several management practices have been proven effective for storing C in dryland soils. Most of them are driven from indigenous knowledge and improved by scientific evidence, others were adapted from newly developed technologies. Some of these practices are discussed below as key examples of storing carbon in the context of dryland farming in Africa.

11.7.3.1 Fertilizer Use and Soil Nutrient Status

Fertilizer use in Africa is among the lowest worldwide, which contributes to nutrient mining (Roy and Nabhan 1999; Henao and Baanante 2006; Bouabid et al. [Chapter 13 in this book]). Crop fertilization is a key to enhancing crop yields. There is overwhelming evidence for the need for increased nutrient inputs in the region, and for the necessity of fertilizer use (mainly N) if crop yields are to be enhanced (Henao and Baanante 2006; Sanchez et al. 2007). Any increase in crop production means an increase in biomass, which represents a potential residue source for C sequestration. Therefore,

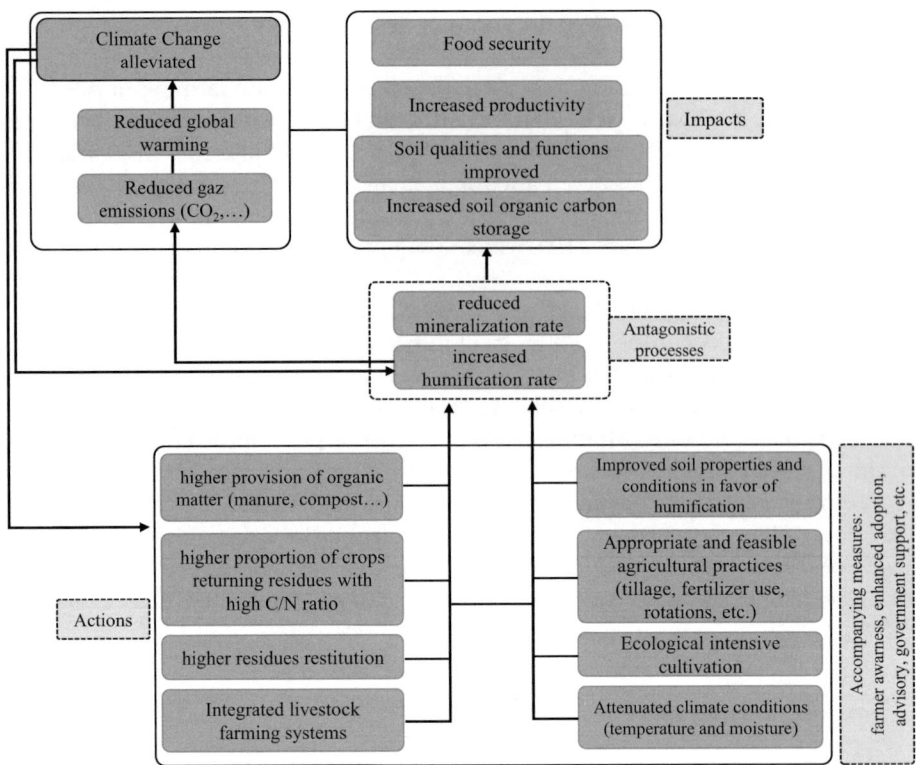

FIGURE 11.9 Carbon storage solutions tree: converting the constraints (Figure 11.8) into solutions.

adequate fertilizer use combined with soil water management (rainfed or under irrigation) is an effective way for increasing biomass production and SOC stock. A recent review by Oladele and Braimoh (2014) showed that the average impact of fertilizer application on the carbon stock in Africa was estimated to be about 626 kg C ha^{-1}.yr^{-1}.

Han et al. (2016) conducted a global meta-analysis of SOC changes under different fertilizer managements, including unbalanced application of chemical fertilizers (UCF), balanced application of chemical fertilizers (CF), chemical fertilizers with straw application (CFS), and chemical fertilizers with manure application (CFM). Their results show that topsoil organic carbon increased by about 0.9, 1.7, 2.0 and 3.5 g kg^{-1} under UCF, CF, CFS and CFM, respectively, corresponding to relative SOC increases of 10%, 15.4%, 19.5% and 36.2% respectively. The C sequestration durations were estimated to be 28–73 years under CFS and 26–117 years under CFM, but with high variability across climatic regions. At least 2.0 Mg ha^{-1} yr^{-1} C input is needed to maintain the SOC in about 85% of the cases examined. They highlighted a great C sequestration potential of applying CF, and that adopting CFS and CFM is highly important for either improving or maintaining SOC stocks across all agroecosystems. Liu and Zhou (2017) reported that SOC builds up quickly in manure and manure plus N-P fertilizer treatments compared to the sole use of N-P fertilizers in newly built terraced land in the semi-arid highlands of northern China. The rate of increase of SOC decreased with time and tended to stabilize at about 3.4 g kg^{-1} yr^{-1} after a cumulative input of manure equivalent to 14 Mg C ha^{-1} over a six year period.

11.7.3.2 Fallow

Fallow plays multiple roles in the sequestration of soil C, especially where fallow is associated with good residue management. When the soil is left uncultivated, a protective spontaneous vegetation cover takes place. This is particularly beneficial in dry areas where the soil is exposed to erosion.

Soil moisture is preserved, especially in the deep layers, and improves biomass production. The vegetation cover represents a sink for photosynthetically fixed C that becomes available for potential return and sequestration in the soil after plant senescence. In Nigeria, a study showed that deforestation caused a decline in soil carbon from 25 to 13.5 Mg ha^{-1} in seven years, but after 12–13 years of bush fallow, the soil restored its total carbon content (Juo et al., 1995).

11.7.3.3   Crop Rotation

The importance of crop rotation in agricultural systems is well established and this practice has become an integral part of many soil conservation technologies. The adoption of rotation has many benefits, such as improving soil structure, combating weed build-up, improving soil fertility and diversifying crops and thus crop residue types. In many dryland areas of Africa, crop rotations are used as alternatives to fallow. The use of different crops with deep rooting facilitates the distribution of OM through the soil profile and allows C storage in deeper horizons, where it is less subject to degradation.

Crop rotation with nitrogen-fixing (N-F) legume crops is used especially to improve soil N status, especially in low N fertilizer farming systems. The main concern with rotations involving N-F crops is their important nitrogen contribution, which can lower the C/N ratio of the residues and consequently favors OM mineralization at the expense of humification. Although legume crops are of paramount importance in many African countries, continuous rotation with such crops may not be in favor of storing SOC. Therefore, it is important that rotation involves crops that leave residues of high C/N ratios in an alternating manner with those involving legumes.

11.7.3.4   Conservation Agriculture

Conservation agriculture (CA) consists of farming practices that contribute to maintaining good soil cover and include, among others, practices with minimum mechanical soil disturbance, no-tillage, organic mulch cover and crop diversification. CA is nowadays widely considered suitable for smallholders' farming systems in marginal land in general (FAO 2018), and in drylands in particular (ICARDA 2012).

Tillage is one of the practices that has the most influence on the mineralization of SOM. When it is too intensive, it speeds up the process. Soil tillage promotes soil aeration and increases the activity of microorganisms involved in the mineralization of SOM. It can also make the soil more susceptible to losses of OM through increased erosion (both wind and water). Therefore, the adoption of reduced tillage or no-till (direct seeding), with the maintenance of significant surface residues, is a measure that is advocated to contribute significantly to reducing the rapid decomposition of SOM in drylands, while offering other favorable conditions for crop growth, such as improved soil moisture conditions and soil fertility status (Moussadek et al. 2011; Mrabet 2012; Bayala et al. 2012; Mrabet and Wall 2015).

Mulching plays an important role in preserving a permanent soil cover that contributes to soil temperature and moisture conditioning. Mulching with residues of high C/N ratio is likely to contribute to the accumulation of stable humus and to C sequestration (FAO 2004). Even if the amounts are low, their influence under dry conditions has been shown to be significant. The favorable conditions created with mulching also have indirect positive effects on the soil moisture status, nutrient cycling and crop growth.

11.7.3.5   Agroforestry and Tree Cropping

Agroforestry is a practice of introducing trees and shrubs into production and land management systems. In many areas of Africa, agroforestry systems are known to store larger amounts of C than tree-free farming systems (Somarriba et al. 2013; Unruch et al. 1993; Lorenz and Lal 2010; Nair et al. 2010). The establishment of trees on agricultural land is relatively efficient and cost-effective compared to other mitigation strategies, and it provides a range of important ancillary benefits for improving the livelihoods of small farms and for contributing to the adaptation to climate change (FAO 2010).

While it is widely accepted that agroforestry has great potential for C sequestration and for establishing buffers between primary forests and areas under anthropic pressure, the concern for many agroforestry ecosystems in Africa lies in the competition between the extraction of wood for household needs and carbon sequestration (Wise and Cacho 2005). Driving the population to minimize the removal of wood to meet their needs, while satisfying the goals of C sequestration projects based on agroforestry parks, may result in adverse effects from increased cutting of primary forests (Chomitz 1999). Therefore, in order to implement a sustainable agroforestry project in socioeconomically vulnerable areas, it is necessary to consider the factors of adoption and success (Mercer 2004). The main challenge of any project of this type is to find solutions to financial and socioeconomic constraints and to understand the importance of the cultural context (Nair 1993). The socioeconomic context is a critical determinant of C sequestration that needs to be taken into account beyond the scientific evidence of the benefits of agroforestry in carbon sequestration.

According to the World Bank (2012), tree crop farming has increased in many African countries and contributes significantly to C sequestration, with a rate of about 1.4 t C ha^{-1} yr^{-1}. These estimates covered cocoa (*Theobroma cacao*) in Ghana and Cameroon, coffee (*Coffea* L.) in Burkina Faso, indigenous fruit trees in South Africa, oil palm (*Elaeis guineensis*) in Côte d'Ivoire, exotic tree species in Ethiopia, rubber plantations (*Hevea brasiliensis*) in Nigeria and Ghana, and cashew (*Anacardium occidentale*) and teak (*Tectona grandis*) plantations in Nigeria. Planted surface areas went from 7 million hectares (Mha) in 1965 to 21 Mha in 1981, 43 Mha in 1990 and 187 Mha in 2000. The planting of olive orchards in Mediterranean North Africa, an ancient practice, is a good example of soil C sequestration, especially because of olive trees' longevity. They are often associated with other crops (intercropping), which increases their benefits for high C/N residue returns as well as for additional economic value. In Morocco, the olive plantations increased from 600,000 ha in 2005 to over 1 million ha in 2017 (MAPMDREF 2018). Oladele and Braimoh (2013) estimated that C sequestration from various African cases of tree crop farming and associated alley farming varied from 1,359 and 1,458 kg C ha^{-1} yr^{-1}, respectively, and proposed that these practices be scaled up to more farmers due to their multiple positive effects on the environment. Ogunkunle and Eghaghara (1992) assessed the SOM contents in the surface layer (0–15 cm) of soils under different land-use types, some of which were mixed cropping. The soil under the original secondary forest had the highest SOM content (61 g kg^{-1}) and the cultivated soil under the 4-year-old cassava (*Manihot esculenta*) had the lowest content, 36 g kg^{-1}. However, the soil under the 10-year-old cacao (*Theobroma cacao*) plantation mixed with yam (*Dioscorea rotunda, bulbifera*), plantain (*Musa paradisiaca*) and cocoyam (*Colocasia esculenta*) presented intermediate SOM contents, 43 g kg^{-1}.

11.7.3.6   Mixed- and Inter-Cropping

The use of mixed- or intercropping with various field or tree crops has been reported to promote C sequestration in the soil compared to sole cropping. A well-managed mixed- or intercropping may result in greater surface and below-ground organic residues than sole crops, and sequester more soil carbon over time due to greater input of root litter (Cong et al. 2014, Cong et al. 2015). Dyer et al. (2012) found that maize and wheat intercropping enhanced soil C sequestration and reduced C emissions significantly compared to the corresponding monocultures. Beedy et al. (2010) indicated that the use of intercropping had positive effects on soil fertility reconstruction as well as on C sequestration. Hu et al. (2015) found that intercropping of wheat and maize produced better yields compared to sole crops, and generated less C emissions, especially with reduced tillage practice. Cong et al. (2014) demonstrated a divergence in soil organic carbon and nitrogen content over a seven-year field experiment, in which they compared rotational strip intercrop systems and ordinary crop rotations. Soil organic C content in the top 20 cm was 4% greater in intercrops than in sole crops, corresponding to an average difference in C sequestration rate among the two systems of 184 kg C ha^{-1} yr^{-1}. Soil organic N content in the top 20 cm was 11%

greater in intercrops than in sole crops, indicating also an average difference in N sequestration rate of 45 kg N ha^{-1} yr^{-1}.

Other studies reported that the mix of residues from intercropping can accelerate or inhibit decomposition depending on the nature of the resulting residue mixture, their biochemical composition and their C/N ratio. Cong et al. (2015) reported different decomposition rate trends depending on the crops involved in the intercropping. SOM in strip intercrop plots decomposed faster than in monocrop plots with lower soil C/N ratio of the mixed residues. Root litter mixtures of maize (*Zea mays*) and wheat (*Triticum aestivum*) decomposed as expected from single litters, but litter mixtures of maize and faba bean (*Vicia faba*) decomposed faster than expected.

11.7.3.7 Organic Amendments and Composting

In low-input agricultural systems, the return of organic materials such as manure to the soil represents an important source of nutrients and contributes to enhancing SOM contents and improving soil physical, chemical and biological properties. The use of manure is a traditional practice in many smallholder farming systems of Africa, but the amounts used are insufficient to maintain or increase SOC to satisfactory levels. Studies have shown that biomass production and SOC levels are increased in African systems receiving manure alone or in combination with chemical fertilizers (Woomer et al. 1998; Kapkiyai, 1999; Tiessen et al. 1998; Roose and Bathès 2001; Ringius 2002; Vågen et al. 2005; Bationo 2007).

Composting is an important organic practice and an excellent means for adaptation to climate change. Composting involves a good mix of OM of various sources (manure, crop residues, and other organic waste) so that it can be broken down into a compost that can be used as a soil amendment. Composting can have a rapid effect on SOC increase (Calderón et al. 2017; Demelash et al. 2014), and has the advantage of ensuring the removal of phytopathogenic agents. Compost composition varies widely depending on the source of organic materials involved, as their humic compounds and their organic C and N content affect their decomposition in the soil.

Most farming systems in Africa are mixed crops and livestock, which often generate substantial amounts of manure and other organic wastes. Despite the low production of such outputs, their composting with residues from other sources can provide a valuable product for improving soil fertility and enhancing soil C storage. According to Lal (1999), soil amendment with compost would sequester from 0.1 to 0.3 Mg ha^{-1} yr^{-1} in dryland soils.

11.7.3.8 Afforestation and Rangeland Rehabilitation

Deforestation and itinerant cultivation causes major damage to soils in Africa. Loss of SOC from forested land can occur rapidly after deforestation. Vågen et al. (2005) compiled changes in SOC for different land-use conversions in different ecosystems of sub-Saharan Africa (Table 11.3). The results show that the shifts toward cultivation, especially from forest and savannah dramatically affects the SOC stock and can attain −0.9 Mg C ha^{-1} yr^{-1}.

Restoring appreciable levels of SOC after afforestation may take years, especially in drylands. When consequent degradation is still at a reversible stage, afforestation associated with conservation techniques can help restore the degraded land and bring back SOC to a satisfactory level. Nosetto et al. (2006) reported that after 15 years of afforestation with pine (*Pinus ponderosa*) trees on degraded drylands in Argentina, more than 50% C was added to the initial ecosystem C pool, with annual sequestration rate ranging 0.5–3.3 Mg C ha^{-1} yr^{-1}. The C gains in afforested stands were higher above- than below-ground (150% vs. 32%).

Liniger et al. (2011) reported that afforestation has high potential for C sequestration and is comparable to the use of conservation agriculture. Silvopasture systems with 50 trees per hectare, can store 110 to 147 tons of CO$_2$eq per hectare. Lal (2004) underlined that afforestation, through the establishment of various types of tree plantations, has great potential for C sequestration in the tropics. The SOC accumulation rate under 18-year plantation of acacia in northern Senegal was about 0.03% yr^{-1} under the tree canopy and 0.02% yr^{-1} in open ground, corresponding to

TABLE 11.3

Changes in SOC for Different Land-Use Conversions in Sub-Saharan Africa, Based on Selected Long-Term Studies

Region	Change in land use From	to	Mean	Minimum	Maximum	SOC gains*	N
				Mg C ha⁻¹ yr⁻¹			
HS	Natural forest	Cultivated	−0.90	−1.00	−0.80		2
		Fallow	−0.57	−1.14	0.00		2
	Cultivated	Fallow	1.06	0.23	2.77	100.0	9
		Afforestation	0.12	−0.29	0.56	71.4	21
WSS	Cultivated	Cultivated CR	0.19	−	−		1
		Cultivated F	0.05	−	−		1
		Cultivated NT	0.33	−1.00	1.30	66.7	6
		Fallow	1.37	0.10	5.30	100.0	10
		Afforestation	0.07	−0.98	0.57	50.0	9
	Cultivated M	Fallow	−0.14	−	−		1
	Fallow	Cultivated	−0.11	−0.18	−0.06		2
	Savannah	Cultivated	0.05	0.00	0.15	33.3	5
		Cultivated C	−0.12	−0.15	−0.09		3
		Cultivated CR	−0.06	−0.13	0.15	20.0	2
		Cultivated F	0.09	0.00	0.20	66.7	5
		Cultivated M	0.04	−0.52	0.94	33.3	3
ESS	Savannah	Cultivated	−2.77	−5.30	−0.77		3
	Woodland	Cultivated	0.36	−	−		4
		Cultivated M	0.55	−	−		1
SA	Savannah	Cultivated	−0.82	−1.26	−0.40		4
		Fallow	0.00	−	−		1
		Pasture	0.05	−0.31	0.40	50.0	2
	Pasture	Afforestation	−0.16	−0.19	−0.13		2
	Fallow	Cultivated	−0.75	−	−		1

Studies sources: Agbenin and Goladi (1997); Aweto (1981); Bationo et al. (2000); Dominy and Haynes (2002); Drechsel et al. (1991); Feller et al. (1981); Glaser et al. (2001); Hartemink (1995, 1997); Impala (2001); Juo et al. (1995); Lal (2000); Manlay (2000); Materechera and Mkkabela (2001); Morris and Gray (1984); Onim et al. (1990); Pieri (1989); Solomon et al. (2000; 2002); Trouve et al. (1994).

Source: Adapted from Vågen et al. (2005)

Notes: HS = Humid and sub-humid; WSS = West Sudanian Savannah; ESS = East Sudanian Savannah; SA = Southern Africa; cultivated M = w/manure; cultivated C = w/cover crops; cultivated CR = w/ crop residues; cultivated F = NPK fertilizers only; cultivated NT = no-till; N = number of observations. * Percentage of observations with net gain (> 0).

SOC sequestration rates of 420 and 280 kg C ha⁻¹ yr⁻¹. Afforestation actions, mainly those involving participatory approaches, also have the indirect effects of providing alternative income and substitutes for wood fuel for the local population (Lepetu et al. 2015). Garrity et al. (2010) indicated that introducing the "evergreen approach" of afforestation (integration of particular wood tree species into the annual food crop systems) in Africa can contribute significantly to the carbon sequestration potential of agroforestry systems, enhance the quality of degraded soils and improve the livelihood of the populations. It is estimated that "evergreen agriculture" systems can accumulate C both above- and below-ground in the range of 2–4 Mg C ha⁻¹ yr⁻¹. Given the vast areas of deforested land that could be used for afforestation in Africa, the potential of

SOC from such a practice is huge, assuming the necessary investment funds could be deployed (Garrity et al. 2010).

Rangelands in Africa, mainly in the savannahs, represent vast ecosystems with a large biodiversity and a high potential for C cycling and storage. The effect of species diversity on soil C and N stocks in these natural grasslands has been attributed to the positive interactions among plant communities. However, grazing and cultivation are two key factors that can disturb the functioning and the services of these ecosystems, and affect their potential for C sequestration. Conant and Paustian (2002) reported that the total grassland in Africa amounts to about 838.2 million ha, of which 87.7 million ha (10.4%) have been subjected to overgrazing. They estimated a total potential carbon sequestration of 16.7 Tt C yr^{-1} for the African continent through the cessation of overgrazing and rehabilitation, of which 16.7 Tt C yr^{-1} lies in the moderate to highly degraded lands. These figures represent about 37% of global rangeland areas.

Lipper et al. (2010) recognize two main reasons for looking into the potential of sequestering carbon from West African rangelands: (i) the degradation and depletion of carbon stocks in these systems has resulted in declining rangeland and agricultural productivity, in turn reducing the livelihood of the local population and leading to their impoverishment (Batjes 2004; Tieszen et al. 2004); and (ii) increasing carbon stocks in the system can be not only a way of improving the ecological health and productivity of the livestock systems, but also a significant and low-cost way of mitigating climate change (Woomer et al. 2004). The per hectare amounts in rangelands of western Africa are low, but aggregate potential is high. Avoiding degradation and slightly rehabilitating degraded lands are the least costly methods and can generate significant reductions in carbon storage and emissions. Bazin (2010) considers that the rangelands of the African savannahs have great potential for SOC if proper animal load and rehabilitation measures are taken.

11.7.4 COMPARATIVE BENEFITS OF SUSTAINABLE LAND MANAGEMENT TECHNOLOGIES

Figures regarding carbon sequestration from different management practices vary largely across agroecosystems. It is very difficult to make direct comparisons of their specific benefits as they are affected by many physical, environmental and socioeconomic factors. The changes for a given practice in various areas can vary considerably.

A comprehensive assessment by the World Bank (2012) on carbon sequestration by agricultural soils in Africa, Asia and Latin America compared the climate benefits of a series of sustainable land management technologies as measured by the net rate of carbon sequestration adjusted for emissions, a measurement referred to as the "abatement rate" (AR) and is expressed in tons of carbon dioxide equivalent t CO_2e ha^{-1} yr^{-1}.

In the case of Africa (Figure 11.10), AR estimates vary from 0.29 t CO_2e ha^{-1} yr^{-1} for chemical fertilizer use to 10.3 t CO_2e ha^{-1} yr^{-1} for the use of biochar. The figure also shows that management techniques such as improved fallow, alley farming, afforestation and tree crop farming, have high AR rates (around 7.5 t CO_2e ha^{-1} yr^{-1}), compared to mulching, rotations, no-till and use of manure (1.3 to 1.8 t CO_2e ha^{-1} yr^{-1}). Intercropping and mixed cropping involving trees have moderate effects on AR (4 to 5.3 t CO_2e ha^{-1} yr^{-1}).

11.8 CONCLUSIONS

Storing C in soils or enriching it with OM is not only an option for mitigating or offsetting GHG emissions, but also for improving soil quality (water retention, aggregate stability and structure, biological activity, chelating capacity, attenuation of the alkalinity and sodicity, etc.) and soil productivity. All aspects converge toward the ultimate goal of contributing to food security in the short and long term. However, the current situation in the African drylands is characterized by a low SOC content and a trend toward its decline if no measures are taken. This C decline is of particular

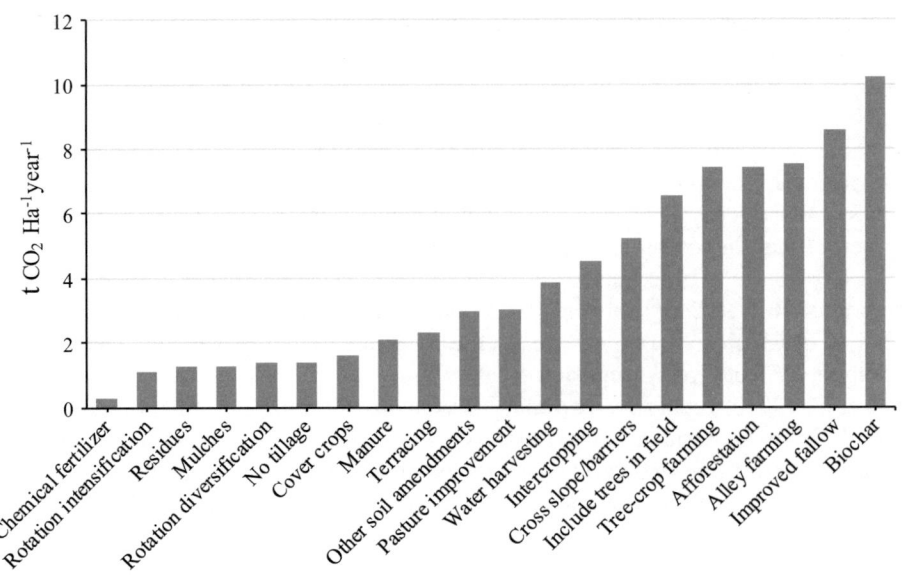

FIGURE 11.10 Abatement rates of the land management practices on carbon sequestration in Africa. (Adapted from World Bank, 2012.)

concern in the arid and semi-arid areas that are more vulnerable and prone to decrease in productivity. Reversing this situation is synonymous with "triggering major changes in the farming systems" with sustained supporting measures.

It is incontestable that the agricultural soils of Africa are "victims" of climate change, but they are also potential "savers" in the fight against global warming, given their potential for colossal C storage at the continental scale. It is obvious that climate change, especially through the rise in temperature, can drastically modify the behavior of the soil on this continent, particularly by increasing the rate of mineralization of SOM. In addition, soil microbial dynamics can be disturbed, as extreme droughts inhibit the biological activity for C humification, while soil engorgement following heavy rain events may favor the emission of N_2O by denitrification and the loss of nitrate-nitrogen needed by crops. The alternating wetting and drying phenomena that occur with climatic disturbance can also modify the geochemical processes in the soil and lead to disturbance of its vital functions. Arid conditions can lead to salinity and therefore affect negatively crop production and SOC accumulation. Other indirect effects of climate change on soils are reduced biomass production, and therefore reduced vegetation cover and OM, both of which play an important part in the protection against soil erosion.

The storage of C in the soil, as a form of C sequestration, is an irrefutable solution to compensate for GHG emissions. Theoretical mass balances are established; however, putting this concept to practice, especially in the context of the African drylands, would require drastic modifications of current farming practices as well as introducing adapted innovations that take into consideration the multiple climatic, physical and socioeconomic situations that can be encountered in this vast continent.

Considering the African "soil potential," two initiatives are worth reflecting on regarding SOC sequestration and agricultural adaptation in the context of climate change. The "4 PM" initiative proposed by France at the COP 21, which focuses on increasing soil carbon content annually by 0.4%. This is a very optimistic initiative that would require special conditions for African soils and their stakeholders. The "3A initiative" (Adaptation of African Agriculture), proposed by Morocco at the COP 22, considers agricultural adaptation and food security as a prerequisite and as an indirect way to contributing to SOC sequestration. Synergies can be established between these two initiatives, and eventually other initiatives, if the specific contexts for the various agroecological zones of Africa are taken into consideration.

Improving SOC sequestration in African dryland soils remains a challenge as it requires answers to a series of questions, some of which were previously addressed by the FAO (2012), but deserve readdressing based on the aspects and issues discussed in the present paper:

- What are the conditions that can guarantee the sustainability of C storage when agricultural practices are changed, considering the antagonistic effects of the processes of humification and mineralization?
- What is the maximum and feasible level of SOC sequestration taking into account the natural balance of C in the conditions of dryland soils of Africa?
- How to act, in an effective way, on humification in the African drylands in the present warming trends (favorable to mineralization), the extreme phenomena (droughts, floods, etc.), that limit biomass production and consequently C sequestration?
- Which cropping systems to advocate and what is their socioeconomic feasibility?
- Which monitoring systems to adopt and at what scale?
- What is the perception of small African farmers (stakeholders) with respect to proposed agroecological innovations?
- What emissions associated with increased fertilizer use, irrigation or livestock need to be integrated into carbon sequestration estimates?
- What is the added value for farmers of applying new farming systems in favor of carbon sequestration, and what support measures are required (access to inputs, technical advisory, access to market, policy, etc.)?

REFERENCES

Agbenin, J.O., Goladi, J.T. 1997. Carbon, nitrogen and phosphorus dynamics under continuous cultivation as influenced by farmyard manure and inorganic fertilizers in the savanna region of northern Nigeria. *Agriculture, Ecosystems and Environment* 63:17–24.

Allison, F.E. 1955. Does nitrogen applied to crop residues produce more humus? *Soil Science Society of America Journal* 19:210–211.

Aweto, A.O. 1981. Secondary succession and soil fertility restoration in southwestern Nigeria. II. Soil fertility restoration. *Journal of Ecology* 69:609–614.

Ayanlaja, S.A., Sanwo, J.O. 1991. Management of soil organic matter in the farming systems of the low land humid tropics of West Africa: A review. *Soil Technology* 4(3):265–279.

Bahri, H., Dignac, M., Rumpel, C., Rasse, D.P., Chenu, C., Mariotti, A. 2006. Lignin turnover kinetics in an agricultural soil is monomer specific. *Soil Biology and Biochemistry* 38(7):1977–1988.

Bationo, A., Kihara, J., Vanlauwe, B., Waswa, B., Kimetu, J. 2007. Soil organic carbon dynamics, functions and management in West African agro-ecosystems. *Agricultural Systems*, 94(1):13–25.

Bationo, A., Wani, S.P., Bielders, C.L., Vlek, P.L.G., Mokwunye, A.U. 2000. Crop residue and fertiliser management to improve soil organic carbon content, soil quality and productivity in the desert margins of West Africa. In: *Global Climate Change and Tropical Ecosystems*, R. Lal, J.M. Kimble, B.A. Stewart (eds.). CRC Press, Boca Raton, FL, pp. 117–145.

Batjes, N.H. 2004. Estimation of soil carbon gains upon improved management within croplands and grasslands of Africa. *Environment, Development and Sustainability* 6:133–143.

Bayala, J., Sileshi, G.W., Coe, R., Kalinganire, A., Tchoundjeu, Z., Sinclair, F., Garrity, D. 2012. Cereal yield response to conservation agriculture practices in drylands of West Africa: A quantitative synthesis. *Journal of Arid Environments* 78:13–25.

Bazin, F. 2010. Contribution de l'agriculture africaine au changement climatique et potentiel d'atténuation. *Grain de sel*, 49(January–March):22–24. www.inter-reseaux.org/IMG/pdf_p22_24_Bazin.pdf.

Beedy, T.L., Snapp, S.S., Akinnifesi, F.K., Sileshi, G.W. 2010. Impact of *Gliricidia sepium* intercropping on soil organic matter fractions in a maize-based cropping system. *Agriculture, Ecosystems and Environment* 138(3–4):139–146.

Bengtsson, G., Bengtson, P., Månsson, K.F. 2003. Gross nitrogen mineralization-, immobilization-, and nitrification-rates as a function of soil C/N ratio and microbial activity. *Soil Biology and Biochemistry* 35(1):143–154.

Berdai, H., Aghzar, N., Cherkaoui, F.Z., Soudi, B. 2002. Azote minéral résiduel et son évolution pendant l'été en fonction du précédent cultural en climat méditerranéen. *Étude et Gestion des Sols* 9:7–23.

Biancalani, R., Petriet, M., Bunning, S. 2015. *Land Use, Land Degradation, and Sustainable Land Management in the Drylands of Sub-Saharan Africa*. Unpublished paper. FAO, Rome, Italy.

Bot, A., Bernites, J. 2005. *The Importance of Soil Organic Matter: Key to Drought-Resistant Soil and Sustained Food Production*. FAO Soils Bulletin 80. FAO, Rome, Italy.

Bouajila, K., Jeddi, F.B., Taamallah, H., Jedidi, N., Sanaa, M. 2014. Effets de la composition chimique et biochimique des résidus de cultures sur leur décomposition dans un sol Limono-Argileux du semi aride. *Journal of Materials and Environmental Science* 5:159–166.

Bremer, E., Ellert, B.H., Janzen, H.H. 1995. Total and light-fraction carbon dynamics during four decades after cropping changes. *Soil Science Society of America Journal* 59:1398–1403.

Calderón, F.J., Vigil, M.F., Benjamin, J. 2017. Compost input effects on dryland wheat and forage yields and soil quality. *Pedosphere* 28(3):451–462.

Campbell, C.A., Jame, Y.W., Winkleman, G.E. 1984. Mineralization rate constants and their use for estimating nitrogen mineralization in some Canadian prairie soils. *Canadian Journal of Soil Science* 64:333–343.

Campbell, E.E., Paustian, K. 2015. Current developments in soil organic matter modeling and the expansion of model applications: A review. *Environmental Research Letters* 10(12):123004.

Chen, B., Liu, E., Tian, Q., Yan, C., Zhang, Y. 2014. Soil nitrogen dynamics and crop residues. A review. *Agronomy for Sustainable Development* 34(2):429–442.

Chomitz, K.M. 1999. Evaluating carbon offsets from forestry and energy projects: How do they compare? World Bank Policy Research Working Paper No. 2357. The World Bank, Washington DC. https://op enknowledge.worldbank.org/handle/10986/19838.

Conant, R.T., Paustian, K. 2002. Potential soil carbon sequestration in overgrazed grassland ecosystems. *Global Biogeochemical Cycles* 16(4):90–99.

Condron, L.M., Hopkins, D.W., Gregorich, E.G., Black, A., Wakelin, S.A. 2014. Long-term irrigation effects on soil organic matter under temperate grazed pasture. *European Journal of Soil Science* 65(5):741–750.

Cong, W.F., Hoffland, E., Li, L., Janssen, B.H., van der Werf, W. 2015. Intercropping affects the rate of decomposition of soil organic matter and root litter. *Plant and Soil* 391:399–411.

Cong, W.F., Hoffland, E., Li, L., Six, J., Sun, J.H., Bao, X.G., Zhang, F.S., Van der Werf, W. 2014a. Intercropping enhances soil carbon and nitrogen. *Global Change Biology* 21(4):1715–1726.

Cong, W.F., Hoffland, E., Li, L., Six, J., Sun, J.H., Bao, X.G., Zhang, F.S., Van der Werf, W. 2014b. Intercropping enhances soil carbon and nitrogen. *Global Change Biology* 21(4):1715–1726.

Corbeels, M., Cardinael, R., Naudin, K., Guibert, H., Torquebiau, E. 2018. The 4 per 1000 goal and soil carbon storage under agroforestry and conservation agriculture systems in sub-Saharan Africa. *Soil and Tillage Research*, 188:16–26.

Davi Davidson, E.A., Ackerman, I.L. 1993. Changes in soil carbon inventories following cultivation of untilled soils. *Biogeochemistry* 20:161–193.

Demelash, N., Bayu, W., Tesfaye, S., Ziadat, F., Sommer, R. 2014. Current and residual effects of compost and inorganic fertilizer on wheat and soil chemical properties. *Nutrient Cycling in Agroecosystems* 100:357–367.

Dominy, C.S. and Haynes, R.J. 2002. Influence of agricultural land management on organic matter content, microbial activity and aggregate stability in the profiles of two oxisols. *Biology and Fertility of Soils* 36:298–305.

Drechsel, P., Glaser, B., Zech, W. 1991. Effect of four multipurpose tree species on soil amelioration during tree fallow in central Togo. *Agroforestry Systems* 16:193–202.

Dyer, L., Oelbermann, M., Echarte, L. 2012. Soil carbon dioxide and nitrous oxide emissions during the growing season from temperate maizesoybean intercrops. *Journal of Plant Nutrition and Soil Science* 175(3):394–400.

Elliott, E.T., Coleman, D.C. 1988. Let the soil work for us. Ecological Bulletins 39, Ecological Implications of Contemporary Agriculture. Proceedings of the 4th European Ecology Symposium, 7–12 September 1986, Wageningen, The Netherlands, pp. 23–32.

FAO (Food and Agriculture Organization). 2002. Rapport sur les ressources en sols du monde – La séquestration du carbone dans le sol pour une meilleure gestion des terres: Options de gestion du sol pour la séquestration du carbone. FAO, Rome, Italy.

FAO (Food and Agriculture Organization of the United Nations). 2004. Carbon sequestration in dryland soils. World Soils Resources Reports No. 102. FAO, Rome, Italy .

FAO (Food and Agriculture Organization of the United Nations). 2005. *The Importance of Soil Organic Matter: Key to Drought-Resistant Soil and Sustained Food Production.* FAO Soils Bulletin 80, A. Bot, J. Benites (eds.). FAO, Rome, Italy.

FAO (Food and Agriculture Organization of the United Nations). 2010. Pour une agriculture intelligente face au climat: Politiques, pratiques et financements en matière de sécurité alimentaire, d'atténuation et d'adaptation. FAO, Rome, Italy.

FAO (Food and Agriculture Organization of the United Nations). 2012. Identifying opportunities for climate-smart agriculture investments in Africa. FAO, Rome, Italy.

FAO (Food and Agriculture Organization of the United Nations). 2018. Conservation agriculture. www.fao. org/conservation-agriculture/en/. Accessed May 2018.

Feller, C., Ganry, F., Cheval, M. 1981. Décomposition et humification des reésidus végétaux dans un agro-système tropical. I. Influence d'une fertilisation azotée (urée) et d'un amendement organique (compost) sur la reépartition du carbone et de l'azote dans différents compartiments d'un sol sableux. *Agronomie Tropicale* 36(1):9–25.

Fog, K. 1988. The effect of added nitrogen on the rate of decomposition of organic matter. *Biological Reviews* 63:433–462.

Garrity, D.P., Akinnifesi, F.K., Ajayi, O.C., Weldesemayat, S.G., Mowo, J.G., Kalinganire, A., Larwanou, M., Bayala, J. 2010. Evergreen Agriculture: A robust approach to sustainable food security in Africa. *Food Security* 2(3):197–214.

Glaser, B., Lehmann, J., Führböter, M., Solomon, D., Zech, W. 2001. Carbon and nitrogen mineralization in cultivated and natural savanna soils of Northern Tanzania. *Biology and Fertility of Soils* 33:301–309.

Guggenberger, G. 2005. Humification and mineralization in soils. In: *Microorganisms in Soils: Roles in Genesis and Functions,* F. Buscot, A. Varma (eds.), Soil Biology, Springer, Berlin/Heidelberg, pp. 85–106.

Han, P., Zhang, W., Wang, G., Sun, W., Huang, Y. 2016. Changes in soil organic carbon in croplands subjected to fertilizer management: A global meta-analysis. *Scientific Reports* 6:27199.

Hans, H.J., Evans, C.E. 1957. *Nitrogen and Carbon Changes in Great Plains Soils as Influenced by Cropping and Soil Treatments.* Technical Bulettin No 1164. United States Department of Agriculture, Washington DC.

Hartemink, A.E. 1995. Soil fertility decline under sisal cultivation in Tanzania. Technical Paper No. 28. ISRIC, Wageningen, The Netherlands.

Hartemink, A.E. 1997. Soil fertility decline in some major soil groupings under permanent cropping in Tanga Region, Tanzania. *Geoderma* 75:215–229.

Hassink, J. 1994. Active organic matter fractions and microbial biomass as predictors of N mineralization. *European Journal of Agronomy* 3(4):257–265.

Henao, J., Baanante, C. 2006. *Nutrient Mining in Africa: Implications for Resource Conservation and Policy Development.* Summary Report. International Fertilizer Development Center (IFDC), Alabama.

Hénin, S., Dupuis, M. 1945. Bilan de la matière organique des sols. *Annals of Agronomy* 1:17–29.

Hénin, S., Monnier, G., Turc, L. 1959. Un aspect de la dynamique des matières organiques du sol. *Comptes Rendus Hebdomadaires des Séances de l'Académie des Sciences* 248:138–141.

Hu, F., Chai, Q., Yu, A., Yin, W., Cui, H., Gan, Y. 2015. Less carbon emissions of wheat– maize intercropping under reduced tillage in arid areas. *Agronomy for Sustainable Development* 35(2):701–711.

ICARDA (International Center for Agricultural Research in Dry Areas). 2012. Conservation agriculture: Opportunities for intensified farming and environmental conservation in dry areas. Research to Action 2. www.icarda.org/publications-and-resources/research-to-action.

Impala. 2001. First Annual Report of Impala Project, Covering the Period October 2000–December 2001. ICRAF, Nairobi, Kenya.

Janssen, B.H. 1996. Nitrogen mineralization in relation to C:N ratio and decomposability of organic materials. *Plant and Soil* 181:39–45.

Janzen, H.H., Campbell, C.A., Ellert, B.H., Bremer, E. 1997. Soil organic matter dynamics and their relationship to soil quality. In: *Soil Quality for Crop Production and Ecosystem Health*, E.G. Gregorich, M.R. Carter (eds.), Developments in Soil Science, Vol. 25 . Elsevier, Amsterdam, The Netherlands, pp. 277–291.

Jarecki, M.K., Lal, R. 2003. Crop management for soil carbon sequestration. *Critical Reviews in Plant Sciences* 22(5):471–502.

Jenny, H. 1941. *Factors of Soil Formation*. Me Graw-Hill, New York.

Jensen, H.L. 1929. On the influence of the carbon: Nitrogen ratios of organic material on the mineralization of nitrogen. *Journal of Agricultural Science* 19:71–82.

Jones, A., Breuning-Madsen, H., Brossard, M., Dampha, A., Deckers, J., Dewitte, O., Gallali, T., Hallett, S., Jones, R., Kilasara, M., Le Roux, P., Micheli, E., Montanarella, L., Spaargaren, O., Thiombiano, L., Van Ranst, E., Yemefack, M., Zougmore, R., (eds.) 2013. *Soil Atlas of Africa*. European Commission, Publications Office of the European Union, Luxembourg .

Ju, X.T., Li, S.X. 1998. The effect of temperature and moisture on nitrogen mineralization in soils. *Plant Nutricao Fertilitatis Science* 4(1):37–42.

Juo, A.S.R., Franzluebbers, K., Dabiri, A., Ikhile, B. 1995. Changes in soil properties during long-term fallow and continuous cultivation after forest clearing in Nigeria. *Agriculture, Ecosystems and Environment* 56:9–18.

Kapkiyai, J.J., Karanja, N.K., Qureshi, J.N., Smithson, P.C., Woomer, P.L. 1999. Soil organic matter and nutrient dynamics in a Kenyan nitisol under long-term fertilizer and organic input management. *Soil Biology and Biochemistry* 31:1773–1782.

Kha, N. 1982. Mise en évidence, par etude microstructurale et microanalyse X, de l'effet stabilisateur des argiles gonflantes au cours des processus de biodegradation. *Pedologie* 32:175–192.

Ladd, J.N., Van Gestel, M., Jocteur Monrozier, L., Amato, M. 1996. Distribution of organic ^{14}C and ^{15}N in particle-size fractions of soils incubated with ^{14}C, ^{15}N-labelled glucose/NH_4, and legume and wheat straw residues. *Soil Biology and Biochemistry* 28(7):893–905.

Lal, R. 1999. Global carbon pools and fluxes and the impact of agricultural intensification and judicious land use. In: Prevention of Land Degradation, Enhancement of Carbon Sequestration and Conservation of Biodiversity through Land Use Change and Sustainable Land Management with a Focus on Latin America and the Caribbean. World Soil Resources Report 86. FAO, Rome, Italy, pp. 45–52.

Lal, R. 2000. Restorative effects of Mucuna utilis on soil organic C pool of a severely degraded Alfisol in western Nigeria. In: *Global Climate Change and Tropical Ecosystems*, R. Lal, J.M. Kimble, B.A. Stewart (eds.). CRC Press, Boca Raton, FL, pp. 147–165.

Lal, R. 2002. Carbon sequestration in dryland ecosystems of West Asia and North Africa. *Land Degradation and Development* 13(1):45–59.

Lal, R. 2004. Soil carbon sequestration impacts on global climate change and food security. *Science* 304:1623–1627.

Lal, R. 2005. Soil carbon sequestration for sustaining agricultural production and improving the environment with particular reference to brazil. *Journal of Sustainable Agriculture* 26:23–42.

Lal, R. 2016. Soil health and carbon management. *Food and Energy Security* 5(4):212–222.

Laudelout, H. 1993. Bilan de la matière organique du sol : le modèle de Hénin (1945). In: *Mélanges offerts à Stéphane Hénin. Sol-agronomie-environnement*. ORSTOM, Paris, France, pp. 117–123.

Laudelout, H., Meyer, J. 1951. Temperature characteristics of the microflora of Central African soils. *Nature* 168:791.

Laudelout, H., Meyer, J., Peeters, A. 1960. Les relations quantitatives entre la teneur en matière organique du sol et le climat. *Agricultura* 8:103–140.

Lepetu, J., Nyoka, I., Oladele, O.I. 2015. Farmers' planting and management of indigenous and exotic trees in Botswana: Implications for climate change mitigation. *Environmental Economics* 6(3):20–30.

Liao, Y., Wu, W., Meng, F., Li, H. 2016. Impact of agricultural intensification on soil organic carbon: A study using DNDC in Huantai County, Shandong Province, China. *Journal of Integrative Agriculture* 15(6):1364–1375.

Liniger, H.P., Mekdaschi Studer, R., Hauert, C., Gurtner, M. 2011. *Sustainable Land Management in Practice – Guidelines and Best Practices for Sub-Saharan Africa*. TerrAfrica, World Overview of Conservation Approaches and Technologies (WOCAT) and Food and Agriculture Organization of the United Nations (FAO), Rome, Italy.

Lipper, L., Dutilly-Diane, C., McCarthy, N. 2010. Supplying carbon sequestration from West African rangelands: Opportunities and barriers. *Rangeland Ecology and Management* 63(1):155–166.

Liu, C.A., Zhou, L.M. 2017. Soil organic carbon sequestration and fertility response to newly-built terraces with organic manure and mineral fertilizer in a semi-arid environment. *Soil and Tillage Research* 172:39–47.

Lorenz, K., Lal, R. 2010. *Carbon Sequestration in Forest Ecosystems*. Springer, Dordrecht, The Netherlands.

Manlay, R.J. 2000. Dynamique de la matière organique à l'échelle d'un terroir agro-pastoral de savane ouest-africaine (Sud-Sénégal). PhD thesis, Ecole Nationale du Génie Rural, des Eaux et Forêts, Centre de Montpellier, France.

Mann, L.K. 1986. Changes in soil carbon storage after cultivation. *Soil Science* 142:279–288.

MAPMDREF (Ministère de l'Agriculture, de la Pêche Maritime, du Développement Rural et des Eaux et Forêts). 2018. *L'Agriculture du Maroc en chiffre : Edition 2017.* Ministère de l'Agriculture, de la Pêche Maritime, du Développement Rural et des Eaux et Forêts. www.agriculture.gov.ma/sites/default/files/AgricultureEnChiffre2017VAVF.pdf.

Martin, J.P., Haider, K. 1986. Influence of mineral colloids on turnover rates of soil organic carbon. In: *Interactions of Soil Minerals with Natural Organics and Microbes*, P.M. Huang, M. Schnitzer (eds.), SSSA Special Publication 17, Soil Science Society of America, Madison, WI, pp. 283–304.

Mercer, D.E. 2004. Adoption of agroforestry innovations in the tropics: A review. *Agroforestry Systems* 61–62(1–3):311–328.

Monties, B. 1980. *Les polymères végétaux. Polymères pariétaux et alimentaires non azotés.* Gauthier-Villars, Paris, France.

Morris, A.R., Gray, D.C. 1984. A comparison of soil nutrient levels under grassland and two rotations of Pinus patula in the Usutu Forest, Swaziland. Proceedings of the 'International Union Forestry Reasearch Organisation' Symposium on Site and Productivity of Fast Growing Plantations, Voluntary Papers, Vol. 2, A.P.G. Schonau, C.J. Schutz (eds.), Pretoria, South Africa. pp. 881–892.

Moussadek, R., Mrabet, R., Dahan, R., Douaik, A., Verdoodt, A., Van Ranst, E., Corbeels, M. 2011. Effect of tillage practices on the soil carbon dioxide flux during fall and spring seasons in a Mediterranean vertisol. *Journal of Soil Science and Environmental Management* 2(11):362–369.

Mrabet, R., Moussadek, R., Fadlaoui, A., van Ranst, E. 2012. Conservation agriculture in dry areas of Morocco. *Field Crops Research* 132:84–94.

Mrabet, R., Wall, P. 2015. *Practical Guide for Conservation Agriculture in West Asia & North Africa.* International Center for Agricultural Research in the Dry Areas (ICARDA), Beirut, Lebanon .

Mustin, M. 1987. *Le compost. Gestion de la matière organique.* Editions François Dubusc, Paris, France.

Nair, P.K.R. 1993. *An Introduction to Agroforestry.* Kluwer Academic Publishers, Dordrecht, The Netherlands.

Nair, P.K.R., Nair, V.D., Kumar, B.M., Showalter, J.M. 2010. Carbon sequestration in agroforestry systems. *Advances in Agronomy* 108:237–307.

Naman, F., Soudi, B. 1999. Problématique de gestion de la matière organique des sols en zones irriguées. Proceedings of the 4th African Crop Science Conference, 11–14 October, Casablanca, Morocco, pp. 161–165.

Naman, F., Soudi, B., Chiang, C.N. 2001. Impact de l'intensification agricole sur le statut de la matière organique des sols en zones irriguées semi-arides au Maroc. *Etude et gestion des sols* 8:269–277.

Naman, F., Soudi, B., Chiang, C.N., Adlouni, E.L., C. 2018. Evolution of carbon and nitrogen biomass of vertisol and fersiallitic soil after previous cultivation of wheat and sugar beet in the irrigated perimeter of the Doukkala in Morocco. *Journal of Materials and Environmental Science* 9(5):1544–1550.

Naman, F., Soudi, B., El Adlouni, C., Chiang, C.N. 2015. Humic balance of soils under intensive farming: The case of soils irrigated perimeter of Doukkala in Morocco. *Journal of Materials and Environmental Science* 6:3574–3581.

Nosetto, M.D., Jobbágy, E.G., Paruelo, J.M. 2006. Carbon sequestration in semi-arid rangelands: Comparison of Pinus ponderosa plantations and grazing exclusion in NW Patagonia. *Journal of Arid Environments* 67:142–156.

Ogunkunle, A.O., Eghaghara, O.O. 1992. Influence of land use on soil properties in a forest region of Southern Nigeria. *Soil Use and Management* 8(3):121–124.

Oladele, O.I., Braimoh, A.K. 2013. Climate change mitigation potential of tree crop and alley farming practices in Africa. *Asia Life Sciences* 9(20):185–119.

Oladele, O.I., Braimoh, A.K. 2014. Potential of agricultural land management activities for increased soil carbon sequestration in Africa—A review. *Applied Ecology and Environmental Research* 12(3):741–751.

Onim, J.F.M., Mathuva, M., Otieno, K., Fitzhugh, H.A. 1990. Soil fertility changes and response of maize and beans to green manures of Leucaena, Sesbania and pigeonpea. *Agroforestry Systems* 12:197–215.

Pieri, C. 1989. *Fertilité´ des terres de savane. Bilan de trente ans de recherche et de développement agricoles au Sud du Sahara. Ministeere de la Coopeeration.* CIRAD, Paris, France.

Plaza-Bonilla, D., Arrúe, J.L., Cantero-Martínez, C., Fanlo, R., Iglesias, A., Álvaro-Fuentes Agron, J. 2015. Carbon management in dryland agricultural systems: A review. *Agronomy for Sustainable Development* 35:1319–1334.

Power, J.F., Legg, J.O. 1978. Corp residue management systems. In: W.R. Oshwald (ed.), ASA Special Publication, Vol. 31, Madison, WI, Am. Soc. Agron., pp. 85–100.

Quattara, B., Quattara, K., Sepantite, G., Mando, A., Sedogo, M.P., Bationo, A. 2006. Intensity cultivation induced effects on soil organic carbon dynamic in the western cotton area of Burkina Faso. *Nutricao Cycl. Agro-Ecosystems* 76:331–339.

Ringius, L. 2002. Soil carbon sequestration and the CDM: Opportunities and challenges for Africa. *Climatic Change* 54:471–495.

Roose, E., Barthès, B. 2001. Organic matter management for soil conservation and productivity restoration in Africa: A contribution from Francophone research. *Nutrient Cycling in Agroecosystems* 61:159–170.

Roy, R.N., Nabhan, H. 1999. Soil and nutrient management in sub-Saharan Africa in support of the soil fertility initiative. Proceedings of the Expert Consultation Lusaka, Zambia, 6–9 December. FAO, Rome, Italy.

Sanchez, P., Palm, C., Sachs, J., et al. 2007. The African millennium villages. *Proceedings of the National Academy of Sciences* 104:16775–16780.

Solomon, D., Fritzsche, F., Lehmann, J., Tekalign, M., Zech, W. 2002. Soil organic matter dynamics in the sub-humid agro-ecosystems of the Ethiopian Highlands: Evidence from natural 13C abundance and particle size fractionation. *Soil Science Society of America Journal*, 66: 969–978.

Solomon, D., Lehmann, J., Zech, W. 2000. Land-use effects on soil organic matter properties of chromic luvisols in semi-arid northern Tanzania: Carbon, nitrogen, lignin and carbohydrates. *Agriculture, Ecosystems and Environment* 78:203–213.

Soltner, D. 2003. Les bases de la production végétale. *Tome I: Le Sol et Son Amélioration*. 23ème Edition, Collection Sciences et Techniques Agricoles.

Somarriba, E., Cerda, R., Orozco, L., Cifuentes, M., Dávila, H., Espin, T., Mavisoy, H., Ávila, G., Alvarado, E., Poveda, V., Astorga, C., Say, E., Deheuvels, O. 2013. Carbon stocks and cocoa yields in agroforestry systems of Central America. *Agriculture, Ecosystems and Environment* 173:46–57.

Soudi, B., Chiang, C.N., Zraouli, M. 1990. Seasonal variation of mineral nitrogen and the combined effect of soil temperature and moisture content on mineralization. *Actes de l'institut Agronomique et vétérinaire Hassan II* 10(1):29–38.

Soudi, B., Naâman, F., Chiang, C.N. 2000. *Problématique de gestion de la matière organique des sols: Cas des périmètres irrigués du Tadla et des Doukkala (Maroc)*, B. Soudi (ed.). Séminaire 'Intensification agricole et qualité des sols et des eaux', Rabat, Marocco, 2–3 November, pp. 25–30.

Swift, M.J., Heal, O.W., Anderson, M.J. 1979. *Decomposition in Terrestrial Ecosystems*. Blackwell Scientific Publications, Oxford, UK.

Tate, R.L. 1995. *Soil Microbiology*. Wiley, New York, NY.

Theng, B.K.G., Curchman, G.J., Newman, R.H. 1986. The occurrence of interlaycr clay/-organic complexes in two New Zealand soils. *Soil Sciences*, 1–42:262–266.

Tiessen, H., Feller, C., Sampaio, E.V.S.B., Garin, P. 1998. Carbon sequestration and turnover in semiarid savannas and dry forest. *Climatic Change* 40:105–117.

Tieszen, L.L., Tappan, G.G., Touré, A. 2004. Sequestration of carbon in soil organic matter in Senegal: An overview. *Journal of Arid Environments* 59:409–425.

Traoré, O., Somé, N.A., Traoré, K., Somda, K. 2007. Effect of land use change on some important soil properties in cotton-based farming system in Burkina Faso. *International Journal of Biological and Chemical Science* 1(1):7–14.

Trouve, C., Mariotti, A., Schwartz, D., Guillet, B. 1994. Soil organic carbon dynamics under *Eucalyptus* and *Pinus* planted on savannas in the Congo. *Soil Biology and Biochemistry* 26:287–295.

Unruh, J.D., Houghton, R.A., Lefebvre, P.A. 1993. Carbon storage in agroforestry: An estimate for sub-Saharan Africa. *Climate Research* 3:39–52.

Vågen, T.-G., Lal, R., Singh, B.R. 2005. Soil carbon sequestration in sub-Saharan Africa: A review. *Land Degradation and Development* 16(1):53–71.

Vigil, M.F., Kissel, D.E. 1991. Equations for estimating the amount of nitrogen mineralized from crop residues. Soil Sci. *Società Amer.* 5:757–761.

Waksman, S.A. 1924. Influence of microorganisms upon the carbon-nitrogen ratio in soil. *Journal of Agricultural Science* 14:555–562.

West, O., Post, W.M. 2015. Soil organic carbon sequestration rates by tillage and crop rotation: A global data analysis. *Soil Science Society of America Journal* 66:1930–1946.

Wise, R., Cacho, O. 2005. A bioeconomic analysis of carbon sequestration in farm forestry: A simulation study of Gliricidia sepium. *Agroforestry Systems* 64:237–250.

Woomer, P.L., Palm, C.A., Qureshi, J.N., Kotto-Same, J., 1998. Carbon sequestration and organic resource management in African smallholder agriculture. In: *Management of Carbon Sequestration in Soils, Advances in Soil Science (series)*. Lal, R., Kimble, J.M., Follett, R.F., Stewart, B.A. (eds.). CRC Press, Florida, pp. 158–173.

Woomer, P.L., Touré, A., Sall, M. 2004. Carbon stocks in Senegal's Sahel transition zone. *Journal of Arid Environments* 59:499–510.

World Bank. 2012. Carbon sequestration in agricultural soils. Economic and Sector Work. Report No.: 67395-GLB. Washington DC, World Bank.

Zech, W., Senesi, N., Guggenberger, G., Kaiser, K., Lehmann, J., Miano, T.M., Miltner, A., Schroth, G. 1997. Factors controlling humification and mineralization of soil organic matter in the tropics. *Geoderma* 79(1–4):117–161.

12 Degradation and Climate-Smart Options for Restoring the East African Soils

Kennedy Were, Bal Ram Singh, and George Ayaga

CONTENTS

If soils are not restored, crops will fail even if rains do not; hunger will perpetuate even with emphasis on biotechnology and genetically modified crops; civil strife and political instability will plague the developing world even with sermons on human rights and democratic ideals; and, humanity will suffer even with great scientific strides.

—Rattan Lal

12.1 INTRODUCTION

Soil is a basic natural resource that supports all forms of life on Earth, productive landscapes and societal development. It underpins provision of essential ecosystem goods and services, such as food, fodder, fuel wood and fiber; habitat for flora and fauna; the physical matrix, chemical environment and biological setting for water, nutrients, air and heat exchange for organisms; water purification; recycling of materials; physical support to organisms and structures; source and sink for pollutants; climate regulation; and regulation of hydrological processes, including infiltration, percolation, drainage, stream flow and water storage (Osman 2014). Due to this generosity, judicious management of soils is of paramount significance for the sustainability of ecosystem goods and services that they provide and for overall human welfare and development.

Despite their essential ecological functions, soils are unappreciated and have been politically and physically neglected leading to extensive degradation in most terrestrial biomes and agro-ecologies globally (Agriculture for Impact 2014). The major degradation threats from which soils need protection include: erosion, decline in organic matter and biodiversity, soil contamination, surface sealing, soil compaction, salinization, floods and landslides. The process of soil degradation per se is induced by natural or anthropogenic factors that alter the structure and quality of soils leading to adverse changes in their properties (e.g., nutrient content) and lowering their current and future capacity to produce ecosystem goods and services (Gomiero 2016; Bouma and Batjes 2000). Even though soil degradation is considered a global pandemic, the African continent is the most affected. The Global Assessment of Soil Degradation (GLASOD) project commissioned by the United Nations Environment Programme (UNEP) estimated that 65 per cent of agricultural land, 31 per cent of permanent pastureland and 19 per cent of forest and woodland in Africa was degraded (Oldeman 1994; Bouma and Batjes 2000). Similarly, UNEP (2006) reported that about 14 per cent of the total land area in East Africa was suffering from severe to very severe degradation. This is inimical to food security, vibrant rural livelihoods, ecosystem sustainability and the achievement of sustainable development goals (SDGs) in the region. As in other parts of the world, chemical degradation (especially nutrient depletion) and physical degradation (especially water and wind erosion) are the principal forms of soil degradation that threaten the productivity of arable lands in Africa. For example, at country level, Stoorvogel and Smaling (1990) reported that nitrogen (N), phosphorus (P) and potassium (K) were being mined at the rates of 42 to 46, 1 to 3 and 29 to 36 kg ha^{-1} year^{-1}, respectively, in Kenya, and at the rates of 27 to 32, 4 to 5 and 18 to 21 kg ha^{-1} year^{-1}, respectively, in Tanzania, while at the regional level, Smaling (1997) estimated the rates of NPK outflows at 22, 2.5 and 15 kg ha^{-1} year^{-1}, respectively. These large negative nutrient balances are attributed to the removal of harvested crops, leaching, soil erosion and overexploitation of soil nutrient stocks without sufficient compensation of the losses using fertilizers (Zingore et al. 2015).

Although soil degradation is a natural process, it is accelerated by a complexity of interacting biophysical and socioeconomic factors; for example, population growth and climate change. According to the official United Nations population estimates and projections, the world population, growing at the rate of 1.10 per cent per year, reached 7.6 billion as of mid-2017, and is projected to be 8.6 billion in 2030, 9.8 billion in 2050 and 11.2 billion in 2100 (United Nations 2017). Several scholars have argued that as population grows, the degradation of the soil resource base increases (Muchena et al. 2005). Essentially, a growing population triggers a cobweb of processes, such as deforestation, diminishing landholdings, increasing pressure on agricultural land, population shift from high potential agricultural areas to marginal lands, and food shortages, which then fuel soil degradation. In addition, changing climate has an intricate linkage with soil degradation because of the feedback between soil degradation and the elements of climate. Climate change aggravates soil degradation by altering the spatial and temporal patterns of temperature, rainfall, solar radiation and winds. For instance, it is expected that climate change will increase soil erosion in parts of sub-Saharan Africa (SSA) through heavy rainfall and increased wind speed. Soil properties and processes, including the decomposition of soil organic matter (SOM), leaching and soil water regimes, will also be affected by rising temperatures. The resultant soil degradation will, in turn, reinforce the detrimental effects of temperature rise on the actual agricultural yields (Were et al. 2016). Therefore, combating soil degradation is indispensable to guarantee food security, ecological health and sustainable development in Africa. This calls for climate-smart soil management implying that the strategy should be to: (i) increase agricultural productivity and incomes in environmentally and socially sustainable ways; (ii) adapt and build farmers' resilience to climate change; and (iii) contribute to climate change mitigation by reducing the emission of greenhouse gases (GHGs) (i.e., CO_2, CH_4 and N_2O) and increasing soil organic carbon (SOC) storage on farmlands (FAO 2013).

This chapter focuses on the degradation of soils across East Africa, including Kenya, Uganda, Tanzania, Rwanda, Burundi, Ethiopia and Somalia. We begin with a brief description of the East African setting, followed by details of the soil resources and status of soil degradation in the region. We then discuss the nexus between climate change and soil degradation in East Africa, and finalize by describing the best-bet climate-smart options for restoring, managing and building resilience of the degraded soils in the region.

12.2 THE EAST AFRICAN SETTING

The seven East African countries, comprising Kenya, Uganda, Tanzania, Rwanda, Burundi, Ethiopia and Somalia (Figure 12.1) cover a total land area of about 3.3 million km^2 (The World Bank Group 2017) with a total population of about 293 million (United Nations 2017). The arable land area accounts for between 2 and 37 percent, permanent cropland area between 0 and 14 per cent, and forest land area between 8 and 52 per cent of the total land areas of the countries. The East African landscape is a rich mosaic of resources, including ecoregions, soils, climates, water, flora and fauna. For example, the climate varies from the warm tropics with very low rainfall in northeastern Kenya and Somalia to the

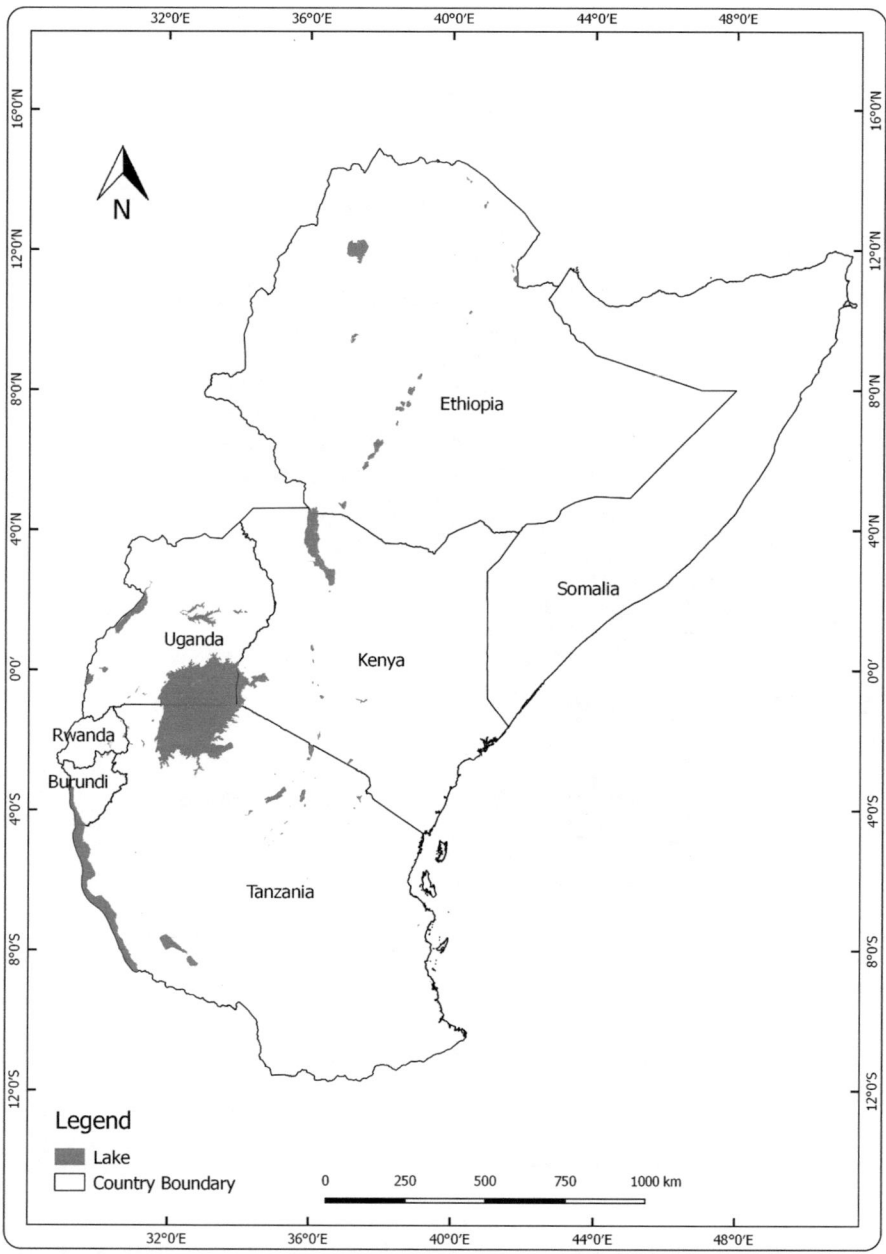

FIGURE 12.1 Location map of the East African countries.

cool tropics in the highland areas, while the vegetation ranges from the woodlands, savannahs, grasslands, bushlands and thickets, to the Afro-alpine heath and moorlands, to the desert scrublands and shrublands. In addition, the ecoregions, defined by temperature and moisture, include the tropic warm/arid, tropic warm/semi-arid, tropic warm/humid, tropic warm/sub-humid, tropic cool/arid, tropic cool/semi-arid, tropic cool/humid, and tropic cool/sub-humid zones (Figure 12.2) (HarvestChoice 2015).

Agriculture, nature-based tourism and extraction of other land-based natural resources (e.g., minerals) are the main land uses that drive economic growth in East Africa. Like in most parts

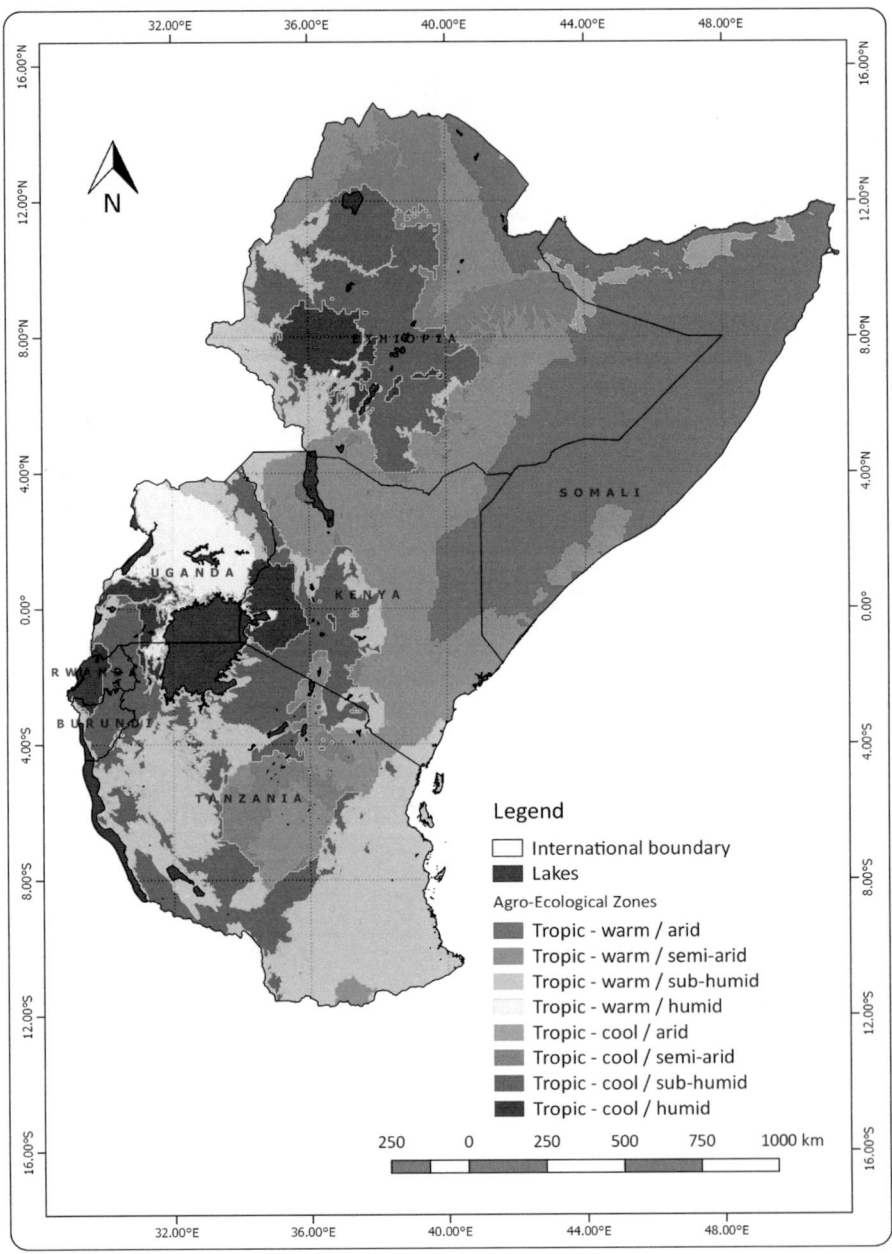

FIGURE 12.2   (See color insert.) The ecoregions of East Africa. (Data from HarvestChoice, 2015.)

TABLE 12.1

Total Population, Rural Environment and Land Use in East Africa

Country	Total population ('000s)	Rural population (% of total)	Land area ('000 km²)	Forest area (% of land area, 2015)	Permanent cropland (% of land area, 2014)	Arable land (% of land area, 2000)
Kenya	49,700	74	569.1	7.8	0.9	2.6
Uganda	42,863	84	200.5	10.4	11.0	26.5
Tanzania	57,310	68	885.8	52	2.4	9.7
Rwanda	12,208	71	24.7	19.5	10.1	36.5
Burundi	10,864	88	25.7	10.7	13.6	37.4
Ethiopia	104,957	81	1,000.0	12.5	1.1	10
Somalia	14,743	60	627.3	10.1	0.0	1.7
Total	**292,645**		**3,333.1**			

Source: Data from United Nations (2017) and The World Bank Group (2017).

of SSA, East African agriculture is characterized by small-scale subsistence crop and livestock production under rainfed conditions. The smallholder farmers, mostly women, cultivate small parcels of degraded land, and often have poor access to irrigation, affordable inputs (e.g., fertilizer), financial credit services, technology, input and output markets, agricultural extension services and agricultural information. This explains the low yields for most crops. Cropping systems, including maize (*Zea mays* L.), beans (*Phaseolus vulgaris*), sorghum (*Sorghum bicolor*), finger millet (*Eleusine coracana*) and teff (*Eragrostis tef*), are common, in addition to extensive grazing and intensive dairy, vegetable, coffee (*Coffea*) and tea (*Camellia sinensis*) production.

The challenges facing East African agriculture are compounded by the high population growth in the region. Parts of East Africa support some of the densest populations in the world (Muchena et al. 2005). For example, the population density of Burundi and Rwanda is approximately 390 and 460 persons per km², respectively (United Nations 2017). The attendant increase in demand for food, feed, fiber, fuel and shelter has had a profound impact on food security and soil resources. The FAO (2015) reported that about 124 million people in East Africa were hungry and unable to consume enough food to lead active and healthy lives in 2014–2016. Widespread land degradation in the region, largely driven by conversion of marginal lands and fragile natural ecosystems (e.g., forests) to agroecosystems, has also been documented (UNEP 2006). The degradation occurs mainly through soil nutrient depletion and accelerated soil erosion. Bekunda et al. (2002) observed that nutrient mining in East Africa is among the highest in SSA, with an estimated annual nutrient depletion rate of 41 kg N, 4 kg P, and 31 kg K per hectare. (Table 12.1)

12.3 SOIL RESOURCES AND STATUS OF SOIL DEGRADATION IN EAST AFRICA

As mentioned in Section 12.2, East Africa is generously endowed with soil resources, which, coupled with climatic diversity, have allowed for the domestication of several crops (e.g., maize, beans, teff, finger millet and sorghum). Maintaining the health of these soil resources underpins ecological stability and sustainability, food security, viable rural livelihoods and agricultural resilience under a changing climate in the region.

12.3.1 Soil Classification

According to *Soil Taxonomy*, the soils of East Africa are classified as Ultisols, Oxisols, Entisols, Alfisols, Aridisols, Andisols, Vertisols, Mollisols and Inceptisols (Eswaran et al. 1996). Interested readers are referred to Osman (2014) for detailed description of these soil orders. Vertisols are distributed mainly along the rift valley, but large contiguous areas are also found in Kenya, Tanzania and Ethiopia. The upper slopes of Mt Kilimanjaro, Mt Kenya and parts of the Ethiopian and Kenyan Highlands have Andisols, while the countries adjacent to the Congo basin, including Uganda, Rwanda and Burundi, are dominated by SOM-rich Oxisols. Tanzania has a very large extent of Ultisols, which have a close association with Oxisols. Aridisols are confined to the northeastern and coastal provinces of Kenya and the southeasternmost parts of Ethiopia and Somalia, while Alfisols occur in Ethiopia and Tanzania with a few pockets in Kenya. Similarly, Entisols occupy parts of Ethiopia and Kenya with sporadic occurrences in Tanzania, and Inceptisols are largely located in Ethiopia. Lastly, Mollisols occur in Kenya and Tanzania, but their extent is rather small.

12.3.2 Soil Degradation

Just to reiterate, soil degradation is a widespread problem throughout East Africa, the causes, types, extent and impact of which are varied. This is undermining the capacity of the soils to perform their ecological functions, as well as to produce plant materials of good quality and quantity. Details of the causes, types, extent and impact of soil degradation follow.

12.3.2.1 Causes of Soil Degradation

Although soil resources in East Africa are mostly managed at local level, their condition and behavior is a consequence of complex and cumulative interactions between biophysical, socioeconomic and political factors operating in specific spatial and temporal contexts. The interactions either trigger, or exacerbate soil degradation. Typical proximate sources of soil degradation in the East African region include overcultivation (extractive farming), overgrazing, deforestation (agricultural expansion), shifting cultivation and overexploitation of vegetation (Muchena et al. 2005; Tully et al. 2015; Nkonya et al. 2016) (Figure 12.3). These proximate factors are underpinned by fundamental sociopolitical and economic structures and processes (e.g., population pressure, poverty, land tenure,

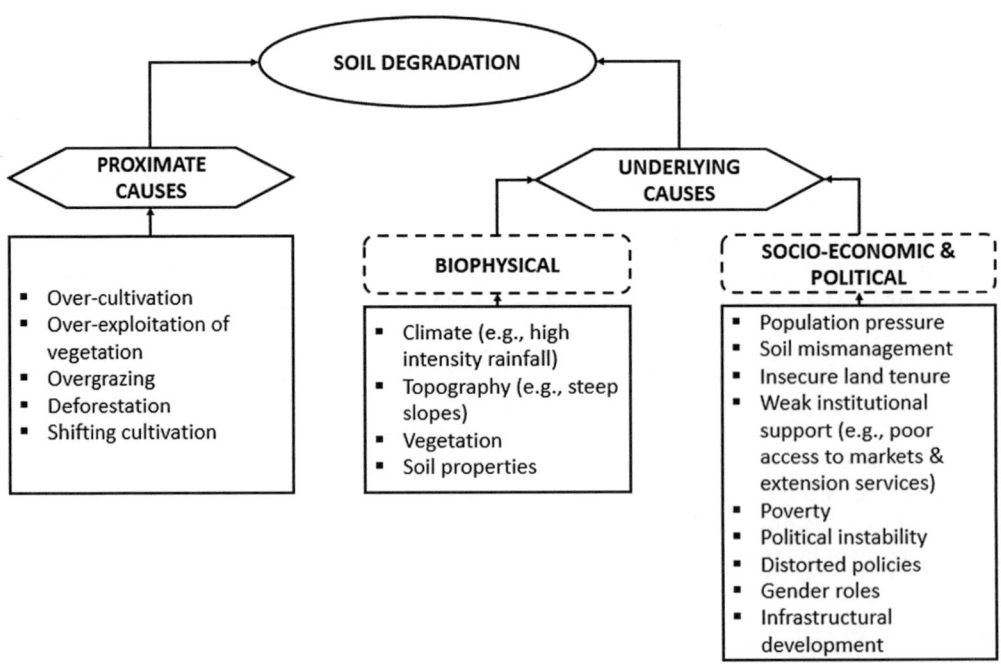

FIGURE 12.3 Driving factors of soil degradation in East Africa.

institutional support, policies and political stability) and biophysical factors (e.g., terrain, soils and climate), which influence the type of and the effectiveness of degradation processes. For example, Woomer and Muchena (1993) noted that soil degradation in East Africa was becoming severe owing to the rising population. The increasing pressure on agricultural land was resulting in much higher nutrient outflows and the subsequent breakdown of traditional soil fertility management strategies, such as fallowing. Some of the drivers of soil degradation are gradual and operate over a long period of time (i.e., progressive), while others are rapid and abrupt (i.e., episodic). Moreover, the proximate factors operate mainly at the local levels, while the underlying factors operate diffusely at regional and global levels. It is also worth noting that the impact of soil degradation and the attendant socio-ecological responses may reinforce, or suppress, the drivers leading to new conditions.

12.3.2.2 Types of Soil Degradation

Over time, the soils of East Africa have undergone tremendous physical, chemical and biological alterations as a result of both natural and human-induced processes (Figure 12.4). The dominant form of physical degradation is erosion by water (Figure 12.5), although other forms, such as decline in soil structure and tilth, surface sealing and crusting, compaction and hard setting, waterlogging and desertification also exist. For example, the gross soil loss from the Ethiopian highlands, one of the severest cases of land degradation in the world, has been estimated at up to 1,900 million Mg per year, 80 per cent (i.e., 1,520 million Mg) of which is from the croplands, which constitute only 22 per cent of the total highland area (Tolcha 1991). In the entire country, the average annual soil loss rates vary between 42 and 300 Mg/ha for croplands (Gebreselassie et al. 2016; Hurni et al. 2015). Other studies have also shown that loss of soil due to erosion ranges from 35 to 246 Mg/ha per year in Rwanda (Olson and Berry 2003). The major process that sets in motion chemical degradation is nutrient depletion, as shown by large negative nutrient balances in Table 12.2, although other processes, including acidification, salinization, leaching and chemical pollution, also occur (Lal 1997). Biological degradation is primarily manifested through loss of soil fauna (e.g., earthworms), decline in soil organic matter (SOM) and increase in soil-borne pathogens. Most of these soil degradation

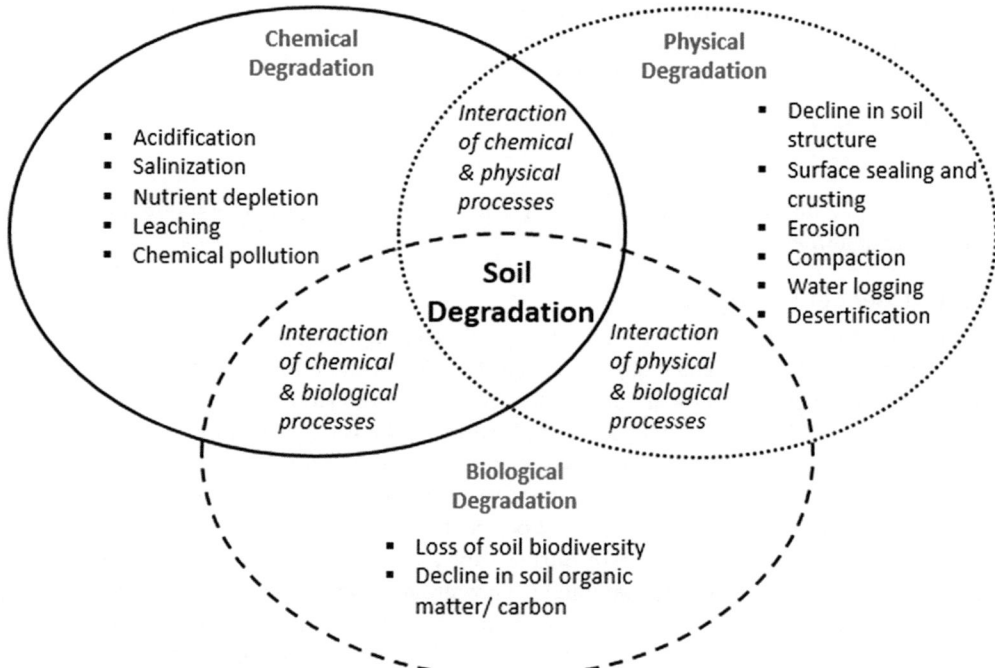

FIGURE 12.4 Types of soil degradation in East Africa.

FIGURE 12.5 **(See color insert.)** Illustrations of water and wind erosion in Kenya. (Photo Credit: authors.)

TABLE 12.2
Nutrient Balances in East Africa

		Nutrient balances (kg/ha^{-1} year^{-1})		
Site	**Study site**	**N**	**P**	**K**
East African highlands	Burundi, Ethiopia, Kenya, Rwanda	–36	–5	–25
Kenya	Aggregate at country level	–46 to –42	–3 to –1	–36 to –29
Tanzania	Aggregate at country level	–32 to –27	–5 to –4	–21 to –18
Uganda	Aggregate at country level	–1	–8	–43

Source: Muchena et al. (2005).

processes occur simultaneously with a severe impact on the functions and potential productivity of the East African agroecosystems.

12.3.2.3   Extent of Soil Degradation

Minimal assessments of the extent and severity of land degradation in East Africa have been conducted with variable outcomes depending on the source and computation methods. Based on qualitative expert judgement, Global Assessment of Human-Induced Soil Degradation (GLASOD) estimated that about 65 per cent of agricultural land in Africa is degraded, three-quarters of which is severely degraded (Figure 12.6). This partly explains the high levels of hunger, malnutrition and poverty on the continent despite the advances in agricultural research, including biotechnology. More recently, UNEP (2006) reported that 14 per cent of the total land area in East Africa suffers from severe to very severe degradation. In Burundi and Rwanda, 76 and 71 per cent of the respective country's total area are very severely degraded. They are followed by Uganda, Kenya and Ethiopia, where areas with severe to very severe degradation constitute about 53, 30 and 28 per cent of total land area, respectively. Bai et al. (2008) also estimated that about 23 and 30 per cent of Kenya's total land area was subject to very severe degradation in 1997 and a better part of the 2000s, respectively, and about 64 per cent to moderate degradation. Specifically, land degradation is more pronounced

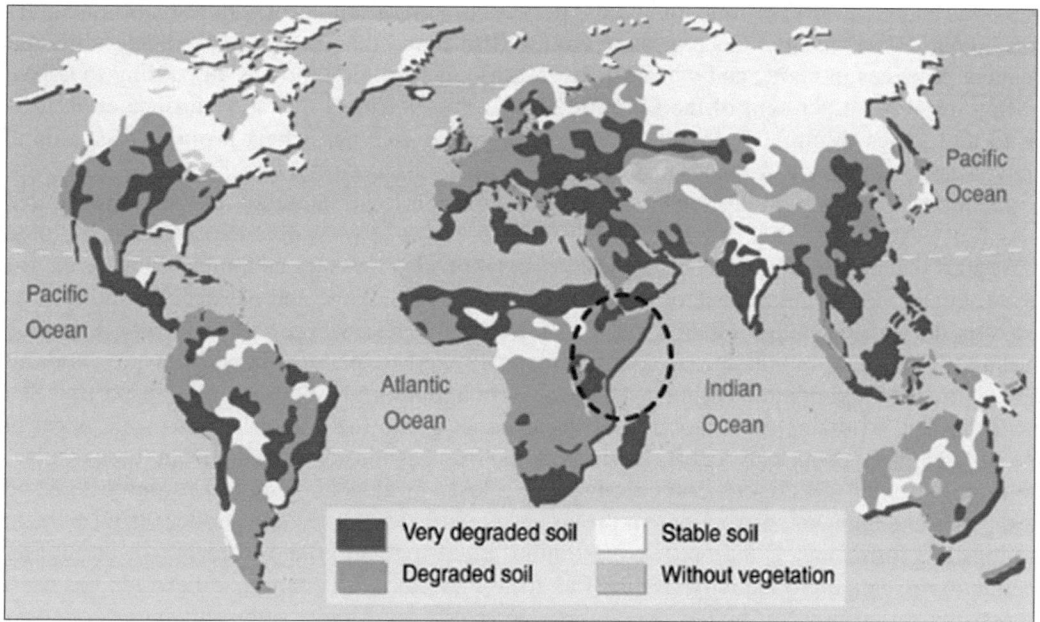

FIGURE 12.6 (See color insert.) Human-induced soil degradation in East Africa and the world. (Source: UNEP, International Soil Reference and Information Centre (ISRIC), and *World Atlas of Desertification*, 1997.)

in the eastern and northeastern parts of Kenya where 12.3 per cent of the land suffers from severe degradation, 52 per cent from moderate degradation and 33 per cent is vulnerable to land degradation (Muchena 2008; UNEP 2006). Similarly, Gebreselassie et al. (2016) reported that over 85 per cent of Ethiopia's total land area was moderately to very severely degraded, and about 75 per cent was affected by desertification. Soil erosion is particularly serious in the high and low potential cereal zones of the north-central highlands in Ethiopia. Lastly, Kirui (2016) found that the hot spots of land degradation covered about 51 per cent of Tanzania's total land area. These figures clearly show the unabated spread of soil degradation in East Africa, which does not augur well for food security and rural livelihoods.

12.3.2.4 The Impact and Cost of Soil Degradation

The socioeconomic and environmental ramifications of soil degradation in East Africa are severe because over 65 per cent of the population is rural, relies on natural resources for livelihood and has low capacity to restore degraded soils. The immediate and most significant socioeconomic consequence of soil degradation is the reduction in crop and livestock yields, which translates to food scarcity and hunger, low income, poverty and increased cost of soil fertility management. Physical degradation processes (e.g., erosion and compaction) impinge on the growth and yields of crops and grazing resources by reducing the rooting depth, available water and nutrient reserves. In particular, erosion detaches the colloidal fractions of soils (i.e., clay and humus) – which are requisite for soil fertility, aggregation, structural stability and favorable pore size distribution – and transports them in runoff. Severe erosion-induced losses in crop productivity have been observed in the highlands of Ethiopia, Kenya and Uganda (Lal 1995, Lal and Singh 1998). This is in tandem with the observations documented in literature that cereal availability per capita in East Africa was on the decline (i.e., from 136 kg/year in the 1980s to 118 kg/year in the 2000s) owing to land degradation (Kirui 2016).

Furthermore, Berry (2003) and Bojö and Cassells (1995) reported that the annual costs of land degradation related to soil erosion and loss of essential nutrients from agricultural and grazing lands was about US $106 million, or about 3 per cent of agricultural gross domestic product (GDP) in Ethiopia. More recently, Gebreselassie et al. (2016) found that the use of land degrading management practices in maize and wheat farms in Ethiopia resulted in losses amounting to US $162 million (or about 2 per cent of the GDP in 2007), while Kirui (2016) and Mulinge et al. (2016) found that losses resulting from such practices in maize, rice and wheat farms in Tanzania and Kenya were equivalent to US $1.8 million and US $270 million, respectively. In other words, soil degradation is impacting negatively, not only on agronomic productivity, but also on economic growth in East Africa.

Besides the socioeconomic impacts, soil degradation also has serious implications for environmental quality. For example, one of the consequences of biological and physical degradation processes is the transformation of soils into net sources of GHGs with a positive feedback to global warming (Lal 2014a, 2014b, 2015a). Olson et al. (2016) aptly explained that erosion processes alter the dynamics of soil organic carbon (SOC) stocks in agricultural land units by transporting SOC-rich sediment, oxidizing SOC stocks and releasing CO_2 into the atmosphere, as well as causing loss of SOC through surface runoff. Environmental impacts mostly emanate from the interplay of the three forms of soil degradation culminating in loss of valuable biological resources (i.e., species, genes and habitats), disruptions in provision of ecosystem goods and services (e.g., elemental cycling, soil formation, C (carbon) sequestration, flood regulation and water purification) and a decline in net biome productivity (NBP) (Lal 2014b, 2015a). For instance, soil erosion has caused excessive siltation in water bodies (e.g., lakes and rivers) resulting in reduced volumes of surface water and hydroelectric generation capacity of dams, eutrophication and pollution of water reservoirs. In Ethiopia, the hydroelectric generation capacity of the Koka dam has been curtailed by a 30 per cent loss of its total storage volume to sedimentation (Gebreselassie et al. 2016).

12.4 CLIMATE CHANGE AND SOIL DEGRADATION IN EAST AFRICA

Climate change is, undoubtedly, the greatest environmental challenge of the 21st century and its impacts will increase the vulnerability of agricultural systems in East Africa. Numerous models have predicted rising temperatures and changes to the duration of rainy seasons, with more erratic conditions for growing crops throughout the region (Thompson et al. 2015). Observed and projected figures for precipitation in East Africa indicate variable changes at both spatial and temporal scales due to location-specific physical processes (IPCC 2014). For example, the rapid warming of the Indian Ocean has reduced the amounts of rainfall and increased the frequency of drought spells in the months of March to June (William et al. 2012). Overall, the short rains in the months of December to February (DJF) will increase by 5–20 per cent as the Indian Ocean warms, while the long rains in the months of June to August (JJA) will decrease by 5–10 per cent by 2050 (Christensen et al. 2007). With respect to temperature, greater warming is expected across all seasons in the 21st century with the temperature increases exceeding the global mean increase of 2.5 °C by 2099. As such, heat extremes will be more frequent during the hot season and hyper-arid and arid areas will grow by 3 per cent (IPCC 2014).

The foregoing climatic changes will adversely affect East African soils considering the strong links between the pedosphere and the atmosphere. Essentially, climate change will aggravate soil degradation by altering the spatial and temporal patterns of temperature, rainfall, solar radiation and winds. Already, Lal (2015b) reported that water and wind will increase soil erosion by 36 per cent in Africa between 1980 and 2090 under a changing climate. Climate change will also affect soil erosion through changes in the rates of decomposition and evapotranspiration, soil erodibility, land-use changes and net primary production (NPP). The warming will increase the rates of NPP (which inputs detritus to the soils) and decomposition (which removes C from the soils). However, in general, the warming will deplete SOC stocks by stimulating the rates of decomposition more than of NPP. This is because of the greater sensitivity of SOM to temperature than NPP (Mengistu et al. 2015). Depletion of SOC will, in turn, affect soil quality by undermining the physical, chemical and biological properties that determine the fertility and functions of the soil. For example, depletion of SOC will reduce the activity and diversity of soil biota, nutrient supply, cation exchange capacity (CEC), aggregation and water-holding capacity, as well as increase CO_2 emissions causing a positive feedback to global warming. In view of the last, it suffices to say that soil degradation is both a victim and a culprit in the climate change discourse.

12.5 CLIMATE-SMART OPTIONS FOR RESTORING AND MANAGING DEGRADED SOILS IN EAST AFRICA

Although the East African soils are severely degraded, they can still be restored, conserved and enhanced through conversion to judicious land use and the adoption of proven climate-smart soil management practices (CSMPs), which also lower GHG emissions. The restoration of degraded ecosystems can generate private and public benefits; hence, it constitutes a potentially important means of generating "win-win" solutions to address poverty, food insecurity and environmental issues (Scherr et al. 2012). For example, restoration of degraded ecosystems enhances human livelihoods and resilience of local communities by providing food, livestock feeds and fuelwood (Milder et al. 2011). It also repairs ecosystems and improves ecological resilience in terms of watershed functions, habitat for wildlife, biodiversity conservation, reducing soil erosion and C sequestration (Bernazzani et al. 2012; Scherr and Shames 2009).

Numerous CSMPs for soil restoration exist, yet none can be said to be the panacea for soil degradation in East Africa. This implies that soil restoration projects should objectively assess, select and implement appropriate CSMPs for a given locality taking into consideration the biophysical and socioeconomic characteristics. An objective assessment would also facilitate an understanding of the merits and demerits of the CSMPs, which is crucial for identifying synergies and trade-offs. According to Lal (2015a), the most desirable technologies and practices should aim at: (i) minimizing losses of water and nutrients out of the ecosystem; (ii) creating positive ecosystem C, nutrient and water budgets; (iii) enhancing biodiversity; (iv) strengthening plant-soil feedback; and (v) minimizing soil disturbances and risks of erosion (Figure 12.7). Some examples of CSMPs that have been documented in literature include conservation agriculture (CA), precision agriculture, integrated nutrient management (INM), soil C sequestration, residue retention (mulching), soil biological management, agro-forestry, soil and water conservation, controlled grazing at optimal stocking rates and use of organic amendments (e.g., manure), crop residues, cover crops, biochar, improved plant varieties with greater root mass, crop rotations, and soil and water conservation structures (e.g., terraces, tied ridges and windbreaks). The following paragraphs expound on some of these CSMPs.

To begin with, the CA strategy is based on the principles of minimum soil disturbance, maintenance of at least 30 per cent permanent organic mulch on soil surface and a diversified cropping system through intercropping or crop rotation (Aune and Coulibaly 2015; FAO 2013) (Figure 12.8). If implemented properly on suitable soil types, CA has a wide range of co-benefits, including: (i) increased infiltration capacity, soil water content, SOM (C storage), soil biological activity and flexibility in planting and harvesting; (ii) reduced weeds, soil erosion, soil surface temperatures, soil compaction, emission of soil C to the atmosphere and fossil fuel consumption by avoiding ploughing; and (iii) improved soil tilth and fertility. However, concerns have been raised with regard to the labor demand for weed control and limited availability of mulching material and input.

The use of INM to address nutrient depletion and soil fertility constraints is another effective approach for achieving soil restoration. INM provides a framework where the best of organic and conventional practices are combined in a way that is environmentally appropriate and sustainable (Agriculture for Impact 2014). Further, INM recognizes the importance of knowledge on how to adapt the practices to local conditions with a view to maximizing nutrient and water-use efficiency, as well as improving crop productivity. Among the common INM practices are: intercropping and rotating cereals with legumes; using crop residues, deep-rooted cover crops and biosolids

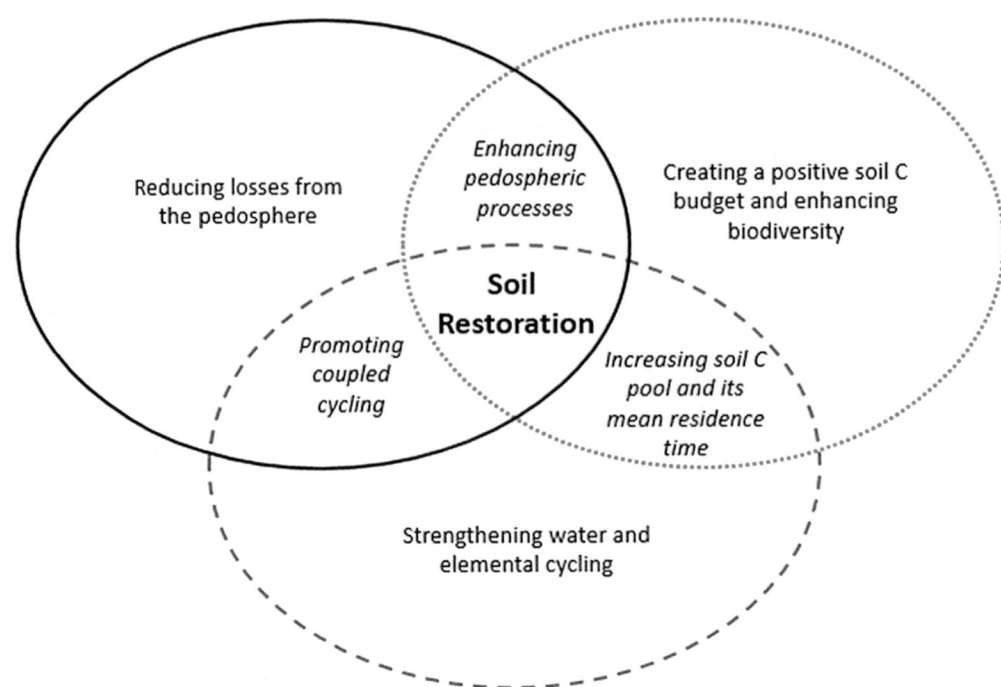

FIGURE 12.7 Three strategies for restoring soils in East Africa.

(Source: adapted from Lal, 2015a).

(e.g., manure and compost); using mycorrhizal inoculation; and applying both organic and inorganic materials to the same crops (Zingore et al. 2015; Lal 2009). The inclusion of legumes in cereal systems allows cereals to benefit from the N that is fixed by legumes (i.e., biological nitrogen fixation). These are useful strategies to enhance soil fertility, improve soil structural stability (or aggregates), increase productivity and sequester C in an SOC pool. It is evident that INM strategy

FIGURE 12.8 Bean (*Phaseolus vulgaris*) production under CA in Laikipia, Kenya. (Photo Credit: African Conservation Tillage Network.)

is in accord with organic farming because it integrates elements, such as nutrient recycling using compost and manure.

Similar to the developed world, where farmers practice precision agriculture, the East African smallholder farmers also need to know the exact nutrient and water requirements, as well as the specific sites in need, in order to minimize the amount of inputs (i.e., fertilizers and water) and costs (Agriculture for Impact 2014). Microdosing of mineral fertilizer – the application of small amounts of fertilizer in the planting pocket – has been found to be a promising approach that can restore the soils, increase productivity and give good returns for a small investment in fertilizer. Microdosing of water can also be achieved through precision or micro-irrigation technologies (e.g., drip irrigation) that ensure the right amount of water is placed close to the growing plant at the right time, or that fewer drops of water are used per crop. The agroecological merits of microdosing are varied. They include the creation of a positive effect on soil C and soil quality by increasing the root biomass and straw production, the promotion of a more efficient resource use and the alleviation of problems associated with parasitic weeds, particularly *Striga hermonthica* (Aune and Coulibaly 2015). However, some critics have argued that the returns from microdosing are too low to make up for the costs of investing in fertilizer, and that it contributes to nutrient mining. These are substantial comments, which warrant further research to obtain conclusive answers. Fine resolution digital maps of soil fertility constraints can also be used to help farmers take spatially targeted, or site-specific remedial measures to address nutrient deficiencies. For example, under the guidance of Ethiopia's Agricultural Transformation Agency (ATA), farmers growing hybrid maize were able to achieve 6–8 Mg/ha when they applied an appropriate balance of NPK coupled with boron (B) that was determined to be deficient in the country using digital soil maps (Agriculture for Impact 2014).

The SOC sequestration through the uptake and transfer of atmospheric CO_2 into the soil reservoir by plants, plant residues and other organic solids also holds the key to soil restoration in East Africa. This is because SOC is a major determinant of soil quality and functions. For instance, it fosters the storage of water and nutrients available to the plant, the activity and diversity of soil biota, soil structure and tilth, and the susceptibility of soil to erosion. When the concentration of SOC falls below a certain threshold (i.e., 1.0–1.5 per cent, depending on the soil type, climate and land use), fundamental soil properties are adversely affected inhibiting plant growth. Thus, CSMPs that replenish the SOC pool and maintain it above the critical level are essential for reducing soil degradation risks and reversing degradation trends. Most of these CSMPs have already been highlighted above (e.g., agro-forestry and the use of biochar, manure and crop residues). Basically, the strategy is to create a positive ecosystem C budget through: (i) increasing the soil application of biomass C; (ii) decreasing losses of SOC by soil erosion; (iii) moderating soil temperature and reducing the rates of mineralization; and (iv) enhancing the mean residence time of SOC by increasing soil aggregation and stability (Lal 2014a). The recarbonization of soil in order to address SOC loss offers a myriad of benefits, such as improved soil and water quality, increased water and nutrient retention capacity, efficiency in the use of inputs, decreased vulnerability to extreme climatic events and susceptibility to soil degradation, as well as sustained agronomic productivity and food security.

Other proven options that can reduce the risk and trends of soil degradation in East Africa include improving activity and species diversity of soil fauna and flora (e.g., earthworms and termites), adoption of soil conservation measures on eroded landscapes, and optimal stocking rates. The reduction in the risk of soil degradation through enhancing soil biological resources can be ascribed to the critical roles in key ecosystem functions that soil biota perform, such as biomass decomposition, nutrient cycling, moderating atmospheric CO_2, improving aggregation and creating disease suppressive soils (Lal 2015a). With regard to optimal stocking, it is widely known that high stocking rates accounts for much of the degradation of rangelands. High stocking rates can reduce plant cover due to overgrazing and trampling effects, as well as adversely affecting soil physical qualities (e.g., structure and water infiltration), increasing soil erosion risks and reducing biomass productivity. Therefore, rehabilitation of degraded rangelands through exclosures can be effective in restoring vegetation, reducing soil erosion hazard and improving soil fertility. Finally, soil and water conservation structures and measures (e.g., terracing, tied ridges, shelter belts, contour ploughing, reforestation and afforestation) can be instrumental in maintaining pedological processes, reducing loss of soil and water, strengthening the cycling of C and other vital elements. e.g., N, P and S), and, ultimately, building resilient soils in East Africa.

12.6 SUMMARY AND CONCLUSIONS

Put in a nutshell, the focus of this chapter has been soil degradation in East Africa and the potential climate-smart options to restore and increase its resilience. It is widely recognized that soil degradation is a serious environmental issue in the region, with about 14 per cent of the total land area suffering from severe to very severe degradation. This manifests itself mainly in the form of erosion by water and depletion of nutrients and SOM. Although soil degradation is a natural process, it has been exacerbated by a complexity of cumulative interactions between biophysical, socioeconomic and political factors operating in specific spatial and temporal contexts in East Africa. One of the immediate socioeconomic consequences of degraded soils is the reduction in agronomic productivity translating to food scarcity, low income, poverty and stunted economic growth. For example, the annual costs of land degradation related to soil erosion and loss of essential nutrients from agricultural and grazing lands is about US $106 million (about 3 per cent of the agricultural GDP) in Ethiopia. Additionally, degraded soils have implications for environmental quality. For example, one of the environmental consequences of the biological and physical degradation processes is the transformation of soils into net sources of GHGs with a positive feedback to global warming. The environmental costs mostly emanate from the interplay of the different forms of soil degradation leading to loss of biodiversity, disruptions in the provision of ecosystem services and a decline in net biome productivity. We have also explained that the degraded soils can be restored through a conversion to judicious land use and the adoption of recommended CSMPs, such as CA. The CSMPs have been underscored because of the intricate linkages between climate change and soil degradation. Such practices are manifold, but none can be said to be a panacea for soil degradation. Thus, soil restoration projects initiated to reverse the downward spiral of soil degradation and strengthen socioeconomic development should objectively select and scale out the best fit CSMPs for a given locality, considering the biophysical and socioeconomic contexts. Such projects would also gain immensely from policies and approaches, which foster partnerships between governments, the private sector and non-governmental organizations.

REFERENCES

Agriculture for Impact. 2014. No ordinary matter: Conserving, restoring and enhancing Africa's soils. A Montpellier Panel Report. https://www.mamopanel.org/media/uploads/files/NO_ORDINARY_MATTER-_CONSERVING_RESTORING_AND_ENHANCING_AFRICAS_SOILS_2014.pdf (Accessed June 15, 2016).

Aune, J.B., and Coulibaly, A. 2015. Microdosing of mineral fertilizer and conservation of agriculture in sustainable agricultural intensification in sub-Saharan Africa. In: *Sustainable Intensification to Advance Food Security and Enhance Climate Resilience in Africa*, ed. Lal, R., Singh, B.R., Mwaseba, D.L. et al. Springer, Berlin, Germany, pp. 223–234.

Bai, Z.G., Dent, D.L., Olsson, L. et al. 2008. *Global Assessment of Land Degradation and improvement 1: Identification by Remote Sensing. Report 2008/01.* FAO/ISRIC, Rome, Italy/Wageningen, Netherlands.

Bekunda, M.A., Nkonya, E., Mugendi, D., et al. 2002. Soil fertility status, management, and research in East Africa. *East African Journal of Rural Development* 20: 94–112.

Bernazzani, P., Bradley, B.A., and Opperman, J.J. 2012. Integrating climate change into habitat conservation plans under the U.S. Endangered Species Act. *Environmental Management* 49: 1103–1114.

Berry, L. 2003. *Land Degradation in Ethiopia: Its Extent and Impact. A Study Commissioned by the Global Mechanism with Support from World Bank.* Technical Report 29pp.

Bojö, J., and Cassells, D. 1995. Land degradation and rehabilitation in Ethiopia: A reassessment. Working, AFTES Working Paper No. 17, World Bank, Washington DC.

Bouma, J., and Batjes, N.H. 2000. Trends of worldwide soil degradation. In: Hohenheimer Umwelttagung, Tagungsband 32. Verlag Günter Heimbach, Stuttgart, Germany. pp. 33–43.

Christensen, J.H., Hewitson, B., Busuioc, A. et al. 2007. Regional climate projections. In: *Climate change 2007: The Physical Science Basis – Contribution of Working Group I to the Fourth Assessment Report of the Intergovernmental Panel on Climate Change*, ed. Solomon, S., Qin, D., Manning, M. et al. Cambridge University Press, Cambridge, UK and New York, NY.

Eswaran, H., Almaraz, R., van den Berg, E., and Reich, P. 1996. An assessment of the soil resources of Africa in relation to productivity. *Geoderma* 77(1): 1–18.

FAO. 2013. *Climate-smart Agriculture Sourcebook*. FAO, Rome, Italy. www.fao.org/docrep/018/i3325e/i3325e.pdf (Accessed March 13, 2016).

FAO. 2015. *The State of Food Insecurity in the World 2015: Meeting the 2015 International Hunger Targets: Taking Stock of Uneven Progress*. FAO, Rome, Italy. www.fao.org/3/a-i4646e.pdf (Accessed March 13, 2016).

Gebreselassie, S., Kirui, O., and Mirzabaev, A. 2016. Economics of land degradation and improvement in Ethiopia. In: *Economics of Land Degradation and Improvement. A Global Assessment for Sustainable Development*, ed. Nkonya, E., Mirzabaev, A., and von Braun, J. Springer International Publishing AG Switzerland, pp. 401–430.

Gomiero, T. 2016. Soil degradation, land scarcity and food security: Reviewing a complex challenge. *Sustainability* 8: 281. doi: 10.3390/su8030281.

HarvestChoice. 2015. *AEZ (16-class, 2009)*. International Food Policy Research Institute, Washington DC and University of Minnesota, St. Paul, MN. http://harvestchoice.org/data/aez16_clas (Accessed June 15, 2017).

Hurni, K., Zeleke, G., Kassie, M., et al. 2015. Economics of land degradation (ELD). Ethiopia case study: Soil degradation and sustainable land management in the rainfed agricultural areas of Ethiopia: An assessment of the economic implications. Report for the Economics of Land Degradation Initiative. ELD, Bonn, Germany.

IPCC. 2014. *Climate Change 2014: Impacts, Adaptation, and Vulnerability. Report from the Intergovernmental Panel on Climate Change*. www.ipcc.ch/report/ar5/ (Accessed June 15, 2017).

Kirui, O. 2016. Economics of land degradation and improvement in Tanzania and Malawi. In: *Economics of Land Degradation and Improvement – A Global Assessment for Sustainable Development*, ed. Nkonya, E., Mirzabaev, A., and von Braun, J. Springer International Publishing AG Switzerland, pp. 609–649.

Lal, R. 1995. Erosion-crop productivity relationships for soils of Africa. *Soil Science Society of America Journal* 59: 661–667.

Lal, R. 1997. Degradation and resilience of soils. *Philosophical Transactions of the Royal Society of London. Series B: Biological Sciences* 352: 997–1010.

Lal, R. 2009. Soil degradation as a reason for inadequate human nutrition. *Food Security* 1: 45–57.

Lal, R. 2014a. Climate strategic soil management. *Challenges* 5: 43–74.

Lal, R. 2014b. Soil conservation and ecosystem services. *International Soil and Water Conservation Research* 2(3): 36–47.

Lal, R. 2015a. Restoring soil quality to mitigate soil degradation. *Sustainability* 7: 5875–5895.

Lal, R. 2015b. Sustainable intensification for adaptation and mitigation of climate change and advancement of food security in Africa. In: *Sustainable Intensification to Advance Food Security and Enhance Climate Resilience in Africa*, ed. Lal, R., Singh, B.R., Mwaseba, D.L. et al. Springer, Berlin, Germany, pp. 3–17.

Lal, R., and Singh, B.R. 1998. Effects of soil degradation on crop productivity in East Africa. *Journal of Sustainable Agriculture* 13(1): 15–36.

Mengistu, D., Bewket, W., and Lal, R. 2015. Soil erosion hazard under the current and potential climate change induced loss of soil organic matter in the Upper Blue Nile (Abay) River Basin, Ethiopia. In: *Sustainable Intensification to Advance Food Security and Enhance Climate Resilience in Africa*, ed. Lal, R., Singh, B.R., and Mwaseba, D.L., et al. Springer International Publishing AG Switzerland, pp. 137–163.

Milder, J.C., Majanen, T., and Scherr, S.J. 2011. Performance and potential of conservation agriculture for climate change adaptation and mitigation in sub-Saharan Africa. Ecoagriculture Discussion Paper No. 6. EcoAgriculture Partners, Washington DC. http://ecoagriculture.org/wp-content/uploads/2015/08/PerformancePotentialofConservationAg.pdf (Accessed March 13, 2016).

Muchena, F.N. 2008. *Indicators for Sustainable Land Management in Kenya's Context. GEF Land Degradation Focal Area Indicators*. ETC-East Africa, Nairobi, Kenya.

Muchena, F.N., Onduru, D.D., Gachini, G.N., and de Jager, A. 2005. Turning the tides of soil degradation in Africa: Capturing the reality and exploring opportunities. *Land Use Policy* 22: 23–31.

Mulinge, W., Gicheru, P., and Murithi, F. 2016. Economics of land degradation and improvement in Kenya. In: *Economics of Land Degradation and Improvement- A Global Assessment for Sustainable Development*, ed. Nkonya, E., Mirzabaev, A., and von Braun, J. Springer International Publishing AG Switzerland, pp. 471–498.

Nkonya, E., Mirzabaev, A., and von Braun, J. (eds) 2016. *Economics of Land Degradation and Improvement- A Global Assessment for Sustainable Development*. Springer International Publishing AG Switzerland.

Oldeman, L.R. 1994. The global extent of soil degradation. In: *Soil Resilience and Sustainable Land Use*, ed. Greenland, D.J., and Szabolcs, I. CAB International, Wallingford, UK, pp. 99–118.

Olson, J., and Berry, L. 2003. Land degradation in Rwanda. In: *Its Extent and Impact. A Study Commissioned by the Global Mechanism with Support from World Bank*. Technical Report, 21pp.

Olson, K.R., Al-Kaisi, M., Lal, R., and Cihacek, L. 2016. Impact of soil erosion on soil organic carbon stocks. *Journal of Soil and Water Conservation* 71(3): 61A–67A.

Osman, K.T. 2014. *Soil Degradation, Conservation and Remediation*. Springer, Dordrecht/Heidelberg/New York/London.

Scherr, S.J., and Shames, S. 2009. Mitigating climate change through food and land use. Worldwatch Report No. 179. Worldwatch Institute, Washington DC. www.worldwatch.org/system/files/179%20Land%20 Use.pdf (Accessed March 13, 2016).

Scherr, S.J., Shames, S., and Friedman, R. 2012. From climate-smart agriculture to climate-smart landscapes. *Agriculture and Food Security* 1:12. doi: 10.1186/2048-7010-1-12.

Smaling, E.M.A., Nandwa, S.M., and Janssen, B.H. 1997. Soil fertility is at stake. In *Replenishing soil fertility in Africa*. SSSA Special Publication No. 51, ed. Buresh, R.J., Sanchez, P.A., and Calhoun, F. Soil Science Society of America & American Society of Agronomy, Madison, WI.

Stoorvogel, J.J., and Smaling, E.M.A. 1990. Assessment of soil nutrient depletion in sub-Saharan Africa: 1983–2000. Report 28. DLO Winand Staring Centre for Integrated Land, Soil and Water Research (SC-DLO), Wageningen, Netherlands.

The World Bank Group 2017. *World Bank Open Data: Free and Open Access to Global Development Data*. https://data.worldbank.org/ (Accessed August, 2017).

Thompson, K., Chidawanyika, F., Kruszewska, I., et al. 2015. Building resilience in East African agriculture in response to climate change. Greenpeace research laboratories. Technical Report 04–2015. Greenpeace Africa, Johannesburg, South Africa.

Tolcha, T. 1991. Aspects of soil degradation and conservation measures in Agucho catchment, western Harerge. Soil Conservation Research Project, Research Report 19. University of Berne, Switzerland, Ministry of Agriculture and Environmental Protection, Ethiopia, and the United Nations University.

Tully, K., Sullivan, C., Weil, R., and Sanchez, P. 2015. The state of soil degradation in sub-Saharan Africa: Baselines, trajectories, and solutions. *Sustainability* 7: 6523–6552.

UNEP. 2006. *Africa Environment Outlook 2*. Our Environment, Our Wealth. Division of Early Warning and Assessment, United Nations Environment Programme, Nairobi, Kenya. www.earthprint.com/ (Accessed October 14, 2016).

United Nations, Department of Economic and Social Affairs, Population division. 2017. *World Population Prospects: The 2017 Revision, Key Findings and Advance Tables*. Working Paper No. ESA/P/WP/248. https://esa.un.org/unpd/wpp/publications/Files/WPP2017_KeyFindings.pdf (Accessed May 27, 2018).

Were, K.O., Gelaw, A.M., and Singh, B.R. 2016. Smart strategies for enhanced agricultural resilience and food security under a changing climate in sub-Saharan Africa. In: *Climate Change and Multi-Dimensional Sustainability in African Agriculture*, ed. Lal, R., Kraybill, D., Hansen, D.O et al. Springer International Publishing AG Switzerland, pp. 431–453.

Williams, A.P., Funk, C., Michaelsen, J., et al. 2012. Recent summer precipitation trends in the Greater Horn of Africa and the emerging role of Indian Ocean sea surface temperature. *Climate Dynamics* 39: 2307–2328.

Woomer, P.L., and Muchena, F.N. 1993. Overcoming soils constraints in crop production in Tropical Africa. In: *Sustaining Soil Productivity in Intensive African Agriculture: Seminar Proceedings*, 15–19 November, ed. Ahenkorah, Y., Owusu-Bennoah, E., and Dowuona, G.N.N., Technical Centre for Agricultural and Rural Cooperation, ACP-EU, Accra, Ghana.

Zingore, S., Mutegi, J., Agesa, B., et al. 2015. Soil degradation in sub-Saharan Africa and crop production options for soil rehabilitation. *Better Crops* 99: 24–26.

13 Nitrogen Dynamics and Management in Rainfed Drylands
Issues and Challenges

Rachid Bouabid, Brahim Soudi, and Mohamed Badraoui

CONTENTS

13.1 DRYLAND CHARACTERISTICS

Dryland agriculture occurs in many parts of the world. It is quite difficult to give a precise definition to dryland agriculture. For the purpose of the present chapter, dryland agriculture or dryland farming is referred to as that occurring in areas where precipitation is insufficient, meaning that water is the key limiting factor for crop production. Dryland agriculture, dryland farming and rainfed farming are often used interchangeably. The term "dry" is used to describe areas characterized by low amounts of rainfall and where the shortage of water affects vegetation growth, productivity and other related agricultural activities. Limitation of water is not only in terms of total amount of precipitation, but also in terms of its unpredicted distribution with regard to the crop growth cycle, mainly the critical growth periods. Dryland agriculture occurs in regions where evapotranspiration exceeds precipitation, and, therefore, is associated to arid, semi-arid and sub-humid temperate regions. The limited rainfall in conjunction with its erratic distribution often induces drought periods that can vary in terms of intensity and frequency; these drought periods are becoming even more amplified and structural with climatic changes. These conditions call for particular soil management practices especially for water and nutrients.

Drylands are defined based on the ratio of average annual precipitation (Pr) to potential evapotranspiration (PET) UNEP-WCMC (2007). Land where this ratio lies between 0.05 and 0.65 are considered drylands, while land where the ratio is less than 0.05 are not included in the dryland category but are considered to be desert areas, where no crop growth occurs without irrigation (Table 13.1, Figure 13.1). Aridity is also assessed on the basis of the number of days for which the soil water balance is favorable to plant growth (growing season). The delineation of drylands has been updated since 2007 to consider additional dryland areas of relevance to the Convention for Biological Diversity (CBD) with dry features despite their Pr/PET greater than 0.65 (Figure 13.1).

Unlike the common perception that drylands are limited to major parts of Africa, Latin America and the Middle East, drylands are actually present in many regions around the world, such as Central Asia, Russia, the northwest of the United States and Canada, and Australia. Drylands occur on all continents and occupy about 6.31 Bha (47.2% of the global land area) (Bha = 10^9 ha). They

TABLE 13.1
Dryland Categories According to UNEP-WCMC (2007)

Classification	Aridity index (Pr/PET)	Rainfall (mm)
Hyper-arid	< 0.05	< 200
Arid	0.05 < Pr/PET < 0.20	< 200 (winter) or < 400 (summer)
Semi-arid	0.20 < Pr/PET < 0.50	200–500 (winter) or 400–600 (summer)
Dry sub-humid	0.50 < Pr/PET < 0.65	500–700 (winter) or 600–800 (summer)

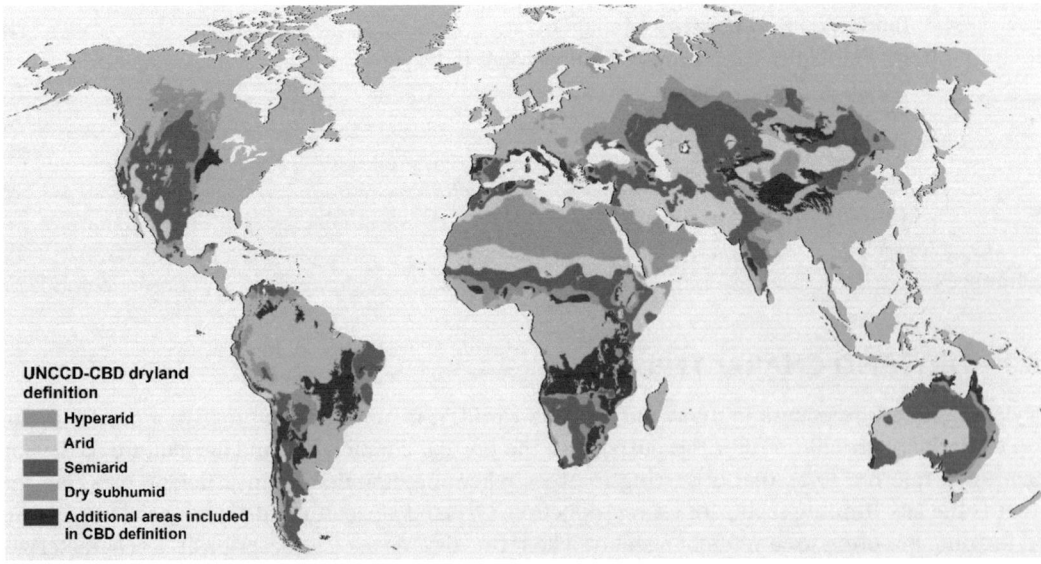

FIGURE 13.1   (See color insert.) World distribution of drylands areas. (From UNEP-WCMC (2007); drawn using GIS data available at www.unep-wcmc.org.)

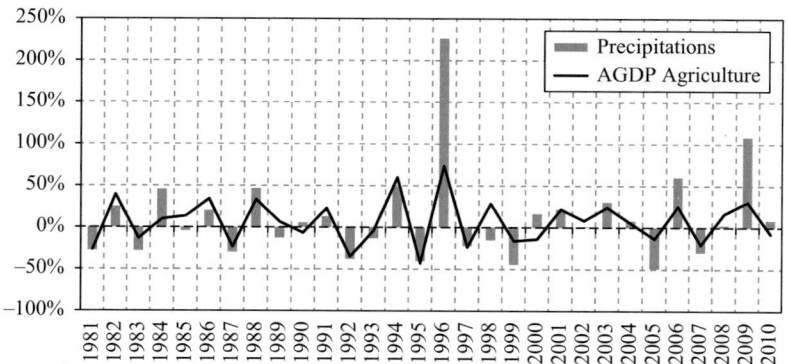

FIGURE 13.2 Variation of relative agriculture gross domestic product and annual precipitation (the value for each year is normalized with that of the previous year). (From Balaghi (2015), unpublished data.)

cover about 2.0 Bha in Africa, 2.0 Bha in Asia, 0.68 Bha in Australasia, 0.76 Bha in North America, 0.56 Bha in South America and 0.3 Bha in Europe (UNEP 1992).

Dryland agriculture encompasses a diversity of cropping systems, varying from subsistence farming to intensive and high cash crops and it is fundamental to the economy of the regions where it prevails. Although irrigation often occurs, rainfed agriculture generally predominates and represents the primary production sector. In the example of Morocco, a typical Mediterranean dryland country, the economy is highly dependent on agriculture, the country's agriculture gross domestic product (AGDP) fluctuations follows the same trend as that of annual precipitations (Figure 13.2).

Enhancing dryland productivity is a lesser constraint in developed countries where agricultural practices benefit from advanced research and technologies. These difficulties are yet to be overcome in developing countries, where, despite valuable indigenous knowledge and satisfactory research, farmers are still struggling with basic needs to cope with subsistence farming and to bring their practices (especially nutrient inputs) to a level that guarantees the expression of the production potentials of their lands and crops. The challenges are multiple if drylands are to meet the needs of a vulnerable population, especially with the rising issues of climate changes.

Many factors of soil degradation threaten the potential production capacity in these areas. Nutrient depletion, wind and water erosion, and salinity are very common issues. Although dryland farming has known significant changes in the past few decades in many parts of the world, mainly with a shift from traditional agriculture to more intensive cropping, nutrient inputs and management, especially nitrogen (N), remains the main concern for crop production. The use of fertilizers is at low levels in many countries. In China, deficiency of nitrogen is reported to be almost everywhere, and that of phosphorus (P) affects at least one-third of arable land (Li et al. 2009b). About 68 Mha of dryland soils in Pakistan with less than 300 mm of rainfall have a negative nutrient balance (Shah and Arshad 2006). Nitrogen use efficiency (NUE) is considered more critical than that of other nutrients as it is highly affected by the amount and distribution of rain during a crop season. Nitrogen is also related to organic matter dynamics, and this latter also affects water status necessary for nutrient bioavailability and uptake under water stress conditions. Therefore, this chapter will put more focus on the nitrogen dynamic as it is the major critical factor in dryland agriculture.

13.2 MAJOR SOILS OF DRYLANDS

Although soil fertility and its evolution are significantly dependent on cropping systems and the degree of crop intensification, its overall status is also determined by soil type. Soil fertility cannot be dissociated from soil pedogenesis. As drylands are characterized in general by restraining soil forming factors (mainly climate, vegetation and living organisms), the soils are usually low to moderately differentiated. The rate of soil formation and the degree of differentiation are inversely related to aridity (Nettleton and Peterson 1983). In low rainfall areas, the variation in parent materials and topography are the main driving factors for differences of soil types and their properties. Soils are, in general, not very deep and have low organic matter (OM) and high base saturation due to low leaching. In many cases, they are affected by calcium carbonates inherited from calcareous parent materials or from secondary precipitation, which is responsible for high alkalinity. A variety of soils can be encountered in drylands depending on the other soil-forming factors, mainly Aridisols (2.12 Bha), Entisols (2.33 Bha), Mollisols (0.8 Bha), Alfisols (0.38 Bha), Vertisols (0.21 Bha) and others (0.47 Bha) (Dregne 1976; Noin and Clark 1997). In Mediterranean-type climates, red soils are common, due to the differentiation of iron oxides, while Spodosols, Ultisols and Oxisols are rarely encountered in these kind of environments. Soils in drylands are not necessarily the result of pedogenesis of present climates, but may have developed under ancient climates. Soils of tropical climates found presently in many dryland areas are the result of paleoclimates. Well- differentiated Alfisols and Oxisols in North Africa, India and Australia are examples of such conditions.

Although vegetation and crop production in drylands are controlled by climate and water, the soil characteristics determine how much of that water will be stored and consequently be available for plants during their growth stages, mainly during the dry or drought periods. Soil moisture is a primary condition for organic matter and nutrient cycling and for nutrient bioavailability for crops, and therefore becomes the sine qua non condition for production. Depth, texture, clay type and organic matter content are important factors that determine the storage capacity and the retention forces of soil available water. Organic matter and, therefore, organic carbon (C) cycles are dramatically affected by dry conditions, as a result of natural and anthropic factors. The low amounts of residues returned and the high rates of decomposition are often not in favor of carbon sequestration.

13.3   NITROGEN DYNAMICS AND USE EFFICIENCY

Besides the shortage and the sporadic distribution of water, the low productivity potential of soils in drylands is commonly attributed to low and imbalanced use of fertilizers, which does not satisfy crop nutrient requirements. Some of the soil characteristics in drylands are also in favor of particular conditions affecting nutrient cycling and availability (high pH, calcium carbonates, swelling-shrinking clays, high base saturation). Drylands are fragile in most areas and vulnerable to degradation factors that affect the physical, chemical and biological settings of the soils, which in turn affect the essential nutrients pool. Nutrient sources in poor drylands are from organic sources and fertilizer use in developing countries is of appreciable use only in the more favorable rainfed areas.

Under conditions of low moisture, the majority of mineral nutrients (when present) are less mobile, are often fixed or precipitated in the soil and are therefore less available to plants. A long-term experiment conducted in India (Co-ordinated Research Projects of Dryland Agriculture) recognized that "dry soils are as hungry as they are thirsty" (Singh and Kumar 2009). Nitrogen and phosphorus availability are of universal concern in all dryland soils. Potassium (K) is more variable as some parent materials may contain significant amounts of potassium-bearing minerals, but coarse textured soils are usually potassium deficient (Sivanappan 1995; Ryan and Sommers 2010; Ryan et al. 2012). In rainfed agriculture, phosphorus and potassium are usually applied as deep fertilizers at the start of the season. As they are of relatively low mobility in the soil, they present fewer constraints in terms of management compared to nitrogen. However, since nitrogen is more mobile, its dynamic along the crop cycle is more intricate and more related to organic matter.

Fertilizer input in dryland agriculture varies with soil and climate, but is, in general, very low in rainfed farming in developing countries. For instance, in North Africa and the Middle East, fertilizer use is less than 20 kg ha^{-1}, all nutrients considered. The soils of the drylands of Africa undergo important nutrient loss (Roy and Nabhan 1999) that may reach 30 kg ha^{-1} yr^{-1} of NPK (nitrogen, phosphorus and potassium) (Henao and Baanante 2006). In Pakistan, dryland soils are reported to be almost entirely deficient in nitrogen (Shah and Arshad 2006). Similar situations were stated for India (Singh and Venkateswarlu 1985) and Kenya (Murage et al. 2000). When nutrients are not replenished to the soil, a progressive mining takes place. The low nutrient status in drylands in many parts of the world is actually a vicious circle that involves low nutrient cycling, low productivity, low residue return, low soil conditioning and protection, and therefore affects the sustainability of production systems as well as food security in the regions concerned.

13.3.1 NITROGEN AND ORGANIC MATTER

Organic matter is an important component of the soil and is an important source of nutrients. It plays a major role in supplying the soil with nitrogen and other nutrients as a result of microbial decomposition. It acts also as a sink when immobilization takes place. Organic matter is the main cycling component of soil fertility in low-input farming. The soils of drylands, in general, have low organic matter content resulting from low biomass productivity and low returns (Ryan et al. 1997; Lal 2002; Lal 2004). The nitrogen supply from organic matter depends on the importance of the two processes of organic matter decomposition, that is, humification and mineralization. The former results in stable humus, and the latter releases CO_2 and mineral nutrients to the soils. While both processes in general co-occur, the dominance of one over the other depends on various abiotic and biotic factors, mainly soil temperature and moisture, the nature of organic matter, and the pool of microorganisms. The rates of these processes have been widely studied under controlled incubation conditions and under humid temperate field conditions. However, only limited knowledge is available for dry or rainfed conditions. Various models for organic matter humification and mineralization have been proposed (Henin and Dupuis 1945; Kolenbrander 1969; Campbell et al. 1984; Jenkinson and Rayner 1977; Sauerbeck and Gonzalez 1977; Biederbec et al. 1994). Janssen (1984), Mary and Guérif (1994). Most studies agree that organic matter mineralization depends on the C/N ratio. Low C/N ratio leads to higher mineralization of organic matter with high net N mineralized, while high C/N leads to more humification and high N immobilization. Vigil and Kissel (1991) reported that the break point between net N mineralization and net-immobilization was at a C/N ratio of 40. Under warm dryland conditions, moderate temperatures favor rapid decomposition of organic matter by microbial activity during the periods of adequate soil moisture.

Organic matter cycling and nitrogen release from its mineralization are issues that can be looked at from different angles (Soudi and Bouabid 2018, present volume). While the accumulation of organic matter (i.e., humus) in the soil is highly recommended for soil conditioning and carbon sequestration, rapid mineralization on a season basis is of concern to low-input smallholder farmers as it is the main source of nutrients for their crops. In warm drylands or Mediterranean type climates, annual organic matter depletion can be very important and may exceed 60% (Corbeels 1998). Applications of nitrogen fertilizers (even in small quantities), or the presence of residual nitrogen from previous legume crops, can lead to rapid decomposition of organic matter. Singh and Singh (1994) found that the combined nitrogen fertilizer with straw resulted in a higher nitrogen mineralization compared to straw and fertilizer applied singly (Figure 13.3). The balance of mineral nitrogen can be appreciable for the season crop, but the accumulation of stable humus that plays other important roles (structure, moisture retention, aggregate protection, etc.), will be negatively affected.

Farming systems favoring residues return or the addition of farm manure with high C/N ratio may contribute to some build-up of stable humus in the soil. However, if low or no nitrogen fertilizer is used, a negative nitrogen balance will be carried over and yield potential will not be achieved.

The predilection of the work of humification and mineralization is a hidden issue in crop management systems in dryland and raises the question about which of the processes the farming practices should aim for. Where soil nutrient status is very low, the management of organic matter is usually directed toward getting higher mineralization rates, as farmers rely on the nutrients pool resulting from this process. Despite the appreciable annual additions of organic matter of different

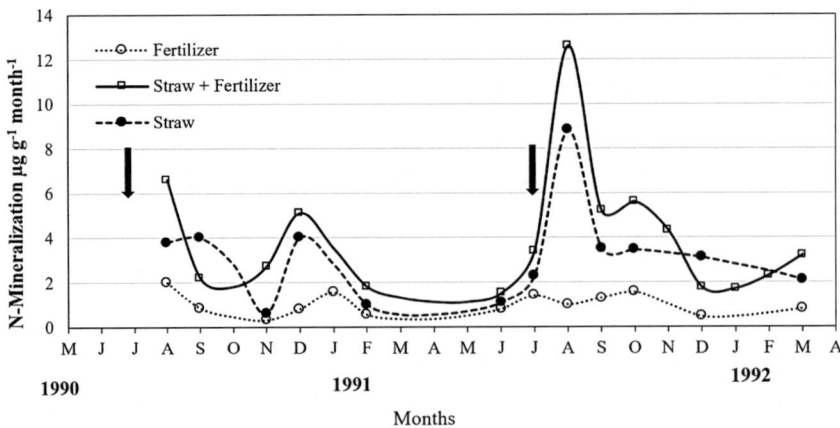

FIGURE 13.3 Effect of applications of straw and fertilizer singly and in combination (arrows indicate time of application), on N mineralization ($\mu g\ g^{-1}\ month^{-1}$). (Redrawn with permission from Singh and Singh 1994).

forms (leftover residues, farm manure), we tend to see rapid depletion of organic matter. This is even more marked in systems adopting rotations with legume crops. The residual nitrogen from biological nitrogen fixation and the recycling of the crop residues with low C/N ratio lead to higher mineralization rates and therefore to rapid loss of soil organic matter. The lack of humic substances may also reduce the low chelation potential of the soil with respect to micronutrients, especially in coarse textured or high pH calcareous soils. In farming systems favoring the return of residues of high C/N ratio, some storage of stable humus can occur, but would be too small to evolve to a significant build-up; on the contrary, the mineral nutrient pool, especially of nitrogen, will not be able to guarantee good crop growth and yield potential will not be achieved.

Therefore, in drylands, especially those of warm climates, the turnover of soil organic matter affects, and is affected by, nitrogen. Complex interactions drive the mineralization of organic matter and the bioavailability of mineral nitrogen, and therefore require good management practices to achieve the compromise goals of nitrogen supply while maintaining sufficient stable organic matter in the soil.

13.3.2 Nitrogen and Soil Water Status

Nitrogen is a major yield-limiting nutrient for crop production in drylands. In arid and semi-arid rainfed regions, soil nitrogen content is closely related to soil moisture regime. Appreciable amounts of residual mineral nitrogen are usually found soon after the start of the rainy season (Chiang et al. 1983; Warren et al. 1997), but will change depending on the amount and type of organic matter available. When dry conditions persist, the top soil may not provide enough nitrogen to the crop. If the root system is well developed, uptake may occur more in deeper soil layers (Campbell et al. 1977; Strong and Cooper 1980).

In dryland soils, nutrient use efficiency is closely related to water use efficiency (Shepherd et al. 1987; Palta and Fillery 1995). Nutrient dynamics and availability have strong interactions with soil available water (He and Dijkstra. 2014). These interactions can affect crop growth and yield positively or negatively, depending on the crop growth stages. Water limitations, low use of fertilizers and low organic matter return to soils make nutrient management a challenge in drylands. Amounts, forms, timings and methods of application of fertilizers need to be considered taking into consideration not only soil nutrient dynamics and supply capacity but also soil water status.

Important year-to-year rainfall variations significantly influence soil water and nutrient status, which in turn causes important repercussions on nutrient inputs. For the same initial soil nutrient status, the same nutrient input may lead to differences in yield when rainfall differs, or when rainfall is irregular from one year to another. As the climatic conditions of the year progress, nutrient inputs need to be adjusted accordingly. A reduction or increase of the amount of nitrogen initially planned is necessary depending on whether the year is dry or relatively wet. Dry or wet conditions affect the soil pool as well, and the adoption of a different balance approach is necessary in order to predict crop needs with the changing season conditions. In the case of drought conditions, water stress not only affects nutrient mobility and availability, but also negatively affects the growth of the crop itself.

When soil moisture is adequate, moderate to high temperatures can promote nitrification of ammonium. Several studies reported that nitrification is high when soil water content is within an adequate range of water-holding capacity (40% to 70%), but it becomes low or is even inhibited beyond this range (at extremes), as the soil become dry or waterlogged (Justice and Smith 1962; Malhi and McGill 1982; Flower and Challagha 1983; Klemedtsson et al. 1988; Shelton et al. 2000; Sahrawat 2008). The nitrification process can transform the entire $N-NH_4^+$ from organic matter mineralization or from ammonium-N-based fertilizer within two to three days. This is also true for the transformation of urea. Since most crops absorb more nitrogen in the form of nitrate-N compared to ammonium-N, this transformation is advantageous if leaching risks are minimal. However, with high risks of leaching following storm events during the rainy season, nitrates may leach out of the rhizosphere. The loss of ammonium-N by volatilization in high pH conditions is less when nitrification is important. Since nitrate absorption by plants occurs mainly by passive mechanisms, the mobility of nitrogen in the form of nitrate is important for plant uptake. Abdel Moumen et al. (2010) evaluated nitrogen loss using ^{15}N mass balance under two rainfall conditions (270 mm and 340 mm yr^{-1}) in calcareous soils using different forms of urea. The results showed that N loss by volatilization was relatively low (11% and 18%) for both rainfall conditions. They also reported that urea hydrolysis is delayed by dry conditions and that nitrification is delayed by cold conditions following urea hydrolysis.

In semi-arid areas, the nitrogen requirements of crops are dictated by the amount of seasonal precipitation (Myers 1984). The input of mineral nitrogen for a crop may undergo important depletion in one year, while in another year large amounts of residual nitrogen can stay in the soil as a result of limited rainfall, low leaching and poor crop growth (Mor-Neir and Harpaz 1987; Seligman et al. 1986). In the latter situations, N recovery is expected to be important for the subsequent crop, whether in the same year or the following year. Coorbeels et al. (1998) studying N recovery by sunflower (*Helianthus annuus*) after wheat (*Triticum aestivum*) in a Typic Haploxerert in Morocco found that the residual labelled ^{15}N fertilizer represented 3.6% of the fertilizer added to the previous winter wheat crop. This low plant recovery was attributed to microbial immobilization of N fertilizer and to the limited N uptake due to drought conditions during the sunflower season. After the sunflower harvest, 49.7% of added labelled fertilizer was left in the soil profile. Total recovery of the ^{15}N fertilizer over the two growing seasons was 83%. Seligman et al. (1986) reported low N recovery from residues of high C/N ratio, and Wood et al. (1996) reported low nitrogen recovery of various crops using ^{15}N labelled urea and ammonium sulfate under different rotations.

Generally, in dry years, there are small or non-significant differences in terms of N uptake among unfertilized and fertilized crops. However, important differences in N uptake occur during rainy years. For this latter case, N mineralization can account for a large part of available N for the crop when sufficient organic matter is present. Figure 13.4 illustrates important differences in terms of N uptake and dry matter accumulation by wheat in three years with different rainfall conditions in northern Morocco (Coorbeels et al. 1997).

Nitrogen use efficiency is not only affected by the total amount of rainfall, but also by rainfall patterns. The occurrence of rainfall during vegetative and reproductive stages is a determinant for water use as well as for nutrient use by crops. The amount and distribution of rain in drylands affects the efficacy of applied nitrogen fertilizers. However, the unpredicted occurrence of rain makes it difficult to adopt a unique strategy for the choice of the form, the timing and the method of application, especially when splitting is a common practice.

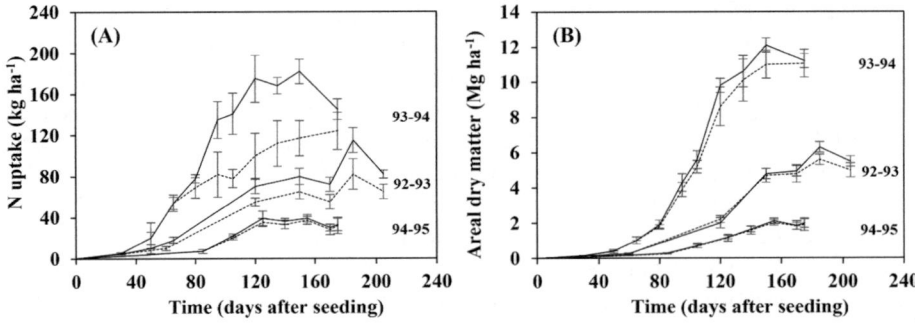

FIGURE 13.4 N uptake (A) and areal dry matter accumulation (B) of winter wheat during three years with contrasting rainfall: 401, 332 and 264 mm for 1993/94, 1992/93 and 1994/95 respectively (A: anthesis, PM: physiological maturity). (Redrawn with permission from Coorbels et al. 1999.)

Crop nutrient uptake depends on the solubility of mineral nutrients and their movement in the soil. Under dry conditions, not only the form and the amount of mineral nutrients are affected by the two processes of mass flow and/or diffusion, but also the mobility. Under the action of water movement and roots suction, different forms of nitrogen behave differently as a result of complex interactions. It is generally accepted that nitrogen is more influenced by mass flow compared to phosphorus and potassium (Barber 1984). However, limited knowledge is available on the behavior of nitrogen mobility under dry soil conditions. Song and Li (2006) studied root function in uptake of nitrogen of maize and the effect of soil water on the transfer and distribution of nitrates and ammonium under irrigated and non-irrigated regimes. Their findings showed that both root growth and water supply had a significant effect on nitrate-N transfer. Under irrigation, the difference of nitrate-N concentrations at different distance points from the maize plant were smaller, while clear differences of nitrate-N concentrations were observed under conditions of limited root growth space without water supply. They concluded that nitrate-N is transferred as solute to plant root systems with water uptake by plants. However, the transfer and distribution of ammonium-N differed from that of nitrate-N and were not influenced by root growth or by soil water supply. This implies that in dry conditions, nitrate-N placement (especially at early stages) is as critical as for other less mobile nutrients. Localized fertilizer placement in the seed rows for cereals in drylands shows better nutrient use compared to broadcast, especially when rain occurs after sowing (McKenzie et al. 2001; Kelley and Sweeney 2005, Zhang et al. 2013). The effect of roots on nitrogen absorption will also depend on the root density and specific area. Any factors that favor root growth would favor root absorption as well.

The application of fertilizers in general, and nitrogen in particular, followed by dry conditions can result in increased soil solute concentrations in the topsoil. High amounts of fertilizers may even reduce the uptake. The interactions between fertilizer use and crop uptake is a dynamic process that can have different effects depending on stages of crop growth. There is a strong need in rainfed dryland agriculture to have a good understanding of the interactions between soil water status, nutrients dynamics and crop behavior in order to achieve the best nutrient use efficiency.

13.3.3 Nitrogen Mineralization and Immobilization

Organic nitrogen mineralization is an important process that drives nitrogen changes in the soil. It represents a major process of nitrogen-supply potential, and an important pool of nitrogen that needs to be taken into consideration in the N balance and N fertilization, especially in low-input systems. N mineralization is intimately related to organic matter decomposition, and their processes are consequently affected by the same factors. Soil moisture and temperature status are the main

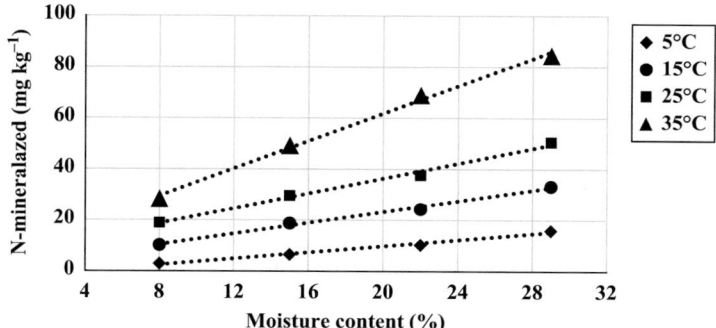

FIGURE 13.5 Effect of water and temperature on N mineralization. (Drawn from data reported by Li et al. 2009c.)

factors driving N mineralization in general, and under dry conditions in particular. The populations of living organisms and microorganisms responsible for organic matter and nitrogen transformations decline with aridity due to the limited moisture needed for their survival.

N mineralization results from the interaction between soil moisture and temperature. It often follows a linear process (Figure 13.5) with increasing soil moisture content at different temperatures (Stanford and Epstein 1974; Ju and Li 1998; Li et al. 2009c). Within a range of water and temperature conditions suitable for biological activity, N mineralization occurs and varies with these two factors. However, under extreme moisture or temperature regimes N mineralization is very low or totally stopped. In most dryland soils, with the exception of those of North America (dry and cold), high temperatures and low moisture conditions are the main concerns.

N mineralization is the highest during wetter years compared to drier years, especially during the periods of optimal moisture and temperature conditions for microbial activity (i.e., spring in Mediterranean climates), and varies with the type of cropping systems. N mineralization was found to be the highest under fallow soils compared to cropped soils. This trend is attributed to the fact that fallow practices are in favor of soil moisture conservation. Figure 13.6 shows significant differences in terms of mineralized N under different cropping for two years with contrasting rainfall in rainfed conditions of Morocco (Coorbeels et al. 1999). There are major differences between the two years, but also important differences between fallow and cultivated plots within the "normal" year (401 mm) compared to the dry year (264 mm). In addition, the study underlines the constraint

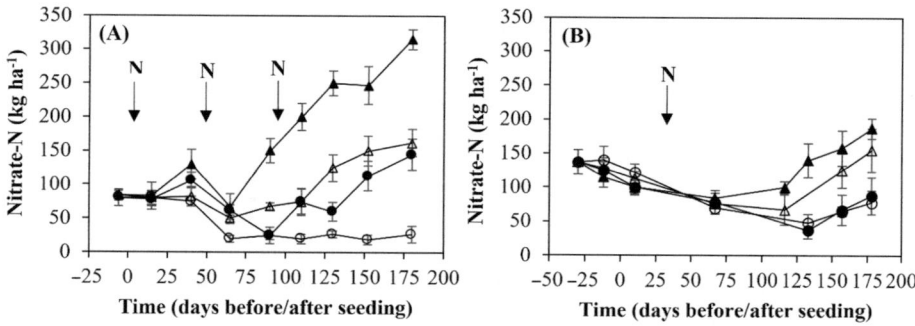

FIGURE 13.6 Change in soil nitrate-N from mineralization in the total rooting depth: (○) unfertilized fallow plot; (●) fertilized fallow plot: (△) unfertilized wheat cultivated plot; and (▲) fertilized wheat cultivated plot; (N) dates of N applications. (A) year with 401 mm (B) year with 264 mm. (Redrawn with permission from Coorbeels et al. 1999.)

relating to N fertilizer applications between the two years. The 401 mm rainfall allowed three splits of nitrogen, while the 264 mm rain imposed only one N application.

When compaction is not a concern, soils in drylands are rarely water saturated for long periods and are well aerated most of the time. These conditions are in favor of rapid organic N mineralization as well as nitrification of ammonium, either from fertilizers or from organic matter mineralization. Unless there is an important fixation by high charge 2:1 clays, nitrogen is present in the soil mainly in the form of nitrates. In such cases, assuming no major rain events at the start of the season to cause nitrate leaching from the topsoil, nitrate-N is the dominant form of nitrogen potentially available for crops and can be used as a reliable, quick soil test index reflecting soil nitrogen supplying capacity.

Various models were used to assess nitrogen mineralization by combining the potentially mineralizable nitrogen (N_0) with functions representing the effect of temperature and soil moisture on the mineralization rate constant (k) (Cabrera et al. 2005). Such models can be adopted to better predict nitrogen mineralization under limited water conditions, such as those of dryland soils. Campbell et al. (1988) compared two versions of such models using data obtained with different treatments: (i) summer fallow, and wheat following wheat grown on summer fallow; (ii) on dryland; and (iii) with irrigation. While the models showed close results for estimated and measured values under irrigation, they tended to underestimate nitrogen values under dryland conditions. Since the model was considered not dynamic, as it does not allow for N_0 to be replenished continuously by nitrogen derived from decomposition of fresh residues and rhizosphere microbial biomass, net nitrogen mineralized from this source might explain the underestimation predicted under dry conditions, which was always obtained whenever the soil became very dry and was re-wetted by rainfall: a phenomenon attributed to the possible flush of mineral nitrogen that can occur after re-wetting but not taken into consideration in the model. Soudi et al. (1990a) and El Harradi et al. (2003) observed similar flushes of initial mineral nitrogen in the first few days of incubation, with amounts ranging from 26% to 40% of total mineralized N. This flush was attributed to the pretreatment effect of soil samples by drying and re-wetting before incubation. This phenomenon is important to consider because it simulates the mineral-N increase observed under field conditions of rainfed drylands following drastic seasonal changes upon first season rain. The early mineralization flush can be attributed to a partial decay of the microbial biomass following summer desiccation. Dead microbial tissues have a low C/N ratio that makes them easily biodegradable by the surviving microbial biomass when soil moisture becomes favorable. High amounts of mineral-N ($N-NO_3 + N-NH_4$) varying from 58% to 73% of the total mineralized-N supply were reported under field conditions (Soudi et al. 1990b) (Figure 13.7). This shows that early season N flush can be beneficial for crops (such as cereals) if the re-wetting of the soil after dry or drought periods does not cause significant

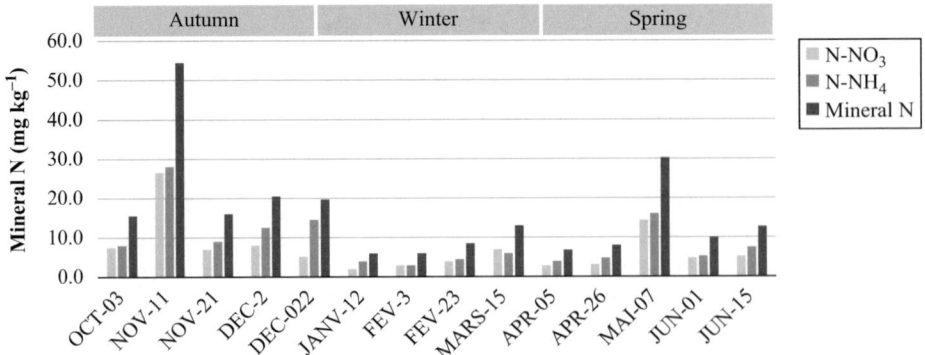

FIGURE 13.7 Seasonal variation of mineral-N in a Calcixeroll in Morocco. (Adapted from Soudi et al. 1990b.)

leaching of the mineralized nitrogen, and needs to be taken into consideration in the estimation of crop nitrogen fertilization.

Nitrogen availability to crops, mainly at early stages, can be significantly affected by nitrogen immobilization before remineralization takes place. The incorporation of residues of high C/N from previous crops or from the addition of organic amendments induces a net N immobilization during initial decomposition. However, the amount and rate of immobilization and subsequent remineralization vary largely depending on the nature of the residues as well as on several biotic and abiotic factors, mainly soil moisture conditions and N fertilizer use. Understanding the immobilization-remineralization trends upon the incorporation of residues in dryland soils is of great importance in relation to the synchronization of N supply and crop demand during critical growth stages, especially those occurring during early season decomposition (Vigil and Kissel 1991; Jensen 1997; Reinertsen et al. 1984; Recous et al. 1995 ; Vanlauwe et al. 1996; Corbeels et al. 2000; Li et al. 2009b).

Although the C and N dynamics during decomposition of plant residues in the soil are well understood, differences may be important among soils in rainfed drylands and those in humid regions. N immobilization by microbial activity can account for important impounding of N from the soil when residues are of high C/N ratio, which adds up to its reduced availability because of low soil moisture storage. Immobilization can happen at different periods of the growing cycle, but is more likely to occur in early season when organic matter from previous crops has not yet undergone much decomposition. Immobilization can affect the overall available mineral N, as well as its use efficiency, and depends on the nature of the organic residues incorporated in the soil. In a study on N cycling in response to the decomposition of residues in a typical wheat/sunflower rotation in a Vertisol in Morocco, Coorbeels et al. (2000) reported that N availability may be severely restrained during the growing season when past wheat straw residues are incorporated into the soil. The amount of inorganic N immobilized due to such incorporation may enter into competition with the plant demand, engendering low N efficiencies. Under dryland conditions, the rate of release of the immobilized N is usually very slow, and affects the overall recovery of nitrogen by the crop. Practices that increase organic matter of high C/N (wheat straw) may contribute to reducing organic matter mineralization, but can result in important microbial immobilization. However, the incorporation of organic matter of low C/N ratio (legume residues) is in favor of high N mineralization, but would lead to rapid organic matter loss.

13.3.4 Nitrogen and Nutrient Balance

Crop nutrition is driven by a balanced nutrient absorption. Nutrients need to be available and absorbed in adequate quantities required by the crop. When a nutrient is lacking in the soil to the required threshold, it becomes limiting to the absorption and functioning of the others. Imbalanced nutrient use has been reported to be a major factor affecting nutrient use efficiency and nutrient depletion in low-input regions in general, and in dryland regions in particular (Murage et al. 2000; Rashid et al. 2004; Ryan 2008b; Li et al. 2009c; Ryan 2011). Fertilizer use is rarely based on soil testing. Furthermore, in many countries, several fertilizer formulae distributed in the market are not appropriate for the soil fertility status and/or for the requirements of the crops. This situation can lead to low response of the crops to the fertilizers used. In low-input situations, this may contribute to the depletion and imbalance of one or more nutrients.

Combined use of N and phosphorus (P) has been reported to improve N use efficiency, reduce N loss and improve crop yield (Sharma et al. 2007; Ma et al. 2010; Duan et al. 2014). Phosphorus is critical to root growth, which is related to water and N use efficiency. Jin et al. (2006) found that the addition of P enhanced the concentration and accumulation of N and P in shoots and seeds of soybean cultivars. The addition of P alleviated the effect of drought stress on plant growth, P accumulation, and grain yield. They suggested that phosphorus fertilization could mitigate drought stress at the reproductive stage, resulting in less yield effects and improved grain quality. Studer et al. (2017) reported the interactive effect of drought and the application of nutrient NPK individually or in combination under drought stress for maize. Al Karaki et al. (1996) found that high proline accumulation in leaves of water-stressed crops at high P levels might be an adaptive response to drought for sorghum compared to bean. Several other studies in drylands corroborate the combined positive effect of N and P on N use efficiency and crop production (Sharma 2007; Ma et al. 2010).

Potassium is an essential nutrient that affects many physiological processes involved in biotic and abiotic stresses resistance (Kant 2002; Cakmak 2005; Wang et al. 2013). It is critical in plant tolerance to dry and drought conditions as it is involved in stomatal regulation, osmotic adjustment, membrane stability, cell elongation, and photosynthesis (Benlloch-Gonzalez et al. 2010; Mengel 2007; Marschner 2012). Crops with adequate K fertilization show greater adaptation to water stress (Lindhauer 1985; Pervez et al. 2004; Premachandra et al. 2009). Asgharipour et al. (2011) reported that adequate K nutrition increased N uptake and improved drought resistance of sorghum. Zahoor et al. (2017) also found that K application regulates N metabolism and osmotic adjustment in cotton functional leaf under drought stress.

13.3.5 BIOLOGICAL NITROGEN FIXATION AND WATER STRESS

Biological nitrogen fixation (BNF) is an important process for N supply in low-input cropping systems. BNF is even more critical for the sustainability of farming systems in dryland soils. Water stress, as a result of dry or drought conditions, is one of the most influencing factors that affects both the legume plant as well as nodule formation and functioning. BNF declines under dry conditions and thus affects the legume crop N uptake, plant growth and subsequent role for N enrichment of the soil awaited from the rotation system. Although the sensitivity to water stress varies among legume crops, some crops (soybean [*Glycine max*], cowpea [*Vigna unguiculata*] and black gram [*Vigna mungo*]) have been reported to be more vulnerable to modest dry conditions (Sinclair and Serraj 1994).

The inhibition of BNF under drought has been widely reported and is related to various factors such as reduced nitrogenase activity, reduced nodule permeability, reduced CO_2 flux, low O_2 conductance, inhibition of photosynthetate, and shoot nitrogen-feedback signaling (Sinclair and Serraj 1984; Duran et al. 1987; Serraj et al. 1999; Serraj et al. 2001; Gonzales et al. 1998; Gonzales et al. 2001; Valentine et al. 2011). Growth and nitrogen fixation are depressed under water stress. The ability of plants to recover and the time for recovery are both related to the duration of the stress period. Weisz et al. (1985) found that drought resulted in decreased nodule conductance, which was detected as early as three days following water stress application. The effect on BNF is more important as drought is prolonged. The loss of nodule activity is also related to nodule water potential (Bennett and Albrecht 1984; Sprent 1971; Durand et al. 1987). Nodule respiration increases with increase in soil water potential and declines upon water rehydration as a result of decaying (Nandwal et al. 1991). Under water stress, reduced BNF is attributed to depressed oxygen uptake and therefore to respiration, as a result of either a loss of oxygen conductance through the nodule or inhibition of oxygen requiring reactions, or probably both (Pankhurst and Sprent 1975a; Pankhurst and Sprent 1975b; Sinclair and Gourdiaan 1981; Serraj and Sinclair 1996a,b).

Water stress sensitivity and tolerance are attributed to several molecules in the nodules, including ureides (Atkins et al. 1992; Sinclair and Serraj 1995b; Serraj et al. 1999; Vadez et al. 2000; Serraj et al. 2001; King and Purcell 2005; Coleto et al. 2014), glutamine (Neo and Layzell 1997; Curioni et al. 1999) and asparagine (Bacanamwo and Harper 1997; Vadez et al. 2000; Lima and Sodek 2003). An extensive review by Arrese-Igor et al. (2011) addresses the various physiological processes involved in nodule BNF under drought. In general, BNF is a more drought-sensitive mechanism when ureides are the main nodule exporters, which is the case for soybean, and it is more drought-tolerant when amides are the main nodule exporters, which is the case of pea (*Pisum sativum*), chickpea (*Cicer arietinum*) and faba bean (*Vicia faba*) (Sinclair and Serraj 1984; 1995; Serraj and Sinclair 1996a,b; Vadez et al. 2000; Purcell et al. 2000; Vadez and Sinclair et al. 2001; Serraj 2003). Such traits can be used in the segregation of legume genotypes as to their BNF tolerance to water stress under dryland conditions. Despite the available research on the mechanisms involved in the sensitivity or tolerance of BNF under water-limiting conditions, soybean (a ureide type exporter) is the most studied crop (Arrese-Igor et al. 2011), and only limited research is available on other crops.

Although the sensitivity of BNF is important among legumes, the recovery capacity is an even more fundamental trait for crop adaptation and productivity after periods of water limitation. The rapidity of recovery of both transpiration and nodule nitrogen fixation activity after soil water deficits is an important criterion for rainfed dryland conditions, especially at the establishment stage of the crop. Legume genotypes having such a capacity are of great value to sustain nitrogen supply and ensure good yields under water deficit conditions (Cerizini et al. 2016; Cerizini et al. 2017).

A meta-data analysis study by Daryanto et al. (2015), using research on legume yield responses to drought under field conditions between 1980 and 2014, showed that the amount of water reduction was positively related with yield reduction, but the extent of the impact varied with legume species and the phenological stage during which drought occurred. Overall, lentil (*Lens culinaris*), groundnut (*Arachis hypogaea*), and pigeon pea (*Cajanus cajan*) were found to experience lower drought-induced yield reduction compared to other legumes such as cowpea and green gram (*Vigna radiate*). Yield reduction was generally more severe when legumes experienced drought during their reproductive stage. Higher yield reductions are also observed for legumes grown on soils with medium texture compared to those cultivated on soils of either coarse or fine texture. In contrast, legume yield reductions were not related to regions and their associated climatic factors.

13.3.6 ROOT SYSTEMS AND NUTRIENT USE AND EFFICIENCY

Crop roots are the hidden part of the plant. A prolific root system is essential to uptake of water and nutrients in drought prone and nutrient limiting situations to ensure good growth and productivity. The effect of water and nutrients on crop growth and productivity is often observed and assessed based on how the shoot parts of the crop are progressing. It is only rarely that the rooting system is examined to see how it is behaving under dry conditions. Nutrient use efficiency, mainly N, is closely related to the root system density and depth (Zhang et al. 2013). It is even more important in drylands where topsoil can be subject to important water deficit by evapotranspiration. Plants may adapt by extending their rooting system deep into the soil to reach for water and nutrients. In rainfed dryland farming, crop adaptation and yield are correlated to root mass (Passioura 1983; Palta et al. 2011). Nutrient use efficiency is improved by root system development. In North African countries where olive production is common, farmers consider from experience that when a young tree is subject to water stress, it develops a denser and deeper root system that allows it to become vigorous with age, tolerate dry conditions and make better use of nutrients. All conditions that can improve root development will increase water and nutrient use efficiency. Good soil structure becomes a determining factor. Porous soils will allow root growth, but may not favor upward capillary water evaporation. Therefore, appropriate soil tillage and surface residue management are needed. Crops have varying rooting systems, and the essential nutrients have differing impacts on root growth and physiological functions. The addition of organic matter is a factor that can contribute to ensuring healthy structure. Nitrogen and phosphorus are the key elements for root development, and K is a key element for regulating water uptake and osmotic pressure of the roots. The complex interactions among soil conditions and plant behavior affect tremendously the ability of the crop to express its potential under favorable conditions or to adapt and survive under stressful conditions.

13.4 NITROGEN MANAGEMENT AND AGRICULTURAL PRACTICES

Nutrient- and water-management practices are a major challenge for increasing crop production under rainfed dryland soils. The techniques include the use of organic and inorganic fertilizers, recycling crop residues, rotations including legumes, fallow and soil and water conservation. The adaptation and the degree of technological development of any of these techniques depend on the local physical and socioeconomic context. Many practices have evolved from indigenous knowledge of the farmers, and others were adopted based on successful experience in similar areas. Nutrient use efficiency is improved with the use of selected cultivars, planting date, weed control and pest management.

13.4.1 Combined Use of Organic and Inorganic N Sources

Organic matter is considered a good soil conditioner, especially for its favorable impact on water status. Soil humus improves aggregate formation and soil structure. The increase of humus can significantly increase the soil available water capacity (AWC) (Bouyoucos 1939; Russel et al. 1952; Hollis et al. 1977). Hudson (1994) found good relationships between SOM and AWC. He reported that for the example of silt loam soils, the AWC for a soil with 4% OM (by weight) was more than twice that of a soil with 1% OM. The use of manure improves SOM content, which in turn improves soil AWC (Salter and Howarth 1961; Salter and Williams 1963). Bauer (1974) reported that a change of 1% of soil humus from the addition of manure resulted in a change of AWC in the range of 5.0 to 9.2 mm m^{-1} in soils with coarse texture and in the range of 0.83 to 2.5 mm m^{-1} in soils with moderately fine and fine textures.

Combined use of inorganic and organic fertilizers is adopted to sustain soil fertility and improve crop productivity in drylands (Palm et al. 1997; Sharma et al. 2002; Place et al. 2003; FAO 2004; Hadda and Arora 2006; Bationo et al. 2007; Li et al. 2009a; Kihara et al. 2011; Harraq et al. 2016; Calderon et al. 2017). In dryland farming, the combined use of inorganic and organic fertilizer is more beneficial than their separate use. Bouraima (2015) reported that application of manure reduced surface runoff and soil erosion, and N and P loss were reduced by 41% and 33%, respectively, in the case of combined use of manure and chemical fertilizers.

The use of manure is indeed important as an organic amendment for soils in general, and for dryland soils in particular. However, its use can carry some problems depending on the origin of the organic materials (e.g., infested fields or barns). Insects, pathogens and weed seeds can be propagated through the use of manure. When manure travels from one area to another, problems associated with the use of manure can be disseminated at different spatial scales (Petit et al. 2013). Therefore, proper management of manure (e.g., composting) is needed to avoid such problems.

13.4.2 Management of Fertilizer Applications

One of the challenges for soil nutrient uptake by plants is soil moisture availability. When soil water is limited, not only is there water stress for the plant, but also the nutrient movement and absorption by the crops are slowed down or even stopped. Fertilizers can only be applied during the periods of the season when there is a good chance of rain. This is particularly true for nitrogen, which is split in two or more dressing fractions. For instance, in most Mediterranean areas, the practice for cereal fertilization is that P, K and a fraction of N are applied as basal application at sowing. N is usually split in two or three dressing applications that target tillering and heading stages. The risk of shortage of adequate moisture in mid-season can jeopardize crop yields (Nageswara Rao et al. 1985; Mattheus et al. 1990; Manyowa 1994; Gimenez et al. 1997; Ben Naceur et al. 1999). In many situations, farmers are reluctant to apply nitrogen when there is no upcoming rain, fearing that nitrogen may be wasted and avoiding additional salinity stress to the crops.

IAEA (2005) reported on nutrient and water management project cases in rainfed dominated agriculture (<300 mm) in eleven countries with a wide range of cropping systems, characterized by low nutrient and organic matter status. These included wheat/maize systems on the Loess Plateau of China (Cai et al. 2005), sorghum/castor (*Ricinus communis*) rotations in Andhra Pradesh of India (Ramana et al. 2005), maize (*Zea mays*)-based systems in the Machakos district of Kenya (Sijali and Kamoni 2005), peanut (*Arachis hypogaea*) production in the Senegal (Sene and Badiane 2005), and wheat/vetch (*Vicia sativa*) rotations in the Safi-Abda region of Morocco (El Mejahed and Aouragh 2005). The results from all the studies showed that irrespective of the management practices, an important portion (20% to 60%) of applied fertilizer N was lost. These losses were attributed mainly to alkaline soil pH conditions, and occurred essentially at the beginning of the crop vegetative period. On the other hand, the residual value of applied N available to subsequent crops was very low, rarely exceeding 9% of the crop N requirement. Their findings underlined the importance of fertilizer management practices to minimize losses, especially during the early part of the cropping season. Split application during the dry season allowed an important increase of wheat yields, but the amount of N to be applied at each split needs to be based on the soil N status and crop demand for N. The combined manure with mineral fertilizers in correct proportions could provide 10% to 15% of the crop-N requirement and contribute to increasing observed yields.

Another example of constraining fertilizer use is the case of rainfed fruit tree production. Olive orchards are one of the main crops in many dryland areas, especially in the Mediterranean. The cycle of the olive tree starts with blooming, which usually occurs in mid or late spring (the start of the dry season). Most of the active growing cycle occurs in summer and early fall. Farmers usually apply P and K and a first fraction of N early spring. The splitting of the remaining N becomes contingent of the rainfall conditions during the rest of the season. The most important vegetative growth and fruit setting occur in summer when no water is available for N mobility in the soil. The P and K mobility is even more critical under such conditions.

Methods of application and placement of fertilizers are critical to achieve good distribution of nutrients in the rhizosphere and their uptake by the roots. They should be adapted depending on the planting density and root system (structure and depth). For instance, the wheat root system is shallow compared to that of maize. The latter is often row planted and tends to have a deep and pivoted root. When soil moisture is deficient, the farther the nutrients are from the root, the less chance there is for their uptake, regardless of the absorption mechanism. Interception is reduced due to limited root growth, and mass flow is hindered by the lack of water needed for the nutrients to be drawn by the water movement exerted by the roots in response to transpiration.

13.4.3 TILLAGE AND SURFACE RESIDUES MANAGEMENT

Tillage practices significantly affect soil properties that are related to soil moisture, organic matter and nutrient dynamics in the soil. Under dryland conditions, conventional tillage can expose the soil to high evaporation, rapid loss of organic matter, and potential wind and water erosion. Water use efficiency is generally low with inappropriate tillage practices. Hatfield et al. (2001) reported that it is possible to improve water use efficiency by 25% to 40% through soil management practices that involve tillage. When adequate amounts of crop residues are present, conservation tillage is highly effective for conserving soil and water, achieving favorable crop yields, maintaining soil organic carbon contents, and enhancing soil and water quality (Unger et al. 1997).

Conservation tillage and reduced tillage have gained increasing attention in drylands (Unger 2002; Vere 2005; Cai et al. 2006; Wang 2006; Avci 2011; Mrabet et al. 2012). They are defined as any tillage and planting practices that can leave, respectively, 15% to 30% and 30% or more crop residues on the soils after planting (CTIC 2018). Among the benefits of such a management approach is the maintenance and accumulation of organic matter, as well as water and nutrient use efficiencies in the soil. Results of long-term experiments in the drylands of China (Wang 2006) showed that the positive effect of reduced tillage on soil water availability, nutrient balance and soil fertility indices were strongly improved by the use of crop residue, either incorporated or applied as surface mulch. Simulations using the CENTURY model revealed a positive effect on carbon sequestration under reduced tillage. Johnson et al. (2018) reported that conservation tillage systems are much better options for the cultivation of different drought-tolerant common bean varieties in semi-arid areas of Kenya due to their soil moisture conservation ability compared with conventional tillage systems, which cannot be sustainable practices in such marginal areas. Sainju et al. (2012), Sainju et al.

(2016) and Sainju et al. (2016) studied N balance in a dryland agroecosystem in response to till-age, crop rotations (involving cereals and legumes), and cultural practices (involving planting date, seeding rates and spacing, application of N fertilization), over several years in the northern Great Plains of the United States. They reported that surface residue N was 30–34% greater in no-till (NT) than in conventional tillage (CT). They also found that N sequestration rate at 0–20 cm from 2004 to 2011 varied from 29 to 89 kg N ha^{-1} year^{-1} under CT and NT with spring wheat/pea rotations. Long-term experiments conducted using NT under Mediterranean conditions showed significant increase of SOM and N (Mrabet et al. 2001; Bessam and Mrabet 2003). Lal (2004) suggested that conservation tillage practices are among the appropriate management practices that can increase soil carbon sequestration.

Surface residue management is recognized as a water conservation practice and has received important research attention in arid and semi-arid regions (Duley and Russel 1939; Van Doren 1978; Unger 1984; Smika et al. 1986; Peng et al. 2015). Surface residue management and residue incorporation greatly affect the nutrient cycling, especially that of N (Sauerbeck and Gonzalez 1977; Rescous 1995; Seligman et al. 1986; Schomberg 1994; Vanlauwe 1996; Jensen 1997; Liwang 1999; Schomberg and Steiner 1999). Mulching, as a surface residue management, is of great importance in water conservation in drylands. Mulch farming, when it is adopted in dryland farming, contributes by modulating soil temperature, reducing water evaporation, protecting soil from erosion and contributing to the sequestration of C (FAO 2004). As stated earlier, any practice that contributes to improving soil moisture, will also contribute to better conditions for nutrient bioavailability and uptake.

Several studies conducted in the drylands of the Loess Plateau of China reported on the importance of mulching for water and N dynamics. Jin et al. (2008) showed that minimum tillage and mulching over a period of seven years consistently increased the yield of winter wheat, primarily by better water harvest. N uptake by grain and straw and N export, as well as residual soil N, were the highest compared to conventional tillage with no mulching. Bayalaa et al. (2011) evaluated the effects of several different conservation agriculture studies in the drylands of Africa (Burkina Faso, Mali, Niger and Senegal) for various crops. They also found that mulching was among the best practices for improving soil fertility and crop yields, especially when rainfall is less than 600 mm. Qin et al. (2015) reviewed the effects of mulching on wheat and maize, using data from 74 studies conducted in 19 countries. They indicated that mulching significantly increased yields, water use efficiency and N use efficiency by up to 60%, compared with no mulching. The effects were larger for maize than for wheat, and larger for plastic mulching than for straw mulching. Plastic mulch-ing performed better at relatively low temperatures, while straw mulching had the opposite trend. The effects of mulching tended to decrease with increasing water input. Mulching effects were not related to soil organic matter content. They concluded that soil mulching can significantly increase maize and wheat yields and water and nitrogen use efficiencies, and thereby may contribute to closing the yield gap between attainable and actual yields, especially in dryland and low nutrient input agriculture. Wang and Xing (2016) reported that mulching significantly improved soil water content, soil nitrate-N content and its vertical distribution in maize root-zone. The results of a three-year field experiment by Gao et al. (2009) assessing the effect of mulching with different materials (straw and plastic) showed that N uptake, NUE and yield of wheat were higher with mulching. In addition, after three years, residual nitrate-N in the 0–200 cm soil averaged 170 kg ha^{-1}, which was equivalent to about 40% of the total N uptake by wheat in the three growing seasons.

Although plastic mulching has been evaluated in many studies and demonstrated efficacy for water and N conservation, organic residue mulching is a much preferred practice because it is the most appropriate for smallholder farmers and has limited or no environmental impacts. Plastic mulching, used for a variety of crops including row vegetables and fruit crops, despite its multiple uses, is slowly degradable and represents a threat to the environment. Organic mulches, on the other hand, provide a source of organic matter and nutrients, and their cycling in the soil contrib-utes to improving soil properties. When previous crop or fallow residue mulching is practiced with conservation tillage in dryland agriculture, the benefits are not only on soil water and nutrients but also on the soil health as a whole (Schomberg et al. 1994; Schomberg and Jones 1999; Fuentes et al. 2003).

13.4.4 Legume Inclusion in Crop Rotations

Monoculture is a common practice in many dryland areas. After harvest, the soil is left bare for several months, and is often subject to grazing for the leftover residues. In summer time, wind erosion can be very significant, while water erosion can happen during the fall before the next season's sowing. Grazing on crop residues causes depletion of organic matter, reduced topsoil protection and nutrient imbalance. Such practice can lead to nutrient mining when inputs are lower than outputs.

The N status of soils can be improved by the integration of legumes in the crop rotations. This is especially important in drylands where N from previous legume crops is not subject to much leaching, as in humid conditions. It is common practice in Mediterranean arid and semi-arid regions to use food legumes, such as chickpea, lentil and faba bean as alternatives to fallow and continuous cropping (Ryan et al. 2008a; Ryan and Sommers 2010). The adoption of legume crops in the rotations can contribute to increasing organic matter and therefore the potential release of N from mineralization (Ryan et al. 2008; Ryan et al. 2010). When the residues left in the soil are of low C/N ratio, they are susceptible to higher rates of mineralization, and therefore contribute to releasing more N, P and other nutrients in the soil during the cycle of the following crop. Depending on the crop, biological fixation by legume crop can provide variable amounts of mineral nitrogen during a growth cycle (Hardarson 1987). Montanez (2000) reported net nitrogen supplies from legume crops varying from 50 to more than 300 kg N ha^{-1} yr^{-1}.

The contribution of N from BNF of legumes in a rotation depends on many factors, among which are the availability and compatibility of the rhizobium strain, soil water and temperature conditions, soil aeration and the amount of initial nitrogen in the soil. In drylands, soil water status is often a determinant factor in the early stages of crop growth, along with the availability of the starter nutrients. Molybdenum is a key element for rhizobia N fixation. This element is not very mobile in the soil if soil moisture is limited and it can be affected by soil pH conditions as well. In calcareous soils, common in drylands, the high pH is not a limiting factor to molybdenum compared to other micronutrients; however, this is not the case in acid soils.

Legume crops are also used in intercropping systems, either with field crops or fruit trees (Vandermeer 1989), mainly in smallholder farming. Intercropping is a common practice in many dryland areas (especially in India and China), and has been reported to improve crop yields and soil nutrient status compared to sole crops (Ramana 2005; Zang et al. 2015; Rekha 2017; Singh 2017). Legume crops have different BNF capacities that can be taken into consideration for efficient intercropping (Montañez 2000). Using ^{15}N techniques, Bationo et al. (2003) reported that cowpea fixed more than twice the amount of atmospheric N compared with groundnuts, and that the inclusion of cowpea in millet-based cropping systems improved nitrogen use efficiency by about 30%. Ramirez-Garcia (2015) found that barley intercropped with vetch had improved root growth and N uptake. Song et al. (2007) investigated crop yield and various chemical and microbiological properties in rhizosphere of wheat, maize and faba bean grown in the field solely and intercropped (wheat/faba bean, wheat/maize and maize/faba bean). They found that intercropping increased crop yield, changed N and P availability, and affected the microbiological properties in the rhizosphere of the three species compared to sole cropping. Bouhafa et al. (2015) and Daoui et al. (2012) reported that legumes improve soil N and P and contribute indirectly to improving soil nutrient status for the olive trees (*Olea europea*) in Mediterranean rainfed olive orchards. The question remains whether intercropping should be adapted to all situations. It may be feasible in smallholder farming systems, but would be of controversial practicability in the more intensive, large farms of favorable rainfed areas.

13.4.5 INTEGRATED SOIL FERTILITY MANAGEMENT

Integrated soil fertility management (ISFM) is defined as the set of sound soil fertility management practices that include both the use of fertilizers as well as organic inputs in combination with
the knowledge of how to adapt these practices to local conditions, in order to maximize nutrient
use efficiency and optimize crop productivity (Vanlauwe et al. 2010; Vanlauwe et al. 2015). The
approach of ISFM is of particular interest in farming systems that use low chemical fertilizers
and rely more on organic sources of nutrients. As discussed earlier, nitrogen dynamics are closely
related to organic matter, and therefore are of particular interest in the context of ISFM. Improving
soil organic matter content also aims at favoring other soil characteristics, such as water status,
which in return affects nitrogen bioavailability and use efficiency under dry conditions. ISFM is
seen also as a soil conservation measure that can contribute to carbon sequestration.

The perception and adoption of soil fertility management in relation to agronomic efficiency
need to be considered from multiple socioeconomic angles with changes that can occur in steps and
with increasing knowledge (Figure 13.8) (Vanlauwe et al. 2010; Vanlauwe et al. 2015). Innovations
adapted to large market-oriented farms in developed countries may not be suitable for smallholders
in developing countries, where farmers are still struggling with subsistence agriculture. The move

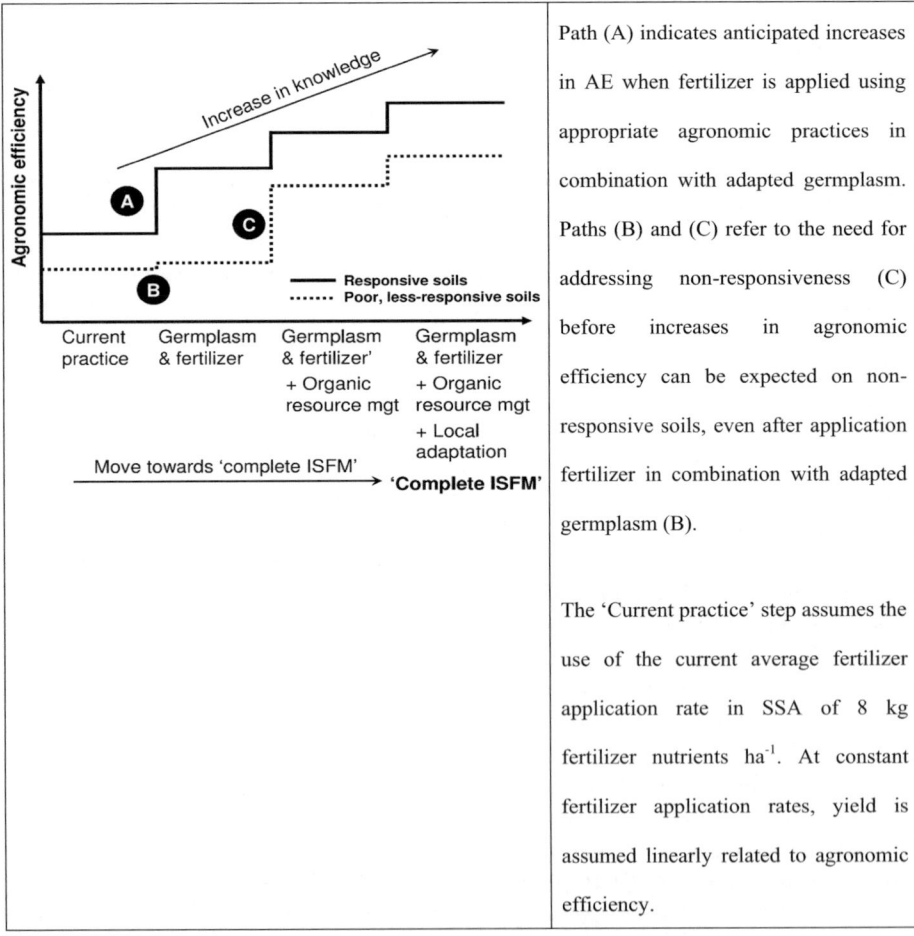

FIGURE 13.8 Conceptual relationship between the agronomic efficiency of fertilizers and organic resource
and the implementation of various components of ISFM. (From Vanlauwe et al. 2010.)

through the various steps of ISFM depends also on the land's degree of responsiveness. The conditions might be the same, but the concerns, the practices, and the means to face these conditions are widely different. For instance, land ownership status is a major factor toward investing in the long-term improvement of soil fertility. Itinerant farming as a result of drought and soil degradation causes a continuous decline of soil fertility in many countries in the Sahel and sub-Saharan Africa (Reich 2001; Darkoh 2003; FAO 2005).

13.4.6 Rs Principles for Nitrogen Management in Drylands

FAO (2004) recommended that soil fertility management practices can be grouped according to the movement of nutrients "into", "within," and "out of" a system and that the practices can be sorted into four groups as suggested by Hilhorst and Muchena (2000):

- Adding nutrients to the soil
- Reducing losses of nutrients from the soil
- Recycling nutrients
- Maximizing the efficiency of nutrient uptake

A number of management practices were drawn from several case studies in Senegal and Sudan (FAO 2004). Any soil fertility management approach to be adopted or adapted to drylands needs to comply with the principles of one or more of these categories of practices.

N management also needs to comply with "Rs" stewardship principles, commonly referred to as the four Rs: "Right amount," "Right time," "Right place" and "Right form" (Ryan et al. 2012). In the context of drylands, since N availability to crops is highly dependent on several management approaches, especially those related to organic matter and water, we suggest considering an additional, fifth "R," which is the "Right approach," as explained below:

13.4.6.1 Right Amount
- The amount of N needs to be based on crop requirement, target optimum yields, soil N balance. The knowledge of initial N stock as well as the potential of N mineralization from organic matter are key factors in dryland soils. The apparent N use efficiency indices for specific conditions are required to make reasonable estimations of the amount to apply to reach potential yields. Since leaching is generally in rainfed early season conditions, nitrate-N can be used as a good soil test.
- Since rainfall can change from year to year, N rates should be adjusted to the actual rainfall conditions that coincide with the critical stages of the target crop. A drier year will require reducing the N rates initially planned, while a rainier season would impose increasing the N amounts, in order to achieve higher yields and to compensate for the loss by leaching.
- Volatilization is a common N loss process in dryland soils, especially in calcareous soils with high pH and lime content. Appropriate adjustments based on assessment studies are needed.
- In dryland agriculture, nitrogen is accorded greater importance by farmers compared to other nutrients. Caution is needed to avoid excess N that may affect crop growth and production.
- The response of N depends on other nutrients and also influences the uptake of other nutrients. Understanding N interactions with other macro- and micronutrients, especially under water stress conditions is essential.

13.4.6.2 Right Time
- Due to limited rainfall, N applications need to be reasoned according to the probabilities of the occurrence of rain and should follow the critical growth stages of the crop.

- Basal N fertilization at the start of the season would depend on the stock of mineral N, as well on SOM content. The quality of organic matter (C/N) will determine the importance of N mineralization vs. N immobilization, and the resulting N balance to be considered for subsequent N topdressings.
- N topdressing is intimately linked to rain occurrence in order to increase the N use efficiency and avoid temporary stress to roots for water and nutrient uptake.
- Root development is important for water and nutrient uptake; and N is important for root development and for the synergies with other nutrients involved, such as phosphorus. N needs to be available to the crop at an early stage for the crop's root development is a prerequisite for water use efficiency in later stages.

13.4.6.3 Right Place
- Geographically, N management depends on the degree of aridity from one area to another. Responses to N are higher in the more favorable areas that are less prone to sporadic rainfall occurrence. N rates and time of application will depend ultimately on the amounts and spatial variability of rainfall in a given region.
- At the parcel level, N is more efficient where roots are dominantly present. Where banding is possible, local applications of N give a better response. The incorporation of N with seeders at sowing are found to trigger better initial growth compared with broadcast application followed by tillage that usually returns N to depths beyond the zone of initial root development.

13.4.6.4 Right Source
- The forms of N need to be considered according to the limited soil water conditions, which determine N mobility, as well as to the soil properties, which determine both the mobility as well as the chemistry of N.
- Different forms of N can be adopted at different growing stages of the crop.
- Ammonium-N can be subject to significant loss under dry conditions, high pH and high lime content. Nitrification of ammonium can contribute significantly to buffer soil pH.
- Urea is a dominant N fertilizer. Its use under cool, moist conditions can give good responses. When soil moisture is favorable, urea can be transformed rapidly to ammonium, which in turn goes through rapid nitrification (a few days). However, urea is prone to volatile loss under dry conditions, and can cause injuries on crop leaves when broadcasted as topdressing.
- Loss by denitrification is very rare in dryland soils.
- The use of organic matter is a good source of N and a good soil conditioned for N cycling. The knowledge of N mineralization rates needs to be taken into consideration in the N balance.

13.4.6.5 Right Approach
- Nutrient balance: the right amount of all essential elements is crucial to guarantee positive interactions for plant growth, water and nutrient use and efficiency.
- Surface residue management and mulching play important roles in protecting the soil and improving its fertility, mainly the N and C cycling.
- Integrated soil fertility management is particularly beneficial in dryland soils.
- A good knowledge of the trends of mineralization vs. immobilization of OM residues and amendments (manure, compost, etc.) is necessary to make compromises between nutrient release (mainly N) and SOM build-up. Different strategies are required depending on the nature of the OM used/recycled (mainly the C/N ratio). Mineralization is often sought in low fertilizer use systems, which is the dominant situation in drylands. However, the need for increasing SOM in such systems is well justified.

- Early inorganic N applications help prevent N immobilization when the C/N ratio of SOM is high.
- Crop rotations with N fixing legume crops, an ancient practice to be sustained to improve soil conditions and the status of N and other nutrients. Rotations are also a good alternative to fallow under the favorable dryland conditions.
- Conservation agriculture practices including reduced tillage, no-till or direct seeding can improve organic matter, water status and N bioavailability.
- Weed control reduces competition for water and N especially under constrained dryland agriculture conditions.

13.5 CONCLUSIONS

Dryland agriculture is constrained by many interacting factors, but the main one remains the limitation of water, which in turn affects soil nutrient bioavailability, organic matter dynamics and the overall productive capacity. Limitation of water is not only in terms of total amount of rainfall, but also in terms of its erratic distribution with regard to crop growth stages. Dryland agriculture encompasses a diversity of cropping systems, varying from subsistence farming to intensive agriculture. Many factors of soil degradation threaten the potential production capacity in these areas. Nutrient depletion, wind and water erosion, and salinity are very common issues. However, nutrient inputs and management, especially N, remain the main concern for crop production. Most drylands around the world suffer from low fertilizer use and inappropriate management practices. N use and use efficiency is considered more critical than that of other nutrients, as it is highly affected by the amount and distribution of rain during the crop season. N and organic matter dynamics are intimately interrelated, and are both affected by soil water status, which is a determinant of nutrient bioavailability and uptake in water stress conditions. Mineralization of organic matter is a process favored by warm conditions. It is regarded as a good source of N in low-input farming systems. However, it tends to deplete SOM, which is needed in dryland soils for other positive traits. Biological nitrogen fixation represents a good source of N especially in rotations and intercropping systems involving legume crops, but can be significantly affected by water stress in dry conditions. Combining organic and mineral fertilizer sources can enhance mineral N availability as well as crop growth better than either form alone. Synergies among nutrients are important under dryland soil conditions. N uptake and use by plants is highly affected by other nutrients such as potassium, a key element for water stress tolerance, and phosphorus, a crucial element for root development. Enhancing N use and use efficiency is more of a challenge under dryland conditions. It requires a good understanding of the processes governing N dynamics under water-limited conditions as well as the choice of appropriate management practices (fertilizer amounts and forms, crop rotations, reduced tillage, mulching, intercropping, etc.). These practices have interactive effects on soil properties, water status, N cycling, organic matter dynamics, etc. They need to be adapted to the multitude of situations than can be encountered in dryland areas as a result of the interaction of variable climatic, physical and socioeconomic factors, in order to guarantee effective N strategies, which may impact also on C strategies. It is recommended that N management needs to obey to "5Rs" stewardship principles, which consider an additional "R" for "Right approach," a set of practices and methods that would ensure sustainable and optimal N dynamics under dryland conditions.

REFERENCES

Abdel Monem, M., Lindsay, W.L., Sommer, R., and Ryan, J. 2010. Loss of nitrogen from urea applied to rainfed wheat in varying rainfall zones in northern Syria. *Nutr. Cycl. Agroecosys.* 86(3): 357–366.

Abdel Wahab, A.M., and Abd-Alla, M.H. 1995. The role of potassium fertilizer in nodulation and nitrogen fixation of faba bean (*Vicia faba L.*) plants under drought stress. *Biol. Fertil. Soils.* 20: 147–150.

Al-Karaki, G.N., Al-Karaki, R.B., and Al-Karaki, C.Y. 1996. Phosphorus nutrition and water stress effects on proline accumulation in sorghum and bean. *J. Plant Physiol.* 148: 745–751.

Arrese-Igor, C., González, E.M., Marino, D., Ladrera, R., Larrainzar, E., and Gil-Quintana., E. 2011. Physiological response of legume nodules to drought. *Plant Stress* 5(Special Issue 1): 24–31.

Asgharipour, M.R., and Heidari, M. 2011. Effect of potassium supply on drought resistance in sorghum: Plant growth and macronutrient content. *Pak. J. Agri. Sci.* 48(3): 197–204.

Atkins, C.A., Fernando , M., and Hunt, S. 1992. A metabolic connection between nitrogenase activity and the synthesis of ureides in nodulated soybean. *Physiol. Plant.* 84: 441–447.

Avci, M. 2011. Conservation tillage in Turkish dryland research. *Agron. Sust. Develop* 32(2): 299–307.

Bacanamwo, M., and Harper, J.E. 1997. The feedback mechanism of nitrate inhibition of nitrogenase activity in soybean may involve asparagine and/or products of its metabolism. *Physiol. Plant.* 100: 371–377.

Balaghi, R. 2015. Changement climatique, environnement et risques naturels au Maroc. Paper presented at Atelier de formation et d'appui CEDRIG pour les projets ASAP-M, PAMPAT et PAMPAT. Oriental, Rabat, Marocco, 20–22 January (unpublished).

Bationo A., Mokwunye U., Vlek P.L.G., Koala S., Shapiro B.I., and Yamoah C. 2003. Soil fertility manage-ment for sustainable land use in the West African Sudano-Sahelian zone. In: *Soil fertility management in Africa: A regional perspective*. Gichuru M.P., Bationo A, Bekunda M.A., Goma P.C., Mafongoya P.L., Mugendi D.N., Murwira H.M., Nandwa S.M., Nyathi P., and Swift M.J. (eds.). TSBF-CIAT.

Bationo, A., Kihara, J., Vanlauwe, B., Waswa, B., and Kimetu, J. 2007. Soil organic carbon dynamics, func-tions and management in West African agro-ecosystems. *Agric. Syst.* 94: 13–25.

Barber, S.A. 1984. *Soil Nutrient Bioavailability, A Mechanistic Approach.* John Wiley and Sons, New York.

Bauer, A. 1974. Influence of soil organic matter on bulk density and available water capacity of soils. *Farm. Res.* 31: 44–52.

Bayala, J., Sileshib, G.W., Coec, R., Kalinganirea, A., Tchoundjeud, Z., Sinclaire, F., and Garritye, D. 2012. Cereal yield response to conservation agriculture practices in drylands of West Africa: A quantitative synthesis. *J. Arid Environ.* 78: 13–25.

Ben Naceur, M., Naily, M., and Selmi, M. 1999. Effet d'un deficit hydrique, Survenant a differents stades de developpement du ble, sur l'humidite du sol, la physiologie de la plante et sur les composantes du rende-ment. *New Medit* 10(2): 53–60.

Benlloch-Gonzalez, M., Romera, J., Cristescu, S., Harren, F., Fournier, J.M., and Benlloch, M. 2010. K^+ star-vation inhibits water-stress-induced stomatal closure via ethylene synthesis in sunflower plants. *J. Exp. Bot.* 61: 1139–1145.

Bennett, J.M., and Albrecht, S.L. 1984. Drought and flooding effects on N2 fixation, water relations, and dif-fusive resistance of soybean. *Agron. J.* 76: 735–740.

Bessam, F., and Mrabet, R. 2003. Long-term changes in soil organic matter under conventional and no-tillage systems in semiarid Morocco. *Soil Use Manage.* 19: 139–143.

Biederbeck, V.O. Janzen, H.H., Campbell, C.A., and Zentner, R.P. 1994. Labile soil organic matter as influ-enced by cropping practices in an arid environment. *Soil Biol. Biochem.* 26(12): 1647–1656.

Bohner, A. 2014. Faktoren der Nährstoffverfügbarkeit im Boden des Dauergrünlandes. In: Fachtagung für Biologische Landwirtschaft, 06. November 2014, Report HBLFA Raumberg-Gumpenstein, Irdning, Austria pp. 15–22.

Bouhafa, K., Moughli, L., Daoui, K., Douaik, A., and Taarabt, Y. 2015. Soil properties at different distances of intercropping in three olive orchards in Morocco. *Int. J. Plant Soil Sci.* 7(4): 238–245.

Bouraima, A.K., He, B., and Tian, T. 2015. Runoff, nitrogen (N) and phosphorus (P) losses from purple slope cropland soil under rating fertilization in Three Gorges Region. *Environ. Sci. Pollut. Res.* 23(5): 4541–4550.

Bouyoucos, G.J. 1939. Effect of organic matter on the water-holding capacity and wilting point of mineral soils. *Soil Sci.* 47(5): 377–384.

Bussière, F., and Cellier, P. 1994. Modification of the soil temperature and water content regimes by a crop residue mulch – Experiment and modelling. *Agric. Forest Meteorol.* 68(1–2): 1–28.

Cabrera, M.L., Kissel, D.E., and Vigil, M.F. 2005. Nitrogen mineralization from organic residues: Research opportunities. *J. Environ. Qual.* 34: 75–79.

Cai, G. Dang, T., Guo, S., and Hao, M. 2005. Effects of nitrogen and mulching management on rainfed wheat and maize in northwest China. In: *Nutrient and water management practices for increasing crop production in rainfed arid/semi-arid area.* Proceedings of a coordinated research project. IAEA-TECDOC-1468, 209-2016, 77–88.

Cai, D.X., Ke, J., Wang, X.B., Hoogmoed, W.B., Oenema, O., and Perdok, U.D. 2006. Conservation tillage for dryland farming in China. In: Proceedings of the 17th ISTRO Conference, Kiel, Germany, August 28–September 3, International Soil Tillage Research Organization, Kiel, Germany, pp. 1627–1633.

Cakmak, I. 2005. The role of potassium in alleviating detrimental effects of abiotic stresses in plants. *J. Plant Nutr. Soil Sci.* 168: 521–530.

Calderón, F.J., Vigil, M.F., and Benjamin, J. 2017. Compost input effects on dryland wheat and forage yields and soil quality. *Pedosphere* 28: 451–462.

Campbell, C.A., Jame, Y.D., and De Jong, R. 1988. Predicting net nitrogen mineralization over a growing season: Model verification. *Can. J. Soil Sci.* 68: 537–552.

Campbell, C.A., Jame, Y.W., and Winkleman, G.E. 1984. Mineralization rate constants and their use for estimating nitrogen mineralization in some Canadian prairie soils. *Can. J. Soil Sci.* 64: 333–343.

Campbell, C.A., Nicholaichuk, W., Davidson, H.R., and Cameron, D.R. 1977. Effects of fertilizer N and soil moisture on growth, N content and moisture use by spring wheat. *Can. J. Soil Sci.* 57: 289–310.

Cerezini, P., dos Santos, D., Fagotti, L., Pípolo, A.E., Hungria, M., and Nogueira, M.A. 2017. Water restriction and physiological traits in soybean genotypes contrasting for nitrogen fixation drought tolerance. *Scicia Agricola (Piracicaba, Braz.)*, 74(2): 110–117.

Cerezini, P., Riar, M.K., and Sinclair, T.R. 2016. Transpiration and nitrogen fixation recovery capacity in soybean following drought stress. *J. Crop Improve.* 30(5): 562–571.

Chiang, C., Soudi, B., and Moreno, A. 1983. Soil nitrogen mineralization and nitrification under Moroccan conditions. In: *Nutrient balances and the need for fertilizers in semi-arid and arid regions.* 17th Colloquium of the International Potash Institute, Rabat and Marrakech. International Potash Institute, Bern Switzerlab, 129–139.

Christensen, B.T., and Johnston, A.E. 1997. Soil organic matter and soil quality—Lessons learned from long-term experiments at Askov and Rothamsted. *Dev. Soil Sci.* 25: 399–430.

Coleto I., Pineda M., Rodino A.P., De Ron A.M., and Alamillo J.M. 2014. Comparison of inhibition of N_2 fixation and ureide accumulation under water deficit in four common bean genotypes of contrasting drought tolerance. *Ann. Bot.* 113: 1071–1082.

Corbeels, M. 1997. *Nitrogen availability and its effect on water limited wheat growth on Vertisols in Morocco.* PhD thesis.

Corbeels, M., Hofman, G., and Van Cleemput, O. 1997. Response of rainfed bread wheat to nitrogen fertilizer on Moroccan Vertisols. 11th World Fertilizer Congress. Ghent, Belgium, September 7–13, 1997.

Corbeels, M., Hofman, G., and van Cleemput, O. 1998. Residual effect of nitrogen fertilization in a wheat-sunflower cropping sequence on a vertisol under semi-arid Mediterranean conditions. *Eur. J. Agron.* 9: 109–116.

Corbeels, M., Hofman, G., and van Cleemput, O. 1999. Soil mineral nitrogen dynamics under bare fallow and wheat in vertisols of semi-arid Mediterranean Morocco. *Biol. Fertil. Soils* 28: 321–328.

Corbeels, M., Hofman, G., and Van Cleemput, O. 2000. Nitrogen cycling associated with the decomposition of sunflower stalks and wheat straw in a Vertisol. *Plant Soil.* 218: 71–82.

CTIC (Conservation Technology Information Center). 2018. *Tillage Type Definitions.* www.ctic.purdue.edu/resourcedisplay/322/. Accessed July 2018.

Curioni, P.M.G., Hartwig, U.A., Nosberger, J., and Schuller, K.A. 1999. Glycolytic flux is adjusted to nitrogenase activity in nodules of detopped and argon-treated alfalfa plants. *Plant Physiol.* 119: 445–454.

Daoui, K., Fatemi, Z., Bendidi, A., Razouk, R., Chergaoui, A., and Ramdani, A. 2012. Olive tree and annual crops association's productivities under Moroccan conditions. In: First European Scientific Conference on Agriforestry, Book of Abstracts, 9–10 October, Brussels, Belgium. www.eurafagroforestry.eu/action/conferences/I_EURAFConference.

Darkoh, M.B.K. 2003. Desertification in the drylands: A review of the African situation. *Ann. Arid Zone* 42(3–4): 289–307.

Daryanto, S., Wang, L., and Jacinthe, P.A. 2015. Global synthesis of drought effects on food legume production. *PLOS ONE* 10(6): e0127401.

Devi, M.J., and Sinclair, T.R. 2013. Fixation drought tolerance of the slow-wilting soybean PI 471938. *Crop Sci.* 53(5): 2072–2078.

Dregne, H.E. 1976. *Soils of Arid Regions.* Elsevier, Amsterdam, The Netherlands.

Duan, Y., Shi, X., Li, S., Sun, X., and He, X. 2014. Nitrogen use efficiency as affected by phosphorus and potassium in long-term rice and wheat experiments. *J. Integr. Agric.* 13(3):588–596.

Duley, F.L., and Russel, J.C. 1939. The use of crop residues for soil and moisture conservation. *J. Am. Soc. Agron.* 31: 703–709.

Durand, J.L., Sheehy, J.E., and Minchin, F.R. 1987. Nitrogenase activity, photosynthesis and nodule water potential in soybean plants experiencing water-deprivation. *J. Exp. Bot.* 38: 311–321.

El Mejahed, K. and Aouragh, L. 2005. Green manure and nitrogen fertilizer effects on soil quality and profitability of a wheat-based system in semi-arid Morocco. In: *Nutrient and water management practices for increasing crop production in rainfed arid/semi-arid area.* Proceedings of a coordinated research project. IAEA-TECDOC-1468. 209-2016, 89–106.

Elherradi, Z., Soudi, B., and Elkacemi, K. 2003. Evaluation de la minéralisation de l'azote de deux sols amendés avec un compost d'ordures ménagères. *Etude et Gestion des Sol* 10(3): 139–153.

Engin, M., and Sprent, J.I. 1973. Effects of water stress on growth and nitrogen-fixing activity of Trifolium repens. *New Phytol.* 72: 117–126.

FAO (Food and Agriculture Organization of the United Nations). 2002. Land degradation assessment in drylands. LADA project. World Soil Resources Reports No. 97. FAO, Rome, Italy.

FAO (Food and Agriculture Organization of the United Nations). 2004. Carbon sequestration in dryland soils. World Soils Resources Reports No. 102. FAO, Rome, Italy.

FAO (Food and Agriculture Organization of the United Nations) 2005. The importance of soil organic matter: Key to drought-resistant soil and sustained food production. In: *FAO Soils Bulletin 80*, A. Bot and J. Benites (Eds.), FAO, Rome, Italy.

FAO (Food and Agriculture Organization of the United Nations) 2005. Land and environmental degradation and desertification in Africa: Issues and options for sustainable economic development with transformation. In: *UN-ECA/FAO Agricultural Division, Monograph No 10*, S.C. Nana-Sinkam (Ed.), FAO, Rome, Italy.

Finn, G.A., and Brun, W.A. 1980. Water stress effects on CO_2 assimilation, photosynthate partitioning, stomatal resistance, and nodule activity in soybean. *Crop Sci.* 20: 431–434.

Fischer, K.S., Lafitte, R., Fukai, S., Atlin, G., and Hardy, B. 2003. *Breeding Rice for Drought-Prone Environments*. International Rice Research Institute, Los Baños, Philippines. http://books.irri.org/971 2201899_content.pdf.

Flowers, T.H., and O'Callaghan, J.R. 1983. Nitrification in soils incubated with pig slurry or ammonium sulphate. *Soil Biol. Biochem.* 15(3): 337–342.

Fuentes, J.P., Flury, M., Huggins, D.R., and Bezdicek, D.F. 2003. Soil water and nitrogen dynamics in dryland cropping systems of Washington State. *Soil Till. Res.* 71: 33–47.

Gálvez, L., González, E.M., and Arrese-Igor, C. 2005. Evidence for carbon flux shortage and strong carbon/nitrogen interactions in pea nodules at early stages of water stress. *J. Exp. Bot.* 56: 2551–2561.

Gao, Y., Li, Y., Zhang, J., Liu, W., Dang, Z., Cao, W., and Qiang, Q. 2009. Effects of mulch, N fertilizer, and plant density on wheat yield, wheat nitrogen uptake, and residual soil nitrate in a dryland area of China. *Nutr. Cycling Agroecosyst* 85(2): 109–121.

Gimenez, C., Orgaz, F., and Fereres, E. 1997. Productivity in water-limited environments: Dryland agricultural systems. In: *Ecology in Agriculture*. Jakson L.E. (ed). Academic Press.

González, E.M., Aparicio-Tejo, P.M., Gordon, A.J., Minchin, F.R., Royuela, M., and Arrese-Igor, C. 1998. Water-deficit stress effects on carbon and nitrogen metabolism of pea nodules. *J. Exp. Bot.* 49(327): 1705–1714.

González, E.M., Gálvez, L., Royuela, M., Aparicio-Tejo, P.M., and Arrese-Igor, C. 2001. Insights into the regulation of nitrogen fixation in pea nodules: Lessons from drought, abscisic acid and increased photoassimilate availability. *Agronomie, EDP Sciences* 21(6–7): 607–613.

Haas, H.J., Evans, C.E., and EMiles, E.F. 1957. *Nitrogen and Carbon Changes in Great Plains Soils As Influenced by Cropping and Soil Treatments*. U.S. Department of Agriculture Technical Bulletin 1164.

Hadda, M.S., and Arora, S. 2006. Soil and nutrient management practices for sustaining crop yields under maize-wheat cropping sequence in sub-mountain Punjab, India. *Soil Environ.* 25(1): 1–5.

Hardarson, G., Danso, S.K.A., and Zapata, F. 1987. Biological nitrogen fixation in field crops. In: *Handbook of Plant Science in Agriculture*, B.R. Christie (ed.), CRC Press Inc., Boca Raton, FL, pp. 165–192.

Harraq, A., Bouabid, R., Bahri, H., and Boumchita, H. 2016. Dynamique de minéralisation de l'azote des fertilisants organiques et contribution des différentes formes d'azote à l'absorption de cet élément par la pomme de terre conduite en mode biologique. *Eur. J. Sc. Res.* 137(2): 196–213.

Hatfield, J.L., Sauer, T.J., and Prueger, J.H. 2001. Managing soils to achieve greater water use efficiency: A review. *Agron. J.* 93: 271–280.

He, M., and Dijkstra, F.A. 2014. Drought effect on plant nitrogen and phosphorus: A meta analysis. *New Phytol.* 204: 924–931.

Henao, J., and Baanante, C. 2006. *Agricultural Production and Nutrient Mining in Africa: Implications for Resource Conservation and Policy Development*. International Fertilizer Development Center (IFDC), Muscle Shoals, AL, www.ifdc.org.

Hénin, S., and Dupuis, M. 1945. Essai de bilan de la matière organique du sol. *Ann. Agron.* 11: 17–29.

Hilhorst, T., and Muchena, F. 2000. Managing soil fertility in Africa: Diverse settings and changing practice. In: *Nutrients on the Move: Soil Fertility Dynamics in African Farming Systems*, T. Hilhorst and F. Muchena (eds.), IIED Drylands Programme, London.

Hollis, J.M., Jones, R.J.A., and Palmer, R.C. 1977. The effects of organic matter and particle size on the water retention properties of some soils in the West Midlands of England. *Geoderma* 17: 225–238.

Hudson, B.D. 1994. Soil organic matter and available water capacity. *J. Soil Water Conserv.* 49(2): 189–194.

IAEA. 2005. *Nutrient and water management practices for increasing crop production in rainfed arid/semi-arid area.* Proceedings of a coordinated research project. IAEA-TECDOC-1468.

IFPRI (International Food Policy Research Institute. 1996. *Feeding the World, Preventing Poverty and Protecting the Earth: A 2020 Vision.* International Food Policy Research Institute, Washington DC. www.ifpri.org/publication/feeding-world-preventing-poverty-and-protecting-earth.

Irshad, M., Honna, T., Yamamoto, S., Eneji, A.E., and Yamasaki, N. 2005. Nitrogen mineralization under saline conditions. *Commun. Soil Sci. Plant Anal.* 36(11–12): 1681–1689.

Janssen, B.H. 1984. A simple method for calculating decomposition and accumulation of 'young' soil organic matter. *Plant Soil.* 76: 297–304.

Jenkinson, D.S., and Rayner, J.H. 1977. The turnover of soil organic matter in some of the Rothamsted classical experiments. *Soil Sci.* 123: 298–305.

Jensen, E.S. 1997. Nitrogen immobilization and mineralization during initial decomposition of ^{15}N labeled pea and barley residues. *Biol. Fert. Soils* 24: 39–44.

Jin, K., De Neve, S., Moeskops, B., Lu, J., Zhang, J., Gabriels, D., Cai, D., and Jin, J. 2008. Effects of different soil management practices on winter wheat yield and N losses on a dryland loess soil in China. *Aust. J. Soil Res.* 46(5): 455–463.

Johnson, Y.K., Ayuke, F.O., Kinamac, J.M., and Sijali, I.V. 2018. Effects of tillage practices on water use efficiency and yield of different drought tolerant common bean varieties in Machakos County, Eastern Kenya. *Am. Sci. Res. J. Eng. Tech. Sci.* 40(1): 217– 234.

Ju, X.T., and Li, S.X. 1998. The effect of temperature and moisture on nitrogen mineralization in soils. *Plant Nutr. Fertil. Sci.* 4(1): 37–42.

Justice, J.K., and Smith, R.L. 1962. Nitrification of ammonium sulfate in a calcareous soil as influenced by combinations of moisture, temperature, and levels of added nitrogen. *Soil Sci. Soc. Am. J.* 26: 246– 250.

Kant, S., and Kafkafi, U. 2002. Potassium and abiotic stresses in plants. In: *Potassium for Sustainable Crop Production.* N.S. Pasricha and S.K. Bansal (eds.), Potash Institute of India, Gurgaon, India, pp. 233–251.

Kelley, K.W., and Sweeney, D.W. 2005. Tillage and urea ammonium nitrate fertilizer rate and placement affects winter wheat following grain sorghum and soybean. *Agron. J.* 97: 690–697.

Kihara, J., Bationo, A., Mugendi, D.N., Martius, C., and Vlek, P.L.G. 2011. Conservation tillage, local organic resources and nitrogen fertilizer combinations affect maize productivity, soil structure and nutrient balances in semi-arid Kenya. *Nutr. Cycl. Agroecosyst.* 90: 213–225.

King, C.A., and Purcell, L.C. 2005. Inhibition of N_2 fixation in soybean is associated with elevated ureides and amino acids. *Plant Physiol.* 137: 1389–1396.

Klemedtsson, L., Svensson, B.H., and Rosswall, T. 1988. Relationships between soil moisture content and nitrous oxide production during nitrification and denitrification. *Biol. Fertil. Soils* 6(2): 106–111.

Kolenbrander, G.J. 1969. *De bepaling van de waarde van verschillende soorten organische stof ten aanzien van hun effect op her humusgehalte bij bouwland.* Inst. Bodemvruchtbaarheid, Haren, Netherlands, C 6988.

Lal, R. 2002. Carbon sequestration in dryland ecosystems of West Asia and North Africa. *Land Degrad. Dev.* 13(1): 45–59.

Lal, R. 2004. Carbon sequestration in dryland ecosystems. *Environ. Manage.* 33(4): 528–544.

Li, P., Dong, H., Liu, A., Liu, J., Sun, M., Li, Y., Liu, S., Zhao, X., and Mao, S. 2017. Effects of nitrogen rate and split application ratio on nitrogen use and soil nitrogen balance in cotton field. *Pedosphere* 27(4): 769–777.

Li, S.X. Wang, Z.H., Malhi, S.S., Li, S.Q., Gao, Y.J., and Tian, X.H. 2009a. Nutrient and water management effects on crop production, and nutrient and water use efficiency in dryland areas of china. *Adv. Agron.* 102: 223–265.

Li, S.X., Wang, Z.H., Gao, Y.J., and Stewart, B.A. 2009b. Nitrogen in dryland soils of China and its management. *Adv. Agron.* 101: 123–181.

Li, S.X., Wang, Z.H., Hu, T.T., Malhi, S.S., Li, S.Q., Gao , Y.J., and Tian, X.H. 2009c. Nutrient and water management effects on crop production and nutrient and water use efficiency in dryland areas of China. *Adv. Agron.* 102: 223–265.

Lima, J.D., and Sodek, L. 2003. N-stress alters aspartate and asparagine levels of xylem sap in soybean. *Plant Sci.* 165: 649–656.

Lindhauer, M.G. 1985. Influence of K nutrition and drought on water relations and growth of sunflower (*Helianthus-annuus L.*). *J. Plant Nutr. Soil Sci.* 148: 654–669.

Liwang M., Peterson G.A., Ahuja L.R., Sherrod L., Shaffer M.J., and Rojas K.W. 1999. Decomposition of surface crop residues in long-term studies of dryland agroecosystems. *Agron. J.* 91: 401–409.

Ma, L., Ahuja, L.R., Shaffer, M.J., Rojas, K.W., Peterson, G.A., and Sherrod, L. 1999. Decomposition of surface crop residues in long-term studies of dryland agroecosystems. *Agron. J.* 91: 401–409.

Ma, Q., Yu, W.T., Shen, S.M., Zhou, H., Jiang, Z.S., and Xu, Y.G. 2010. Effects of fertilization on nutrient budget and nitrogen use efficiency of farmland soil under different precipitations in Northeastern China. *Nutr. Cycl. Agroecosyst.* 88: 315–327.

Malhi, S.S. and McGill, W.B. 1982. Nitrification in three Alberta soils: Effect of temperature, moisture and substrate concentration. *Soil Biol. Biochem.* 15: 397–399.

Manyowa, N.M. 1994. Maize production in Zimbabwe: Coping with drought stress in marginal agroecological zones. In: *Bilan hydrique agricole et sécheresse en Afrique tropicale: vers une gestion. Universités francophones*, F.N. Reynier and L. Netoyo (eds.), John Libbey Eurotext, Paris, France.

Marino, D., Frendo, P., Ladrera, R., Zabalza, A., Puppo, A., Arrese-Igor, C., and González, E.M. 2007. Nitrogen fixation control under drought stress. Localized or systemic? *Plant Physiol.* 143: 1968–1974.

Marschner, P. 2012. *Marschner's Mineral Nutrition of Higher Plants*, 3rd edition. Academic Press, London, pp. 178–189.

Mary, B. and Guérif, J. 1994. Intérêts et limites des modèles de prévision de l'évolution des matières organiques et de l'azote dans le sol. *Cahiers Agricultures*. 3: 247–257.

Matthews, R.B., Reddy, D.M., Rani, A.U., Azam-Ali, S.N., and Peacock, J.M. 10. 1990. Response of four sorghum lines to mid-season drought. I. Growth, water use and yield. *Field Crops Res.* 25: 279–296.

McKenzie, R.H., Middleton, A.B., and Zhang, M. 2001. Optimal time and placement of nitrogen fertilizer with direct and conventionally seeded winter wheat. *Can. J. Soil Sci.* 81: 613–622.

Mengel, K. 2007. Potassium. In: *Handbook of Plant Nutrition*, A.V. Baker and D.J. Pilbeam (eds.), CRC Press, Taylor and Francis, Boca Raton, FL.

Middleton, N., and Thomas, D. 1997. *World Atlas of Desertification*, 2nd edition. United Nations Environment Programme (UNEP), Arnold, London.

Montañez, A. 2000. Overview and case studies on biological nitrogen fixation: Perspectives and limitations. *Sci. Agric.*: 1–11.

Mrabet, R., Ibno-Namr, K., Bessam, F., and Saber, N. 2001. Soil chemical quality changes and implications for fertilizer management after 11 years of no-tillage wheat production systems in semiarid Morocco. *Land Degrad. Dev.* 21: 1–13.

Mrabet, R., Moussadek, R., Fadlaoui, A., and van Ranst, E. 2012. Conservation agriculture in dry areas of Morocco. *Field Crops Res.* 132: 84–94.

Murage, E.W., Karanja, N.K., Smithson, P.C., and Woomer, P.L. 2000. Diagnostic indicators of soil quality in productive and non-productive smallholders' fields of Kenya's Central Highlands. *Agric. Ecosyst. Environ.* 79: 1–8.

Myers, R.J.K. 1984. A simple model for estimating the nitrogen fertilizer requirement of a cereal crop. *Fert. Res.* 5: 95–108.

Myers, R.J.K., C.A. Campbell, and K.L. Weier. 1982. Quantitative relationship between net nitrogen mineralization and moisture content of soils. *Can. J. Soil Sci.* 62: 111–124.

Nageswara Rao, R. C., Sardar Singh, Sivakumar, M. V. K., Srivastava, K. L., and Williams, J. H. 1985. Effect of water deficit at different growth phases of peanut. I. Yield response. *Agronomy J.* 77: 782–786.

Nandwal, A.S., Bharti, S., Sheoran, I.S., and Kuhad, M.S. 1991. Drought effects on carbon exchange and nitrogen fixation in Pigeonpea (*Cajanus cajan L.*). *J. Plant Physiol.* 138: 125–127.

Neo, H.H., and Layzell, D.B. 1997. Phloem glutamine and the regulation of O_2 diffusion in legume nodules. *Plant Physiol.* 113: 259–267.

Nettleton, W.D., and Peterson, F.F. 1983. Aridisols. In: *Pedogenesis and Soil Taxonomy*, L.P. Wilding, N.E. Smeck, G.F. Hall (eds.), Elsevier, Amsterdam, Netherlands, pp. 165–216.

Noin, D., and Clarke, J.I. 1997. Population and environment in arid regions of the world. In: *Population and Environment in Arid Regions*, J. Clarke and D. Noin (eds.). MAB/UNESCO, Vol. 19, The Parthenon Publishing Group, New York, pp. 1–18.

Noy-Meir, I., Harpaz, Y., 1978. Agro-ecosystems in Israel. In: Cycling of Mineral Nutrients in Agricultural Ecosystems. Frissel, M.J. (ed.). *Agro-Ecosystems* 4:143–167.

Palm, C.A., Myers R.J.K., and Nandwa S.M. 1997. Combined use of organic and inorganic nutrient sources for soil fertility maintenance and replenishment. In: *Replenishing Soil Fertility in Africa*. Buresh R.J., Sanchez, P.A., and Calhoun, F. (eds.). Soil Science Society of America, Madison, WI, 193–217.

Palta, J.A. and Fillery, I.R.P. 1995. N application enhances remobilization and reduces losses of pre-anthesis N in wheat grown on a duplex soil. *Aust. J. Agric. Res.* 46(3): 519–531.

Palta, J.A., Chen, X., Milroy, S.P., Rebetzke, G.J., Dreccer, M.F., and Watt, M. 2011. Large root systems: Are they useful in adapting wheat to dry environments? *Funct. Plant Biol.* 38: 347–354.

Place, F., Barret, C.B., Freeman, H.A., Ramisch, J.J., and Vanlauwe B. 2003. Prospects for integrated soil fertility management using organic and inorganic inputs: evidence from smallholder African agricultural systems. *Food Policy.* 28: 365–378.

Pankhurst, C.E., and Sprent, J.I. 1975a. Surface features of soybean root nodules. *Protoplasma* 85: 85–98.

Pankhurst, C.E., and Sprent, J.I. 1975b. Effects of water stress on the respiratory and nitrogen-fixing activity of soybean root nodules. *J. Exp. Bot.* 26: 287–304.

Passioura, J.B. 1983. Roots and drought resistance. *Agric. Water Manage.* 7: 265–280.

Peng, Z., Ting, W., Haixia, W., Min, W., Xiangping, M., Siwei, M., Rui, Z., Zhikuan, J., and Qingfang, H. 2015. Effects of straw mulch on soil water and winter wheat production in dryland farming. *Sci. Rep.* 5: 10725.

Pervez, H., Ashraf, M., and Makhdum, M.I. 2004. Influence of potassium nutrition on gas exchange characteristics and water relations in cotton (*Gossypium hirsutum L.*). *Photosynthetica* 42: 251–255.

Petit, S., Alignier, A., Colbach, N., Joannon, A., Le Cœur, D., and Thenail, C. 2013. Weed dispersal by farming at various spatial scales. A review. *Agron. Sustain. Dev.* 33(1): 205–217.

Premachandra, G.S., Saneoka, H., and Ogata, S. 2009. Cell membrane stability and leaf water relations as affected by potassium nutrition of water-stressed maize. *J. Exp. Bot.* 42(6): 739–745.

Purcell L.C., King C.A., Ball R.A. 2000. Soybean cultivar differences in ureides and the relationship to drought tolerant nitrogen fixation and manganese nutrition. *Crop Sci.* 40: 1062–1070.

Ramana, M.V., Khadke, K.M., Rego, T.J., Kumar Rao, J.V.D.K., Myers, R.J.K., Pardhasaradhi, G., and Venkata Ratnam, N. 2005. Management of nitrogen and evaluation of water-use efficiency in traditional and improved cropping systems of the southern Telangana region of Andhra Pradesh, India. In: *Nutrient and water management practices for increasing crop production in rainfed arid/semi-arid area.* Proceedings of a coordinated research project. IAEA-TECDOC-1468, 139–154.

Ramirez-Garcia, J., Martens, H.J., Quemada, M., and Thorup-Kristensen, K. 2015. Intercropping effect on root growth and nitrogen uptake at different nitrogen levels. *J. Plant Ecol.* 8(4): 380–389.

Rashid, A. and Ryan, J. 2004. Micronutrient constraints to crop production in soils with Mediterranean-type characteristics: A review. *J. Plant Nutr.* 27: 959–975.

Recous, S., Robin, D., Darwis, D., and Mary, B. 1995. Soil inorganic N availability: Effect on maize residue decomposition. *Soil Biol. Biochem.* 27: 1529–1538.

Reich, P.F., Numbem, S.T., Almaraz, R.A., and Eswaran, H. 2001. Land resource stresses and desertification in Africa. In: *Responses to Land Degradation*, E.M. Bridges, I.D. Hannam, L.R. Oldeman, F.W.T. Pening de Vries, S.J. Scherr, and S. Sompatpanit (eds.). Proc. 2nd. International Conference on Land Degradation and Desertification, Khon Kaen, Thailand. Oxford Press, New Delhi, India.

Reinertsen, S.A., Elliott, L.F., Cochran, V.L., and Campbell, G.S. 1984. Role of available carbon and nitrogen in determining the rate of wheat straw decomposition. *Soil Biol. Biochem.* 16: 459–464.

Rekha, R.G., Desai, B.K., Satyanarayan, R., Umesh, M.R., and Shubha, S. 2017. Effect of intercropping system and nitrogen management practices on nutrient uptake and post-harvest soil fertility 8(2): 236–239.

Roy, R.N., and Nabhan, H. 1999. Soil and nutrient management in sub-Saharan Africa in support of the soil fertility initiative. Proceedings of the Expert Consultation Lusaka, Zambia, 6–9 December. FAO, Rome, Italy.

Russell, M.B., Klute, A., and Jacob, W.C. 1952. Further studies on the effect of long-time organic matter additions on the physical properties of Sassafras silt loam. *Soil Sci. Soc. Amer. Proc.* 16: 156.

Ryan, J. 1998. Changes in organic carbon in long-term rotation and tillage trials in northern Syria. In: *Management of Carbon Sequestration in Soil*, R. Lal, J. Kimble, R. Follett, and B.A. Stewart (eds.). Advances in Soil Science, CRC, Boca Raton, FL, pp. 285–295.

Ryan, J. 2008a. Crop nutrients for sustainable agricultural production in the drought-stressed Mediterranean region. *J. Agric. Sci. Technol.* 10: 295–306.

Ryan, J. 2008b. A perspective on balanced fertilization in the Mediterranean region. *Turk. J. Agric. For.* 32: 79–89.

Ryan, J., Masri, S., Garabet, S., Diekmann, J., and Habib, H. 1997. *Soils of ICARDA's Agricultural Experiment Stations and Sites: Climate, Classification, Physical and Chemical Properties, and Land Use.* ICARDA Tech. Bull., ICARDA, Aleppo, Syria.

Ryan, J., Masri, S., Pala, M., and Bounejmate, M. 2002. Barley-based rotations in a typical Mediterranean agroecosystem: Crop production trends and soil quality. In: *International meeting on Soils with Mediterranean Type of Climate (selected papers)*, P. Zdruli, P. Steduto, and S. Kapur (eds.), vol. 7. Options Méditerranéennes: Série A. Séminaires Méditerranéens. n. 50, pp. 287–296.

Ryan, J., Monem, M.A., Azzaoui, A., El Mejahed, K., El Gharous, M., and Mergoum, M. 1992. A current perspective on dryland cereal fertilization in Morocco. In: *Fertilizer Use Efficiency under Rain-Fed Agriculture in West Asia and North Africa*, J. Ryan and A. Matar (eds.), Proceedings of the Fourth Regional Workshop, 5–10 May 1991, ICARDA, Agadir, Morocco. pp. 106–115.

Ryan, J., Pala, M., Harris, H., Masri, S., and Singh, M. 2010. Rainfed wheat-based rotations under Mediterranean-type climatic conditions: Crop sequences, N fertilization, and stubble grazing intensity in relation to cereal yields parameters. *J. Agric. Sci.* 48: 205–216.

Ryan, J., Singh, M.,. and Pala, M. 2008. Long-term cereal-based rotation trials in the Mediterranean region. Implications for cropping sustainability. *Adv. Agron.* 97: 276–324.

Ryan, J., Sommer, R., and Ibrikci, H. 2012. Fertilizer best management practices: A perspective from the dryland West Asia–North Africa region. *J. Agron. Crop Sci.* 198: 57–67.

Ryan, J., and Sommers, R. 2010. Fertilizer best management practices in the dryland Mediterranean area – Concepts and perspectives. 19th World Congress of Soil Science, Soil Solutions for a Changing World, 1–6 August, Brisbane, Australia.

Sahrawat, K.L. 2008. Factors affecting nitrification in soils. *Commun. Soil Sci. Plant Anal.* 39(9–10): 1436–1446.

Sainju, U.M., Lenssen, A.W., Caesar-TonThat, T., Jabro, J.D., Lartey, R.T., Evans, R.G., and Allen, B.L. 2012. Dryland soil nitrogen cycling influenced by tillage, crop rotation, and cultural practice. *Nutr. Cycl. Agroecosyst.* 93(3): 309–322.

Sainju, U.M., Lenssen, A.W., Allen, B.L., Stevens, W.B., and Jabro, J.D. 2016. Nitrogen balance in response to dryland crop rotations and cultural practices. *Agric. Ecosyst. Environ.* 233: 25–32.

Salter, P.J., and Howarth, F. 1961. The available-water capacity of a sandy loam soil: II. The effects of farmyard manure and different primary cultivations. *J. Soil Sci.* 12(2): 335–342.

Salter, P.J., and Williams, J.B. 1963. The effect of farmyard manure on the moisture characteristics of a sandy loam soil. *J. Soil Sci.* 14: 73–81.

Sauerbeck, D.R., and Gonzalez, M.A. 1977. Field decomposition of carbon-14-labelled plant residues in various soils of the Federal Republic of Germany and Costa Rica. In: *Soil Organic Matter Studies.* Vol. 1. International Atomic Energy Agency (IAEA). Vienna, Austria, pp. 159–170.

Schomberg, H.H., and Jones, O.R. 1999. Carbon and nitrogen conservation in dryland tillage and cropping systems. *Soil Sci. Soc. Am. J.* 63: 1359–1366.

Schomberg, H.H., and Steiner, J.L. 1999. Nutrient dynamics of crop residues decomposing on a fallow no-till soil surface. *Soil Sci. Soc. Am. J.* 63: 607–613.

Schomberg, H.H., Steiner, J.L., and Unger, P.W. 1994. Decomposition and nitrogen dynamics of crop residues: Residue quality and water effects. *Soil Sci. Soc. Am. J.* 58: 372–381.

Seligman, N.G., Feigenbaum, S., Feinerman, D., and Benjamin, R.W. 1986. Uptake of nitrogen from high C-to-N ratio 15N-labeled organic matter residues by spring wheat grown under semi-arid conditions. *Soil Biol. Biochem.* 18: 303–307.

Seligman, N.G., Loomis, R.S., Burke, J., and Abshahi, A. 1983. Nitrogen nutrition and phenological development in field-grown wheat. *J. Agric. Sci.* 101: 691–697.

Sene, M. and Badiane, A.N. 2005. Optimization of water and nutrient use by maize and peanut in rotation based on organic and rock phosphate soil amendments. In: *Nutrient and water management practices for increasing crop production in rainfed arid/semi-arid area.* Proceedings of a coordinated research project. IAEA-TECDOC-1468, 197–208.

Serraj, R., Sinclair, T.R., and Purcell, L.C. 1999. Symbiotic N2 fixation response to drought. *J. Exp. Bot.* 50: 143–155.

Serraj, R. and Sinclair, T.R. 1996a. Inhibition of nitrogenase activity and nodule oxygen permeability by water deficit. *J. Exp. Bot.* 47: 1067–1073.

Serraj, R. and Sinclair, T.R. 1996b. Processes contributing to N_2-fixation insensitivity to drought in the soybean cultivar Jackson. *Crop Science.* 36: 961–968.

Serraj, R., Vadez, V., and Sinclair, T.R. 2001. Feedback regulation of symbiotic N_2 fixation under drought stress. *Agronomie* 21: 621–626.

Shah, Z.H., and Arshad, M. 2006. *Land Degradation in Pakistan: A Serious Threat to Environments and Economic Stability.* Eco Services International. www.eco-web.com/edi/060715.html. Accessed in April 2018.

Sharma, K.L., Srinivas, K., Das, S.K., Vittal, K.P.R., and Kusuma, G.J. 2002. Conjunctive use of inorganic and organic sources of nitrogen for higher yield of sorghum in dryland Alsol. *Indian J. Dryland Agric. Res. Dev.* 17: 79–88.

Sharma, K.L., Vittal, K.P.R., Ramakrishna, Y.S., Srinivas, K., Venkateswarlu, B., and Kusuma Grace, J. 2007. Fertilizer use constraints and management in rainfed areas with special emphasis on N use efficiency. In: *Agricultural Nitrogen Use & Its Environmental Implications*, Y.P. Abrol, N. Raghuran, and M.S. Sachdev (eds.), IK International Publishing House, New Delhi, India.

Shelton, D.R., Sadeghi, A.M., and McCarty, G.W. 2000. Effect of soil water content on denitrification during cover crop decomposition. *Soil Sci.* 165: 365–371.

Shepherd, K.D., Cooper, P.J.M., Allan, A.Y., Drennan, D.S.H., and Keatinge, J.D.H. 1987. Growth, water use and yield of barley in Mediterranean-type environments. *J. Agric. Sci.* 108: 365–378.

Sijali, I.V. and Kamoni, P.T. 2005. Optimization of water and nutrient use in rain-fed semi-arid farming through integrated soil-, water- and nutrient-management practices. In: *Nutrient and water management practices for increasing crop production in rainfed arid/semi-arid area*. Proceedings of a coordinated research project. IAEA-TECDOC-1468, 209–2016.

Sinclair, T.R. and Serraj, R. 1995. Legume nitrogen-fixation and drought. *Nature.* 378: 344.

Sinclair, T.R., and Goudriaan, J. 1981. Physical and morphological constraints on transport in nodules. *Plant Physiol.* 67: 143–145.

Sinclair, T.R., Purcell, L.C., Vadez, V., and Serraj, R. 2001. Selection of soybean (Glycine max) lines for increased tolerance of N_2 fixation to drying soil. *Agronomie, EDP Sciences* 21(6–7): 653–657.

Sinclair, T.R., and Serraj, R. 1995a. Dinitrogen fixation sensitivity to drought among legume species. *Nature* 1995: 378–344.

Sinclair, T.R., and Serraj, R. 1995b. Legume nitrogen fixation and drought. *Nature* 378: 344–344.

Singh, R.P. and Venkateswarlu, P. 1985. Role of all India coordinated research project for dryland agriculture in research development. *Fert. News.* 30(4): 43–55.

Singh, H. and Singh, K.P. 1994. Nitrogen and phosphorus availability and mineralization in dryland reduced tillage cultivation: Effects of residue placement and chemical fertilizer. Soil Biol. Biochem. 26(6): 695–702.

Singh, A.K., and Kumar, P. 2009. Nutrient management in rainfed-dryland agro-ecosystems in the impending climate change scenario. *Agric. Situation India* 66(5): 265–269.

Singh, M. 2017. Production potential and profitability of pigeonpea [*Cajanus cajan (L.) Millsp.*] as influenced by intercropping with blackgram [*Vigna mungo (L.) Hepper*] and integrated nutrient management. PhD Thesis. Faculty of Agriculture, Indira Gandhi Krishi Vishwavidyalaya, Raipur, India. http://krishiko sh.egranth.ac.in/bitstream/1/5810038158/1/thesis%20Madhulika%20singh.pdf.

Sivanappan, R.K. 1995. Soil and water management in the dry lands of India. *Land Use Policy* 12(2): 165–175.

Smika, D.E., and Unger, P.W. 1986. Effect of surface residues on soil water storage. In: *Advances in Soil Science*, B.A. Stewart (ed.), Vol. 5. Springer, New York.

Smith, S.J., Young, L.B., and Miller, G.E. 1977. Evaluation of soil nitrogen mineralization potentials under modified field conditions. *Soil Sci. Soc. Am. J.* 41: 74–76.

Song, H.X., and Li, S.X. 2006. Root function in nutrient uptake and soil water effect on NO_3–N and NH_4-N migration. *Agric. Sci. China* 5(5): 377–383.

Song, Y.N., Zhang, F.S., Marschner, P., Fan, F.L., Gao, H.M., Bao, X.G., Sun, J.H., and Li, L. 2007. Effect of intercropping on crop yield and chemical and microbiological properties in rhizosphere of wheat (*Triticum aestivum L.*), maize (*Zea mays L.*), and faba bean (*Vicia faba L.*). *Biol. Fertil. Soils* 43: 565–574.

Soudi, B. 1988. Etude de la dynamique de l'azote dans les sols marocains: Caractérisation et pouvoir minéralisateur. PhD Thesis. es-Sciences Agronomiques, Institut Agronomique et Vétérinaire Hassan II, Rabat, Morocco.

Soudi, B., Sbai, A., and Chiang, C.N. 1990a. Nitrogen mineralization in semi-arid area of Morocco: Rate constant variation with depth. *Soil Sci. Soc. Am. J.* 54: 756–761.

Soudi, B., Sbai, A., and Chiang, C.N. 1990b. Variations saisonnières de l'azote minéral et effet combiné de la température et de l'humidité du sol sur la minéralisation. *Actes Inst. Agron. Vet.* 10(1): 29–38.

Sprent, J.I. 1971. The effects of water stress on nitrogen fixing root nodules. I. Effects on the physiology of detached soybean nodules. *New Phytol.* 70: 9–17.

Stanford, G., and Epstein, E. 1974. Nitrogen mineralization and water relations in soils. *Soil Sci. Soc. Am. J.* 38: 103–106.

Strong, W.M., and Cooper, J.E. 1980. Recovery of nitrogen by wheat from various depths in a cracking clay soil. *Aust. J. Exp. Agric.* 20: 82–87.

Studer, C., Hu ,Y., and Schmidhalter, U. 2017. Interactive effects of N-, P- and K-nutrition and drought stress on the development of maize seedlings. *Agriculture* 7(11): 90.

UNEP. 1992. *World Atlas of Desertification*, First Edition. Edward Arnold, London, UK.

UNEP-WCMC. 2007. A spatial analysis approach to the global delineation of dryland areas of relevance to the CBD Programme of Work on Dry and Subhumid Lands. Dataset based on spatial analysis between WWF terrestrial ecoregions (WWF-US, 2004) and aridity zones (CRU/UEA; UNEPGRID, 1991). Dataset checked and refined to remove many gaps, overlaps and slivers (July 2014).

Unger, P.W. 1984. Tillage and residue effects on wheat, sorghum, and sunflower grown in rotation. *Soil Sci. Soc. Am. J.* 48: 885–891.

Unger, P.W. 2002. Conservation tillage for improving dryland crop yields. *Cienc. Suelo* 20(1): 1–8.

Unger, P.W., Schomberg, H.H., Dao, T.H., and Jones, O.R. 1997. Tillage and crop residue management practices for sustainable dryland farming systems Paul W. *Ann. Arid Zone* 36(3): 209–232.

Vadez, V., and Sinclair, T.R. 2000. Ureide degradation pathways in intact soybean leaves. *J. Exp. Bot.* 51: 1459–1465.

Vadez, V., Sinclair, T.R., and Serraj, R. 2000. Asparagine and ureide accumulation in nodules and shoots as feedback inhibitors of N_2 fixation in soybean. *Physiol. Plant.* 110: 215–223.

Valentine, A.J., Benedito, V.A., and Kang, Y. 2011. Legume nitrogen fixation and soil abiotic stress: From physiology to genomics and beyond. *Annu. Plant Rev.* 42: 207–248.

Van Doren, D.M., Jr., and Allmaras, R.R. 1978. Effect of residue management practices on the soil physical environment, microclimate, and plant growth. In: *Crop Residue Management Systems*, W.R. Oschwald (ed.). Spec. Publ. No. 31, American Society of Agronomy, Madison, WI, pp. 49–83.

Vandermeer, J. 1989. *The Ecology of Intercropping*. Cambridge University Press, Cambridge, UK.

Vanlauwe, B., Bationo, A., Chianu, J., Giller, K.E., Merckx, R., Mokwunye, U., Ohiokpehai, O., Pypers, P., Tabo, R., Shepherd, K.D., Smaling, E.M.A., Woomer, P.L., and Sanginga, N. 2010. Integrated soil fertility management: Operational definition and consequences for implementation and dissemination. *Outlook Agric.* 39: 17–24.

Vanlauwe, B., Descheemaeker, K., Giller, K.E., Huising, J., Merckx, R., Nziguheba, G., Wendt, J., and Zingore, S. 2015. Integrated soil fertility management in sub-Saharan Africa: Unravelling local adaptation. *Soil* 1: 491–508.

Vanlauwe, B., Nwoke, O.C., Sanginga, N., and Merckx, R. 1996. Impact of residue quality on the C and N mineralization of leaf and root residues of three agroforestry species. *Plant Soil* 183: 221–231.

Vere, D. 2005. Research into conservation tillage for dryland cropping in Australia and China. ACIAR's Impact Assessment Series; Project Report Number 33; Australian Centre for International Agricultural Research. www.aciar.gov.au.

Vigil, M.F., and Kissel, D.E. 1991. Equations for estimating the amount of nitrogen mineralized from crop residues. *Soil Sci. Soc. Am.* 5: 757–761.

Wang, M., Zheng, Q., Shen, Q., and Guo, S. 2013. The critical role of potassium in plant stress response. *Int. J. Mol. Sci.* 14: 7370–7390.

Wang, X. 2006. Conservation tillage and nutrient management in dryland farming in China. Doctorate Thesis. Wageningen University, Wageningen, Netherlands.

Wang, X., and Xing, Y. 2016. Effects of mulching and nitrogen on soil nitrate-N distribution, leaching and nitrogen use efficiency of maize (Zea mays L.). *PLOS ONE* 11(8): e0161612.

Warren, G.P., Atwal, S.S., and Irungu, J.W. 1997. Soil nitrate variations under grass, sorghum and bare fallow in semi-arid Kenya. *Exp. Agric.* 33: 321–333.

Weisz, P.R., Denison, R.F., and Sinclair, T.R. 1985. Response to drought stress of nitrogen fixation (acetylene reduction) rates by field-grown soybeans. *Plant Physiol.* 78(3): 525–530.

Wood, M., McNeill, A.M., Pilbeam, C.J., Swift, R.S., Harris, H.C., and Mugane, P.G. 1996. Sustainability of nitrogen use in two dryland farming systems. In: *Progress in Nitrogen Cycling Studies*, O. Van Cleemput, G. Hofman, and A. Vermoesen (eds.), Kluwer Academic Publishers, pp. 303–306.

Zahoor, R., Zhao, W., Dong, H., Snider, J.L., Abid, M., Iqbal, B., and Zhou, Z. 2017. Potassium improves photosynthetic tolerance to and recovery from episodic drought stress in functional leaves of cotton (*Gossypium hirsutum* L.). *Plant Physiol. Biochem.* 119: 21–32.

Zhang, X., Huang, G., Bian, X., and Zhao, Q. 2013. The effect of root interaction and nitrogen fertilization on the chlorophyll content, root activity, photosynthetic characteristics of intercropped soybean and microbial quantity in the rhizosphere. *Plant Soil Environ.* 50(2): 80–88.

Zhang, Y., Liu , J., Zhang, J., Liu , H., Liu, S., Zhai, L., Wang, H., Lei, Q., Ren, T., and Yin, C. 2015. Row ratios of intercropping maize and soybean can affect agronomic efficiency of the system and subsequent wheat. *PLOS ONE* 10(6): e0129245.

14 The Nuclear Option

Darryl D. Siemer

CONTENTS

14.1 INTRODUCTION

Because today's politically correct (non-nuclear) renewable energy sources would fail to support even the near future's (2050 AD) anticipated human population without severe environmental consequences (Springer and Duchin 2014; Sims 2011), this chapter invokes a future similar to that envisioned by Oak Ridge National Laboratory's (ORNL's) H. E. Goeller and Alvin Weinberg four decades ago (Goeller and Weinberg 1976). In this scenario, by 2100 AD, a "sustainable nuclear renaissance" – not an "all of the above" mix of politically correct energy sources – would address the root causes of mankind's perpetual squabbling over resources responsible for most human misery both then and now. It would simultaneously reverse environmental degradation and render food production sustainable by enabling the mining, grinding, shipping, and distribution of powdered basalt over farmland thereby affecting Mother Nature's too slow, notoriously unreliable, and sometimes catastrophic (volcanic) approach to both soil-building and CO_2 sequestration. Wherever possible this chapter will make its points with simple calculations utilizing reasonable assumptions and readily obtained data, not sweeping generalizations or wishful thinking. Finally, since it currently seems unlikely that Africa or any of its Western world "helpers" will develop/implement an appropriate nuclear renaissance, it also discusses the most likely way that this scenario might be realized – China will take the lead.

14.2 AFRICA'S SPECIAL ISSUES

Most of Africa's 54 countries continue to exhibit alarmingly high rates of both population growth and poverty (United Nations 2015). Approximately 380 million of its 1.2 billion people are extremely poor – often hungry – and ten of the world's most underdeveloped countries – Mozambique, Guinea, Burundi,

Burkina Faso, Eritrea, Sierra Leone, Chad, Central African Republic, Democratic Republic of Congo, and Niger – are located therein. Although considered exceptionally underdeveloped, none of them are among the 20 recognized to have the world's lowest living costs (Leffel 2017) meaning that Africa's poor people are considerably poorer than are those in more technologically advanced, but poor by OECD standards, nations such as Romania. Africa's people are plagued by a lack of basic infrastructure due to dysfunctional and, often, self-serving governance, further complicated by long festering civil/tribal/religious conflicts, and, therefore, most of them face bleak futures. The fact that such countries are ill-equipped to deal with any sort of natural disaster, possess economies comprised primarily of subsistence farming on progressively poorer-quality land, and have grossly underfunded public health services, education services and physical infrastructure, constitute only some of the factors considered in compiling quality-of-life rankings. The United Nation's measures of Human Development (United Nations Development Programme 2018) also considers the "fairness" of income/wealth distribution, for which factor Africa's countries are also low ranking (CIA World Factbook 2018). Cambridge's Sir Partha Dasgupta, recipient of almost every award that economists can bestow, has pointed out that most of the recent gross national product (GNP) increases of second/third world countries have come at the expense of their average citizens' personal assets (Dasgupta 2003).

Africa's burgeoning population exacerbates all of its problems. As of 2015, the UN's mid-range population growth projections predicts that it will have ~4.5 billion inhabitants by 2100 AD – about three times that anticipated for the world's currently most populous nation, China. The populations of 28 African countries are supposed to more than double between 2015 and 2050 and, by 2100, those of Angola, Burundi, Democratic Republic of Congo, Malawi, Mali, Niger, Somalia, Uganda, United Republic of Tanzania and Zambia are to increase at least fivefold.

Frankly, I consider these projections unrealistic. Both the Western world's current political trends and most of human history suggest that another "world war" is apt to kill far more people than did the 20th century's and thereby curb population growth, especially within poorer counties. US Pentagon studies (Gunn 2017) concluded that the majority of such deaths will be due to disease and starvation caused by the disintegration of technology-dependent societies contending for increasingly limited and degraded resources (land, food, fuel, high-grade ores, etc.). In his seminal book, *Small is Beautiful: A Study of Economics as if People Mattered*, Ernst Schumacher (1973) observed that today's technological economy is unsustainable because the finite resources driving it are treated as inventory (income) rather than capital. The "sustainability" of today's economic systems therefore requires continued growth of both population and total wealth (gross domestic product, GDP), which is impossible in a finite world. This volume's ultimate goal is to devise a genuinely sustainable economy in which every individual person regardless of where they live does indeed matter. Until we acknowledge those facts, accept that goal, and begin to act accordingly, we are just spinning our wheels.

Since food represents humanity's most basic need and its source is farmland, I'll begin by describing what's been happening along those lines in Africa. A recent Brookings Institute report (McArthur 2013) points out that "no matter how effectively other conditions are remedied, per capita food production in Africa will continue to decrease unless soil fertility depletion is effectively addressed." It goes on to say that a second major problem with the oft-assumed African "land abundance" hypothesis is its inconsistency with evidence that its soils are being simultaneously depleted and eroded by current agricultural practices, including a decline in fallowing. While some African politicians, along with the management of "land grabbing" (?) international agribusiness concerns, seem to feel that Africa still has plenty of as yet undeveloped arable land, many of Africa's poorest people (mostly subsistence farmers) can't afford to let any of theirs lie fallow: some families live on 0.9 US acre (not ha) farms, yielding under 1 t of grain/ha, while the first-world's farmers routinely produce 3 to12 t/ha of whatever "cash crop" they choose to plant on several orders of magnitude larger farms.

The key differences between the agricultural practices of developed nations and most of Africa's include:

1. Developed nations heavily fertilize their croplands – most of Africa's farmers can't afford artificial fertilizers and often have to burn any manures or crop residues they can gather to cook their food.
2. Developed nations' farmers can afford to irrigate their croplands – most of Africa's farmers can't, an issue compounded by the fact that much of Africa's nominally arable land doesn't get enough rain to reliably support anything other than skeletal cows or goats.
3. Most developed-nation farms are large enough and productive enough to enable their owners to buy/utilize specialized machinery, which renders their labor both far less exhausting and much more rewarding. Africa's poorest farmers work themselves to death with primitive tools.

14.3   WHY EVERYTHING BOILS DOWN TO ENERGY INPUTS

Of course, those differences simply reflect the relative amounts of raw/primary energy supporting the lifestyles of rich vs. poor people, which measure in today's world boils down to their relative per capita fossil fuel consumptions. On-farm agricultural energy consumption in rich countries entails the burning of diesel oil, gasoline, and/or LP gas in engines plus the use of electricity, generally produced by burning more fossil fuel, usually coal. It is considerably higher in high-GDP countries (around 20.4 GJ/ha) than it is in low-GDP countries (around 11.1 GJ/ha) and far greater than on Africa's poorer subsistence farms (Giampietro 2002). For example, most of the energy input to a subsistence farm consists of human labor, which, throughout an 8-hour work day, amounts to about 75 watts per person (Wikipedia 2018e). If 100% of such useful (mechanical) energy ($75 \times 8 \times 3,600 = 2.16E+6$ J/day) is devoted to cultivating a 0.9 acre (0.36 ha) plot throughout a 6-month growing season, the area-normalized "energy services" devoted to that effort is 1.08 GJ/ha/a ($2.16E+6$ J/day $\times 365$ days $\times 6/12/2.47$ acre/ha/$1E+9$), which is about 10% of the raw/primary food energy required to keep each person so occupied alive throughout the year (2,500 kcal/day $\times$ 365 days/a). Energy-wise that's not very efficient – today's tractors generate about one-third of a joule's worth of useful mechanical energy from a joule's worth of fuel heat energy and don't consume any energy when not running.

According to a Food and Agriculture Organization of the United Nations (FAO) report (Sims 2011), the raw/primary energy consumed by the world's "food sector" amounts to ~95 exa (10^{18}) Joules – approximately 20% of current total global raw/primary energy consumption (~570 EJ exa (10^{18}) Joules) – and generates over 20% of anthropogenic greenhouse gas emissions. Land-use changes, particularly those linked to deforestation brought about by the expansion of agricultural lands to raise food crops and biofuels constitutes another ~15% of anthropogenic greenhouse gas (GHG) emissions. Only about 5% of that energy, ~6 EJ, directly supports on-farm activities such as cultivating and harvesting crops, pumping water, housing livestock, heating protected crops, drying, and short-term storage. The rest/majority of the agricultural sector's energy demand is devoted to transport, fertilizer, and pesticide production, food processing, packaging, storage, and distribution.

All of the world's developed countries have adopted Dr. Borlaug's fossil-fueled "Green Revolution," which enabled ~90% of the world's ~7.5 billion people to consume as much food – both basic necessities along with some luxury items – as they want. Approximately one-half of the world's current population would quickly starve if that hadn't happened. However, the fossil fuels enabling both it and most of the 20th century's other technological revolutions generated environmental impacts, including the greenhouse gas emissions responsible for global warming/climate change. From 1870 to the present, fossil fuel burning dumped about 580 Gt (580 Pg C) into the atmosphere in the form of ~2,100 Gt of CO_2. That gas – partitioned between the atmosphere, oceans, and land, warming all three – is causing increasingly severe and frequent weather events, including "Super El Niños" (Hong 2016), ocean acidification, drought, biofuel production-driven food cost escalation, air pollution, deforestation, potable/irrigation water shortages, sea-shoreline erosion/flooding, and relentless increases in the cost of living in the world's poorer regions. Those effects constitute threat multipliers that have aggravated stressors – poverty, environmental degradation, hunger, political instability, and social tensions – and have engendered mass migrations, as well as terrorist activity and other forms of violence. During that same period, two world wars and numerous smaller ones were fought over natural resources – primarily "lebensraum" (land) and fossil energy, usually petroleum – some of which resulted in the Christian country "winners" creating new countries in the oil rich Islamic Persian Gulf, which, of course, eventually engendered more conflict.

Securing those resources has proven to be expensive for those winners. A Princeton University report concluded that simply keeping the US Navy's fifth fleet within the Persian Gulf from 1976 to 2006 had cost its taxpayers ~$6.8 trillion (2008 dollars) and would probably cost them another $0.5 trillion during 2007 (Stern 2010), which figures didn't include the costs of actual conflicts. Since that fleet remains on station, the total cost of "maintaining presence" therein has now probably reached about $12 trillion. A 2013 Kennedy School of Government report (Maass 2010) concluded that the total cost of the USA's recent wars in the Middle East and Northern Africa would probably be $4 trillion–$6 trillion and had accounted for roughly 20% of its increase in national debt between 2001 and 2012 (modern wars are fought with borrowed money).

Concerted international effort to address fossil fuel's environmental impacts began with the UN's 1997 Kyoto Protocol to which many, mostly small and not particularly impactful, countries signed up. While the billions of dollars spent on climate science research since then has generated thousands of papers/reports and paid for hundreds of other conferences both large and small, neither that science/knowledge nor the policy changes implemented by many countries that favor/subsidize politically correct renewable energy sources have had much effect upon mankind's GHG emissions. As this is being written (December 2018) representatives from 195 countries have again gathered (in Katowice, Poland) for this year's United Nations Climate Change Conference, COP 24 (COP = Conference of the Parties). This meeting is focused upon producing rules to flesh out the "details" of the 2015 Paris climate accord (COP 21), the landmark agreement signed by all of its attendees except Nicaragua and Syria, to battle climate change and hopefully limit global warming to <1.5°C, one-half a degree under the 2°C limit set earlier at COP 15 (Copenhagen conference). Since 2015, the International Panel on Climate Change's (IPCC's) leadership has been trying to breathe new life into that accord amid backsliding from several key nations, most notably the United States, over commitments made when they signed it. To date, the IPCC's efforts have not really accomplished much because key "parties" refuse to agree upon a mechanism ensuring that they honor their commitments with respect to either GHG emissions or contributions to a $100 billion/a climate mitigation fund.

The latest version of the British Petroleum company's annual *Statistical Review of World Energy* (Anonymous 2018a) contains not only information from the preceding year, but also historic data on both consumption and production of all forms of energy during the last several decades. Its principal conclusion is that humanity is not fulfilling the goals set by the Paris Agreement regarding the threats posed by climate change. In 2017, mankind took a step backward with respect to the timid advances made during the preceding two years: the use of fossil fuels had grown, increasing CO_2 emissions by ~1.6%. Apparently that trend continues – anthropogenic CO_2 emissions rose another 2.7% in 2018 (Jackson et al. 2018). Worse, most climate models indicate that by 2100 AD, even if the "commitments" made by COP 21–24's attendees were to be honored, in toto they would likely cause global warming of between 2.7°C and 3.2°C, well above the 1.5–2°C threshold that many climate modeling experts consider a tipping point beyond which Nature's positive feedback mechanisms will render catastrophic impact inevitable (Hansen 2008).

The most important thing mankind must do is to replace today's fossil fuel-dominated energy system with something that is both "clean" (no GHG emissions) and sufficiently reliable (not "intermittent') to power an even bigger, more interconnected, more prosperous, cleaner and fairer future world economy.

14.3.1 How Much Such Energy Would be Needed?

The exceptionally "rich" lifestyle of the USA's ~320 million people is supported by about 99.5 EJ (98 quads) of raw/primary energy per annum, ~80% of which is provided by fossil fuels (LLNL 2015). That translates to a mean per capita raw/primary energy consumption rate (power) of 9,860 watts (99.5E+18J/3,600/24/365/320E+6) or about eight (99.5/3.2e + 8/570/7.5e + 9) times that of the world's average person today. Since 1 joule's worth of raw/primary (heat) energy provides about 0.4 joules' worth of useful "energy services" (the efficiency of most of fossil fuel's applications is Carnot-limited) and Europeans apparently can live almost as well consuming one-half that much raw/primary energy per capita, let's assume that supporting the lifestyles of each of the future's equally rich people would require ~2(9,860 × 0.5 × 0.4 = 1,972 ≈ 2,000) kW's worth of energy services (electricity). Consequently, an egalitarian world with 11.2 billion inhabitants must possess power plants able to supply an average power of about 22(11.2E+9 × 2,000 × 3,600 × 24 × 365/1E+12/3,600/24/365 = 22.4) TW$_e$ (terawatt electrical). Finally, assuming that each individual region's peak power demand is ~40% higher than its average and that no worldwide, zero-loss, "super grid" exists, our descendants would need ~30,000 (22.4 × 1.4 × 10^{12}/10^9) one-GW$_e$ power plants to live that well.

That power could not be generated with fossil fuel, because, even if there were enough of it (there isn't), burning it would have catastrophic consequences. For example, the raw/primary (heat) energy represented by the world's remaining 1,139 billion tons of coal, 187 trillion m^3 of natural gas, and 1.707 trillion barrels of petroleum (Anonymous 2018b) is about 5.0E+22 J's, which, if consumed by 40% Carnot efficient power plants, could generate 22 TW$_e$ for just 29 years – about one-third of a typical first-world human life span. Additionally, those reserves collectively contain about 1,200 Gt of carbon, which, if converted to CO_2 and dumped into the atmosphere, would push global warming well past any of the tipping points suggested by the ICPP's climate modeling experts.

Because the "half-life" of CO_2 already in the atmosphere is about a half century (Moore and Braswell 1994), achieving the goals of the Paris climate accord (limiting maximum temperature rise to 1.5°C) at this late point in time would require an almost immediate switch to clean (no GHG emissions) energy sources *and* the removal of enough carbon dioxide already in the atmosphere to reduce its concentration to ~350 ppm (Hansen 2008). Consequently, some of the ICPP's more optimistic post-COP 21 scenarios/reports assume that "bio-energy with carbon capture and storage" (BECCS) represents a magic bullet that could address both global warming and the future's energy supply conundrum in a politically correct (no nuclear) fashion (Martin 2015). All such scenarios are unrealistic because raising sufficient switch grass, palm oil, wood, etc. to do so would require vast amounts of land, water, and fertilizer that most people (especially the hungrier ones) would consider better utilized if applied to producing something other than fuel. It is also unrealistic because carbon capture and sequestration (CCS) is both difficult and expensive, which is why, after several decades and many billions of dollars' worth of "study" and "demonstrations," only about 0.08% of anthropogenic CO_2 is currently so managed (Wikipedia 2018a). BECCS-based "save the world" scenarios are also impossible because they don't scale. For example, burning 100% of the world's current annual grain (about 2.5 Gt, see Statista 2017) plus "bone dry wood" (about 1.9 Gt, see Wikipedia 2018k) harvests in "clean" (CCS equipped) 40%-efficient (optimistic) heat-to-electricity power plants would generate useful energy services (electricity) equivalent to the output of ~935 one-GW$_e$ ("full sized") nuclear reactors – ~3% of the number required to render 11.2 billion people one-half as energy rich as the USA's people are now. Any backup system for low-capacity factor energy sources (solar and wind) must be able to satisfy most, not just 3%, of total demand. Furthermore, the carbon (about 1.8 Gt) in that much biofuel (carbohydrate) represents only about 0.3% of mankind's total anthropogenic carbon emissions to date, which means that even if 100% of the CO_2 that it represents were to be captured and sequestered, it wouldn't actually accomplish much. Consequently, because it could not achieve either of the IPCC's goals, and would surely compete with food, fiber, and construction-type wood production, all rosy BECSS-based scenarios are hopelessly unrealistic and therefore do not "deserve further study" – a conclusion common to the majority of reports in any scientific field. Additionally, because growing biofuel removes inorganic nutrients and soil organic carbon, it represents another extractive technology that would further degrade the environment and compromise human food production (Lal 2008).

The most alarming thing about how things have been going recently (see the Fourth National Climate Assessment – US Global Change Research Program 2018) is that civilization remains absolutely dependent upon a resource that will inevitably become prohibitively expensive when most of the cheaper/easier-to-access coal, oil, and natural gas has been consumed, which is likely to occur well before 2100 AD. Unless the world's decision makers have already developed/implemented a simultaneously "clean," reliable (not intermittent), and affordable alternative by then, civilization is likely to collapse heralding the onset of a dark ages akin to that depicted in Mad Max movies.

14.4   THIS "TECHNOLOGICAL FIX'S" SPECIFICS

Let's do some ball-park calculations that demonstrate how Weinberg and Goeller's vision could address Africa's issues.

14.5   ASSUMPTIONS

First, I'm going to assume that the UN's population projection for Africa circa 2100 (about 4.5 billion) turns out to be right.

Next, since I'm primarily concerned with showing what a nuclear renaissance should be able to accomplish with respect to assuring that continent's (and the world's) food security, I'm going to assume that part of the useful energy (electricity) it provides is devoted to doing so – in other words, nuclear reactors would provide the necessary water, fertilizer, and soil-building minerals plus the power required to properly utilize them.

Finally, I'm going to assume that Africa's decision makers, along with who/whatever else chooses to help/enable them, collectively decide that its citizens should be able to enjoy the same standard of living enjoyed by the EU's today.

14.6   WHY POLITICALLY CORRECT RENEWABLES WON'T WORK

Assuming that "energy services" means electricity (reasonable because most of its applications are nearly 100% efficient) and that no worldwide power grid exists circa 2100, generating Africa's share of the world's total energy supply would require about 12,000 ($30 \times 4.5/11.2 \times 10^{12}/10^9$) full-sized (~one-GW$_e$) power plants generating an average of 9 TW$_e$ ($4.5E+9 \times 2$ kW).

The following shows why today's most popular alternative/renewable energy sources could not meet that demand.

First, let's see how many of today's nominally "cheap," rooftop-type, solar panels would be required to produce 9 TW$_e$'s worth of useful energy. At Home Depot today one can purchase four, *real* state-of-the-art (19% efficient), 265 watt-rated, 1.61 m^3 (39″ by 65″) solar panels for $1,412. If they were to be employed in Nigeria, which purportedly exhibits an average solar irradiance of 5.5 kWh/m^2/day (Ojuso 1990), each of those panels would generate a time-averaged power of 70.1 watts ($1.61 \times 0.19 \times 5,500 \times 3,600/(24 \times 3,600)$), which means that their "capacity factor" (average energy/power generated/nameplate rating) when so situated would be 26.4% (70.1/265) – about two-and-a-half times what it would be if installed in northern Europe instead. This suggests that powering Africa's 4.5 billion future citizens with them would require 128 billion (900E+9/70.1), such panels costing about $45 trillion. Since such power is intermittent, they would also have to buy with enough batteries to keep everything running during the roughly 73.6% (100 – 26.4) of the time that those panels would not produce much. How many batteries would that be? Assuming that Africa's inhabitants could get by with just one day's worth of energy storage (unconservative – widespread cloudy and windless periods often last longer than one day) they would have to build/buy about 200 billion kWh's worth of storage capacity (2,000 J/s × 4.5E+9 × 3,600 s/hr × 24 hr/day/3.6 E6 J/kWh = 2.16E+11 ≈ 200 billion kWh), which, if implemented with Tesla's equally real and state-of-the-art 13.5 kWh, ~$7,000, lithium ion battery "Power Walls," would cost the hopefully more prosperous descendants of today's subsistence farmers another $100 trillion ($7,000/13 kWh × 200E+9 kWh/3.6E+6 J/kWh) of today's dollars to purchase the *first* time (batteries only last for a few years). Real world windmills exhibit similar capacity factors, meaning that if they were to be used instead, a similar amount of energy storage capacity or some other sort of clean backup power would be required. Natural gas – today's backup for (and thereby justification for financial investment in (Pfotenhauser 2014)) wind and solar power plants – will probably be prohibitively expensive by then because all the "cheap" natural gas will have already been fracked/consumed. Leaving it in the ground along with most of the world's remaining coal would be a good idea because burning them would otherwise add to the already excessive amounts of anthropogenic GHGs responsible for climate change – coal as CO_2 and methane both as is and after it has been oxidized to CO_2 – methane is a far "stronger" GHG than is CO_2 and several percent of that currently fracked leaks directly into the atmosphere (Alvarez et al. 2012).

14.7   WHICH CROPS AND HOW MUCH LAND?

Next, we must decide what crops these people should grow and how much land would be required to do it. For simplicity's sake, let's assume that the everyone will be consuming 2,500 kcal/day (1.05E+7 J) of food, most of which would be provided by two especially productive crops raised upon the minimum amount of soil capable of providing yields currently achieved in the United States.

Table 14.1 lists the amount of land required on a modern farm to produce one gram of protein with various crops. Picking the outstanding crop candidates in this list is not too difficult – primarily maize (corn) because it's exceptionally productive, nutritious, and already widely produced/consumed/accepted in Africa, plus some sort of pulse (legume) to complement its unbalanced mix of amino acids (not enough lysine) (Lal 2017). Of the likely candidates, peanuts seem to make the most sense because they are a "hot weather crop," taste considerably better, contain more fat/oil, also already widely produced/consumed/accepted in Africa, and would not extract as much phosphorus (P) and potassium (K) (key macronutrients) from its soil per food-calorie as would the next runner-up, soybeans (see Table 14.3).

Assuming zero waste, providing 2,500 kcal/day of food for 4.5 billion people translates to 4.1E+15 kcal's (1.72E+19 J) worth of those foodstuffs per year. If we also assume that 75% of their food-type calories are to be provided by maize, the total amount of land required to feed every African circa 2100 AD adds up to 9.71E+7 ha, of which 7.28E+7 ha would be devoted to maize and 2.43E+7 ha to peanuts.

That combination would provide everyone with 59 grams of "complete" protein per day along with virtually everything else that humans need to first grow and then remain healthy. Africa's folks would probably also want to (and should) devote perhaps an additional 5–10% of similarly productive/managed land to raising the lower calorie/protein but tastier fruits, vegetables, and spices that render vegetarian diets far more palatable than most people realize. Additionally, if Africa's much more prosperous (on the average) future inhabitants were to decide that chicken should provide 20% of their food calories (500 kcal/day/person – about fourteen times more than they currently consume), similar calculations suggest that roughly 10% additional land would be required to raise the additional peanuts and maize needed to feed those birds. The substitution of cricket "meat" (Van Huis 2012) for that much chicken would require only about 5% more peanuts/corn/land than would a strictly vegetarian dietary.

9.71E+7 ha is about the same amount of land that the USA currently devotes to producing all of the foodstuffs listed in Table 14.1 to support its ~320 million people (7.2% of the number assumed

TABLE 14.1
What Should They Be Growing and Raising?

Food	Land use (m²/g protein)	Food	Land use (m²/g protein)
Beef/mutton	1.0243	Dairy	0.0440
Pork	0.1299	Wheat	0.0354
Fresh produce	0.0982	Rice	0.0229
Poultry	0.0751	Maize	0.0144
Eggs	0.0514	Pulses	0.0104

(Clark and Tilman 2017)

herein for Africa). Since Africa has a total area of 29.6E+8 ha, 7.4% of which is purportedly arable (Worldstat 2018), 9.71E+7 ha represents only 44% of its arable land, which, hopefully, means that Africa's future citizens would continue to share some of its still useful land with its iconic suite of wild animals.

14.8   THE WHYS, HOWS, AND COSTS OF DESALINATION

Africa's 4.5 billion future inhabitants wouldn't be able to feed themselves with ~50% of its arable land unless they become able to irrigate it. Because irrigated land almost always produces higher yields than do rainfed farms and permits double and sometimes even triple cropping in warmer regions, such lands currently provide around 40% of global cereal supply (Simms 2011). Only ~4% of Africa's cropland is currently irrigated, due primarily to prohibitive costs, insufficient water, and a general lack of commitment to infrastructure-related investment in things like power plants and the fuel needed to operate them. Consequently, it's unlikely that this chapter's cornucopian scenario would be implemented by either Africans themselves or the institutions/people that have provided most of their help to date.

Pumping water onto approximately 10% of the world's total arable land (around 300 Mha) currently consumes around 0.225 EJ/yr. Another 0.05 EJ/yr of indirect energy is devoted to the manufacture and delivery of irrigation equipment (Smil 2008). Around two-thirds of the water currently used for irrigation is drawn from underground aquifers. Energy-intensive electricity-powered pumping from deep wells accounts for about two-thirds of that and projections suggest that that fraction will become ~90% by 2050, when shallow reserves are almost totally depleted. Current aquifer water extraction rates exceed recharge rates – grossly so in many places. Additionally, global warming is rapidly melting the glaciers that feed the riverine systems providing much of the world's cheap-to-deliver irrigation water. Global warming has caused Mount Kilimanjaro's "snows" to disappear along with most of those within the USA's Glacier National Park. Building more dams won't solve the problem because they don't create water. Additionally, a comprehensive review of Nigeria's outside-funded dam projects (Tomlinson 2018) concluded that while they do make money for local promoters and the outsiders that fund/support them, they decrease net agricultural productivity by turning fertile downstream floodplains into deserts. In addition to killing wetland-dependent wildlife, those dams have served to lower, not raise, the incomes of far more people than have benefited. Most such dams don't generate nearly as much electrical power as "promised," due to inadequate maintenance and low (water-limited) capacity factors. Finally, at best, dams represent a temporary fix for the problems that they are supposed to address because their reservoirs will eventually fill with mud.

Water shortages plus the high cost of desalination – primarily due to high electricity costs – has led some countries rich enough to do so (e.g., China) to reduce their own crop production and rely more heavily upon imported grains.

This situation is unsustainable, which means that we'll next assume that most of the water irrigating Africa's future farmlands would be generated by desalinating seawater – the world's only truly inexhaustible/sustainable water source. Wikipedia's description of Israel's approach to solving its water issues (Wikipedia 2018j), demonstrates how a properly managed future technological society could address the world's water woes. Israel's ~8.5 million people are fed by ~1.045E+9 m^3 of fresh water applied to its mostly irrigated farmland (American-Israeli Cooperative Enterprise 2012). This suggests that the rest of the Middle East's even more water-stressed, mostly poor ~101 million people, including the majority of those living in Palestine, Iraq, Jordan, Lebanon, Oman, Syria, and Yemen (Wikipedia 2018b), could be equally well supported by irrigating their potentially arable land (assuming all of it) with 1.24E+10 m^3 of desalinated seawater. Assuming the ~3 kWh/m^3 energy requirement (Wikipedia 2018i) of today's most popular approach to desalination, reverse osmosis (RO), doing so would require an energy input of 1.34E+17 joules/a, which corresponds to the full-time output of ~4.2 one-GW_e nuclear reactors. The volume of water corresponding to adding 0.51 m (20″) of it over 9.71E+7 ha of African farmland is 4.93E+11 m^3, which, if generated via RO, would require the full-time output of ~169 full-sized nuclear reactors (5.33E+18 J_e/a). In principle at least, Siemen's electrodialysis-based desalination technology would require only about one-half that much energy/reactors and is also less apt to become fouled by seawater's other-than-salt impurities (Hussain and Abolaban 2014).

To continue, Africa's average elevation is about 600 m (2,000 ft) above sea level, roughly the same as that of both North and South America. Assuming that all of Africa's desalinated irrigation water would have to be pumped uphill that far, the energy needed to do so would be 2.90E+18 joules (4.97E+11m^3 × 1,000 kg/m^3 × 600 m × 9.8 m/s) requiring another 92 full-sized power plants.

How much would Africa's desalination equipment cost? The contractual cost of the world's (Saudi Arabia's) biggest (~1 million m^3/day), RO-based desalination plant is $1.89 billion (Wikipedia 2018c). Collectively, these numbers suggest that building enough RO plants to irrigate Africa's future farmlands would require a one-time capital expenditure of $2.55 trillion (4.93E+11 × 0.00189/(365 × 1E+6) – about 11% of the USA's current national debt. (Similarly addressing California's Central Valley's chronic irrigation water problems *should* cost only about $40 billion).

As mentioned earlier, the Western world's recent Middle East military incursions will probably end up costing its people ~thirty times more ($4 trillion to $6 trillion) than it would to build enough nuclear-powered desalination plants to provide sufficient fresh water for everyone living there, and thereby address a root cause of that region's turmoil. Another plus for desalination is that its product does not add additional salts to soil and is also better at remediating already over-salinized soils than is groundwater. A final plus is that because it doesn't already contain near-equilibrium levels of calcium, magnesium, carbonate/bicarbonate, silica, etc., it is a better rock solvent (more corrosive) than is groundwater. The next section will reveal why this is important.

14.9   FERTILIZERS

One of the reasons why the productivity of Africa's farmlands is lower than that of more developed regions is that relatively little fertilizer is used. The three most important components of fertilizers (macronutrients) include nitrogen in either its negative three (ammonia-type) or positive five (nitrate-type) oxidation states, potassium (invariably in its plus one oxidation state), and phosphorus (invariably in its plus five oxidation state). Nitrogen fertilizer production currently accounts for about one-half of the fossil fuel (mostly natural gas) used in primary food production.

14.9.1   NITROGEN AND THE COST OF FIXING IT

We'll start with nitrogen (N). US farmers hoping to produce 13.1 Mg/ha (200 bu/acre) of maize (grain) are advised to add ~258 kg of N/ha (Beegle 2005). Since peanut (legume), can recover/fix its own nitrogen from air, much less nitrogenous fertilizer would be needed for land devoted to its cultivation – let's say 50 kg N/ha. Assuming those application rates, fertilizing Africa's future crop land would require ~2.0E+7 tons of ammonia each year (1 kg N ≈ 1.21 kg of ammonia) A recent review of synthesis technologies (Frattini et al. 2016) concludes that each ton of ammonia made with electrochemically generated hydrogen, pressure swing-generated atmospheric nitrogen, and conventional Haber–Bosch processing equipment would require about 14.2 MWh's (14.2 × 3.6E+9 J) worth of electricity. That, in turn, suggests that satisfying this chapter's scenario's nitrogenous fertilizer requirement would require the full-time output of ~32 one-GW_e power plants.

14.9.2 WHY BASALT SHOULD SUPPLY P&K

Since

- Much of Africa's (and the world's) farmland has already lost a good deal of its topsoil via erosion.
- Much of its remaining topsoil is mineral-depleted.
- Basic (mafic) rock-weathering is how Mother Nature limits the Earth's atmospheric CO_2 concentration (Hartmann et al. 2013).
- Basaltic rocks are both intrinsically basic (contain a good deal of magnesium and calcium) and rapidly weathered by the natural phenomena extant in cultivated soils (Moulton et al. 2000).
- Most of the Earth's crust consists of basaltic rock, a good deal of which is either on or close to the surface of its continents.
- Soils comprised primarily of weathered volcanic ash, most of which originally consisted of molten basalt, are exceptionally productive (Beerling et al. 2018).
- Basaltic rocks contain relatively high concentrations of potassium and phosphorus along with all the other biologically important elements, which is why …

… we will assume that the phosphorus and potassium required to produce Africa's food crops circa 2100 AD will be provided by amending its farmlands with powdered basalt. In order to be effective, any such amendment must weather rapidly enough to release sufficient potassium and phosphorus to support high-yield agriculture, which, in turn, means that the raw rock surfaces must be "fresh" (not already equilibrated with the atmosphere and thereby covered with secondary phases), ground to a considerably smaller particle size than is the quarry waste-type soil amendment rock powder currently being marketed to hobby farmers, and mixed with root-zone topsoil, not just dumped upon the surface of the ground (Campbell 2009; Priyono and Gilkes 2004). Based upon the rather limited amount of scientifically planned/supervised experimentation described in the open-access (not paywalled) technical literature, I'm going to assume that this would require grinding it so that the particles comprising >80% of its mass possess diameters <10 microns. Since rock grinding is highly energy intensive – much more so than simply recovering it from a quarry's rock outcrop or waste dump –the cost estimate for this part of my scenario will be based upon that step's energy demand plus the powder's transport and distribution costs.

First, how much of it must be made? The foodstuff P and K concentrations, land areas, and crop yield figures in Tables 14.1 through 14.3, suggest that the food consumed each year by Africa's 4.5 billion future inhabitants would contain 2.64E+9 kg of potassium and 2.05E+9 kg of phosphorus. Assuming (wrongly I hope, but consistent with the way that things are usually done) that neither nutrient is subsequently recycled back to the soil, both must be replaced each year via basalt weathering. The compositions of flood basalts vary considerably but since all originate from the Earth's fairly well-mixed underlying magma, for the following estimates I'm going to assume a composition with which I'm familiar (Leeman 1982 and Siemer 2019) – that of the basalt comprising Idaho's "Craters of the Moon" National Monument and also covering much of the rest of Idaho's Snake River plain. It contains an average of 0.61 wt% K_2O and 0.55 wt% P_2O_5, which translates to

TABLE 14.2

Candidate Crop Yields USA 2017

Crop	bu/ac*	lbs/bu	lbs/ac	Mg/ha
Barley	72.6	48	3485	4.08
Wheat	46.3	69	3195	3.74
Rye	23.9	56	1338	1.57
Flax	14.1	60	846	0.99
Sorghum	72.1	56	4038	4.73
Rice			7507*	8.79
Oats	61.7	32	1974	2.31
Corn	176.6	56	9890	11.58
Soy	49.1	60	2946	3.45
Peanuts			4074*	4.77

* http://usda.mannlib.cornell.edu/usda/current/Acre/
Acre-06-29-2018.pdf.

TABLE 14.3

Key Characteristics of Candidate Food Crops (data from Wikipedia entries and Their references)

Crop	Mg/ha	MJ/kg	Protein (g/kg)	Fat %	Water %	K %	P %	kcal/ha	kg protein/ha
Sorghum	4.73	13.8	103	3.6	10	0.36	0.29	1.56E+07	487
Maize	11.58	15.3	94	4.7	10	0.28	0.21	4.23E+07	1,043
Peanut*	3.31	23.8	250	48	4.3	0.33	0.33	1.88E+07	828
Soybeans	3.45	18.7	360	20	14	1.8	0.7	1.54E+07	1,242

* My Mg/ha figure for peanuts assumes that the "nut" represents 2/2.88 of the as-harvested crop

requiring 5.21E+8 Mg (tons) of it per year to provide this scenario's potassium and 8.69E+8 Mg to supply its phosphorus. Since phosphorus happens to be the limiting nutrient in this case, at steady-state, we'd be adding 8.95 Mg (8.69E+8/9.71E+7) of powdered basalt/ha/a.

A recent review of rock grinding technologies (Jankovic 2003) suggests that producing 1 Mg of <10 micron basalt powder would require about 100 kWh's worth of electricity. If so, making 8.69E+8 Mg of it would require 3.13E+17 J, which, if done throughout 1 year, would require the full-time output of 9.9 one-GW_e nuclear reactors.

If this powder were to be transported, an average of 1,930 km (1,200 miles) from mine to farm via an electrified rail system as energy-efficient as that currently used to move US coal (185 km/L diesel fuel/short ton), its energy cost would be about 6.80+16 J (assumes 1.1 Mg/short ton, 33% heat-to-mechanical engine efficiency, and 44.5 MJ/kg diesel fuel with a SpG of 0.85). Trebling that figure to account for fuel consumed by trucks and tractors at the railheads, brings the total to 2.04E+17 joules/a, which corresponds to an annual transportation/distribution energy demand requiring another 6.5 one-GW_e nuclear reactors to satisfy.

An application rate of 8.95 Mg/ha/a is not really very large because it represents only about 0.5% of the mineral matter already in a 6-inch deep (root zone) layer of normal density/composition soil and is also considerably less than what current farming practices lose via wind/water erosion (typically about 30 t/ha/a – Pimentel et al. 2009). Consequently, since this scenario's rock grinding/distribution costs are much lower than its irrigation water and nitrogenous fertilizer costs, it would be a good idea to at least start out with a considerably larger application rate, perhaps 40–50 Mg/ha.

Doing so would also reduce the chance of crops being "starved" due to slower-than-I've-assumed weathering rates.

The 309 (169+92+32+9.9+6.5) GW_e's worth of "clean" power plants required to implement the agricultural aspects of my scenario's clean/green future seems to be a rather imposing figure (about the same amount of power currently generated by all of the world's nuclear reactors and about four hundred times more than Africa currently produces/consumes (about 650 TWh, see Wikipedia 2018d), but represents only ~3% of the total energy services consumed by 4.5 billion Africans as "rich" as are today's EU inhabitants.

All of the necessarily huge machinery and manufacturing facilities required to implement this or any other technological fix capable of "saving the world" would be much cheaper to build and would operate much more efficiently, with reliable power, than with that provided by intermittent sources. While it would indeed be "possible" to run desalination/ammonia plants, rock crushers, tractors, locomotives, etc., with windmills and/or solar panels, doing so would be expensive, dangerous, and frustratingly unproductive to such machinery's owner-operators and their customers. Intermittent supplies are suitable for some niche applications, not for powering technological civilizations (Brook 2018).

Again, some of my numbers are rough approximations because the rate and degree to which powdered basalt would release its constituents (weather) under field conditions is affected by a host of factors/variables. A nutrient-specific discussion of some of them may be found in an FAO report describing the use of raw phosphate rock as fertilizer (Zapata and Roy 2004). Thankfully, this subject is beginning to receive a good deal of attention (Taylor et al. 2017) and some hopefully realistic studies have begun (Beerling 2018).

14.10   ADDITIONAL "KILLER APPS"

When completely weathered by the mechanisms collectively responsible for doing so in biologically active soils, each gram of this scenario's basic basalt (10.06 wt% CaO and 7.65 wt% MgO) would release 7.35 milliequivalents' (0.1006 × 2/(40+16)+0.0765 × 2/(24.32+16)) worth of base. If we assume that it would convert soil-gas CO_2 that would otherwise transpire (into the atmosphere) to bicarbonate ion (Hartmann 2013), the application/weathering of 8.95 t/ha of it over 9.71E+7 ha of African farmland would remove/sequester 0.076 Pg (76 million tons) of carbon. That sounds like a lot of "sequestration" but represents only ~0.009% of that currently in the atmosphere (about 3,300 Gt CO_2).

If carbon sequestration is to be a primary goal, another way to go about doing it would be convert corn stover (there's about as much of it produced as grain and approximately 80% of it can be readily collected) and peanut hulls/stems (their leaves would probably end up on the ground) to "biochar." Assuming this scenario's grain crops, that translates to converting about ~11.6 tons of biomass to ~3.1 tons of biochar and 5 tons of bio-oil per ha (Lang 2002 and Garcia-Perez et al. 2010). Because biochar is ~70% elemental carbon, burying it would simultaneously increase Africa's soil's organic matter (SOM) and sequester atmospheric carbon at the rate of ~0.25 Pg (250 million tons) per year. If everyone – not just Africans – were to char their stover and fertilize with Snake River plain basalt, they would collectively sequester the equivalent of about 3 Gt CO_2 per year. However, since the atmosphere already contains about 500 Gt of excess CO_2 ((412ppm–350ppm)/412ppm × 3,300 Gt = 496) and will surely get further out of balance before we kick our addiction to fossil fuels, it would probably take over two centuries for the future's farmers to reduce it to a nominally "safe" (350 ppm) level.

These numbers suggest that the primary rationale for implementing this "enhanced weathering" scheme would be to stop polluting, make everyone "rich," and render agriculture sustainable, not collect/sequester already-dumped atmospheric carbon.

Another plus for "biocharring" is that it should simultaneously produce more than enough "oil" to fuel the machinery required to run the farms, meaning that it could become a profitable sideline for their owners. Figures in a recent report about Nebraska's farm fuel costs suggest that high-input corn farming currently requires about 70 US gallons of diesel fuel/ha/a (Wilson 2015). Five tons of bio-oil purportedly has the energy content of ~37% of that of No. 2 diesel oil (i.e., 2,176 liters, 575 US gallons, or 82 GJ's worth of such fuel) and it should be possible to convert it to a good diesel-type engine fuel (Cataluna 2013).

Another possibility raised by a suitably implemented nuclear renaissance is that other sorts of fuels could be synthesized because electrolytic hydrogen would become much cheaper than it is now. For instance, hydrogenation of the ~11.6 t/ha of stover mentioned previously would produce about three times as much synthetic fuel oil (~4.9E+8 Mg/a) as would making it from biocharring's bio-oil (Agrawal et al. 2007). Similarly, if Africa's 4.5 billion future citizens were to produce/consume as much Portland cement per capita as we do now, Fischer-Tropsch hydrogenation of the CO_2 so generated would produce about 8.2E+8 Mg of transportation fuel/a, which figure, divided by 4.5 billion, represents ~28% of current world per capita petroleum consumption rate. Another possibility, which would not serve to dump such carbon into the atmosphere, would be to make additional ammonia and use it as engine fuel (Kang and Holbrook 2015).

Finally, if Africa were to fully electrify itself via a properly implemented nuclear renaissance, its air and water would be cleaner, its homes and cities more livable, more and better (more interesting, better paying, and more secure) employment opportunities would be available to its young people, population growth rate would drop precipitously (see Tupy 2013), as would the degree of misery/desperation/frustration currently driving young people everywhere (mostly males) to join terrorist gangs and hate groups. In other words, Africa's people would experience the same benefits of radically increased prosperity that South Korea's and China's people have recently enjoyed. An almost never mentioned (too politically incorrect) reason for China's recent success is that its leadership adopted/enforced a one child per family policy at the same time that they decided to encourage/enable its people to become entrepreneurial (Conly 2015). The purpose of that policy was to free up both time and capital, which could then be (and was) devoted to "making China great." It also rendered children born during that period especially "special" to both their parents and society at large, which rendered their lives more enjoyable and successful.

The job of government is not just to give its current citizens anything they want, but to pave the way for a prosperous, stable society for their descendants. Any kind of government-mandated fertility control is unattractive, but unless its goals can be achieved otherwise (by making the lives and futures of already-living people better), it's likely to become necessary in Africa.

14.11 WOULD ANY SORT OF "NUCLEAR RENAISSANCE" WORK?

Regardless of how "small" and "modular" they might become, any "advanced" version of today's light water reactors (LWRs) would consume at least ~160 tons of natural U/GW_e/year, most of which, primarily ^{238}U, would be discarded during its fuel fabrication's U-enrichment step. A genuinely sustainable nuclear renaissance could not be implemented with them or any other sort of burner/converter-type reactor because the uranium industry's estimate of all "affordable proven plus undiscovered" uranium resources likely to be found at concentrations high enough to be worth mining is ~18 million tons (OECD-IAEA 2014). Assuming 2 kW's worth of LWR-generated electrical energy for 11.2 billion people, all of the world's affordable (in that context) uranium would be consumed within about 5 years (1.8E+7/22,000/160 = 5.1).

Fueling burner/converter-type reactors with uranium (U) extracted from seawater couldn't "save the world" either. The country most involved with testing/developing that approach to U mining is Japan (see Tamada 2009 for a slide set and lecture). Dr. Tamada's slides begin with the contention usually prefacing such reports, that is, that "there's 1000× as much U in the oceans as on land." That's incorrect because there's about *350* times as much U in "readily accessible" rock (the first kilometer of the ~3 ppm U crustal rock covering the Earth's land surfaces) as in its seawater (~1.33 billion km³ of ~3 ppb U water). Those slides go on to describe Japan's pilot plant scale demonstrations and end with the conclusion that a 68.7 by 15.2 km (1,030 km²) array of the most promising uranium adsorbent developed (amidioxime-coated, irradiated polyethylene fiber "ropes") might be able to collect enough uranium to fuel six of Japan's almost state-of-the-art LWRs (i.e., ~1,200 Mg U/a) for a "reasonable" cost. That conclusion assumed that the adsorbent would be trapping ~4 g U/kg adsorbent per cycle, roughly three times more than was generally recovered during their demonstrations (Regalbuto 2014). A subsequent US study concluded that an "improved" version of that adsorbent would capture ~3.3 g U/kg adsorbent (Kim et al. 2014). In any case, because those systems' adsorbent arrays must be bottom-anchored, experience significant wave action, and situated where natural currents quickly replenish the water surrounding them, almost 4 million km² of *shallow* (mostly coastal) ocean bottom would have to be covered to fuel 22 TW's worth of conventional-type nuclear reactors. Finally, anyone considering such schemes should be aware that both fishermen and "Rainbow Warriors" are apt to raise objections: the former because their nets and lines would surely become entangled by so situated gigantic synthetic "kelp beds," the latter because those arrays might also entangle/strangle whales and turtles.

Other than for those little details, mining the oceans to fuel conventional reactors is a fine idea.

Realization of Weinberg and Goeller's utopian (but possible) future can happen only if the world's decision makers decide to first develop and then implement an appropriately scaled (big enough) nuclear renaissance and then see to it that untrammeled human nature does not turn their project/vision into another nuclear boondoggle.

A genuinely sustainable nuclear fuel cycle must be implemented with breeder-type reactors that generate as least as much new fuel ("fissile," any readily fissioned actinide isotope) as they consume from "fertile" natural uranium and/or thorium isotopes. Of the ways that that might be accomplished, fast spectrum molten salt reactors (MSRs) seem to offer the most promise (Holcomb et al. 2011). One of the Generation IV International Forum's (GIF's) Advanced Reactor (World Nuclear Association 2019) "Gen IV" concepts, EURATOM's EVOL program's "thorium burning" Molten Salt Fast Reactor (MSFR), is especially attractive (Fiorina 2013) . Siemer 2015 describes the features of an "isobreeding" version (generates only as much new fissile (^{233}U) from "fertile" ^{232}Th as it "burns," no extra) of that concept. Isobreeding is an especially relevant goal/assumption because, at steady-state, the world would not require extra fissile, which would mitigate proliferation concerns and it (isobreeding) would also simplify reactor operation because very little "reprocessing" would be required to keep it running at steady-state.

Another concept that wasn't formally considered by GIF, the Molten Chloride (salt) Fast Reactor (MCFR) would be simpler to build than the MSFR because its ^{238}U -to-^{239}Pu "breeding" cycle's superior neutronics should permit isobreeding without a fertile isotope (^{238}U or ^{232}Th) containing "blanket" surrounding its core. This would also mitigate proliferation concerns because, at steady-state, neither the reactor itself nor its attendant salt cleanup/recycling system would contain "bomb grade" fissile (>90% ^{239}Pu) that might otherwise tempt suicidal terrorists to attempt its "diversion." Another of its practical advantages is that its core could be situated within a tank containing either a molten bismuth/lead "reflector" or a molten salt blanket containing a fertile isotope, which would allow it to breed start-up fissile for other reactors. These are probably some of the reasons why the Bill Gates-backed TerraPower "nuclear start-up" recently decided to split its development efforts between its solid-fueled, liquid-metal cooled, "breed and burn," "Traveling Wave" concept and an unblanketed, chloride salt-based "breed and burn" MSR (Southern 2018). Another start-up, "Elysium," is apparently proposing to build a similar system.

Another especially promising MSR concept, not considered by GIF, is MOLTEX's "stable salt reactor" (Moltex Energy 2018). Its primary technical distinction is that its core consists of a bundle of thin-walled steel tubes containing a fissile isotope (any combination of trivalent ^{239}Pu, ^{235}U, and/or ^{233}U) in a molten chloride-based solvent salt consisting primarily of fertile ^{238}UCl$_3$ and table salt (NaCl). Unlike the others, its fuel salt would be "static," not continuously recirculated between its core and external heat exchangers. Fission-generated heat energy would pass through the walls of those tubes to a rapidly moving (pumped), surrounding, fissile-free, fluoride-based coolant/blanket salt consisting of a ThF_4–NaF eutectic (Scott 2014). That molten salt stream's heat energy would be transferred to a third molten salt stream (e.g., "solar salt"), which, in turn, would exchange its heat with water or CO_2 to generate the pressurized gas driving its power turbines. Both the fuel and primary coolant salt streams would be rendered noncorrosive to conventional stainless steels via redox buffering with divalent Zr – an extremely powerful reducing agent/oxygen scavenger. In this writer's opinion, the MOLTEX concept's chief virtue is that it should be considerably easier/cheaper to implement the "first of a kind" version of it than either the MCFR or MSFR. Another significant plus is that its developers have been willing to reveal more "technical details" than have their competitor – credibility in this hyper-secretive technical field is tough to earn and proportional to the degree to which a concept's champions embrace "openness."

All of these concepts are "fast" because their cores do not contain a moderating material (e.g., liquid water, beryllium oxide, or elemental carbon) that deliberately slow the rapidly moving (fast) neutrons generated by nuclear fission. This enables superior fuel (fissile) regeneration capability, lessens minor actinide (Am, Cm, etc.) build-up, and permits operation with a much higher fission product "ash" salt concentration that, in turn, translates to a much lessened fuel salt reprocessing (clean-up) requirement. It also means that they would not generate large amounts of solid waste comprised of irradiated/contaminated/damaged moderator. To date, the world's graphite-moderated "production" (of weapons-grade plutonium) reactors have generated roughly 250,000 tons of radiologically contaminated graphite, most of which lingers in "temporary" storage (International Atomic Energy Agency 2010). In principle, any of these reactor concepts could convert all actinides, both those introduced and those generated *in situ*, to relatively short-lived and simple-to-manage fission products (FP). All of them would obviate the cost, waste, and safety-related issues inherent to potentially sustainable, solid-fueled reactor concepts, such as the sodium-cooled "Integral Fast Reactor" (IFR), "Traveling Wave" concepts, or General Atomic's helium-cooled "Energy Multiplier Module" (EM2) (Rawls 2010). All of them should be cheaper to build than "advanced" versions of today's ineluctably unsustainable light water reactors because they would operate at much lower/safer pressures and generate more useful (higher temperature) heat energy. And, finally, all of them could be started with fuel comprised of the uranium plus plutonium and minor actinides (collectively called the "transuranic" or TRU elements) extracted from spent LWR fuel assemblies, which would simultaneously simplify/cheapen the long-term management of such "waste."

Since there isn't enough ^{235}U or spent LWR fuel-derived plus "excess" weapons-grade plutonium in the world (roughly 1,900 tons total) to start more than ~400 of any sort of fast breeder reactors, they would have to be configured to breed extra start-up fissile material until enough of them have been built to achieve steady-state (power everything).

All breeder reactor concepts save MOLTEX are at least 50 years old. None except the USA's pet solid-fueled, "liquid metal (cooled) fast breeder reactor" (LMFBR) technology ever received sufficient attention/funding to generate anything but "paper" (conceptual) reactors with, again, one exception – Dr. Weinberg's/ORNL's graphite-moderated "molten salt reactor experiment," which ran for four years during the late 1960s (Wikipedia 2018f), but couldn't "breed" because it was too small (8 MW_t) and wasn't surrounded with a blanket. The reason for this is that circa 1973, a financially strapped US federal government's leadership (the Vietnam war had also been fought with borrowed money ballooning its national debt and triggering inflation) decided to fire the bothersome Dr. Weinberg and study *only* the LMFBR concept (it's a much better plutonium maker) while its nuclear industry was selling much-enlarged versions of Admiral Rickover's enriched-uranium-fueled, light-water-cooled/moderated submarine reactor at cost so that it could then profit by servicing (fueling) them thereafter. Although several large LMFBRs were subsequently built and operated by France and USSR/Russia, the concept never gained much traction with electrical utility owner/operators due to persistent sodium leaks/fires and the complex/expensive nature of both the reactors themselves (Mahaffey 2014) as well as the attendant solid fuel recycling systems that would have been required for sustainable operation. The main reason why genuinely sustainable nuclear fuel cycles still don't get much attention in the US is that most of its political leaders consider nuclear power a stopgap while, with the "temporary" help of fracked natural gas, it "transitions" to an imaginary clean/green world powered by conservation, biofuels, wind turbines and solar panels/ towers (plus super batteries?) linked together with a worldwide grid (Jacobson and Delucchi 2009; Jacobsen et al. 2017). Consequently, the USA's national laboratories' nuclear scientists have helped their "industrial partners" render today's unsustainable nuclear fuel cycle more attractive for temporary niche-filling by championing "small modular" versions of their current reactors even though it is unlikely that they would be as cost-effective as their full-sized forbears. Another reason why the development of a genuinely sustainable nuclear renaissance still gets short shrift is that doing so is impossible without first performing the "hot," hands-on research (not "paper studies") required to develop a practical/affordable system. About 750–900 kg of fission products would be generated within any fast MSR's fuel salt per GW_e year, all of which would have to be properly dealt with to prevent possible damage. Similarly, any tanks, pipes, and heat exchangers in contact with that salt would experience extremely high neutron bombardment rates that could also cause damage. Solutions to problems raised by these facts cannot be discovered until realistic tests are done under realistic conditions. Because the USA's national laboratories have downsized their "hot" experimental facilities/capabilities and replaced them with "modelers," it has become extremely difficult/ expensive to perform such work – most "research" money is currently spent upon buildings and personnel-related overhead.

Finally, and most importantly, successful implementation of any of the concepts mentioned above could satisfy 100% of mankind's power needs with abundant and readily accessible natural actinide fuel – not just the 0.7% of natural uranium (^{235}U) fissionable in a moderated reactor – which would render reliable/steady nuclear power as "renewable" as sunlight (Cohen 1983). For example, the Earth's crust contains an about 3 ppm U by weight. Some rocks (and all uranium ores) contain much more than that – average volcanic rock contains 20–200 ppm, average black shale and phosphate rocks, 50–250 ppm (Ulmer-Scholle 2018). For example, assuming 200 MeV (3.2E–11 J) per fission, an average crustal rock density of 2.7 g/cc, and 50% heat-to-electricity conversion (molten salt reactors run much hotter than do LWRs and would therefore generate more electricity per heat joule), just the U within the topmost kilometer of the Earth's continents (i.e., 3 ppm of ~4.2E+17 tons ≈ 1.2E+12 t U), could continuously generate 22 TW_e for 74 million years. Since the Earth's crust contains three to four times as much thorium as uranium, let's say a total of 12 ppm, breeder reactors could potentially generate ~2.7E+12 J's worth of heat energy from 1 cubic meter of average crustal rock – about 56 times more than that provided by an equal volume of "banked" (*in situ*, not bulk) bituminous coal.

Anyone who feels that such rock mining would be too impactful should consider:

- The USA's mountaintop-removal approach to coal mining.
- The fact that fracking of the relatively "easy" (shallow) 1.6 to 3.2 km deep US shale deposits (Lallanilla 2018), currently being mined, causes lots of mini-earthquakes, occasionally pollutes groundwater, and usually leaks some of its product directly into the atmosphere.
- The fact that Brazil's much heralded, recent offshore oil discoveries are even deeper than that and covered with both water and "rock" (Wikipedia 2018g).
- The fact that nuclear power is far less damaging to both the environment and people than is the coal industry (Kharecha and Hansen 2013).

14.12 REACTOR COSTS

It is difficult to come up with definite cost figures in this particularly contentious arena because, like those of its health care system, there is little correlation between the USA's "should" and "actual" costs, due to a tremendous burden of sometimes unnecessary, usually self-serving, and often litigious "overhead" costs. The effect of those cost drivers are best illustrated by a figure (Figure 14.1) excerpted from a paper compiling historical construction costs for full-sized light water reactors

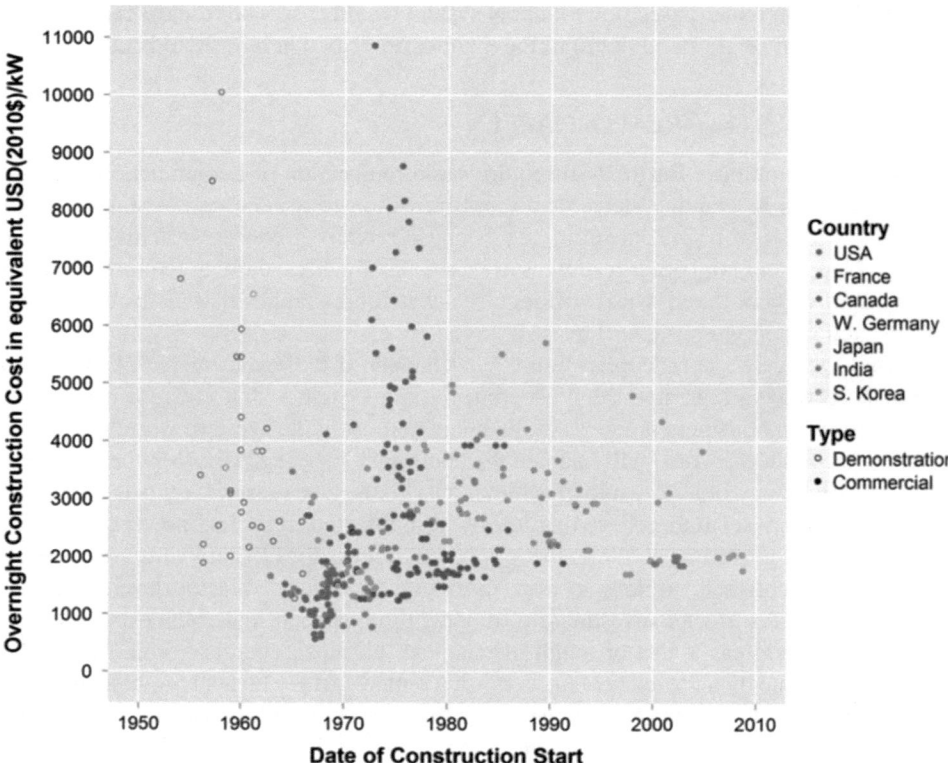

FIGURE 14.1 Reactor build costs around the world (inflation adjusted).

throughout the world (Lovering et al. 2016). The USA's build cost per GW_e started out at under $1 billion 2010-dollars in the late 1960s but quickly ballooned by over an order of magnitude (which eventually killed that enterprise) while those of more sternly disciplined countries (France and South Korea) have remained at about $2 billion for several decades.

Anyway, in 1970, a paper written by two of ORNL's senior-most nuclear engineers (Bettis and Robertson 1970) included an analysis of what power generated by a full-sized molten salt breeder reactor *should* cost. Their figure included the costs of construction (~$0.159 billion × 6.46 or $1.03 billion), fueling, and operating a one-GW_e breeder reactor. Applying the USA's subsequent ~6.46 × inflation factor for most things we purchase to their conclusion ($0.0041/kWh), generates a should-cost figure of $0.0265/kWh – about 40% of the USA's current average wholesale electricity cost. This suggests that if the USA's 320 million citizens set out to power themselves at the rate assumed herein for the future's 11.2 billion people (2 kW average with a 40% peak) with one-GW_e molten salt breeder reactors, building the 857 (30,000 × 0.32/11.2) of them so required should cost about $860 billion in today's dollars. To put that figure into perspective, it is only about 35% (860/634 = 1.35) more than one year's worth of the USA's current "discretionary" military spending (US Office of Management and Budget 2017). In a similar vein, since the USA's national debt is ~$21 trillion and the interest rate on its 10 year treasury bonds is ~3.19%, it seems likely that its taxpayers are paying someone (themselves?) ~$1.85 billion/day to service their debt.

Is $860 billion too much to simultaneously reestablish the USA's "leadership in nuclear power," clean up its atmosphere, restore/maintain the fertility of its farmland, and create millions of worth-while (not just "service") jobs and high-tech business opportunities for its citizens?

Of course, since we're pretending here that Africa's population will be 4.5 billion, the cost of the ~12 thousand full-sized reactors needed to power their part of world would be proportionately higher, ~$12 trillion of today's dollars. That's ~12% of what the backup batteries required by a similarly capable solar panel-based power supply system would cost. The reactors required to power only their agricultural sector (feed them) in the manner described herein should cost ~$350 billion.

14.13   HOW COULD IT *REALLY* HAPPEN?

Avoiding a deadly "hothouse Earth" will require rapid redirection of human actions from selfish, for-profit exploitation to genuine stewardship and transition to a genuinely sustainable economy (Steffen et al. 2018). It won't be cheap or easy because rendering energy systems carbon neutral, ensuring that land, water, and other resources are used sustainably, adapting to climate change, and cleaning up an already polluted world will require substantial changes that many, especially influential people and vested interests, will resist.

Goeller and Weinberg's cornucopian future can't happen if those most able to bear the costs of changing – those who've benefited the most from today's system – don't help to pay for it. Since interfering with current business practices is anathema to most of the Western world's leadership, it's likely that the East, not the West, will become the "outsiders" helping Africa's people to implement either this scenario or any other capable of achieving its goals. One reason for this is that most of the West's decision makers seem to believe that technical innovation in nuclear power should be left to the private sector, not government. While that might make sense for IT (hardware, software, and cell phone app development), it is unlikely to work in this sector, because reactor development is much "riskier" requiring a far greater investment of money, time, special materials, experimental work, and experts able to address a host of tough mechanical, physical, chemical, and, especially, legal issues – it can't be done by a few clever "hackers" in a rented garage. Its goals – achieving long-term sustainability, preventing further environmental damage, and providing a better life for everyone including the poor – are also inconsistent with those of the majority of today's venture capitalists. Getting government "out of the way" might sound nice, but it won't work because every major energy breakthrough in recent history has received public support that moved an idea to proof-of-concept to demonstration *after* which the private sector invested and thereby became rich (Siddiqui 2017).

Of course, the other reason is that many of the West's decision makers believe that the future should/could be powered by an "all of the above" mix of politically correct, renewable technologies (see Jacobsen 2009, Jacobsen 2017; and , for opposing views, Clack et al. 2017 and Brook et al. 2018) and have therefore supported/subsidized efforts consistent with that paradigm and, if not overtly hostile to it, have paid only lip service to the development of a sustainable nuclear fuel cycle. This is understandable because the majority of US politicians were originally lawyers and/or businessmen (not engineers or scientists), and therefore don't think quantitatively about technical issues.

The most accurate measure of a country's commitment to accomplishing anything is how much money it is willing to spend trying. In the world's richest Western nation (USA ~$20 trillion GDP) we hear that "The US Department of Energy (DOE) has selected projects to develop a pebble bed reactor and a molten chloride fast reactor to receive *multi-year* cost-share funding worth *up to* a total of $80 million" (World Nuclear News 2016); and that "Secretary of Energy Rick Perry Announces $60 Million for US Industry Awards in Support of Advanced Nuclear Technology Development," which support is to be split between 13 projects in 10 different states (Harman 2018). Although much of this work will be performed in the government's laboratories by its contractors' employees (it's illegal to do genuinely "hot" research elsewhere in the US), it won't take responsibility for guiding it or establish appropriate goals. Finally, although these projects have accomplished a good deal of modeling, simulation, and "road mapping" (see gain.inl.gov), DOE's nuclear scientists/engineers can't effectively address the future's energy issues until they are able to do the same sorts of risky/messy experimentation performed during the 1950s and 1960s at its "National Reactor Testing Station" (NRTS, now INL). Unfortunately, although those efforts designed, built, operated, and then safely decommissioned ~50 different nuclear reactors, the USA never built a test reactor capable of emulating/evaluating today's front-running MSR concepts and apparently is still not planning to do so.

The good news is that a number of the USA's young people (Breakthrough Institute 2018), environmental groups, and especially distinguished scientists have taken up Goeller and Weinberg's cause (Wikipedia 2018h).

In contrast to its current commitment to reactor development, the USA's DOE/NNSA commits ~$6.9 billion/a to maintaining its stockpile of nuclear weapons and total (DOD+DOE) nuclear weapons-related expenditures are in the order of $20 billion/a to $40 billion/a (Rumbaugh and Cohn 2012).

While the Western world's nuclear power resurgence has slowed to a snail's pace, China is steadily firing up state-of-the-art light water reactors, its new fast reactor went to full power before Christmas 2014, and it has begun construction of a new high-temperature gas-cooled nuclear reactor (Chinese HTR) in April 2015 (Conca 2015). Most importantly, it has also apparently decided to commit 22 billion yuan ($3.3 billion) to developing the technology abandoned by the US 47 years ago, Weinberg and Wigner's molten salt reactors (Chen 2017).

China is not leading that campaign because it has more experience with nuclear power or can operate reactors more efficiently and safely than do the Western world's countries. It's leading because its political leadership has both the will and foresight required to pursue activities likely to serve their country's best interests over the long haul. China's leaders accept the fact that it's their responsibility to provide a safe, comfortable, and clean world for their descendants, and also choose to believe in scientific consensus and that both intra- and international cooperation is beneficial. Consequently they have demonstrated that it is possible to incentivize entrepreneurs to do the things required to address their country's (and the world's) problems without ceding control/choice of those goals. It has been done by lending state support to targeted industries and technologies, particularly those infrastructure-related activities required to address issues such as those serving as the subject of this volume. We should also remember that China is already heavily involved in African development as a part of its 65-nation "Belt and Road" initiative. The opinion of some Western world politicians that such activity is "greedy and evil" is wrong-headed. It is a fine example of help-your-neighbor globalization, not colonization, and African agency, not Chinese rapacity (Bräutigam 2018).

REFERENCES

Agrawal, R., Singh, N. R., Ribeiro, F. H., Delgass, W. N. 2007. Sustainable fuel for the transportation sector. *Proceedings of the National Academy of Sciences of the United States of America*, 104(12), 4828–4833. doi: 10.1073 pnas.0609921104.

Alvarez, R. A., Pacala, S. W., Winebrake, J. J., Chameides, W. L., Hamburg, S. P. 2012. Greater focus needed on methane leakage from natural gas infrastructure. *Proceedings of the National Academy of Sciences of the United States of America*, 109(17), 6435–6440. doi: 10.1073/pnas.1202407109.

American-Israeli Cooperative Enterprise. 2012. Israel overview: Water. In: *The Israel Briefing Book*, M. Bard (ed.). www.jewishvirtuallibrary.org/jsource/brief/Water.html.

Anonymous. 2018a. *BP Statistical Review of World Energy* (June 2018). www.bp.com/content/dam/bp/en/corporate/pdf/energy-economics/statistical-review/bp-stats-review-2018-full-report.pdf.

Anonymous. 2018b. *BP: World Reserves of Fossil Fuels* (July 2018). https://knoema.com/smsfgud/bp-world-reserves-of-fossil-fuels.

Beegle, D. 2005. *Nitrogen Fertilization of Corn. PennState Extension* (accessed 2018). https://extension.psu.edu/nitrogen-fertilization-of-corn.

Beerling, D. J., Leake, J. R., Long, S. P., Scholes, J. D., Ton, J., Nelson, P. N., et al. 2018. Farming with crops and rocks to address global climate, food and soil security. *Nature Plants*, 4(3), 138–147.

Bettis, E. S., Robertson, R. C. 1970. The design and performance features of a single fluid molten salt breeder reactor. *Nuclear Applications and Technology*, 8(2), 190–207.

Bräutigam, D. 2018. U.S. politicians get China in Africa all wrong. *Washington Post*. www.washingtonpost.com/news/theworldpost/wp/2018/04/12/china-africa/?noredirect=on&utm_term=.fe641fe6f38f.

Breakthrough Institute 2018. *Technological Solutions to Environmental Challenges*. https://thebreakthrough.org/.

Brook, B. W., Blees, T., Wigley, M. L., Hong, S. 2018. Silver buckshot or bullet: Is a future 'energy mix' necessary? *Sustainability*, 10(2), 302. doi: 10.3390/su10020302.

Campbell, N. S. 2009. The use of rockdust and composted materials as soil fertility amendments. PhD thesis,
University of Glasgow, Glasgow. http://theses.gla.ac.uk/617/.

Cataluña, R., Kuamoto, P. M., Petzhold, C. L., Caramão, E. B., Machado, M. E., da Silva, R. 2013. Using bio-
oil produced by biomass pyrolysis as diesel fuel. *Energy and Fuels*, 27(11), 6831–6838. doi: 10.1021/
ef401644v.

Chen, S. 2017. China hopes cold war nuclear energy tech will power warships, drones. *Southern China Morning
Post*. www.scmp.com/news/china/society/article/2122977/china-hopes-cold-war-nuclear-energy-tech-w
ill-power-warships.

CIA World Factbook 2018. *Country Comparison: Distribution of Family Income – Gini Index*. https://www.cia
.gov/library/publications/the-world-factbook/rankorder/2172rank.html.

Clack, C. T. M., Qvist, S. A., Apt, J., Bazilian, M., Brandt, A. R., Caldeira, K., et al. 2017. Evaluation of a proposal
for reliable low-cost grid power with 100% wind, water, and solar. *Proceedings of the National Academy
of Sciences of the United States of America*, 114(26), 6722–6727. doi: 10.1073/pnas.1610381114.

Clark, M., Tilman, D. 2017. Comparative analysis of environmental impacts of agricultural production systems, agricultural input efficiency, and food choice. *Environmental Research Letters*, 12(6). doi:
10.1088/1748-9326/aa6cd5/meta.

Cohen, B. L. 1983. Breeder reactors: A renewable energy source. *American Journal of Physics*, 51(1), 75–76.

Conca, J. 2015. Can SMRs lead the U.S. into a clean energy future? *Forbes*. www.forbes.com/sites/jamesconc
a/2015/02/16/can-smrs-lead-the-u-s-into-a-clean-energy-future/#3c70603b31c7.

Conly, S. 2015. Here's why China's one-child policy was a good thing. *Boston Globe*. www.bostonglobe.com/
opinion/2015/10/31/here-why-china-one-child-policy-was-good-thing/GY4XiQLeYfAZ8e8Y7yFycI/s
tory.html.

Dasgupta, P. 2003. World poverty: Causes and pathways. Plenary Lecture Delivered at the World Bank's Annual
Bank Conference on Development Economics (ABCDE), Bangalore, India, 21–22 May. Subsequently
Published in *World Bank Conference on Development Economics*, B. Pleskovic and N.H. Stern (eds.),
2003 (Washington DC: World Bank), 2004. www.econ.cam.ac.uk/people-files/emeritus/pd10000/public
ations/worldpov.pdf.

Fiorina, C. 2013. The Molten salt fast reactor as a fast-spectrum. Candidate for thorium implementation. PhD
thesis, University of Milan, Milan, Italy. www.politesi.polimi.it/bitstream/10589/74324/1/2013_03_
PhD_Fiorina.pdf.

Frattini, D., Cinti, G., Bidini, G., Desideri, U., Cioffi, R., Jannelli, E. 2016. A system approach in energy
evaluation of different renewable energies sources integration in ammonia production plants. *Renewable
Energy*, 99(C), 472–482.

Garcia-Perez, M., Lewis, T., Kruger, C. E. 2010. Methods for producing biochar and advanced biofuels in
Washington state. Part 1: Literature review of pyrolysis reactors. First Project Report. Department of
Biological Systems Engineering and the Center for Sustaining Agriculture and Natural Resources,
Washington State University, Pullman, WA. https://fortress.wa.gov/ecy/publications/documents/1107
017.pdf.

Giampietro, M. 2002. Energy use in agriculture, *Encyclopedia of Life Sciences*. MacMillan Publishers, Nature
Publishing Group. www.els.net.

Goeller, H. E., Weinberg, A. M. 1976. The age of substitutability. *Science*, 191(4228), 683–689, February 20,
1976.

Gunn, L. 2017. National security and the accelerating risk of climate change. *Elem Sci Anth*, 5, 30. doi: 10.1525/
elementa.227.

Hansen, J., Sato, M., Kharecha, P., Beerling, D., Berner, R., Masson-Delmotte, V., et al. 2008. Target atmospheric CO_2: Where should humanity aim? *The Open Atmospheric Science Journal*, 2(1), 217–231. doi:
10.2174/1874282300802010217.

Harman, S. 2018. *Secretary of Energy Rick Perry Announces $60 Million for U.S. Industry Awards in Support of
Advanced Nuclear Technology Development*. U.S. Department of Energy. https://www.energy.gov/article
s/secretary-energy-rick-perry-announces-60-million-us-industry-awards-support-advanced.

Hartmann, J., West, A. J., Renforth, P., Köhler, P., De La Rocha, C. L., Wolf-Gladrow, D. A., Dürr, H. H.,
Scheffran, J. 2013. Enhanced chemical weathering as a geoengineering strategy to reduce atmospheric
carbon, supply nutrients, and mitigate ocean acidification. *Reviews of Geophysics*, 51(2), 113–149.

Holcomb, D. E., Flanagan, G. F., Patton, B. W., Gehin, J. C., Howard, R. L., Harrison, T. J. 2011. *Fast Spectrum
Molten Salt Reactor Options*. ORNL TM-201111/105. https://info.ornl.gov/sites/publications/files/Pub
29596.pdf.

Hong, L.-C. 2016. *Super El Niño*. Springer.

Hussain, A., Abolaban, F. 2014. Nuclear desalination: A viable option for producing fresh water-feasibility and techno-economic studies. *Life Science Journal*, 11(1), 301–307.

International Atomic Energy Agency. 2010. *Progress in Radioactive Graphite Waste Management* (July 2010). https://www.pub.iaea.org/MTCD/Publications/PDF/te_1647_web.pdf.

Jackson, R. B., Quéré, C. L., Andrew, R. M., Canadell, J. G., Korsbakken, J. I., Liu, Z., Peters, G. P., Zheng, B. 2018. Global energy growth is outpacing decarbonization. *Environmental Research Letters*, 13, 120401. doi. 10.1088/1748-9326/af303.

Jacobson, M. Z., Delucchi, M. A. 2009. A plan to power 100 percent of the planet with renewables. *Scientific American*, Nov. issue. www.scientificamerican.com/article/a-path-to-sustainable-energy-by-2030/.

Jacobson, M. Z., Delucchi, M. A., Zack, A. F., Bauer, Z. A. F., Goodman, S. C., Chapman, W. E. et al. 2017. *100% Clean and Renewable Wind, Water, and Sunlight (WWS). All-Sector Energy Roadmaps for 139 Countries of the World* (January 27), http://web.stanford.edu/group/efmh/jacobson/Articles/I/CountriesWWS.pdf.

Jankovic, A. 2003. Variables affecting the fine grinding of minerals using stirred mills. *Minerals Engineering*, 16(4), 337–345.

Kang, D. W., Holbrook, J. H. 2015. Use of NH_3 fuel to achieve deep greenhouse gas reductions from US transportation. *Energy Reports*, 1, 164–168. doi: 10.1016/j.egyr.2015.08.001.

Kharecha, P. A., Hansen, J. E. 2013. Prevented mortality and greenhouse gas emissions from historical and projected nuclear power. *Environmental Science and Technology*, 47(9), 4889–4895. doi: 10.1021/es3051197.

Kim, J., Tsouris, C., Oyola, Y., Janke, C. J., Mayes, R. T., Dai, S., et al. 2014. Uptake of uranium from seawater by amidoxime-based polymeric adsorbent: Field experiments, modeling, and updated economic assessment. *Industrial and Engineering Chemistry Research*, 53(14), 6076–6083. doi: 10.1021/ie4039828.

Lal, R. 2017. Improving soil health and human protein nutrition by pulses-based cropping systems, Chapter Four. In: *Advances in Agronomy*, Donald L. Sparks (ed.), Volume 145, pp. 167–204, Elsevier.

Lal, R. 2008. Sequestration of atmospheric CO_2 in global carbon pools. *Energy and Environmental Science*, 1(1), July issue, 86–100. doi: 10.1039/B809492F.

Lallanilla, M. 2018. *Facts About Fracking*. (accessed 2018). www.livescience.com/34464-what-is-fracking.html.

Lang, B. 2002. *Estimating the Nutrient Value in Corn and Soybean Stover*. Iowa State University Extension, Ames, IA. www.extension.iastate.edu/sites/www.extension.iastate.edu/files/allamakee/stovervalue.pdf.

Leeman, W. P. 1982. Olivine tholeiitic basalts of the Snake River. In: *Cenozoic Geology of Idaho, Idaho Bureau of Mines and Geology Bulletin 26*, I. Plain. B. Bonnichsen and R.M. Breckenridge (eds.), pp. 181–191 (Also characterizes other US basalts). https://www.idahogeology.org/product/b-26.

Leffel, T. 2017. The cheapest places to live in the world −2018. *Cheapest Destinations Blog*. www.cheapestdestinationsblog.com/2017/12/11/the-cheapest-places-to-live-in-the-world-2018/.

LLNL. 2015. Lawrence Livermore National Laboratory's annual energy sanky-diagram overall conversion factors (usually about 40%) are based upon DOE/EIA-0035, 2015-03. https://flowcharts.llnl.gov/content/assets/docs/2014_United-States_Energy.pdf.

Lovering, J. R., Yip, A., Nordhaus, T. 2016. Historical construction costs of global nuclear power reactors. *Energy Policy*, 91, 371–382.

Maass, P. 2010. The ministry of oil defense. *Foreign Policy*. https://www.idahogeology.org/product/b-26.

Mahaffey, J. 2014 In nuclear research, even the goof-ups are fascinating. In: *Atomic Accidents*. Pegasus Books.

Martin, R. 2015. Terrapower quietly explores new nuclear reactor strategy. *MIT Technology Review* (accessed Nov. 30, 2015). www.technologyreview.com/s/542686/terrapower-quietly-explores-new-nuclear-reactor-strategy/.

Martin, R. 2016. The dubious promise of bioenergy plus carbon capture. *MIT Technology Review*. www.technologyreview.com/s/544736/the-dubious-promise-of-bioenergy-plus-carbon-capture/.

McArthur, J. W. 2013. *Good Things Grow in Scaled Packages*. Brookings Institution. www.brookings.edu/research/good-things-grow-in-scaled-packages-africas-agricultural-challenge-in-historical-context/.

Moltex Energy. 2018. *An Introduction to the Moltex Energy Technology Portfolio*. www.moltexenergy.com/learnmore/An_Introduction_Moltex_Energy_Technology_Portfolio.pdf.

Moore III, B., Braswell, B. H. 1994. The lifetime of excess atmospheric carbon dioxide. *Global Biogeochemical Cycles*, 8(1), 23–38, doi: 10.1029/93GB03392.

Moulton, K. L., West, J., Berner, R. A. 2000. Solute flux and mineral mass balance approaches to the quantification of plant effects on silicate weathering. *American Journal of Science*, 300(7), 539–570.

OECD-IAEA. 2014. *Uranium 2014: Resources, Production and Demand.* www.oecd-nea.org/ndd/pubs/2014
/7209-uranium-2014.pdf.

Ojuso, J. O. 1990. Data Bank: The iso-radiation map of Nigeria. *Solar and Wind Technology,* 7, 563–575.

Pfotenhauser, N. 2014. *Big Wind's Bogus Subsidies.* www.usnews.com/opinion/blogs/nancy-pfotenhauer/201
4/05/12/even-warren-buffet-admits-wind-energy-is-a-bad-investment.

Pimentel, D., Marklein, A., Toth, M. A., Karpoff, M. N., Paul, G. S., McCormack, R., Kyriazis, J., Krueger,
T. 2009. Food versus biofuels: Environmental and economic costs. *Human Ecology,* 37(1), 1–12. doi:
10.1007/s10745-009-9215-8.

Priyono, J., Gilkes, R. J. 2004. Dissolution of milled-silicate rock fertilizers in the soil. *Australian Journal of
Soil Research,* 42(4), 441–448. doi: 10.1071/SR03138.

Rawls, J. 2010. *Implications for Waste Handling of the Multiplier Module.* US Nuclear Waste Technical Review
Board (June 29). https://www.nwtrb.gov/docs/default-source/meetings/2010/june/rawls.pdf.

Regalbuto, C. 2014. *Past and Future Efforts to Extract Uranium from Seawater.* Stanford University. March 21,
2014 http://large.stanford.edu/courses/2014/ph241/regalbuto2/.

Rumbaugh, R., Cohn, N. 2012. Resolving ambiguity: Costing nuclear weapons. *Stimson* (September 17). www.
stimson.org/content/resolving-ambiguity-costing-nuclear-weapons.

Schumacher, E. F. 1973. Small is beautiful: A study of economics as if people mattered. HarperCollins, HB171.
S384 1989.

Scott, I. R. 2014. A practical molten salt fission reactor. UK Patent application GB 2508537 A, https://patenti
mages.storage.googleapis.com/pdfs/65e7697f591b41905b81/GB2508537A.pdf.

Siddiqui, F. 2017. No, the private sector won't pick up the slack for federal investments in energy innovation.
Third Way. https://medium.com/third-way/no-the-private-sector-wont-pick-up-the-slack-for-federal-inve
stments-in-energy-innovation-4b3c2029670a.

Siemer, D. 2019. Silicate weathering to mitigate climate change. In: *Soil and Climate,* R. Lal and B.A. Stewart
(eds.). Taylor and Francis, pp. 249–265.

Siemer, D. D. 2015. Why the MSFR is the "best" GEN IV reactor. *Energy Science and Engineering,* 3(2, Feb),
83–97. doi: 10.1002/ese3.59/full.

Sims, R. E. H. 2011. *Energy-Smart Food for People and Climate,* FAO Issue Paper. www.fao.org/docrep/014/
i2454e/i2454e00.pdf.

Smil, V. 2008. *Energy in Nature and Society: General Energetic of Complex Systems.* MIT Press, Cambridge,
MA.

Southern Company. 2018. *Southern Company and TerraPower Prep for Testing on Molten Salt Reactor.* Office
of Nuclear Energy. DOE Office of Nuclear Energy Press release, 9 August 2018. www.energy.gov/ne/
articles/southern-company-and-terrapower-prep-testing-molten-salt-reactor.

Springer, N. P., Duchin, F. 2014. Feeding nine billion people sustainably: Conserving land and water through
shifting diets and changes in technologies. *Environmental Science and Technology,* 48(8), 4444–4451.
doi: 10.1021/es4051988.

Statista 2017. *Total Global Grain Production from 2008/2009 to 2017/2018 (in Million Metric Tons).* www.
statista.com/statistics/271943/total-world-grain-production-since-2008-2009/.

Steffen, W., Rockström, J., Richardson, K., Lenton, T. M., Folke, C., Liverman, D., et al. 2018. Trajectories of
the earth system in the Anthropocene. *Proceedings of the National Academy of Sciences of the United
States of America,* 115(33), 8252–8259. doi: 10.1073/pnas.1810141115.

Stern, R. J. 2010. United States cost of military force projection in the Persian Gulf, 1976–2000. *Energy Policy,*
38(6), 2816–2825. doi: 10.1016/j.enpol.2010.01.013.

Tamada, M. 2009. *Current Status of Technology for Collection of Uranium from Seawater.* Erice seminar 2009.
http://ecolo.org/documents/documents_in_english/uranium-sea-09_Tamada.pdf.

Tamada, M. 2009. *(slides) Collection of Uranium from Seawater.* 2009.11.5TM / IAEA Vienna, Austria.

Taylor, L. L., Beerling, D. J., Quegan, S., Banwart, S. A. 2017. Simulating carbon capture by enhanced weathering with croplands: An overview of key processes highlighting areas of future model development.
Biology Letters, 13(4), 1–8.

Tomlinson, J. 2018. Nigerian briefing – How engineers can help secure a sustainable economy. *Proceedings of
the Institution of Civil Engineers - Energy,* 171(3), 121–128. doi: 10.1680/jener.17.00024.

Tupy, M. 2013. *Prosperity and World Population Growth.* www.cato.org/blog/prosperity-world-population
-growth.

Ulmer-Scholle, D. S. 2018. Uranium – Where is it found? *New Mexico Bureau of Geology & Mineral Resources.*
https://geoinfo.nmt.edu/resources/uranium/where.html.

United Nations Development Programme. 2018. *Human Development Index (HDI).* http://hdr.undp.org/en/co
ntent/human-development-index-hdi.

United Nations, Department of Economic and Social Affairs, Population Division. 2015. *World Population Prospects: The 2015 Revision, Key Findings and Advance Tables*. Working Paper no. ESA/P/WP.241. https://esa.un.org/unpd/wpp/publications/files/key_findings_wpp_2015.pdf.

US Global Change Research Program. 2018. Fourth national climate assessment. *Volume II: Impacts, Risks, and Adaptation in the United States*. https://nca2018.globalchange.gov/.

US Office of Management and Budget. 2017. *OMB Sequestration Update Report to the President and Congress for Fiscal Year 2018*. www.whitehouse.gov/sites/whitehouse.gov/files/omb/sequestration_reports/FY_2018_Sequestration_Update_8-18-17.pdf.

Van Huis, A. 2012. Potential of insects as food and feed in assuring food security. *Annual Review of Entomology*, 1146, 563–583.

Wikipedia, The Free Encyclopedia. 2018a. *Carbon Capture and Storage* (accessed 2018). https://en.wikipedia.org/wiki/Carbon_capture_and_storage.

Wikipedia, The Free Encyclopedia. 2018b. *Demographics of the Middle East* (accessed 2018). https://en.wikipedia.org/wiki/Demographics_of_the_Middle_East.

Wikipedia, The Free Encyclopedia. 2018c. *Desalination* (accessed 2018). https://en.wikipedia.org/?title=Desalination.

Wikipedia, The Free Encyclopedia. 2018d. *Energy in Africa* (accessed 2018). https://en.wikipedia.org/wiki/Energy_in_Africa.

Wikipedia, The Free Encyclopedia. 2018e. *Human Power* (accessed 2018). https://en.wikipedia.org/wiki/Water_supply_and_sanitation_in_Israel.

Wikipedia, The Free Encyclopedia. 2018f. *Molten-Salt Reactor Experiment* (accessed 2018) https://en.wikipedia.org/wiki/Molten-Salt_Reactor_Experiment.

Wikipedia, The Free Encyclopedia. 2018g. *Pre-Salt Layer* (accessed 2018). https://en.wikipedia.org/wiki/Pre-salt_layer.

Wikipedia, The Free Encyclopedia. 2018h. *Pro-Nuclear Movement* (accessed 2018). https://en.wikipedia.org/wiki/Pro-nuclear_movement.

Wikipedia, The Free Encyclopedia. 2018i. *Reverse Osmosis* (accessed 2018). https://en.wikipedia.org/wiki/Reverse_osmosis.

Wikipedia, The Free Encyclopedia. 2018j. *Water Supply and Sanitation in Israel* (accessed 2018). https://en.wikipedia.org/wiki/Water_supply_and_sanitation_in_Israel.

Wikipedia, The Free Encyclopedia. 2018k. *Wood Economy* (accessed 2018). https://en.wikipedia.org/wiki/Wood_economy.

Wilson, R. 2015. Fuel prices and cost of production. https://agecon.unl.edu/cornhusker-economics/2015/fuel-prices-and-cost-of-production.

World Nuclear Association. 2019. *Generation IV Nuclear Reactors*. www.world-nuclear.org/information-library/nuclear-fuel-cycle/nuclear-power-reactors/generation-iv-nuclear-reactors.aspx.

World Nuclear News. 2016. *US Invests in Advanced Reactor Development. World Nuclear News*. (18 January). www.world-nuclear-news.org/NN-US-invests-in-advanced-reactor-development-1801168.html.

Zapata, F., Roy, R. N. 2004. *Use of Phosphate Rocks for Sustainable Agriculture*. A joint publication of the FAO Land and Water Development Division and the International Atomic Energy Agency. Food and Agriculture Organisation of the United Nations, Rome, Italy. www.fao.org/docrep/007/y5053e/y5053e06.htm.

15 Love Songs to Loam
Motivating Youth to Make a Difference by Engaging Science and Religion

Marcia J. Bunge

CONTENTS

15.1 INTRODUCTION

In the face of soil degradation and other serious ethical and environmental problems, a growing number of respected scientists and religious leaders around the world recognize that solutions require collaboration. Pope Francis, the Dalai Lama, and other religious leaders appreciate and rely on scientists to help them address serious problems facing human beings and our planet. Leading scientists, whether religiously affiliated or not, also understand the importance of collaborating with religious leaders and communities to solve ethical challenges of common concern. After all, over 80% of the people on this planet self-identify with some form of religion. Furthermore, motivating people of all worldviews—whether religious or secular—to work together for the common good requires not only stating the facts but also engaging their core values and beliefs.

Although creative alliances between scientists and religious communities are necessary and widespread, many people still assume that religion and science are enemies. Stories in the news tend to focus less on collaborative efforts and more on conflicts between particular scientists who reject religion and religious leaders who reject theories of evolution. Furthermore, even though a host of religious organizations have issued statements on their openness to theories of evolution and on positive relationships between science and religion, many religious leaders are not speaking enough with members of their communities about such statements and more informed perspectives on relationships between religion and science. As Francis Collins, former director of the National Human Genome Research Institute, has famously said, "One of the greatest tragedies of our time is this impression that has been created that science and religion have to be at war" (Swinford 2006).

Thus, even young people who have grown up in religious communities whose leaders welcome and engage scientific discoveries often falsely assume that science and religion are enemies. Several studies now show, for example, that many young people in the United States mistakenly believe that all forms of Christianity are anti-science, and this misconception is one of the of the primary reasons young people brought up in the Church are leaving it (Kinnaman 2011). These studies ring true with my own experience with college students. I am a religious scholar, trained at the University of Chicago, who teaches at

Gustavus Adolphus College (Minnesota). Like other outstanding liberal arts colleges, Gustavus welcomes students from all worldviews, backgrounds, and religious traditions. Each year I am disturbed to find that even students from Christian denominations that publicly engage the sciences believe Christianity is anti-science. For instance, every year a few students will tell me that they used to be a Lutheran or Roman Catholic but had to give up their faith because they were becoming scientists. They are surprised to learn about highly respected scientists and theologians from Catholic, Lutheran, and other Christian traditions who articulate and cultivate positive relationships between science and religion.

15.2   THE GUSTAVUS ACADEMY FOR FAITH, SCIENCE, AND ETHICS

Given both the widespread misconception that religion and science are antithetical and the urgency of environmental and ethical problems that require the cooperation and creativity of both scientists and religious leaders, the Gustavus chaplains and I worked together with other members of our college's staff and faculty to create the Gustavus Academy for Faith, Science, and Ethics. The week-long summer program provides opportunities for high school students to explore their faith and to discover how scientists and people of faith are working together to address some of the world's most pressing problems. In preparation for the Academy, the leadership team of Gustavus staff and faculty offers a five-week training program for college students, who study issues related to science and religion, prepare activities for the Academy, and then help lead it as college mentors. In the Academy, activities with the high school students, called Fellows, include lectures from theologians and scientists in residence, readings, discussions with small mentor groups, recreation, hands-on experiments, worship, student presentations, and many opportunities for participants to reflect on their faith, values, and sense of vocation, that is, how they might cultivate and use their particular strengths and gifts to contribute to the common good whether as scientists or in other professions.

Each year the Academy focuses on a particular theme related to the college's annual Nobel Conference. Held each October, this signature event brings together leading scientists, ethicists, and religious thinkers to explore pressing scientific questions and related ethical issues. Thus, in line with the 2018 Nobel Conference on "Living Soil: A Universe Underfoot," the 2018 summer Academy also focused on soil degradation. During the summer Academy our high school Fellows and college mentors learned about microbiomes, the complexity of soil, and the challenges we face in protecting this fundamental resource. They visited a farm, heard lectures from resident ethicists, theologians, and soil scientists, and carried out soil experiments, while also reflecting on their own faith, ethical commitments, and core values that compel them to help protect the soil. Fellows were then invited back to our campus in October for the 2018 Nobel Conference to reconnect with one another and their mentors and to learn more from leading experts in soil science, ecology, microbiology, and environmental ethics, including Rattan Lal, Jack Gilbert, and David Montgomery.

The Academy was funded by a generous grant from the Lilly Endowment and fits well with the College's heritage, distinctive strengths, and mission. As a liberal arts college affiliated with the Evangelical Lutheran Church in America (ELCA), our mission is to provide students with an excellent education that equips them to use their particular gifts and talents to contribute to the common good. In line with the Lutheran intellectual tradition and long-standing commitment to public education, the college offers excellent programs in the humanities and the sciences to students from all backgrounds and worldviews. The college also offers students many opportunities both inside and outside the classroom to reflect on their own values, ethical and religious commitments, and sense of purpose and to learn about how people from diverse religious and secular worldviews seek to understand and cooperate with one another. Our Academy is one of several programs funded by Lilly's High School Youth Theology Institutes initiative. The Lilly initiative helps church-related colleges and universities create summer programs that seek to cultivate the gifts of young Christian leaders who can positively impact Church and society. Our program focuses distinctively on science and religion. Participants come from across the country and are leaders in their congregations, passionate about issues of social justice, and have a strong interest in the sciences and math.

By engaging both science and religion, the Gustavus Academy dispels the notion that science and religion are enemies, positively impacts students, and motivates them to make a difference in the world. Since the Academy has both leading scientists and theologians in residence, students gain a rich appreciation for both the scientific and ethical dimensions of complex problems such as soil degradation. They also learn firsthand from scientists and theologians about how they are working together to address issues of common concern and how they live out their faith and sense of vocation in their personal and professional lives. By providing a safe and creative space for students to discuss their core values and beliefs alongside their science-related interests and concerns, students also become highly motivated and empowered to make a difference. They understand more clearly not only the nature and scope of problems facing our world but also why they care and how their own values and beliefs both ground and motivate them.

15.3   THE 2018 ACADEMY ON SOIL: REFLECTIONS BY FOUR COLLEGE MENTORS

These and other benefits of the Academy are expressed well by our college mentors, who become ambassadors on- and off-campus for fresh ways to frame positive connections between science and religion. The heartfelt engagement of students with science, faith, and one another is expressed in the following reflections by just four of our college mentors who helped to lead the 2018 Academy on soil:

> The most beneficial part of participating in an academy that explores faith, science, and ethics is that one is able to see connections across the three different subjects. Using Faith and Science together to inform ethical decision-making results in more comprehensive and effective decisions. In the case of the degradation of soil, we were able to build on biblical passages about the care of the earth and our connection to it as well as to use science to create a deeper understanding of the problem and the abundant life that fills the soil. The depth of understanding that comes from using faith and science to address ethical issues helps develop young leaders who not only know what the problem is but also why it is their duty to address that problem.

> **Matthew Ouren (2019 Gustavus Graduate)**

> I appreciated the relationships between both the mentors and Fellows that were developed and strengthened throughout the course of the Academy. The time spent in fellowship, attending lectures, and even eating meals allowed everyone involved in the Academy program to develop close bonds. As a mentor, I appreciated the time I was able to spend in my small group, which consisted of myself and five high school girls. In that small and safe space, I felt more comfortable and completely able to facilitate discussions surrounding faith, science, ethics, and vocation while making sure that the students, who had diverse perspectives and backgrounds, felt comfortable sharing their own reflections and questions.

> **Sophia Gottlick (2020 Gustavus Graduate)**

> At this year's Academy, I discovered deeper affirmation for my spiritual beliefs in humanity's inherent connection to the earth. Through a discussion of biblical passages about soil, especially the Hebrew notion of *Adamah* (found in Genesis), my moral commitment to caring about soil degradation issues grew stronger. Since life is of and from soil, it ought to be respected and honored. Through my Christian lens, since God is depicted as creating from soil in a creative, loving way, all people also ought to approach the earth we tread on with reverence.

> **Alexander Theship-Rosales (2019 Gustavus Graduate)**

> I really appreciate how the Academy allowed the students to be fiercely themselves … I also find that I am more able to answer difficult questions regarding faith and life in general. I am more confident in my own relationship with God, and I have a better understanding of why certain things happen and how God is present in everyday life. I have a newfound love and respect for farming and people who work in

the agricultural industry because of how the Gustavus Academy paired Christianity to cultivating soil. My faith grounds me, and it's clear to me that God is present in the making of our food!

Isabel St. Dennis (2021 Gustavus Graduate)

15.4 THE "LOAMY LOVE SONG": A STUDENT PRESENTATION AT THE 2018 ACADEMY ON SOIL

One of the most moving parts of the Academy is the final presentations by the high school Fellows when they express their own "take away" from the Academy. The presentations take a variety of forms, including PowerPoint presentations, short skits, poems, prayers, or songs. These presentations reveal the students' leadership skills, creativity, and the sheer joy and delight that happens when our concerns for the world, deepest values, passions, and strengths align.

The hit of this year's student presentations was the "Loamy Love Song," The lyrics were written by two high school Fellows, Peter Weiblen and Eric Johnson. They set the words to a familiar and beloved hymn tune, "How Great Thou Art," enabling everyone to join with humor and heart in the chorus. Peter and Eric played the hymn on their guitars and led the chorus with gusto. Even though their song is playful, the hymn tune they chose was especially apt for a song that helps drive home the significance of soil and environmental responsibility. "How Great Thou Art" is translated from a Swedish hymn, "O Store Gud," which was written in 1885 by the Swedish poet Carl Gustav Boberg and set to a beautiful Swedish folk tune. The poet writes of forests, mountains, brooks, and stars, considering "in awesome wonder" God's creation and love. With its attention to the natural world, the hymn is a favorite at many Christian summer camps and environmental workshops.

Although the hymn tune and lyrics of the "Loamy Love Song" speak to Christian young people and would not be appropriate for summer institutes held for students who come from other religious traditions, the song and its enthusiastic reception by participants in this year's Gustavus Academy on soil help illustrate the energy and creativity that comes from connecting scientific understandings of the world with one's core values and religious beliefs.

"Loamy Love Song"

Verse 1
Whether you've got some great corn, soybeans, or wheat
For you I will always be so loyal.
I will love you, and all you give me to eat
Because you're so much more than dirt; you're soil.

Chorus:
Sand, silt, and clay, you're fertile every day!
We love you loam. We love you loam.
Sand, silt, and clay, you're fertile every day!
We love you loam. We love you loam.

Verse 2
I've travelled 'round, and you're in all locations;
for hearty crops, you are fundamental.
I've spread my roots, and I've found my vocation.
You make us all so environmental.
(Sing Chorus)

Verse 3
In Genesis, God made you on the day third.
You save water with all your filtration.

Sympathy to all who have not truly heard
How great you've been, loam, since creation.
(Sing Chorus)

Verse 4
When I feel your texture all through my field
Profits I earn make farming like a heist.
I know for sure, it's going to be a good yield.
I do receive all blessed things through Christ.

(Sing Chorus twice)
Sand, silt, and clay, you're fertile every day!
We love you loam. We love you loam.
Sand, silt, and clay, you're fertile every day!
We love you loam. We love you loam.

15.5   CONCLUSION AND WAYS FORWARD: CREATING SPACES FOR YOUNG PEOPLE OF DIVERSE WORLDVIEWS TO ENGAGE SCIENCE AND RELIGION

Although geared to young Christian leaders, the Gustavus Academy for Faith, Science, and Ethics illustrates the urgent need for and benefits of creating spaces for young people of diverse backgrounds and worldviews to engage religion and science. Such spaces help dispel the dangerous notion that religion and science are at odds, and they allow young people to learn how and why scientists and religious leaders are effectively working together to address serious challenges, such as soil degradation. Furthermore, providing young people with spaces to reflect on their own ethical and religious commitments while exploring creative alliances between science and religion can be highly impactful and motivating. Given that religion is a factor in the lives of over 80% of people on this planet, most young people are shaped by a religious tradition. Since religious beliefs and values often inform ethical decision-making and motivate actions, helping young people reflect on ethical challenges requires paying attention to their religious commitments (Jacobsen 2012). When young people have the opportunity to reflect on their core values and beliefs while learning about scientific and religious responses to serious ethical challenges, they are able to articulate more clearly why they care and are empowered to take action. They come away knowing more about themselves and the world, and they are motivated to make a difference. Providing more opportunities for young people in various parts of the world and from diverse religious backgrounds to engage religion and science will also make a difference in healing the perceived gap between science and religion and motivating all of us—regardless of our particular worldviews—to work together to address ethical issues of common concern.

REFERENCES

Jacobsen, Douglas, and Rhonda Hustad Jacobson. 2012. *No Longer Invisible: Religion in University Education.* New York: Oxford University Press.

Kinnaman, David. 2011. *You Lost Me: Why Young Christians Are Leaving Church … and Rethinking Faith.* Grand Rapids, MI: Baker Books.

Swinford, Steven. 2006. I've found god, says man who cracked the genome. *The Sunday Times.* June 11. Accessed December 8, 2018. www.thetimes.co.uk/article/ive-found-god-says-man-who-cracked-the-genome-qxlhgwjvb0z.

Index